CULTIVATION AND BREEDING OF MEDICINAL AND AROMATIC PLANTS

Dr. Lakshman Chandra De obtained his both M.Sc and Ph. D. Degree in Horticulture-Floriculture from IARI, Pusa, New Delhi. Presently, he is a Principal Scientist working at ICAR-NRC for Orchids, Pakyong, Sikkim. He has 21 years research and extension experiences in production and post-harvest technologies of horticultural crops of NEH Region and 4 years experience as Joint Director at ICAR, A.P. Centre. He has handled Nationally Funded Projects like NATP on 'IPNS for Vegetable Production System', Technology Mission on 'HMNEH', Mega Seed Project on flower crops and DUS Project on orchids. He has significant contribution in preparation of DUS Test Guidelines on Cymbidium, Dendrobium, Vanda, Cattleya, Phalaenopsis, Oncidium and Paphiopedilum orchids. He has published 104 research papers and written 25 book chapters, 24 Technical bulletins, 78 abstracts, 36 popular articles, 29 reports and 9 training manuals. He is the author of 14 popular books titled 'Advanced Commercial Floriculture', 'Medicinal Herbs and Flowers', 'Post-harvest Technology of Flowers and Ornamental Plants', 'Handbook of Edible Fruits', 'Handbook of Vegetable Crops', 'Ornamental Crop Breeding', 'Value Addition in Flowers and Orchids', Handbook of Gardening, 'Nursery and Landscaping' and 'Production Technology of Commercial Flowers', 'Production of Seed and Planting Material of Horticultural Crops' , 'Commercial Orchids' , 'Year Round Production of Orchids' and 'Protected Cultivation of Ornamental Plants' which have been well accepted as reference books in ICAR institutes, SAU's in India and abroad. He was awarded HSI Gold Medal in Floriculture in 2011, HSI Fellowship in 2013, IASR Fellowship in Horticulture (USA) in 2015, Heinrich Hertz Research Award in Horticulture (USA) in 2015, Antonie van Leewenhoek Research Award in Floriculture (USA) in 2016, IASR Fellowship in Floriculture (USA) in 2016, Michael Faraday Research Award in Botany (USA) in 2016, IASR Fellowship in Botany (USA) in 2016, CHAI Fellowship in 2016 and ISOH Fellowship, 2017. He has been included in Biography of Asia/Pacific Who's Who (Vol. XIV) in 2016 and Famous India -Who's Who in 2016. He is a member of Indian Council of Food and Agriculture (ICFA), Board of Studies, Sikkim University, Faculty of Horticulture, UBKV, West Bengal and several scientific societies and National Level Committees and a reviewer of Journal of Botanical Sciences, Journal of Agriculture and Technology, British Biotechnology Journal, Journal of Biological Diversity, Indonesia and Journal of Agriculture and Crop Research and Associate Editor of The Himalayan Review.

CULTIVATION AND BREEDING OF MEDICINAL AND AROMATIC PLANTS

Dr Lakshman Chandra De

Principal Scientist (Horticulture)

ICAR-NRC for Orchids, Pakyong

East Sikkim-737106, Sikkim

Pointer Publishers

Jaipur 302 003 (Raj) India

First Published in 2017 by
Mrs. Shashi Jain for
Pointer Publishers
807, Vyas Building, Chaura Rasta
Jaipur 302 003 (Raj.) India
Phone : 0141-2578159, 2708286
e-mail : pointerpub@hotmail.com
Visit us at : www.pointerpublishers.com

ISBN 978-81-7132-866-6

Printed at
Harish Art Printers
Jaipur-302 003 (Raj.)

PREFACE

Traditional systems of medicine continue to be widely practised on many accounts. Population rise, inadequate supply of drugs, prohibitive cost of treatments, side effects of several allopathic drugs and development of resistance to currently used drugs for infectious diseases have led to increased emphasis on the use of plant materials as a source of medicines for a wide variety of human ailments.The forest in India is the principal repository of large number of medicinal and aromatic plants, which are largely collected as raw materials for manufacture of drugs and perfumery products. About 8,000 herbal remedies have been codified in Ayurveda. Plants, especially used in Ayurveda can provide biologically active molecules and lead structures for the development of modified derivatives with enhanced activity and /or reduced toxicity.Some of the useful plant drugs include vinblastine, vincristine, taxol, podophyllotoxin, camptothecin, digitoxigenin, gitoxigenin, digoxigenin, tubocurarine, morphine, codeine, aspirin, atropine, pilocarpine, capscicine, allicin, curcumin, artemesinin and ephedrine amongothers. Nutraceuticals / functional foods and perfumery/aroma chemicals from natural sources is the fastest growing sector in the global market of medicine.Diseases like diabetes, cancer, arthritis, asthma, depression and heart ailments are associated with both genetic and environmental risk factors.Single ingredient drugs known as new chemical entities (NCEs) often fail to cure such diseases. The plant secondary metabolites offer significant therapeutic opportunities to address such diseases.

Essential oils are highly concentrated, low volume, high value products.Essential oils and aroma chemicals constitute a major group of industrial products. These oils form indispensable ingredients of the necessities in many spheres of human activity. They are adjuncts of cosmetics, soaps, pharmaceuticals, perfumery, confectionery, ice-creams, aerated waters disinfectants,tobacco, agarbathis and a host of related products. Aromatherapy involves the use of essential oils and aromatics derived from plants to cure diseases. Someof the essential oils are reported to be in many ways better than antibiotics due to their safety and wide spectrum of activity. Application of essential oils in agriculture as anti-feedants, repellents, botanical insecticides, natural herbicides and growth boosters are still open to fascinating realms of research.The demand for medicinal plants is increasing everyday and the World Health Organization (WHO) has projected that the global herbal market will grow to $ 5 trillion by 2050 from the current level of $ 62 billion. India and China are the largest producer of more than 70% of the global diversity. The significant global herbal export market include EU, USA, Canada, Australia, Singapore, and Japan. New emerging market (globally) includes Brazil, Argentina, Mexico, China and Indonesia.The market place for medicinal plants used in dietary supplements is growing in the European continent along with Japan and USA.

The top 10 highly traded medicinal plants include Amlaki (*Emblica officinalis* Gaertn.), Satavari (*Asparagus racemosus*Willd.), Aswagandha (*Withania somnifera* Dunal.), Haritaki (*Terminaliac* bebula Retz.), Asoka (*Saraca asoca* DC. Willd.), Bilva (*Aegle marmelos* Corr.), Svarnapatri or Senna leaves (*Cassia angustifolia*Vahl.), Vasa (*Adhatoda vasica* Nees), Pippali (*Piper longum* Linn.) and Brahmi (*Bacopa monnieri* (Linn.) Wettst and other plant material that have huge demand in foreign markets are Isabgol (*Psyllium husk*, Psyllium seeds), Chandan (Sandal wood chips and dust) etc.

Medicinal principles are usually present in different parts of the plant like root, stem, bark, heartwood, leaf, flower, fruit or plant exudates. These medicinal principles are separated by different processes where the most common being extraction. Technology for the manufacture of standardised extracts and phytochemicals is available and there are many extracts already in the international market as drugs.The promotion and development of processing of medicinal and aromatic plants have gained importance recently in many developing countries. Green consumerism and resurgence of interest for plant based products, liberalised and free market economy, increasing awareness about biodiversity conservation and sustainable use of natural resources coupled with poor socio-economic conditions of native populations are ground realities for planning and harnessing the low-cost and purpose oriented process technologies.

North-East India is comprised of eight states namely Arunachal Pradesh, Assam, Manipur, Meghalaya, Mizoram, Nagaland, Tripura and Sikkim and supports 50% of India's biodiversity. Meghalaya is endowed with 850 medicinal plants, of which 377 species are used by majority of people for their primary health care needs. In Assam, herbal plants are used at the household level in a self-help mode by the rural community. Arunachal Pradesh, the treasure house of biological and socio-cultural diversity in the Eastern Himalayan Region of India. The state has 26 major tribes and over 110 sub-tribes who maintain a close relationship with the nature. A number of underutilized minor edible fruits have been identified in the Imphal valley (excluding Jiribam sub-division, Imphal East) of Manipur. The therapeutic application of 39 plant representing 29 genera and 23 families used as medicine by the Meitei community. The rich flora of Sikkim has a number of raw drugs described in Ayurvedic texts. There are about 420 plants are used by the tribal people for various diseases in Sikkim Himalayas region. A total of 25 of such plants were recorded from the medicine men and aged villagers of various villages from Tripura. Some important plants include *Oroxylum indicum, Euphorbia nerifolia, Scoparia dulcis, Jatropha curcas* and *Kaempferia rotunda*.

Trees are erect, branched, long lived plants that are important not only for soil conservation, environmental purification but also useful for their medicinal and aromatic value for human beings. Medicinal shrubs constitute an important natural wealth of a country. They play a significant role in providing primary health care services to rural people. They serve as important therapeutic agents as well as important raw materials for the manufacture of traditional and modern medicines. They are rich sources of bioactive compounds and thus serve as important raw materials for drug production.Climbers and creepers are important components of plant diversity and are also valuable for their medicinal uses, nutrient recycling, etc. Although Western medicine strayed away from herbalism, 75% to 90% of the rural population of the rest world still relies on herbal medicine as their only health care. Medicinal herbs have curative powers and are used in making medicines because of their healing properties as a result of containing active ingredients.

The present book on 'Cultivation and Breeding of Medicinal and Aromatic Plants' has been organized in such a way that it will be useful to everyone interested in the world of medicinal and aromatic plants. One endeavour here is to make people aware of the potential of medicinal and aromatic plants from all angles and to explore the possibilities to develop new vistas for medicinal plants based industries.

The book is written with 15 independent chapters. Almost all the commercially important commercially important medicinal and aromatic plants cultivated in different parts of the world have found a place in this publication. In an introductory chapter on scope and importance of medicinal and aromatic plants has been made to provide useful and complete information on scenario of domestic and international trade. 14 interesting chapters on classification, exports and imports, biodiversity and conservation, ayurvedic medicinal plants, breeding, cultivation of medicinal trees, shrubs, climbers, herbs and aromatic plants, extraction technologies of essential oils, processing, storage and utilization of drugs and IPR issues in medicinal and aromatic plants are discussed in useful way. It may be humbly claimed that this is the first most comprehensive book of its kind on 'Cultivation and Breeding of Medicinal and Aromatic Plants.' Most of the information was gleaned from many libraries of our country. I sincerely hope that this book will be useful to teachers, students, scientists, nurserymen, pharmacists, physicians, amateur and professional gardeners, and above all progressive hobby growers. I am very much thankful to Shri Vipin Jain, Pointer Publishers, Jaipur, Rajasthan for taking keen interest in making the publication as attractive as possible.

Pakyong, Sikkim **Lakshman Chandra De**

CONTENTS

List of Colour Plates

MEDICINAL TREES

MEDICINAL SHRUBS

1

IMPORTANCE AND SCOPE

Of the 2,50,000 higher plant species on earth, more than 80,000 are medicinal. India is one of the world's 12 biodiversity centres with the presence of over 45000 different plant species. India's diversity is unmatched due to the presence of 16 different agro-climatic zones, 10 vegetation zones, 25 biotic provinces and 426 biomes (habitats of specific species). Of these, about 15000-20000 plants have good medicinal value. However, only 7500-8000 species are used for their medicinal values by traditional communities. In India, drugs of herbal origin have been used in traditional systems of medicines such as *Unani* and *Ayurveda* since ancient times. The *Ayurveda* system of medicine uses about 700 species, *Unani* 700, *Siddha* 600, Amchi 600 and modern medicine around 30 species. The drugs are derived either from the whole plant or from different organs, like leaves, stem, bark, root, flower, seed, etc. Some drugs are prepared from excretory plant product such as gum, resins and latex. Even the Allopathic system of medicine has adopted a number of plant-derived drugs which form an important segment of the modern pharmacopoeia. Some important chemical intermediates needed for manufacturing the modern drugs are also obtained from plants (E.g. diosgenin, solasodine, b-ionone). Not only, that plant-derived drug offers a stable market world wide, but also plants continue to be an important source for new drugs.

Traditional systems of medicine continue to be widely practised on many accounts. Population rise, inadequate supply of drugs, prohibitive cost of treatments, side effects of several allopathic drugs and development of resistance to currently used drugs for infectious diseases have led to increased emphasis on the use of plant materials as a source of medicines for a wide variety of human ailments.

Among ancient civilisations, India has been known to be rich repository of medicinal plants. The forest in India is the principal repository of large number of medicinal and aromatic plants, which are largely collected as raw materials for manufacture of drugs and perfumery products. About 8,000 herbal remedies have been codified in Ayurveda. The *Rigveda* (5000 BC) has recorded 67 medicinal plants, *Yajurveda* 81 species, *Atharvaveda* (4500-2500 BC) 290 species, *Charak Samhita* (700 BC) and *Sushrut Samhita* (200 BC) had described properties and uses of 1100 and 1270 species respectively, in compounding of drugs and these are still used in the classical formulations, in the Ayurvedic system of medicine. Unfortunately, much of the ancient knowledge and many valuable plants are being lost at an alarming rate. With the rapid depletion of forests, impairing the availability of raw drugs, Ayurveda, like other systems of herbal medicines has reached a very critical phase.

About 50% of the tropical forests, the treasure house of plant and animal diversity have already been destroyed. *The Red Data Book of India* has 427 entries of endangered species of which 28 are considered extinct, 124 endangered, 81 vulnerable, 100 rare and 34 insufficiently known species (Thomas, 1997). Ayurveda, Siddha, Unani and Folk (tribal) medicines are the major systems of indigenous medicines. Among these systems, Ayurveda is most developed and widely practised in India. Ayurveda dating back to 1500-800 BC has been an integral part of Indian culture. The term comes from the Sanskrit root *Au* (life) and *Veda* (knowledge). As the name implies it is not only the science of treatment of the ill but covers the whole gamut of happy human life involving the physical, metaphysical and the spiritual aspects. Ayurveda recognises that besides a balance of body elements one has to have an enlightened state of consciousness, sense organs and mind if one has to be perfectly healthy. Ayurveda by and large is an experience with nature and unlike in Western medicine, many of the concepts elude scientific explanation. Ayurveda is gaining prominence as the natural system of health care all over the world. Today this system of medicine is being practised in countries like Nepal, Bhutan, Sri Lanka, Bangladesh and Pakistan, while the traditional system of medicine in the other countries like Tibet, Mongolia and Thailand appear to be derived from Ayurveda.Phytomedicines are also being used increasingly in Western Europe. Recently the US Government has established the "Office of Alternative Medicine" at the National Institute of Health at Bethesda and its support to alternative medicine includes basic and applied research in traditional systems of medicines such as Chinese, Ayurvedic, etc. with a view to assess the possible integration of effective treatments with modern medicines.

The development of systematic pharmacopoeias dates back to 3000 BC, when the Chinese were already using over 350 herbal remedies. Ayurveda, a system of herbal medicine in India, Sri Lanka and South-East Asia has more than 8000 plant remedies and using around 35,000-70,000 plant species. China has demonstrated the best use of traditional medicine in providing the health care. China has pharmacologically validated and improved many traditional herbal medicines and eventually integrated them in formal health care system.

Green plants synthesise and preserve a variety of biochemical products, many of which are extractable and used as chemical feed stocks or as raw material for various scientific investigations. Many secondary metabolites of plant are commercially important and find use in a number of pharmaceutical compounds. However, a sustained supply of the source material often becomes difficult due to the factors like environmental changes, cultural practices, diverse geographical distribution, labour cost, selection of the superior plant stock and over exploitation by pharmaceutical industry.

Plants, especially used in Ayurveda can provide biologically active molecules and lead structures for the development of modified derivatives with enhanced activity and /or reduced toxicity. The small fraction of flowering plants that have so far been investigated have yielded about 120 therapeutic agents of known structure from about 90 species of plants.

Some of the useful plant drugs include vinblastine, vincristine, taxol, podophyllotoxin, camptothecin, digitoxigenin, gitoxigenin, digoxigenin, tubocurarine, morphine, codeine, aspirin, atropine, pilocarpine, capscicine, allicin, curcumin, artemesinin and ephedrine among others. In some cases, the crude extract of medicinal plants may be used as medicaments. On the other hand, the isolation and identification of the active principles and elucidation of the mechanism of action of a drug is of paramount importance. Hence, works in both mixture of traditional medicine and single active compounds are very important. Where the active molecule cannot be synthesised economically, the product must be obtained from the cultivation of plant material. About 121 (45 tropical and 76 subtropical) major plant drugs have been identified for which no synthetic one is currently available (table 1). The scientific study of traditional medicines, derivation of drugs through bioprospecting and systematic conservation of the concerned medicinal plants are thus of great importance.

A combination therapy integrating Ayurveda and allopathy whereby the side effects and undesirable reactions could be controlled can be thought of. Studies can show that the toxic effects of radiations and chemotherapy in cancer treatment could be reduced by Ayurvedic medications and similarly surgical wound healing could be accelerated by Ayurvedic medicines. Modern science and technology have an essential role to play in the process. An integrated approach for the cultivation, conservation and preservation of important plant species through plant molecular biology, plant tissue culture; research on the rationale and methodology of Ayurvedic medical practice; isolation of active constituents and their development into new therapeutics; standardisation and validation of known herbal medicines and other related aspects need to be focussed upon.

Despite the diverse nature of crops grown in the country and the existence of a fast growing pharmaceutical sector, the share of India in world trade is quite insignificant considering the large geographical area. However, this is bound to rise rapidly with better research inputs and efficient management of the farm sector. So far, India has been involved in the export of only large volume raw material. To achieve competitive advantage we need to resort to low volume high cost (value) trade through value addition to the raw and unfinished products. It is therefore, necessary to develop genetically superior planting material for assured uniformity and desired quality and resort to organised cultivation to ensure the supply of raw material at grower's end. Post harvest storage and process technologies need to be developed to produce the value added finished products that may be directly utilised by the industry.

Table 1. Major non-synthetic plant drugs

Drugs	Plants	Use
Vinblastine, Vincristine	*Catharanthus roseus*	Anticancer
Ajmalacine	*Catharanthus roseus*	Anticancer, hypotensive
Rescinnamine, Reserpine	*Rauvolfia serpentina*	Tranquilizer
Quinine	*Cinchona sp.*	Antimalarial,amoebic dysentery
Pilocarpine	*Pilocarpus jaborandi*	Antiglucoma
Cocaine	*Erythroxylum coca*	Topical anaesthetic
Morphine	*Papaver somniferum*	Painkiller
Codeine	*Papaver somniferum*	Anticough
Atropine	*Atropa belladonna*	Spasmolytic, cold
Atropine	*Hyoscyamus niger*	Spasmolytic, cold
Cardiac glycosides	*Digitalis sp.*	For congestive heart failure
Artemisinin	*Artemesia annua*	Antimalarial
Taxol	*Taxus baccata*	Breast and ovary cancer,
	T. brevifolia	Antitumour
Berberine	*Berberis*	For leishmaniasis
Pristimerin	*Celastrus paniculata*	Antimalarial
Quassinoids	*Ailanthus*	Antiprotozoal
Plumbagin	*Plumbago indica*	Antibacterial, antifungal
Diospyrin	*Diospyros montana*	
Gossypol	*Gossypium sp.*	Antispermatogenic
Allicin	*Allium sativum*	Antifungal, amoebiasis
Ricin	*Ricinus communis*	
Emetine	*Cephaelis ipecacuanha*	Amoebiasis
Glycyrrhizin	*Glycyrrhizia glabra*	Antiulcer
Nimbidin	*Azadirachta indica*	Antiulcer
Catechin	*Acacia catechu*	Antiulcer
Sophoradin	*Sophora subprostrata*	Antiulcer
Magnolol	*Magnolia bark*	Peptic ulcer
Forskolin	*Coleus forskohlii*	Hypotensive, cardiotonic
Digitoxin, Digoxin	*Digitalis spp.*	Cardio tonic
Thevenerin, Nerrifolin	*Thevetia spp.*	Cardio tonic
Podophyllin	*Podophyllum emodi*	Anticancer
Indicine N-oxide	*Heliotropium indicum*	Anticancer
Elipticine	*Ochrosia*	Anticancer
Homoharringtonine	*Cephalotaxus*	Anticancer
Camptothecine	*Camptotheca acuminata*	Anticancer

Chinese, Indian, Arabian and other traditional systems of medicines make extensive use of about 5000 plants. India is proud to be rich in biological diversity and tenth among the plant rich countries of Asia, sixth as far as centres of diversity especially agrodiversity are concerned. Nearly three fourth of the drugs and perfumery products used in the world are available in natural state in the country. India possesses almost 8% of the estimated biodiversity of the world with around 1,26,000 species. It is one of the 12 mega biodiversity centres with 2 hot spots of biodiversity in western Ghats and north-eastern region. The sacred groves are a miniature ecosystem conserving biodiversity in its pristine form. There are about 400 families in the world of flowering plants, at least 315 are represented in India. According to WHO, around 21,000 plant species have the potential for being used as medicinal plants. About 5000 species have been studied. There are at least 121 major plant drugs of known structure, but none of them is currently produced through synthetic means. For developing phytomedicines as a major area of concern, it would be essential to adopt a holistic interdisciplinary approach, have a scientific basis of the understanding of the plant systems, new innovations and their conservation for utilisation in future on a sustainable basis.

Potential Medicinal Plants for Commercial Cultivation

Acorus calamus, Aconitum sp., Adhatoda vasica, Aloe vera, Ammi majus, Atropa acuminata, Berberis aristata, Carica papaya, Catharanthus roseus, Cassia senna, Cephaelis ipecacuanha, Cinchona spp., Dioscorea spp., Glycyrrhiza glabara, Hedychium spicatum, Heracleum candicans, Hyoscyamus sp.muticus, Inula racemosa, Juglans regia, Juniperus spp., Matricaria chamomilla, Papaver somniferum, Plantago ovata, Podophyllum emodi, Rauvolfia serpentina, Rheum emodi, Saussurea lappa, Swertia chirata, Urginea indica, Valeriana wallichii, Zingiber officinale, Bacopa monnieri, Boerhaavia diffusa, Dudoisia myoporoides, Eclipta alba, Gymnema sylvestre, Phyllanthus amarus, Piper retrofractum Panax quinquefolium, Silybum marialum.

Medicinal Plants having Demand in Indigenous Drug Industry

Bacopa monnieri, Centella asiatica, Andrographis paniculata, Saraca asoca, Terminalia arjuna, Pterocarpus marsupium, Withania somnifera, Asparagus racemosus, Chlorophytum arundinaceum, Desmodium gangeticum, Gymnema sylvestre, Piper longum, Plumbago zeylanica, Tinospora cordifolia.

Important Indian Medicinal Plants of Global Importance

Acorus calamus, Aloe barbadensis, Ammi majus, Ammi vishaga, Atropa acuminata, Atropa belladonna, Berberis spp., Carica papaya, Cassia acutifolia, Catharanthus roseus, Cephalis ipecacuanha, Cinchona spp, Dioscorea deltoidea, Matricaria chamomilla, Plantago ovata, Rauvolfia serpentina, Rheum ausstrale, R. webbianum, Swertia chirayita, Taxus baccata, Urginia indica, Valariana wallichii.

Recent Developments in Botanical Drug Discovery

Nutraceuticals / functional foods and perfumery/aroma chemicals from natural sources is the fastest growing sector in the global market of medicine. Some examples of recent breakthroughs in botanical drug discovering of good impacts are as follows:

Anti-malarial drug: Discovery of Artemisin from *Artemisia annua* as anti-malarial compound, especially for cerebral malaria. One third of world population is exposed to the risk of malaria.

Anti-bird flu respite: The avian-influenza resulting in bird flu caused by HSNI virus can be managed through Tamilflu drug Oseltamivir produced through a 12 steps synthesis, starting from a plant product, the shikimic acid, obtained from the fruits of *Illicium verum* (Chinese star anise) and *Illicium griffithi* (Indian star anise). At present, this is the only drug available for bird flu.

Anti-inflammatory property: Turmeric (*Curcuma longa*) rhizome is being increasingly used for remedy for a number of inflammatory conditions. The pigment curcumin in turmeric rhizome has multiple uses. Non-steroidal, non-aspirin like anti-inflammatory drug is reported from gum resin of *Boswellia serrata*.

Other leads: Oleogum resin from *Commiphora wightii* (*gugul*) for use as *anti-hyperlipidemic* drug (standardized on the basis of bioactive gugulosterone E and Z); some liver protecting drugs form *Picrorhiza kurooa*, *Andrographis paniculata*, *Schizandra chinensis* are showing good promise. Compounds such as 'reserpine' from *Rauvolfia serpentina* and '*paclitaxol*' from Himalaya yew (*Taxus wallichiana*) have important pharmaceutical uses in Europe and USA. Diseases like diabetes, cancer, arthritis, asthma, depression and heart ailments are associated with both genetic and environmental risk factors. Single ingredient drugs known as new chemical entities (NCEs) often fail to cure such diseases. The plant secondary metabolites offer significant therapeutic opportunities to address such diseases. Drug regimes containing several active components have been used in both Indian Ayurvedic and Chinese systems of medicines, where mixtures of plant extract are considered integral to therapy. It should not be overlooked that biological activity of plant extracts often results from additive or synergistic effects of its components.

Aromatic plants possess odorous volatile substances which occur as essential oil, gum exudate, balsam and oleoresin in one or more parts, namely, root, wood, bark, stem, foliage, flower and fruit. The characteristic aroma is due to a variety of complex chemical compounds. The term *essential oil* is concomitant to fragrance or perfumes because these fragrances are *oily* in nature and they represent the *essence* or the active constituents of the plants. They are called *volatile or ethereal oils* as they evaporate when exposed to air at ordinary temperatures. Essential oils are highly concentrated, low volume, high value products. The world of essential oils has since then come out from the narrow field of definition to a wide variety of applications in flavours, disinfectants, oral hygiene, tobacco, pharmaceuticals and in almost all spheres of human activity. In the world wide flavour and fragrance market, essential oils constitute about 17 per cent. The estimate of world production of essential oils varies from 40,000 to 60,000 tonnes per annum. The demand for spice oils is placed at 2,000 tonnes per annum. Out of a total of about 1500 species of aromatic plants which serve as a source of raw materials for the perfumery, information on the chemistry and properties of essential oils of only about 500 species is known in some detail at present. Of these, about 50 species find use as commercial source of essential oils and aroma chemicals, though the number of those having regular and large scale utilization hardly exceeds two dozens.

Essential oils and aroma chemicals constitute a major group of industrial products. These oils form indispensable ingredients of the necessities in many spheres of human activity. They are adjuncts of cosmetics, soaps, pharmaceuticals, perfumery, confectionery, ice-creams, aerated waters disinfectants, tobacco, agarbathis and a host of related products. However, with the recent advances in organic chemistry, the synthetics have outnumbered the naturals in a ratio of about 200:1 due to limitations in the availability of the latter in sufficient quantities at a steady price over a period of time. Naturals are seen as good and safe. The concern for nature and the love for all things which are basic and natural has been spearheading to a green movement of everything natural and nature-based consumer products all over the world. A future shock is awaiting the industry as the already dwindling world resources of coal and petroleum on one side and the philosophy of going back to the nature is gaining acceptance internationally on the other side. This will revert the dependence of the industry on the petrochemicals to the perennial source of naturals. Thus, because of a large spectrum of usage in the everyday life of man, the essential oil and aroma industry has a bright future. In today's world of consumer boom, the role of essential oils has increased many folds. Apart from the hitherto known applications of essential oils, more and more areas are opening up which will benefit the industry. Use of essential oils in therapeutics is becoming popular in Japan and European countries. Aromatherapy involves the use of essential oils and aromatics derived from plants to cure diseases. Some of the essential oils are reported to be in many ways better than antibiotics due to their safety and wide spectrum of activity. Synergistic activity of essential oils needs further probe. Application of essential oils in agriculture as antifeedants, repellents, botanical insecticides, natural herbicides and growth boosters are still open to fascinating realms of research. Production of secondary metabolites in bioreactors under controlled conditions using cell and tissue culture offers exciting frontiers of future research.

❏❏❏

2

CLASSIFICATION OF MEDICINAL AND AROMATIC PLANTS

Of the 2,50,000 higher plant species on earth, more than 80,000 species are reported to have at least some medicinal value and around 5000 species have specific therapeutic value. They are classified according to the part used, habit, habitat, therapeutic value etc, besides the usual botanical classification.

MEDICINAL PLANTS

1. Based on part used

i) **Whole plant** : *Abutilon hirsutum, Boerhaavia diffusa, Phyllanthus neruri, Bacopa monnieri, Andrographis paniculata, Achillea millefolium, Arabidopsia thaliana, Artemisis annua, Borago officinalis, Hydrocotyle asiatica, Pholidota articulata,Reseda odorata, Urtica dioica, Veronica officinalis.*

ii) **Root** : *Acampe papillosa, Aconitum atrex, Butea monosperma, Cassia fistula, Catharanthus roseus, Rauvolfia serpentina, Withania somnifera, Amaranthus spinosus, Angelica glauca, Cimicifuga racemosa, Codonopsis ovata, Lavandula bipinnata, Nardostachys grandiflora, Onosma emodi, Panax ginseng, Plumbago zeylanica, Polygonatum biflorum, Potentilla fulgens, Rheum emodi, Rumex obtusifolia, Saussurea lappa, Trillium erectum, Tylophora indica.*

iii) **Stem** : *Tinospora cordifolia, Ephedra gerardiana, Lactuca sativa, Taraxacum officinale.*

iv) **Bark** : *Saraca asoca, Alstonia scholaris, Cinchona officinalis.*

v) **Leaf** : *Indigofera tinctoria, Lawsonia inermis, Aloe vera, Adhatoda vasica, Eclipta alba, Adiantum capillus veneris, Ammania baccifera, Andropogon citratus, Anisomeles indica, Calendula officinalis, Coleus blumei, Enhydra fluctuans, Euphorbia hirta, Melilotus officinalis, Mentha piperita, Rhyncostylis retusa, Trianthema monogyna, Xanthium strumarium.*

vi) **Flower :** *Biophytum sensityvum, Mimusops elengi, Anthemis nobilis, Jasminum officinale, Jasminum sambac, Rosa alba, Rosa centifolia, Rosa damascena, Tagetes erecta.*

vii) **Fruit :** *Solanum dulcamara,Solanum indicum, Embelia tsjeriam-cottam, Cordia dichotoma, Terminalia bellirica, Capsicum frutescens, Diospyros melanoxylon, Humulus lupulus, Papaver somniferum, Physalis minima, Trapa bispinosa.*

viii) **Seed :** *Datura stramonium, Vernonia anthelmintica, Ammi majus, Anethum sowa, Brassica nigra, Brassica juncea, Foenculum vulgare, Plantago ovata.*

ix) **Rhizome :** *Alpinia galanga, Cephaelis ipecacuanha, Glycyrrhiza glabra, Podophyllum hexandrum, Zingiber officinale, Acorus calamus, Coptis teeta, Curcuma aromatica, Nymphaea alba.*

x) **Pulp :** *Agave sisalana.*

xi) **Bulb :** *Allium cepa, Allium sativum, Urginea indica.*

xii) **Tubers :** *Dioscorea bulbifera, Gloriosa superba, Habenaria acuminata, Habenaria intermedia, Helianthus tuberosus,Kaempferia galanga, Orchis latifolia.*

2. Based on habit

i) **Grasses :** *Cynodon dactylon, Citronella, Cymbopogon.*

ii) **Sedges :** *Cyperus rotundus.*

iii) **Herbs :** *Vernonia cineria, Aconitum heterophyllum, Aconitum nepellus, Acorus calamus, Alpinia galanga, Andrographis paniculata, Atropa acuminata, Bacopa monnieri, Boerhaavia diffusa, Catharanthus roseus, colchicum luteum, Coptis teeta, Costus speciosus, Datura strumonium, Digitalis purpurea, Euphorbia hirta, Ferula narthax, Glycyrrhiza glabra, Hyoscyamus niger, Lobelia nicotianaefolia, Mentha arvensis, Ocimum sanctum, Plantago ovata, Podophyllum hexandra, Rheum emodi, Saussurea lappa, Solanum xanthocarpum, Swertia chirayita, Trianthema monogyna, Tribulus terrestris, Urginea indica, Valerian officinalis, Abutilon hirsutum, Acampe papillosa, Achillea millaefolium, Achyranthes aspera, Allium cepa, Allium sativum, Aloe barbadensis, Aloe vera, Alpinia galanga, Amaranthus tricolor, Ammania baccifera, Ammi majus, Angelica glauca, Anethum sowa, Anthriscum cerefolium, Argemone mexicana, Artemisia vulgaris, Calendula officinalis, Carthamus tinctorius, Chenopodium album, Chlorophytum arundinarea, Crocus sativus, Coriandrum sativum, Dendrobium nobile, Eclipta alba, Enhydra fluctuans, Erigeron canadensis, Erodium cicutarium, Eulophia dabia, Equisetum arvense, Geranium wallichianum, Habenaria accuminata, Malaxis muscifera, Malva parviflora, Marsilea minuta, Melilotus officinalis, Orchis latifolia, Panax ginseng, Papaver somniferum, Pholidota articulata, Polycarpa corymbosa, Polygonum aviculare, Portulaca oleracea, Primula reticulata, Rhyncostylis reticulata, Ruta graveolens, Thymus serphyllum, Tagetes erecta, Trillium erectum, Typha latifolia, Urtica dioica, Vanda tessellata.*

iv) **Shrubs :** *Adhatoda vasica, Aristolochia indica, Berberis aristata, Ephedra gerardiana, Gaultheria fragrantissima, Holarrhenna antidysenterica, Lawsonia inermis, Rauvolfia serpentina, Sida cordifolia, Withania somnifera, Nerium* *oleander, Hibiscus rosa sinensis, Hibiscus mutabilis, Jasminum humile, Jasminum officinale, Jasminum multiflorum, Lavandula officinalis, Plumbago zeylanica, Calotropis gigantea, Woodfordia fruiticosa, Vitex negundo.*

v) **Climbers :** *Asparagus racemosus, Hydrocotyle asiatica, Hemidesmus indicus, Ipomoea hederacea, Piper nigrum, Tinospora cordifolia, Tylophora indica, Coccinia indica, Gloriosa superba, Humulus lupulus, Melothria heterophylla, Dioscorea alata, Basella rubra, Paederia foetida, Cissus quadrangularis, Mucuna pruriens, Ipomoea aquatica, Passiflora edulis, Clitorea ternatea, Merremia hederacea, Antigonon leptopus, Citrullus colocynthis.*

vi) **Trees :** *Azadirachta indica, Aegle marmelos, Alstonia scholaris, Butea monosperma, Cassia fistula, Cinchona officinalis, Cinnamomum zeylanicum, Emblica officinalis, Hydnocarpus kurzu, Madhuca indica, Mallotus philippensis, Pinus roxburghii, Saraca indica, Santalum album, Sterculea urens, Eugenia jambolana, Tamarindus indica, Pterocarpus marsupium, Nyctanthes arbortristis, Bauhinia variegata, Diospyros melanoxylon, Citrus medica, Messua ferrea, Michelia champaka, Moringa oleifera, Acacia nilotica, Olea europea, Terminalia arjuna, Phyllanthus acidus, Averrhoea carambola.*

3. Based on climatic zones

i) **Tropical :** *Andrographis paniculata, Acorus calamus, Azadirachta indica,Adhatoda vesica, Centella asiatica, Ocimum basilicum, Paederia foetida, Withania somnifera.*

ii) **Sub-tropical :** *Mentha arvensis, Costus speciosus, Artemisia maritima, Hydnocarpus kurzii, Lavendula vera, Litsea cubeba, Mucuna pruriens, Pogostemon calelin, Zanthoxylum armatum, Curcuma caesia, Leonotis nepetaefolia.*

iii) **Temperate :** *Atropa belladonna, Coptis tita, Ephedra gerardiana, Geranium nepalensis, Panax pseudoginseng, Swertia chirata, Picrorrhiza kurooa, Satyrium nepalensis, Rubia cordifolia, Taxus baccata, Orchis latifolia.*

iv) **Alpine :** *Aconitum ferox, A. heterophyllum, Illicium griffithii, Berberis* spp., *Podophyllum hexandrum, Rheum emodi, Delphinium subulatum.*

4. Based on life cycle

i) **Annual herbs :** Anise, Basil, Borage, Calendula (Pot Marigold), Chamomile, Chervil, Coriander.

ii) **Biennial herbs :** Caraway seeds, Prime rose, Bai Zhi, Mullein, Teasel, Viper's Bugloss.

iii) **Aromatic foliages :** Allium spp., Artemisia spp, *Foeniculum vulgare, Tagetes minuta.*

iv) **Perennial herbs :** *Aloe vera*, Fennel, mints.

5. Based on therapeutic value

Antimalarial : *Cinchona officinalis, Artemisia annua.*

Anticancer : *Catharanthus roseus, Taxus baccata.*

Antiulcer : *Azadirachta indica, Glycyrrhiza glabra.*

Antidiabetic : *Catharanthus roseus, Momordica charantia, Tinospora cordifolia, Woodfordia fruticosa, Syzygium cumini.*

Anticholesterol : *Allium sativum.*

Antiinflammatory : *Curcuma domestica, Desmodium gangeticum.*

Antiviral : *Acacia catechu.*

Antibacterial : *Plumbago indica.*

Antifungal : *Allium sativum.*

Antiprotozoal : *Ailanthus sp., Cephaelis ipecacuanha.*

Antidiarrhoeal : *Psidium gujava, Curcuma domestica.*

Anti- dysenteric : *Holarrhena antidysenterica, Punica granatum, Syzygium cumini, Grewia hirsuta.*

Anti-Rheumatic : *Plumbago Zeylanica, Withania somnifera, Datura metel, Commiphora wightii, Curcuma longa, Tinospora cordifolia.*

Hypotensive : *Coleus forskohlii, Alium sativum.*

Tranquilizing : *Rauvolfia serpentina.*

Anaesthetic : *Erythroxylum coca.*

Spasmolytic : *Atropa belladona, Hyoscyamus niger.*

Diuretic : *Phyllanthus niruri, Centella asiatica.*

Astringent : *Piper betle, Abrus precatorius.*

Anthelmentic : *Quisqualis indica, Punica granatum.*

Cardiotonic : *Digitalis sp., Thevetia sp.*

Antiallergic : *Nandina domestica, Scutellaria baicalensis.*

Hepatoprotective : *Silybum marianum, Andrographis paniculata, Aloe vera, Glycyrrhiza glabra, Ocimum americanum, Zingiber officinale, Piper longum.*

Anti-fertility drugs : *Berberis aristata,Picrrorrhiza kurroa, Artemisia maritima, Aconitum ferox* and*Podophyllum hexandrum.*

6. Based on Ayurvedic formulations

a) The ten roots of the Dasamoola (*Dasamoolam*)

i) *Desmodium gangeticum (Orila)*

ii) *Uraria lagopoides (Cheria orila)*

iii) *Solanum jacquinii (Kantakari)*

iv) *Solanum indicum (Cheruchunda)*

v) *Tribulus terrestris (Njerinjil)*

vi) *Aegle marmelos (Koovalam)*

vii) *Oroxylum indicum (Palakapayyani)*

viii) *Gmelina arborea (Kumizhu)*

ix) *Steriospermum suaveolens (Pathiri)*

x) *Premna spinosus (Munja)*

b) The ten flowers of the Dasapushpa (*Dasapushpam*)

i) *Biophytum sensitivum (Mukkutti)*

ii) *Ipomoea maxima (Thiruthali)*

iii) *Eclipta prostrata (Kayyuniam)*

iv) *Vernonia cineria (Poovamkurunnil)*

v) *Evolvulus alsinoides (Vishnukranthi)*

vi) *Cynodon dactylon (Karuka)*

vii) *Emelia sonchifolia (Muyalcheviyan)*

viii) *Curculigo orchioides (Nilappana)*

ix) *Cardiospermum halicacabum (Uzhinja)*

x) *Aerva lanata (Cherula)*

c) The four trees of the Nalpamara (*Nalpamaram*)

i) *Ficus racemosa (Athi)*

ii) *Ficus microcarpa (Ithi)*

iii) *Ficus religiosa (Arayal)*

iv) *Ficus benghalensis (Peral)*

d) The three fruits of the Triphala (*Thriphalam*)

i) *Phyllanthus emblica (Nellikka)*

ii) *Terminalia bellerica (Thannikka)*

iii) *Terminalia chebula (Kadukka)*

7. Based upon active constituents

i) Aromatic herbs: Fennel, ginger, garlic, lemon herb.

ii) Astringent herbs: Peppermint, red raspberry.

iii) Bitter herbs: Aloe, cascara, liquorice, pumpkin, senna, yellow dock, yucca, barberry, gentian, safflowers, asparagus, blessed thistle, burdock, butcher's broom, buchu, chaparral, chickweed, corn silk, dandelion, dog grass, grapevine, and parsley.

iv) Mucilaginous Herbs: Althea, aloe, burdock, comfrey, dandelion, Echinacea, fenugreek.

v) Nutritive herbs: Asparagus, banana, broccoli, cabbage, carrot, cauliflower.

8. Botanical classification

Most of the medicinal and aromatic plants belong to the following families:

- Acanthaceae (*Adhatoda vasica*, *Andrographis paniculata*, *Hygrophila auriculata*)
- Agavaceae (*Agave sisalana*)
- Amaranthaceae (*Achyranthes aspera*, *Amaranthus spinosus*)
- Apocynaceae (*Alstonia scholaris*, *Catharanthus roseus*, *Holarrhena antidysenterica*, *Nerium oleander*)
- Araceae (*Acorus calamus*)
- Boraginaceae (*Borago officinalis*, *Onosma echioides*)
- Caryophyllaceae (*Arenaria serpyllifolia*)
- Compositae (Artemisia, Calendula, *Achillea millefolium*, *Anaphalis neelgerriana*, *Anthemis nobilis*, *Arnica sororia*, *Enhydra fluctuans*, *Erigeron canadensis*, *Helianthus tuberosus*, *Lagera alata*, *Saussurea lappa*, *Tagetes erecta*)
- Cruciferae (*Armoracia rusticana* , *Brassica nigra*, *Arabidopsis thaliana*)
- Cucurbitaceae (*Bryonia alba*, *Coccinia indica*, *Cucumis trigonus*)
- Euphorbiaceae (*Emblica officinalis*, *Euphorbia hirta*, *Phyllanthus niruri*)
- Labiatae (*Mentha arvensis*, *Ocimum sanctum*, Lavandula, Pogostemon, *Ajuga bracteosa*, *Anisomeles indica*, *Coleus blumei*)
- Leguminosae (*Cassia fistula*, *Glycyrrhiza glabra*, *Saraca asoca*, *Melilotus officinalis*, *Trifolium pratense*)
- Liliceae (*Colchicum autumnale*, *Asparagus aphyllus*, *Aloe vera*, *Allium cepa*, *Allium sativum*, *Chlorophytum arundinaceium*, *Gloriosa superba*, *Polygonatum biflorum*, *Urginea indica*)
- Malvaceae (*Sida cordifolia*, *Malva sylvestris*, *Abelmoschus moschatus*)
- Meliaceae (*Azadirachta indica*)
- Nyctaginaceae (*Boerhavia diffusa*, Mirabilis jalapa)
- Oleaceae (*Nyctanthes arbortristis*, *Jasminum grandiflorum* , *Jasminum officinale*, *Jasminum sambac*)
- Orchidaceae (*Acampe papillosa*, *Dendrobium nobile*, *Vanda tessellata*, *Coelogyne cristata*, *Cypripedium calceolus*, *Eria muscicola*, *Eulophia dabia*, *Habenaria spp.*, *Malaxis cylindrostachya*, *Orchis latifolia*, *Pholidota articulata*, *Rhyncostylis retusa*, *Satyrium nepalense*)
- Papaveraceae (*Papaver somniferum*, *Argemone mexicana*)
- Phytolaccaceae (*Phytolacca Americana*)
- Poaceae (Cymbopogon, Citronella)
- Ranunculaceae (Aconitum, *Coptis teeta*, *Anemone pulsatilla*, *Cimicifuga americana*, *Hepatica acutiloba*)

- Rosaceae (*Rosa gallica*, *Potentilla fulgens*, *Rosa alba*, *Rosa centifolia*, *Rosa damascena*)
- Rubiaceae(Ipecacuanha, Cinchona, *Rubia cordifolia*)
- Rutaceae (*Aegle marmelos*, wood apple)
- Scrophulariceae (*Bacopa monnieri*, *Digitalis purpurea*, *Picrorhiza kurrooa*, *Scoparia dulcis*, *Veronica officinalis*)
- Solanaceae (*Atropa belladonna*, Datura, *Withania somnifera*, *Hyoscyamus niger*, *Physalis minima*, *Solanum indicum* , *Solanum dulcamara*)
- Umbelliferae (*Hydrocotyle asiatica*, *Ferula narthex*, *Ammi majus*, *Anethum sowa*, *Anthriscus cerefolium*, *Archangelica officinalis*)
- Verbenaceae (*Vitex agnus-castus*)
- Zingiberaceae (*Alpinia galanga*, *Elettaria cardamomum*, *Curcuma amada*, *Curcuma aromatica*, *Curcuma caesia*, *Kaempferia galanga*, *Zingiber officinale*)

AROMATIC PLANTS

Out of a total of about 1500 species of aromatic plants known, only a little over 300 species have been studied in some detail. Of the 50 species which find use as commercial source of essential oils and aroma-chemicals, the number of those having regular and large scale utilization hardly exceeds two dozen.

Based upon growth habit aromatic plants are classified into four categories:

Aromatic grasses : Lemongrass, palmarosa, citronella, vetiver.

Aromatic herbs and shrubs : Mints, ocimums, patchouli, rosemary, clarysage, thyme, celery, coriander, cumin, fennel, ajowan, davana, chamomile, geranium, cardamom, ginger, kacholam.

Aromatic trees : Sandalwood, eucalyptus, clove, camphor, cinnamon, nutmeg, linaloe.

Aromatic flowers : Rose, jasmine, tuberose, marigold, champaca.

□□□

3

Exports and Imports of Medicinal and Aromatic Plants

The demand for medicinal plants is increasing everyday and the World Health Organization (WHO) has projected that the global herbal market will grow to $ 5 trillion by 2050 from the current level of $ 62 billion . India and China are the largest producer of more than 70% of the global diversity. India's share is 2. 5%, against China's 13%. The significant global herbal export market include EU, USA, Canada, Australia, Singapore, and Japan. New emerging market (globally) includes Brazil, Argentina, Mexico, China and Indonesia. The major products as within the herbal sector in an international perspective include pharmaceuticals, which is further divided into medicinal and aromatic plants, plant material extracts, plant materials, spices, herbs and cosmetics. The market place for medicinal plants used in dietary supplements is growing in the European continent along with Japan and USA.

The top 10 highly traded medicinal plants include Amlaki (*Emblica officinalis Gaertn.)*, Satavari (*Asparagus racemosus Willd.)*, Aswagandha (*Withania somnifera Dunal.)*, Haritaki (*Terminalia chebula Retz.)*, Asoka (*Saraca asoca DC. Willd.)*, Bilva (*Aegle marmelos Corr.)*, Svarnapatri or Senna leaves (*Cassia angustifolia Vahl.)*, Vasa (*Adhatoda vasica Nees*), Pippali (*Piper longum Linn.)* and Brahmi (*Bacopa monneri (Linn.) Wettst*) and other plant material that have huge demand in foreign markets are Isabgol (Psyllium husk, Psyllium seeds), Chandan (Sandal wood chips and dust) etc.

Medicinal plants imported to India are *Cuscuta ephithymum*, *Glycyrrhiza glabra*, *Lavandula*, *Operculina turpethum*, *Pimpinella anisum*, *Smilax china*, *Smilax ornata*, *Commiphora weightii*, *Quercus infectoria*, *Garania pictoria*, *Panax pseudoginseng* and *Thymus vulgaris.*

Top 20 medicinal plants have domestic demand are *Emblica officinalis* Gaertn, *Asparagus racemosus* Willd. *Withania somnifera* Dunal. *Terminalia chebula* Retz., *Saraca asoca* [Roxb.], *Aegle marmelos* Carr., *Cassia angustifolia* Vahl., *Adhatoda vasica* Nees *Piper longun* Linn., *Bacopa monnieri* [Linn.], *Sida cordifolia* Linn., *Ocimum sanctum* Linn.

Bambusa bambos Druce., *Boerhaavia difflusa* Linn., *Azadirachta indica* A. Juss. *Solanum nigrum* Linn. *Woodfordia fruticosa* Kurz., *Andrographis paniculata, Syzygium aromaticum and Tinospora cordifolia.*

Indian herbal market is nearly Rs. 5000 crores, which is having annual growth of 14%. Rs. 100 crores worth of herbal product/ingredients/ extracts are being exported.

Modern Pharmacopoeia contains about 25 per cent plant drugs. China has made very good progress in utilizing herbal wealth by promoting its use in the developed world. In China, the annual herbal production is worth US$ 48 billion, with export of about US$ 3.6 billion. Global market for all categories of herbal products estimated at US$ 62 billion. The global market for botanical and plant derived drugs increased from US$ 9.5 billion in 2008 to US$ 32.9 billion in 2013, at an annual growth rate of 11.0 per cent according to a 2009 study by BBC research.

The herbal products industry comprises a number of interrelated sub-section including, i) phytochemicals; ii) functional food; iii) nutraceuticals; iv) herbal teas; v) ethical OTC medicines; vi) flowers and fragrances; vii) aroma therapy; viii) culinary and spices use. Functional food alone enjoys a market of about US$ 2.0 billion in Europe and US$ 5.0 billion in USA, mainly for improvement of gut health, heart health, bone health and immune function. Out of many uses, the most popular ten best selling herbal medicines in USA are Echinacea, *Echinacea species,* Saw palmeto *Serenoa repens,* Garlic, *Allium sativum,* Aloe, *Aloe barbadensis,* Goldenseal *Hydratis Canadensis,* Ephedra *Ephedra species,* Ginseng, *Panax species,* Eleuthero *Eleutherococcus senticosus,* Ginko, *Ginko biloba,* Cranberry *Vaccinium macrocarpon.*

Plant parts of medicinal species include - whole plant, roots, leaves, flowers, fruits, seeds, stem, bark and wood. Destructive harvest of the species in which whole plant and roots are mainly used is an ecological concern, particularly when collected from wild. Many such valuable plants now fall under rare, endangered and threatened (RET) categories.

China and India are the world's leading producing countries, whereas Hong Kong, USA and Germany are the important trade centres. China is a major growing and manufacturing center for herbal extracts and finished products. In addition to wild collection, role of cultivation in meeting pharmaceutical demand is expanding through R&D support in selection of cultivars, rapid propagation of quality planting material and other good agricultural practices. In some areas, MAP species are playing central role in income generation of small and marginal land holders. Of late, wild collections and trade of medicinal plants, mainly in Asia, are being growingly regulated for collection, transit and export. Traditional medicine practitioners prefer wild collected raw materials but R&D efforts in a few species have been able to establish that cultivation rather than wild collection, is a preferred option for sustainable development. Medicinal and aromatic plants, namely, opium poppy (*Papaver somniferum*); isabgol (*Plantago ovata*), senna (*Cassia angustifolia*), mentha spp (*M. arvensis, M. spicata*), aloe (*Aloe barbadensis*), *aonla* (*Emblica officinalis*), kalmegh (*Andrographis painculata)*, long pepper (*Piper longum*), *Cymbopogan* spp. lemon grass; *Citronella* grass and others are cultivated as crops in India.

In China, important cultivated crops are *Penax ginseng*, *Angelica sinensis*, *Coptis chinensis*, and *Paeonia suffriticosa*. It is reported that more than 100 species are cultivated on a commercial scale. In addition, some plant species originated in Asia are cultivated on a commercial scale in Europe and other countries. For example, *Gingko biloba*, native to Southeast China and *Strychnos nux-vomica*, native to South Asia, are cultivated in Europe and China, respectively.

For pharmaceuticals ranging from digitalin to vincristine, the ethnobotanical approach to drug discovery has proved to be successful. In traditional medicine systems, both in India and China, the raw materials are still sourced from the wild. Northeast Himalayan region of India, Nepal, Myanmar and other countries are rich source of wild collections. It is reported that in Nepal, every year more than 15,000 tons of herbs are collected from the wild. The loss of genetic diversity through destructive harvesting and overexploitation may cause enormous problems in the context of sustainable development of MAP.

Global Trends in Use of Essential Oils : Aromatic plants and their products, particularly the essential oils, are now becoming one of the most important export items from many developing countries of Asia. Natural essential oils, the volatile extracts from aromatic plants, are chemical compounds with an odoriferous nature, insoluble in water but soluble in organic solvents. The term 'essential' refers to the presence of an essence or odour and the term 'oil' is used because when these compounds are placed on a transparent paper, they leave behind an oil spot.

Recent advances in chemistry have led to development of technology for preparation of individual odoriferous principles from the aromatic plant materials. Now, there are large essential oil production units which are engaged not only for isolation of the suitable aromas but also blend them to use for perfumes, industry and in making a great variety of products ranging from food, medicine, cosmetics, toiletries, paints, insecticides and a lot of several other products.

World's total production of essential oils is estimated at about 0.10 to 0.11 million tons. Brazil with its production of citrus oil at 40,000 tons is the largest producer of essential oils in the world. In terms of value, India stands in second position with a world share of 21-22 per cent. Among the essential oils, Japanese mint oil, ambrettee seed essence, sandal wood oil, *Citronella* oil, lemon grass oil, palmarosa oil, ginger oil, clove oil, eucalyptus oil, vetiver oil, pepper oil, patchouli oil, etc. India has become largest producer of mint (*Mentha arvensis*) and basil (*Ocimum basilicum*). In recent years, jasmine oil and tube rose oil have made a debut in the world market. Essential oils can be increasingly utilized in diversified areas. In the Northeastern region of India, both patchouli and himalomena aromatic oil are being increasingly produced. Ginger oil and oleoresins are also in use.

Dental care: World's consumption of essential oils and its fractions such as menthol, eucalyptus, carvone, anethola and methyl salicylate in dental care sector is estimated to be 10,000 - 12,000 tons valued at US $ 150 million.

Flavours/Fragrances: The demand for natural flavours is increasing since perception has changed in favour of flavours, which are considered as safe. Consumption of flavours /fragrances was estimated to be of US $ 10,000 million worldwide. Flavours and fragrances, are the two most important sectors covering about 70 per cent within essential oil groups. Out of present market of flavours, 50 per cent comes from natural flavours, the value of which is estimated at US $ 1,200 million. USA and Western Europe together account for 62 per cent of consumption. If Japan is also included, the three major industrial regions with only 15 per cent of world's populations account for 78 per cent of the consumption. Many diverse aroma isolates are sought by the industry for manufacturing of various blended flavours. Essential oils and herbal extracts and their value added products are to be considered as "Commodity" and vigorous marketing and promotional drives are needed. Many countries are making huge investments in promoting their commodities e.g. tea tree oil and eucalyptus oil by Australia, citrus oils and mint oils by USA, citrus oils by Italy, lavender, rosemary, thyme oils by Spain and rose oil by Turkey and Bulgaria.

Challenges

Although there are ample opportunities available in this sector, yet this sector is facing many challenges. Some of these are as follows:-

- Poor agricultural practices and lack of knowledge.
- Irregular availability and inconsistent quality of raw material.
- Lack of R&D of high yielding varieties, product and process development.
- Inefficient processing techniques.
- Poor quality control procedures.
- Lack of regulatory mechanism in trade of medicinal plants
- Traditional medicines also face increasing competition from synthetic drugs.

On export front this sector faces the following challenges.

- Lack of credible documentation of the therapeutic values of medicines and their formulation is a major constraint in export particularly to countries to USA, EU.
- Lack of assumed availability for exports, in terms of quality, quantity and price.
- Lack of information about the total world trade, demand dynamics in the overseas market, marketing techniques etc.
- Lack of knowledge on international and specific regulations of importing countries governing imports of such products, including quality, intellectual property rights (IPR) issues etc.
- Lack of standardization in products, processes and services.

Opportunities

1. Traditional medicines (TM) have been used since ancient times, and empirically passed on from generation. TM has been used to maintain, and promote the human health, prevention as well as to reduce symptomatic disorders and has made great contribution to health.
2. Above table and its analysis shows the export volume of medicinal plants and herbal products is increasing in good percentages.
3. Government has given incentive to drug manufacturers, entrepreneurs, AYUSH institutions etc. for registration of their products for exports to targeted countries.

4. The market for finished formulations in the developed countries should be tapped, instead of exporting crude drug or primary extracts.
5. Identification of specific Herbs for specific disease and successful promotion of these drugs/ or health supplements to developed countries is vital to create interest in the global community. A need-accretion for the Ayurvedic or herbal product from India by this method will help to create a strong footing for the Indian herbal product or medicine. This will ensure sustainable growth of herbal exports from India.
6. Exporting the finest quality of herbal extracts to the Pharmaceuticals, Food processing industries and bulk consumers in the global market to gain value advantage.
7. Proper marketing strategy should be developed to fashion the Indian medicine system and terminologies in the overseas markets and popularize them, considering the mindset of users .

Strategy to Improve Trade of Medicinal Plants and their Products

Following strategies are suggested to improve national as international trade of medicinal plants and their various products :

i) R&D on all aspects of medicinal plants must be intensified.

ii) Materials of proven safety,effectivity and quality need onlybe exported.

iii) Indian plants must be domesticated and cultivated for consistent quality of the produce.

iv) Crop-wise post harvest processing technology for obtaining high quality produce must be developed.

v) Toxicity (safety) and efficacy trials need to be conducted on ayurvedic, unani and sidha medicines and plants drugs.

vi) Cultivation of exotic plants already in extensive usage in USA and Europe must be introduced in India.

vii) Phytomedicinal monographs for each of the medicinal plant, including information on physiological effects, efficacy, quality standard and references etc., are needed to be developed.

viii) Cultivation of medicinal plants should be integrated with that of food crops by devising suitable cropping schedules.

□□□

4

BIODIVERSITY AND CONSERVATION OF MEDICINAL AND AROMATIC PLANTS

India, with its diverse agro-climatic conditions and regional topography, has been considered as the treasure house or botanical garden of plant genetic resources. Hence, India is recognized as one of the world's top 12 mega diversity nations. Our herbal wealth constitutes more than 8,000 species and accounts for around 50 % of all higher flowering plant species of India; around 70 % of the medicinal plants in the country are spread across the tropical forests of Western Ghats. However, available information shows that 1,800 species are used in Classical Indian systems of medicines. Ayurveda uses 1,200, Siddha – 900, Unani – 700, Amchi – 600, Tibetan – 450.

The emerging field of herbal products industry holds a great potential to the economic development of the Indian region. Usage of herbs as a source of food, medicine, fragrance, flavour, dyes and other items in Indian systems of medicine is in increasing trend. It is estimated that, 95 % of the medicinal plants used in Indian herbal industry today are collected from wild. About half a million tonnes of dry material is collected through destructive means indiscriminately and 1.65 lakh ha. of forest is cleaned and felled each year.

With the increase in population, rapid expansion of area under food and commercial crops, deforestation, extension of urban area, establishment of industries in rural areas, etc., there is considerable depletion of plant genetic resources wealth, many of them being in the process of extinction day by day.

BIODIVERSITY – OVERVIEW

Although there are around 8,000 medicinal plant species used by different communities in India across different ecosystems, only around 10% of them (880 species) are in active trade. Among these, around 48 species are exported in the form of raw drugs and extracts, while around 42 species are imported. The wild populations of about 100 of the traded species are known to have declined, thereby making them to be considered threatened. This is the situation of raw drug trade in India that unfolds.

How Many Herbs?

Habit wise analysis of these 880 medicinal plants, indicates that these are well distributed across different life forms with the majority belonging to the herbaceous category. The highest proportion of herbs (41%), including grasses, is followed by trees (26%), shrubs (17%) and climbers (16%).

Botanical Base

It is interesting to note that these 880 traded plants are distributed across 151 families. Of these 79% belong to the Dicots while the Monocots constitutes 11% followed by Pteridophytes 5% and Gymnosperms 3% and one percent each from Fungi and Lichens. This indicates that a very small proportion of raw drugs belong to the category of lower classes in the plant kingdom (Table 1).

Table 1. Botanical representation of traded medicinal plants (Class wise)

Class	Number of Families
Dicots	119
Monocots	17
Pteridophytes	8
Gymnosperms	5
Fungi	1
Lichens	1

Family wise representations of important medicinal species are given in Table 2.

Table 2. Botanical representations of traded medicinal plants (Family wise)

Family	Species
Fabaceae	67
Asteraceae	54
Euphorbiaceae	48
Caesalpiniaceae	41
Apiaceae	37
Lamiaceae	37
Solanaceae	35
Cucurbitaceae	32
Rubiaceae	29
Malvaceae	28

Geographic Distribution

The natural occurrence of traded species is another element of interest here. It is found that these species are distributed across different bio-geographic zones, diverse habitats and landscape elements. About 18% of these species are confined to Himalayan and Trans Himalayan zone including North East India while around 4% is restricted to

Western Ghats and 0.5% is found only in the Desert zone. The rest of the species (around 77%) have a wide range of distribution across the other bio-geographic zones of the country (Table 3).

Table 3. Bio-geographical distribution of traded medicinal plants

Bio-geographic zones	No of medicinal plants identified	Example species
1	2	3
Trans – Himalayan	700	*Ephedra gerardiana, Hippophae rhamnoides, Physochlania praalta, Arnebia euchroma, Ferula jaeschkeana.*
Himalayan	2900	*Aconitum heterophyllum, Arnebia benthamii, Dactylorhiza hatagirea, Podophyllum hexandrum, Picrohiza kurroa, Pistacia chinensis, Nardostachys grandiflora, Rubia sikkimensis, Coptis teeta, Polygonatum cirrhifolium, Swertia chirayata, Valeriana jatamansi, Rhododendron anthopogon, Taxus wallichiana.*
Desert areas	500	*Tecomella undulata, Tribulus rajasthanensis, Citrullus colocynthis, Commiphora wightii, Acacia nilotica.*
Semi-arid areas	1000	*Balanites aegyptiaca, Withania coagulens, Tribulus alatus, Commiphora wightii, Boswellia serrata, Canscora, Acacia nilotica.*
Western ghats	2000	*Myristica malabarcia, Garcinia indica, Coscinium fenestratum, Hydnocarpus pentatandra, Garcinia gummigutta, Vateria indica, Nilgirianthus ciliatus.*
North East India	2000	*Aquilaria malacensis, Smilax glabra, Ambroma augusta, Hydnocarpus kurzii, Vetivaria zizaniodes.*
Deccan peninsula	3000	*Embelia tsjeriam-cottam, Caesalpinia digyna, Screbera sweitenoides, Decalepis hamiltonii, Pterocarpus santalinus.*
Gangetic plains	1000	*Holarrhenaq pubescens, Mallotus philippensis, Pluchea lanceolata, Peganum harmala.*
Andaman & Nicobar	1000	*Claophyllum inophyllum, Adnanthera pavonina,* islands *Barringtonia asiatica, Aisandra butyrace.*
Coastal islands	500	*Rhizophora mucronata* Lam., *Acanthus ilicifolius, Avicennia marina, Sonneratia caseolaris.*

Endangered Medicinal Plants

Plant parts like leaves, bark, roots, fruits, seeds or even whole plant is indiscriminately collected from wild sources without taking care of saving the plants. Many of the important useful species are on the verge of extinction due to over-exploitation and habitat destruction. More than 95% of the medicinal plants are collected from the wild; a number of them have become endangered in their natural habitats. There is need to encourage multiplication and cultivation of these plants. Collection of the following species from wild sources should be prohibited.

Aconitum sp. (Monk's Hood, Bachang), *Acorus* spp.(Sweet Flag, Vekhand), *Anchusa strigosa* (Gaozaban), *Aristolochia bracteata* (Kiramar), *Artemisia annua* (Worm wood), *Atropa acuminata* (Indian belladonna), *Berberis aristata* (Indian Berbery, Daru Haridra), *Brunium persicum* (Kala Zeera), *Chlorophytum* spp (Safed Musli), *Colchicum luteum* (Tara-Tutiya, Suranjan-1- Talah, Golden Collyrium), *Commiophora wightii* (Guggul), *Concinum fariestatum* (Jeevanti), *Coptis teeta* (Halad-Vachnag, Gold Thread, Mamira), *Curculigo orchioides* (Kali Musli), *Didymocarpus pedicellata* (Shila-Pushpa), *Dorsera* sp. (Sundew), *Ephedra gerardiana* (Somlata), *Eulophia campestris* (Saleb Misri), *Ferula jaeskaena* (Indian Hing), *Gentiana kurroa* (Indian Gentian Pashanbheda), *Gloriosa superba* (Glory Lily, Agnisikha), *Gynocardia odorata* (Kadu Bonsha), *Holostemma annualare* (Peet Chandan, Kasturi Mazil), *Hydrocarpus* sp., *Hyoscyamus niger* (Henbane), *Inula racemosa* (Pshkar Mool), *Iphigenia indica* (Nirpani), *Lilium polyphyllum* (Kshira Kankoli), *Microstylis nicifera* (Jeevaka), *Microstylis wallichii* (Rishvakê), *Nardostachys grandiflora* (Jata mansi), *Onosma bracteatum* (Ratnajyot), *Orchis latifolia* (Salam Panja), *Panax pseudo-ginseng* var. *himalicus* (Indian Ginseng), *Physochlaena praelta* (Scholar, Lalthang), *Picrorhiza kurroa* (Kutki), *Piper cubeba* (Kababchini), *Podophyllum hexandrum* (Bankari), *Rauwolfia serpentina* (sarpagandha), *Rheum astrata* (Rewadchini), *Rheum emodi* (Indian Rhubarb, Gandhini), *Saussurea sacra* (Yogispada), *Swertia chirayata* (Chirayata), *Taxus baccata* (Talispatra), *Taxus wallichiana* (Indian Yew).

Loss of Biodiversity of Medicinal Plants

Environmental factors

Rainfall: For the past few years the annual rainfall has decreased resulting in the health of many herbaceous species during summer months.

Deforestations: Deforestations have been reported over the last two decades. The spread of agriculture, logging, fire wood collection, heavy wood collection, heavy grazing, etc., are the main reasons for reduction in area under valuable forests. Many valuable wild medicinal plant species are eradicated or minimized every year due to the deforestation activities.

Siltation of water bodies: Siltation of water bodies in the forests has resulted in the reduction of water holding capacity heading to depletion of underground water.

Lack of pollinators: Honey bee colonies have declined in numbers to the extent of 50- 60%, in forests and other areas. Loss of pollinators has resulted in reduced seed set and dispersal of seeds.

Developmental Activities

Submersion: Loss of many species of medicinal plants has been noticed in forests due to submersion, eg., the Maradavally forest is the catchments of Linganamakki Dam, the main reservoir of Karnataka for irrigation and power generation. Submersion of nearly 10 sq. km of forest area during monsoons has resulted in loss of valuable medicinal plant species.

Infrastructure: Expansion of roads, installation of power lines and construction of buildings has caused extensive damage to forests and medicinal plants, eg., Devanarayandurga forest in Karnataka.

Agriculture and Forestry Methods

Monoculture: There has been a progressive increase in monoculture plantations of economically important indigenous as well as exotic species in forest. Monoculture plantation totally affects the organic productivity and reduces the natural stability and complexity resulting in loss of medicinal plants. eg.,*Eucalyptus* and *Acacia* species in many forests.

Encroachments: Encroachments over forestlands have assumed alarming levels. Apart from felling of trees and clearing vegetation, the cultivation practices followed on high sloppy lands has caused soil erosion, and decline of medicinal plant wealth.

Over-exploitation: Gathering of medicinal plants from the forests are rampant. The collection was by unorganized forest collectors, who, in turn sold the product to a contractor at the price fixed by the latter. But now, due to the awareness created by the members of the 'Local Traditional Medicinal Practitioners Association', illegal gathering has been controlled to a certain extent.

Conservation of Medicinal and Aromatic Plants

Conservation strategy (IUCW, UNEP & WWF, 1980) defines conservation as "the management of human use of the biodiversity so that it may yield the greatest sustainable benefit to present generation while maintaining its potential to meet the needs and aspirations of future generations".

Strategies & Priorities : The primary goals of biodiversity conservation as envisaged in the World Conservation Strategy can be summarized as follows:

- Maintenance of essential ecological processes and life support systems on which human survival and economic activities depend.
- Preservation of species and genetic diversity and sustainable use of species and ecosystems which support millions of rural communities as well as major industries.

Strategies for Conservation of Medicinal Plants

The conservation of the wild medicinal plants or any other such threatened species can be tackled by scientific techniques as well as social actions.

There are basically three scientific techniques of conservation of genetic diversity of these plants.

1. Legislation
2. *In-situ* conservation
3. *Ex-situ* conservation

Legislation: There are no separate policies or regulations for conserving medicinal plants growing in forests in India. There conservation is covered under existing laws pertaining to forestry. Following are the laws formulated by government of India for conservation of forests which directly or indirectly protects the wild herbal flora.

- Forest Act, 1927
- Wildlife (Protection) Act 1972 and Wildlife (Protection) Amendment Act 1991
- Forest (Conservation) Act, 1980
- Environment Protection Act, 1986
- National forest policy, 1988
- National biodiversity act, 2002
- The scheduled tribes and other traditional forest dwellers act, 2006

***In-situ* conservation**

- Conservation of a given species in its natural habitat or in the area where it grows naturally is known as *in-situ* conservation.
- It includes Gene bank / Gene sanction, Biosphere reserves, national parks, sacred sites, Sacred grooves etc.
- It is only in nature that plant diversity at the genetic, species and eco-system level can be conserved on long-term basis.
- It is necessary to conserve in distinct, representative biogeographic zones inter and intra-specific genetic variation.

It is cost-effective way of protecting the existing biological and genetic diversity is the 'in-situ' or on the site conservation wherein a wild species or stock of a biological community is protected and preserved in its natural habitat. The prospect of such a 'ecocentric', rather than a species centred approach is that it should prevent species from becoming endangered by human activities and reduce the need for human intervention to prevent premature extinctions. Establishment of biosphere reserves (Table 4), national parks, wild life sanctuaries, sacred groves and other protected areas forms examples of *'in-situ'* methods of conservation. The idea of establishing protected area network has taken a central place in all policy decision process related to biodiversity conservation at national, international and global level.

The most commonly referred *in situ* conservation methods are highlighted below:

Biosphere Reserves: The Ministry of Environment and Forest, Government of India, had identified 14 biosphere reserves based on survey data and 7 of them have already been made operational by now (Table 4).

National Parks: Out of a total of 91 National Parks in the country, 2 have been established in Himachal Pradesh, that is, Pin Valley National Park and Great Himalayan Park in the districts of Lahaul-Spiti and Kulu, respectively.

In addition to the above wildlife sanctuaries (448), there are World Heritage sites (5), Wetlands (19, including 6 Ramsagar sites), Mangroves (15), Coral Reefs (4) and other areas such as sacred groves, natural monuments, ethno-biological reserves, etc. These will definitely serve in conserving biodiversity in their respective regions.

Table 4: Biosphere reserves in India

Sl. No.	Biosphere reserve	Area (sq. km)	State
1.	Nokrek	80	Meghalaya
2.	Nilgiri	5520	Karnataka, Kerala, Tamil Nadu
3.	Namdapha	4500	Arunachal Pradesh
4.	Nanda Devi	1560	Uttar Pradesh
5.	Sunderbans	9630	West Bengal
6.	Great Nicobar	885	Andaman & Nicobar Islands
7.	Gulf of Manner	555	Tamil Nadu
8.	Manas	2837	Assam
9.	Valley of Flowers	2000	Uttaranchal
10.	Kaziranga	760	Assam
11.	Thar Desert	760	Rajasthan
12	Kanha	760	Madhya Pradesh
13	Rann of Kutch	2000	Gujarat

In-situ conservation of medicinal plants in India can be accomplished through the active support and participation of people who dwell in or near and around the protected forest areas.

Sacred groves

Sacred groves are small or large patches of vegetation protected on the basis of cultural and traditional practices on the religious background (Table 5).

Table 5 : Sacred groves in India

States	Number of documented groves
Andhra Pradesh	750
Arunachal Pradesh	58
Assam	40
Chhattisgarh	600
Gujarat	29
Haryana	248
Himachal Pradesh	5000
Jharkhand	21
Karnataka	1424
Kerala	2000
Maharashtra	1600
Manipur	365
Meghalaya	79
Orissa	322

Rajasthan	9
Sikkim	56
Tamil Nadu	448
Uttaranchal	1
West Bengal	670
Total	13720

The other complementary methods of *in-situ* conservation are:

On-Farm Conservation: On-Farm Conservation involves the maintenance of traditional crop cultivars (land races) or farming systems by farmers within the traditional agricultural system. Traditional farmers use land races, which are developed by the farmer and well adapted to the local environment. This method of conservation has been gaining importance in recent years, though farmers have used it for centuries. In case of agrobiodiversity, the effects of growers practices are of paramount importance. Systemic documentation of farmers' knowledge of diversity and usages is needed.

Home Gardens: Home garden conservation is very similar to on-farm conservation, however scale is much smaller. In rural situations, home gardens tend to contain a wide spectrum of species such as vegetables, fruits, medicinal and spice plants.

Zero energy input based concept of parafrost conservation in the Himalayan region which remains covered with snow.

Outlines for *in-situ* Management

- The Parks Department should prepare a policy at national level on the conservation and utilization of medicinal plants in protected areas.
- The policy should include:
    - Identifying which of the protected areas are most important for medicinal plants;
    - Targets and techniques for recording and monitoring medicinal plants in protected areas;
    - Techniques and procedures for collection of medicinal plants within protected areas.
- The Parks Department should assess the extent to which the protected areas system covers the medicinal plants of the country. It should then create new protected areas and extend existing ones to ensure that all the medicinal plants of the country are conserved.
- The Parks Department should devise economic and social incentives for maintaining natural habitats and wild species.
- Park managers should ensure that the conservation and exploitation of medicinal plants are incorporated into site management plans.
- Species that are heavily depleted by over-collection should be re-introduced into areas where they once grew wild.

Ex-Situ Conservation

Conservation of medicinal plants can be accomplished by the ex-situ i.e. outside natural habitat by cultivating and maintaining plants in botanic gardens, parks, other suitable sites, and through long term preservation of plant propagules in gene banks (seed bank, pollen bank, DNA libraries, etc.) and in plant tissue culture repositories and by cryopreservation).

Field Gene Bank (field repository/clonal repository)

(i). Gene Bank: Storage in the form of seed (Base collection at -20°C; Active collection at +4°C to 10°C). The three national gene banks have been established in India for *ex situ* conservation of medicinal and aromatic plants.

(a) National Bureau of Plants Genetic Resources (NBPGR), New Delhi, under ICAR.

(b) Central Institute of Medicinal and Aromatic Plants (CIMAP), Lucknow, Uttar Pradesh, under the Council of Scientific Industrial Research (CSIR), Ministry of Science and Technology, Government of India.

(c) Tropical Botanical Gardens Research Institute, (TBGRI), Palode, Thiruvananthapuram (Kerala).

The conservation of genetic variability of cultivated plants and their wild relatives is the sole responsibility of the National Bureau of Plant Genetic Resources (NBPGR) that operates under the Indian Council of Agricultural Research (ICAR), Department of Agricultural Research and Education (DARE).

Seed Gene Bank

Germplasm conservation in Seed Gene Bank is more economical. The NBPGR, New Delhi,houses National Gene Bank (NGB) which is primarily responsible for conservation of germplasm of agri-horticultural crops and their wild relatives for long-term seed storage for posterity. These are referred to as "Base Collection" stored in modules maintained at -20°C. The seeds are dried to attain 4-6 per cent moisture content and hermetically sealed in moisture proof aluminium foil packets. These stored seeds remain viable for 50 to 100 years. In most crops, seeds samples with more than 85 per cent seed viability are only processed. The seeds in gene bank are stored preferably as per the gene bank standards recommended by FAO/IPGRI.

National Active Germplasm Sites

The national active germplasm sites (NAGS) are the integral component of the network. There are presently 40 NAGS, which are based at ICAR institutes, (crop-based institutes for a specific crop or a group of crops) and SAUs. These are integral part of national plant biodiversity conservation network. The NAGS are entrusted with the responsibility of multiplication, evaluation, maintenance and the conservation of active collection and their distribution to bonafide users both at the national and international levels. These active/working collections are stored in modules maintained at +4°C and 35-40 per cent relative humidity (RH). Under these temperatures, seeds are expected to remain viable for 15 to 50 years. For medium term storage, seed moisture content is brought down to 8 to 10 per cent. The NBPGR has a network of II regional stations located in different agroclimatic zones of the country to support the active germplasm conservation activities of the regions.

Cryopreservation (in liquid nitrogen at-165°C to-196°C)

Cyropreservation or freeze preservation under liquid nitrogen.

(a) **Seed Preservation:** The seeds have been grouped broadly into two categories, based on their response to dehydration. A majority of them are desiccation tolerant, called 'Orthodox' and hence can be stored for longer durations. The second group of plant species are called 'Recalcitrant', whose seeds suffer injury on their drying and therefore cannot be stored at subzero temperatures.

(b) **Pollen Preservation:** Pollen storage was mainly developed as a tool for controlled pollination of synchronous flowering in plants, especially in fruit tree species. In addition, pollen storage has also been considered as an emerging technology for genetic conservation. Pollen can easily be collected and cryo-preserved in large quantities in relatively small spaces.

Exchange of germplasm through pollen poses fewer plant quarantine problems. In recent years, cryo-preservation techniques have been developed for pollen in a large number of species and cryo-bank of pollen has been established for fruit-tree species in several countries.

In vitro (on Tissue Culture) Conservation

The essential prerequisites for an *in vitro* conservation programme are creation of special. facilities (culture rooms with controlled environment, artificial lights, laminar airflow cabinets, autoclave, etc.) and trained scientists and technicians. Information on the *in vitro* multiplication and/or conservation protocols of those plant species is also desirable. Any *in vitro* conservation programme primarily comprises two stages: (i). *in-vitro* multiplication to build up a large number of plants, and (ii) *in vitro* storage. The material stored *in vitro* may be in the form of meristems, shoot tips, axillary buds, embryos, and even callus and cell suspension. *In vitro* gene banks are easy to maintain and often inexpensive provided effective storage systems are developed.

DNA Storage (Conservation at -20°C)

Storage of DNA is, in principle, simple to carry out and widely applicable. The storage of DNA seems to be relatively easy and cheap. In recent years genetic engineering has resulted in breaking down the species and genus barriers for transferring genes. Transgenic plants have been produced with genes transferred from viruses, bacteria and fungi and even mice. These efforts have led to the establishment of DNA Libraries. Necessary strategies and procedures have to be developed on how to use the material stored in the form of DNA.

Botanical Gardens/ Arboreta

A botanical garden is an institution holding documented collection of living plants for the 'purpose of scientific research, conservation, display and education. They serve as repositories of germplasm collections, specially rare and endangered ones of indigenous and exotic origin. Botanic Garden Conservation International (BGCI), an international organisation with its headquarters in London (U. K.) was established in 1987 for global co-operation and monitoring the conservation programmes of the botanical gardens. The BGCI has 500 member botanical gardens in III countries all over the world. There are about 1,846 botanical gardens worldwide as per the BGCI database.

India has a network of about 140 botanical gardens which include 33 botanical gardens attached to 33 universities botany departments. But hardly 30 botanical gardens have any active programme on conservation. Tropical Botanical Gardens & Research Institute (TGBRI),located in a degraded forest region of Western Ghat Mountains in Kerala has an excellent example in *ex-situ* conservation of plant diversity in India.

❑❑❑

5

MEDICINAL AND AROMATIC PLANTS IN AYURVEDIC SYSTEMS

India has six recognized systems of medicine i.e. Ayurveda, Siddha, Unani, Yoga, Naturopathy and Homoeopathy. Ayurveda is the oldest system of medicine and native to India, but Unani system of medicine originated in Greece and has been introduced by the Arabs in India. Ayurvedic system of medicine is based on three dosas (Vata, pitta and Kapha). The body will be healthy, if all dosas exist in equal quantities. In 1500 B.C to 500 AD, Ayurvedic texts include Charak Samhita, Sushrut Samhita and Astanga Samgraha.

Description of medicinal and cosmetic utility plants viz. *Kustha (Sassurea lappa), Chandana (Santalam album)* and *Kumkum (Crocus sativus)* can be traced from oldest Indian literatures like *Upanishads*, epics, etc. Lord *Dhanvantari* the God of Ayurveda is the Incarnation of Lord Vishnu. Their description denotes the importance and uses of Aromatic material like Kasturi Tilakam, Chandan Lepanam etc. Gandhamadan, a mountain named after so, harboring ample fragrant plants (Ramayan, Vagbhata).

Aroma - as an integral component of *Prithvi Mahabhuta* attributed with *Gandha* Guna (fragrance) and categorized as **Surabhi (compatible)andAsurabhi (incompatible).**

Therapeutic Value: A number of aromatic plant, animal & mineral sources attributed with medicinal value.

- Relevance in **Anatomical applications** (Phana Marma)
- Specific **pharmaceutical dosage** forms to retain the aroma of the drugs. (Arka, Hima, Phanta etc.)
- Numerous **pharmaceutical & clinical applications**
- Special **routes of administration** of dosage forms pertaining to aromatic properties
- Personal, Social & Environmental Hygiene and preventive health care

APPLICATIONS OF AROMA THERAPY IN AYURVEDA

Ayurvedic literatures mention the following wide-ranging applications of Aroma in Health Care System:

- Preventive health
- Environmental Health
- Public health
- Pharmaceuticals
- Therapeutics

APPLICATIONS IN PREVENTIVE HEALTH

- Utility of garlands enriched with aromatic flowers as a part of daily regime
- Inhaling flavors/fumes of aromatic medicinal plants as a daily regime
- Prescribed to stay in gardens having aromatic flowers in summer seasons as a part of seasonal regime
- Fumigation of clothing with aromatic plants/products
- Avoidance of exposure to excessive/improper

APPLICATIONS IN ENVIRONMENTAL & PUBLIC HEALTH

- **Purification of water** with aromatic herbs
- **Fumigating the environment** with aromatic plants/products to prevent epidemics/infectious diseases
- **Sterilizing** the labour rooms, Pediatric wards & OTs with antimicrobial aromatic plants (*Dashang Dhoopa*)

APPLICATIONS IN PHARMACEUTICS

- Addition of aromatic excipients/Treatment of distillates of aromatic medicinal herbs according to targeted indications (*Arka*)

  For Vata Disorders : Guggulu

  For Pitta Disorders : Chandana

  For Kapha Disorders : Jatamamsi
- Use of aromatic excipients for patients' compliance/acceptability
- Treating the oils with aromatic plants/products before utilizing them for manufacture (*taila and ghrita murchana)*

APPLICATIONS IN THERAPEUTICS

- Fumigation for wound-healing
- Fumigation to prevent & manage infectious & contagious diseases
- Induction of emesis (*Madanaphala-Lotus*)
- Aphrodisiac agents

- Nasal Administration of certain medicaments
- Management of headache
- Management of psychotic disorders to restore the consciousness

e.g. Restoring consciousness of *Lakshman* is an example of application of aromatic nasal medicaments as narrated in *Ramayana*.

Aromatic compounds and their uses

Chemical Compounds of Essential Oils	Therapeutic Properties
Aldehydes	Releasing, calming and sedative, Anti-inflammatory, Anti-septic, Anti-microbial
Phenols	Highly anti-septic and anti-microbial, Stimulating immune system
Sesquiterpenol	Mood balancing, Anti-inflammatory, Immune system enhancer
Ketones	Mind relaxer, Mucolytic and expectorantsupporting respiratory system, Stimulates cell regeneration
Esters	Mind relaxer, Anti-spasmodic, Anti-fungal
Alcohol	Anti-bacterial, Anti-viral
Monoterpenes	Detoxifying, Diuretic, Anti-septic, Anti-viral, Anti-inflammatory Analgesic, Anti-spasmodic, Immunity stimulator
Azulene	Anti-histaminic, Highly anti-inflammatory, Promoting healing
Bisabolol	Is the strongest sesquifterpene, Anti-bacterial, Anti-fungal, Anti-inflammatory
Flavonoids	Immunity enhancer, Anti-cancer, Preventing cardio-vascular disorders, Improves circulation, Maintains healthy status
Farnesol	Highly anti-inflammatory, Anti-bacterial
Farnesence	Highly anti-viral
Limonene	Highly anti-viral
Terpene alcohols	Highly anti-bacterial, Immunity enhancer
Terpene hydrocarbons	Highly anti-viral
Thujone	Mucolytic and expectorant, Immunity enhancer

MEDICINAL AND AROMATIC PLANTS IN AYURVEDIC MEDICINE

Scientific name : *Achyranthes aspera* Linn.

Family : Amaranthaceae

Name of Drug : Apamargah

Botanical description: This is an erect herb or undershrub which attains a maximum height upto 1 meter. The stem is stiff and less branched. Branches are pubescent, terete and striate. The leaves are few, thick, elliptic or obovate, pubescent with round apex. The flowers are green or tinged with purple, numerous, appear on a long and woody spike.

Distribution: In different parts of warmer regions of India as weed and also in Sri Lanka, tropical Asia, Africa, Australia and America.

Active ingredients: Achyranthine, aglycone, betaine, pentatriaontane, 6-pentatriacontanone, hexatriacontane and tritriacontane.

Parts used: Seeds, leaves, roots

Medicinal uses: It is used to prepare a special medicine called Kshara, used extensively in surgical procedures to treat fistula, and as oral medicine for obesity, tumors etc. The drug from entire plant is employed in anorexia and urinary diseases, skin diseases, enlargement of cervical glands, piles, hiccups, ascites, pruritus and bronchial asthma.

Scientific name	:	*Achyranthes prostrata* Linn.
Family	:	Amaranthaceae
Name of Drug	:	Apamargah

Botanical description: This is an annual to perennial plant with prostrate or erect stems growing from a long taproot. It can grow 30 - 60cm tall. Spikes are oblong with hooked fascicle on each side.

Distribution: Moist shady places throughout the plains of India, Bengal, Sikkim, Khasi Hills, Bangladesh, Myanmar, Sri Lanka, tropical Asia, Africa, Australia, Oceana and America.

Active ingredients: Achyranthine

Parts used: Whole plants

Medicinal uses: The drug from whole plant is useful in dysentery, arresting bleeding, worms and pathogenic organisma, urinary and skin diseases, asthma. The sap of the plant is used as ear drops to treat otitis and headache. The roots are used as an abortifacient. A decoction of the leaves is applied to snakebites. The sap of the plant is used as ear drops to treat otitis and headache.

Scientific name	:	*Acorus calamus* Linn.
Family	:	Araceae
Name of Drug	:	Vaca

Botanical description: This species possesses bright green, acute thick leaves having wavy margins and creeping and branching rootstocks. Spathes are 15-75cm long, pedicel 3.2-3.8 cm broad. Spadix is obtuse, curved, green. These are naturally aquatic but grow well in ordinary garden soil.

Distribution: Throughout India and Sri Lanka in marshes, wild or cultivated in Sikkim, also distributed in north temperate and warm regions.

Active ingredients: Alpha and beta asarone, acorine, acoradine

Parts used: Rhizome

Medicinal uses: The drug is useful in treatment of bronchial asthma, indigestion, diarrhoea, dysentery, abdominal obstruction and colic, hysteria, insanity. It is also said to be useful in improving intellect and memory power.

Scientific name : *Adhatoda vasica* Nees.

Family : Acanthaceae

Name of Drug : Vasa

Botanical description: It is an evergreen shrub grown upto one metre in height with many long opposite branches. Leaves are large and lance-shaped. Stems are herbaceous above and woody below. Leaves are opposite and exstipulate. Flower are spikes or panicles, small irregular zygomorphic, bisexual, and hypogynous. It has capsular four seeded fruits. The flowers are either white or purple in colour.

Distribution: Sri Lanka, Nepal, India, Pakistan, Indonesia and China

Active ingredients: Vasicine, vasicinone, deoxyvasicine, maiontone, vasicinolone and vasicinol

Parts used: Leaves, flowers, fruits and roots

Medicinal uses: The leaves, flowers, fruit and roots are extensively used for treating cold cough, whooping cough, chronic bronchitis and asthma, as sedative, expectorant and antispasmodic.

Scientific name : *Adenosma indiana* Merr.

Family : Scrophulariaceae

Name of Drug : Sprkka

Botanical description: These are annual herbs grown upto 40cm with angular hairy stems,. Leaves are simple, opposite, ovate to elliptic. Inflorescence is terminal, spikelike and cylindrical. Flowers are bisexual and purple in colour.

Distribution: It is found as weed in South India.

Parts used: Whole plant

Medicinal uses: It is an aphrodisiac. The plant is used as poultice against abdominal pain.

Scientific name : *Adiantum philippense* Linn.

Family : Polypodiaceae

Name of Drug : Hamsapadi

Botanical description: It is a small fern. Stipes are 10-15 cm long, tufted, wiry, naked, polished dark chestnut-brown; fronds are 15-30 cm long and 7.6 cm broad, simply pinnate, often elongated and rooting at the apex; pinnae are subdimidiate, the lower edge nearly in a line or oblique with the petiole, the upper edge rounded; the rachis and both surfaces naked; sori occur in continuous lines along the edge.

Distribution: Southern India

Active ingredients: Triterpene, flavonoids, adiantone

Parts used: Whole plant

Medicinal uses: *Adiantum philippense* is well known for its antioxidant potential. Used in treatment of paralysis, blood diseases, epileptic fits, rabies, dysentery, elephantiasis, pimples, and wounds. The drug is considered diuretic, tonic and carminative.

Scientific name	:	*Aerva lanata* Juss.
Family	:	Amaranthaceae
Name of Drug	:	Bhadra

Botanical description: An annual or perennial herb grown on road side and wastelands. The plant has small and simple leaves, shiny white and red at the base and has two lobes. The stems are about 1.8 m long. The flowers are tiny about 2.5 mm long with green, pink or creamy white color and grow in clusters near the axil of stems.

Distribution: India, Asia and Africa

Active ingredients: Alkaloids, Tannins, Flavonoids, Palmitic acid, Coumestrol, Beta Sitosterol

Parts used: Whole plant

Medicinal uses: The herb is considered diuretic, cooling, lithontripic, anthelmintic, anti-bacterial and analgesic. Employed in diabetes, lithiasis, haematemensis associated with pregnancy, headache and strangury.

Scientific name	:	*Alpinia calcarata* Rosc.
Family	:	Zingiberaceae
Name of Drug	:	Rasna

Botanical description: A rhizomatous perennial herb with a dull dark green leaves, lighter undersides. Foliages are narrow. Flowers are white with yellow and red variegated, fragrant. Inflorescences are upright.

Distribution: Sri Lanka, China

Active ingredients: Protocatechinic acid, 1,8-cineole, quercetin, â-pinene, 4-O-methyl-syringic acid, methyl cinnamate, vanillic acid and diterpenes.

Parts used: Rhizome

Medicinal uses: It helps in purification of blood, improving voice and digesting food. It is employed in bronchitis, asthma and other respiratory ailments.

Scientific name	:	*Alpinia galanga* Sw.
Family	:	Zingiberaceae
Name of Drug	:	Rasna

Botanical description: These are perennial herbs with leafy stem and horizontal rootstock. The leaves are oblong or lanceolate. The flowers are greenish white borne in terminal racemes or panicles.

Distribution: Western Ghats of Southern India, Sri Lanka, Malacca, Malay Islands

Active ingredients: Galangin

Parts used: Rhizome, seeds

Medicinal uses: The drug is used for maintaining strength and youthful vigour and administered in intermittent fever, dyspepsia, cough, rheumatism, intestinal disorders and biliary colic.

Scientific name	:	*Alternanthera sessilis* R.Br. ex. DC.
Family	:	Amaranthaceae
Name of Drug	:	Lonika

Botanical description: It is a prostrate herb with opposite linear or lanceolate or elliptic leaves. Flowers are hermaphrodite, small, white, shining and bracteate.

Distribution: Different parts of India, particularly wet locations, Sri Lanka and all warm countries

Active ingredients: Methylenecycloartanol, cycloeucalenol, campesterol, sitosterol

Parts used: Whole plant

Medicinal uses: The whole plant is considered blood purifier, alleviative, sour and cooling. Fresh leaves are applied over eyes for conjunctivitis and chronic inflammation of the eyelid. The whole plant is also used for increasing breast milk, promoting bile flow, inducing abortion and for curing fever.

Scientific name	:	*Amaranthus viridis* Linn.
Family	:	Amaranthaceae
Name of Drug	:	Tandutiyah

Botanical description: It is an annual herb with erect stems, 10-80cm long. Stems are sparingly or densely branched, channelled. Leaves are triangular-ovate to narrow rhombic, 2-7cm long, 1.5-5.5cm wide, hairless, tip usually narrow and with a small narrow notch., stalks 1-10cm long. Flowers are green, in slender, paniculate spikes, in leaf axils or at the end of branches.

Distribution: It is found as weed in India. Native to America. Distributed in all tropical and subtropical countries.

Active ingredients: Quercetin, rutin, amaranthin

Parts used: Whole plant

Medicinal uses: The entire plant is laxative, diurectic, carminative, acrid and cooling. The drug is used to cure dyspepsia, haemophilic conditions, burning sensations and promote appetite.

Scientific name	:	*Amorphophallus paeniifolius* Nicolson
Family	:	Araceae
Name of Drug	:	Suranah

Botanical description: This plant produces a inflorescence followed by a solitary leaf. The plant is deciduous, dying back to a large underground corm, weighing up to 8kg, after the growing season. It comprises a large spadix crowned with a bulbous purple knob, encircled by a fleshy purple and green-blotched spathe up to 50cm wide. After successful pollination of the female flowers the spadix can extend to 2m tall. The fresh inflorescence emits an odour reminiscent of rotting flesh to attract pollinating carrion flies and beetles. The solitary leaf resembles a small tree. The leaf blade is much divided into hundreds of leaflets and can reach over 1m wide. This blade sits atop a thick fleshy stem up to 13cm diameter and 2m tall. The pustular surface of the stem is attractively blotched with paler shades of green.

Distribution: It is found wild in uplands in India.

Active ingredients: Amylase, betulinic acid, tricontane, lupeol, stigmasterol, betasitosterol, glucose, galactose, rhamnose and xylose.

Parts used: Rhizomes

Medicinal uses: It is anthelmintic, astringent, acrid, carminative and irritant. It is employed in skin diseases, piles and haemophilic conditions.

Scientific name	:	*Andrographis paniculata* Wallich ex. Nees.
Family	:	Acanthaceae
Name of Drug	:	Kiratatiktah

Botanical description: It is a much branched erect annual herb. The branches are four angled. The leaves are lance shaped. The flowers appear in large spreading sparse branches, individual flowers are small, upto 1 cm long, rose coloured. Fruit is capsular upto 2 cm in diameter.

Distribution: In shady places of tropical forests in India and Sri Lanka. Cultivated in East and West Indies.

Active ingredients: Sodium chloride, tannic acid, isoprenaline, andrographolide, penucilides

Parts used: Leaves and stems

Medicinal uses: The drug is used in fever, burning sensation, thirst, cough, cold, diarrhoea, sore throats, oedema, ulcer, worms and difficulty in breathing.

Scientific name	:	*Aristolochia indica* Linn.
Family	:	Aristolochiaceae
Name of Drug	:	Iswari

Botanical description: Leaves are simple, alternate, short petioled, the blade ovate or somewhat wedge shaped, very variable in shape and size. The young leaves are light purplish. The fruit is a capsule, roundish or oblong and hexagonal,

Distribution: India, Sri Lanka, Nepal , Bangladesh

Active ingredients: Aristolochic acid, aristolic acid, aristolic acid methyl ester, methyl aristolochate, aristolamide, aristolinic acid, aristolonitrite.

Parts used: Roots, rhizomes and leaves

Medicinal uses: The plant is used in skin diseases where there is morbidity of vata, pitta and kapha. It is used as an appetiser, aphrodisiac and anthelmintic. The fresh juice of the leaves is a popular antidote to snake poison. The leaves and bark are used in bowel complaints of children, diarrhoea and in intermittent fevers. In traditional medicine the underground parts of the plant are rubbed with honey and given to treat leprosy; and, macerated with black pepper, it is prescribed in diarrhoea.

Scientific name	:	*Argemone mexicana* Linn.
Family	:	Papaveraceae
Name of Drug	:	Svarnaksir

Botanical description: This is an annual herb up to 90 cm tall. Leaves are glaucous, over the veins light blue with conspicuous markings on them. Flowers are yellow, 4 to 6cm across. Capsules are prickly. Commonly known as 'Mexican Poppy'.

Distribution: Found as weeds in different parts of India. The species is native to Mexico.

Active ingredients: Argemixitin, luteolin, eridictyol, sanguinarine, dihydrosanguinarine, berberine, vanillic acid.

Parts used: Seeds, roots

Medicinal uses: It is tonic, anthelmintic, alterative, aperient and bitter. This is antidiabetic, anti-oxidant and hepatoprotective.

Scientific name	:	*Artanema longifolia* Vatke
Family	:	Scrophulariaceae
Name of Drug	:	Kokitakasah

Botanical description: This is an erect, often branched, herbaceous perennial plant with hollow, slightly fleshy stems, growing up to 1 metre tall

Distribution: In the forests of Southern India, Tropical Asia Australia

Parts used: Roots and Seeds

Medicinal uses: It is considered aphrodisiac, demulcent, roborant, cold and also used for the treatment of arthritis, rheumatism, stones in bladder, opthalmia and diarrhoea.

Scientific name	:	*Arundo donax* Linn.
Family	:	Poaceae
Name of Drug	:	Sahah

Botanical description: It is a very tall and robust bamboo-like, perennial grass with large, spreading clumps of thick culms to 6 m tall. The numerous leaves are about 5 cm wide and 30.5-61 cm long, and arranged conspicuously in two opposing ranks on the culms. The leaves look like those of a corn plant. The inflorescence, is 30cm- 60cm long purplish, aging to silver, plume that stands above the foliage. Giant reed spreads from thick, knobby rhizomes.

Distribution: Myanmar, Pakistan, India and North Africa

Active ingredients: Tryptamines

Parts used: Rhizome

Medicinal uses: It is said to cure blood diseases, piles, eye diseases, leucorrhoea, erysipelas, dysuria and thirst.

Scientific name	:	*Asplenium arifolia* Burm.
Family	:	Cheilanthaceae
Name of Drug	:	Akhukarni

Botanical description: The plant is a small herbaceous fern with dark green leaves, blackish stipe and cordiform leathery leaves. The frond are of two types. The sterile fonds are small. The fertile frons are long, stalked and large pinnately lobed. Leaves are heart shaped.

Distribution: Southern part of India

Active ingredients: Flavonoid, phenol

Parts used: Leaf

Medicinal uses: The leaf juice is used in snake bite, headache and migraine.

Scientific name	:	*Azadirachta indica* A. Juss.
Family	:	Meliaceae
Name of Drug	:	Nimbh

Botanical description: A fast-growing tree, can reach a height of about 15 to 20 meters. The trunk is short, straight and has a diameter of 1.2 m, wherein the branches spread out widely. Young leaves are reddish to purple in colour and turn into dark green pinnate leaves on maturity. The white and fragrant flowers are arranged auxiliary in drooping panicles, about 25 cm in length. The smooth olive-like drupe fruit is elongated-oval to nearly roundish in shape.

Distribution: Neem is grown in tropical and semi-tropical regions of India, Myanmar, Bangladesh, Sri Lanka, Malaysia and Pakistan.

Active ingredients: Azadirachtin, nimbinene, 6-desacetylnimbinene, nimbandiol, nimbolide, p-sitosterol, salannin, gedunin, azadirone, nimbin, nimbidine, nimbicidine, nimbinol.

Parts used: Seeds, leaves, flower and bark.

Medicinal uses: Neem products are proved to be anthelmintic, antifungal, anti-diabetic, antibacterial, antiviral, anti-fertility and sedative. Neem tree is generally considered to be an air purifier and preventive against malarial fever, acne, pimples and cholera. The leaves are useful in relieving flatulence, promoting the removal of catarrhal matter and phlegm from the bronchial tubes, and in increasing secretion and discharge of urine. They also acts act as an insecticide. The bark is a bitter tonic and a stimulant. It arrests secretion and bleeding besides counter acting any spasmodic disorders.

Scientific name	:	*Bacopa monnieri* Linn.
Family	:	Scrophulariaceae
Name of Drug	:	Brahmi

Botanical description: This is a glabrous, creeping succulent herb, leaves are oblong, spatulate and obtuse. The flowers are up to 1cm in diameter, pale bluish, pimple veined, fading to white, arises from the axil of the leaves and are borne on the short pedicels. The roots appear from the nodes of the stems.

Distribution: It is found as weed in marshy low lands or wet lands, in plains of India.

Active ingredients: Brahmine, herpestine, hersaponin, monnierin

Parts used: Whole plant

Medicinal uses: The entire plant is highly valued for improving intelligence and memory and revitalization of sense organs. The drug is astringent, laxative, tonic, sweet and cold. It is employed for clearing voice and improving digestion. It is effective in cases of anxiety, nerosis, emanciation and insanity. The drug is useful for diabetes, cough, fever, arthritis, anorexia, dyspepsia, dermatosis, dropsy, heart and nerve trouble.

Scientific name	:	*Barleria prionitis*
Family	:	Acanthaceae
Name of Drug	:	Sahacarah

Botanical description: It is an erect, prickly shrub, usually single-stemmed, reaches to a height of about 1.5 m. The stems and branches are stiff and smooth and light brown to light grey in colour. The leaves are oval-shaped though narrow at both ends (ellipsoid). The base of the leaves is protected by three to five sharp, pale coloured spines. The yellow-orange tubular flowers are found bunched tightly together at the top of the plant, but they also occur singly at the base of leaves. The flowers are tubular, with several long protruding stamens.

Distribution: India, Sri Lanka and Africa

Active ingredients: Hydroxyflavone

Parts used: Whole plant

Medicinal uses: It has numerous medicinal properties including treating fever, respiratory diseases, toothache, joint pains and a variety of other ailments; and it has several cosmetic uses. A mouthwash made from root tissue is said to relieve toothache and treat bleeding gums. The leaves are used to promote healing of wounds and to relieve joint pains and toothache. Extracts of the plant are incorporated into herbal cosmetics and hair products to promote skin and scalp health.

Scientific name	:	*Basella alba* Linn.
Family	:	Basellaceae
Name of Drug	:	Upodika

Botanical description: It is a fast-growing, soft-stemmed vine, reaching 10 metres in length. Its thick, semi-succulent, heart-shaped leaves have a mild flavour and mucilaginous texture. The stem of the cultivar *Basella alba* 'Rubra' is reddish-purple.It bears white or white-pink colour tiny flowers depending upon the species and deep-purple to black colour berries.

Distribution: Asia, America, Australia and Europe

Active ingredients: Beta cyanindin, gomphrenin-I

Parts used: Leaf, stem and seeds

Medicinal uses: The leaves are traditionally used in ayurveda system of medicine to bring sound refreshing sleep when it is applied on head about half an hour before bathing. A paste of the root is applied to swellings and is also used as a rubefacient. Sap is applied to acne eruptions to reduce inflammation. Decoction of leaves used for its mild laxative effects. Pulped leaves applied to boils and ulcers to hasten suppuration. Sugared juice of leaves is useful for catarrhal afflictions. In Ayurveda, it is used for haemorrhages, skin diseases, sexual weakness, ulcers and as laxative in children and pregnant women. The plant is febrifuge, its juice is a safe aperients for pregnant women and a decoction has been used to alleviate labour. The cooked roots are used in the treatment of diarrhoea. The leaf juice is a demulcent, used in cases of dysentery. The Ayurvedic treatment leaves and stem are used for anticancer such as melanoma, leukemia and oral cancer. Root and leaves has been used for the removal of after birth, stomach pains and increase milk production. *Basella alba* is administered orally for the treatment of anal prolapsed or hernia. *Basella alba* are used for the treatment of anemia in women, coughs, cold, cold related infections. Maceration is taken orally for infertility, pelvic inflammatory disease, orchitis, epididymytis, threatened abortion, spurious labour. Leaves are used in constipation, poultice for sores, urticaria and gonorrhoea. It is also used in poultice local swellings, intestinal complaints etc. The mucilaginous liquid obtained from the leaves and tender stalks of plants is popular remedy for headaches.

Scientific name	:	*Bauhinia variegata* Linn.
Family	:	Caesalpinaceae
Name of Drug	:	Kanchanarah

Botanical description: It is a small to medium-sized deciduous tree with a short bole and spreading crown, attaining a height of up to 15 m. The twigs are slender, zigzag; when young, light green, slightly hairy, and angled, becoming brownish grey. Leaves have minute stipules 1-2 mm, early caducous; petiole puberulous to glabrous, 3-4 cm; lamina broadly ovate to circular, often broader than long, 6-16 cm diameter; 11-13 nerved; tips of lobes broadly rounded, base cordate; upper surface glabrous, lower glaucous but glabrous when fully grown. Flower clusters (racemes) are unbranched at ends of twigs. The few flowers have short, stout stalks and a stalklike, green, narrow basal tube (hypanthium).

Distribution: China, Colombia, India, Myanmar, Nepal, Pakistan, Thailand, Vietnam.

Active ingredients: Beflavone, 5,7-dimethoxy-30, 40-methylenedioxy flavone and a new dihydrodibenzoxepin,5,6-dihydro-1,7-dihydroxy-3,4-dimethoxy-methyldibenzoxepin, flavonolglycoside, triterpene saponin, phenanthraquinone, flavonoids.

Parts used: Bark, buds and leaves

Medicinal uses: The buds and barks are alterative and astringent. The decoction of bark is used for treating of leprosy, worm infections, scrofula, ulcers and other skin diseases. Decoction of flower bid is used for the treatment of bleeding piles, blood in urine, and bleeding from mucous surface. The flowers have laxative action. Leaves are nutritive and anti-diabetic.

Scientific name	:	*Benincasa hispida* Cogn.
Family	:	Cucurbitaceae
Name of Drug	:	Kusmandah

Botanical description: It is a climber with much branched stem. Leaf stalks are long and hairy. Leaves are roundish, kidney shaped, base deeply heart shaped. Upper surface is rough, lower surface shortly bristly, blade 10-25cm long, 5-7 lobed, lobes ovate triangular, margin sinuate or toothed. Tendrils are slender, rarely simple. Male flower 5-15cm long, female 2-4 cm long. The fruit is fuzzy at young stage. At maturity, the fruit loses its hairs and develops a waxy coating.

Distribution: India and other parts of tropical Asia, Africa, China, japan and Malayasia

Active ingredients: Alunsenol, mutiflorenol, mucins, aminoacids, starch

Parts used: Fruits, leaves and shoots

Medicinal uses: The fruit is beneficial for the management of a host of medical problems, including epilepsy, lung diseases, asthma, cough, urine retention and internal hemorrhage. It is also an excellent remedy for tapeworms. Ash Gourd is ideal for diabetics and patients suffering from obesity. It combats general debility by stabilizing nerve cells. The cooling properties of its juice are helpful in treating peptic ulcers and for relieving acidity. Ash Gourd provides relief from constipation and tones the gastrointestinal tract.

Scientific name	:	*Biophytum candolleanum* Wt.
Family	:	Oxalidaceae
Name of Drug	:	Alambusa

Botanical description: It is an erect herb with simple hispid stem. Leaves are crowded into the rosette at the tip of the stem, 8-10cm long, paripinnate, leaflets 8-15 pairs, subsessile, glabrous above, pale beneath'; flowers are small, yellow, on long peduncled umbel.

Distribution: Western Peninsula, warmers parts of India in shady, moist places and tropical Indo-Malayan region.

Parts used: Leaves

Medicinal uses: The drug is employed for dysentery, chronic cough, fever, vaginal disorder, urinary calculi, burning sensation and haemorrhoides.

Scientific name	:	*Boerhaavia diffusa* Linn.
Family	:	Nyctaginaceae
Name of Drug	:	Punarnava

Botanical description: It is a deep rooted perennial creeping herb. Leaves appear two in one node, one smaller than the other, upper surface green and lower surface whitish. Base of the leaves is cordate. Flowers arise in short clusters on long axillary stalks, very small in size and reddish in colour. Fruits are glandular with fine ridges.

Distribution: It is found in plains of India during rainy season as weed in sandy tracts, Punjab, Assam and tropical and subtropical Asia.

Active ingredients: Punarnavine, purine-nucleoside, punarnavoside

Parts used: Whole plant

Medicinal uses: It is a rejuvenative medicine and stimulates function of heart and kidneys. It is also anti-inflammatory and antiarthritic. This medicinal plant is used to improve eyesight. It has diuretic properties which is useful in controlling blood sugar levels in diabetic patients.

Scientific name	:	*Boerhaavia erecta* Linn.
Family	:	Nyctaginaceae
Name of Drug	:	Punarnava

Botanical description: It is an erect annual to perennial plant with a stout rootstock, growing up to 1 metre tall. The stems branch mainly from the base and can become woody at their bottoms. Leaves are ovate, cordiform to elliptic. Inflorescence axillary or/and terminal; flowers 2-7 in cymose panicles; peduncle 2-8 cm long. Flowers tubular-campanulate; pedicel 0.5-3 mm long, with 2-3 lanceolate bracteoles less than 1 mm in length.

Distribution: Moist sandy places in South India.

Active ingredients: Phenols, flavonoids and antioxidants

Parts used: root

Medicinal uses: It is laxative, stomachic, expectorant, antipyretic, cardiotonic, astringent, bitter, cooling and diuretic. The drug is administered in diabetes, jaundice, oedema and general debility.

Scientific name	:	*Butea monosperma* (Lam.) Taub.
Family	:	Fabaceae
Name of Drug	:	Palash

Botanical description: It is a medium sized tree, 7 to 15 m tall and the trunk is usually crooked and twisted with irregular branches and rough, grey bark. The leaves are pinnate, with three leaflets, each leaflet 10-20 cm long. Flowers are orange, scentless, are massed along the ends of the stalks—dark velvety green like the cup-shaped calices—and the brilliance of the stiff, bright flowers is shown off to perfection by this deep, contrasting colour. Each flower consists of five petals comprising one standard, two smaller wings and a very curved beak-shaped keel. It is this keel which gives it the name of Parrot Tree.

Distribution: India, Sri Lanka, Myanmar

Active ingredients: Beta-sitosterol, amyrin, bitulinic acid, stigmasterol, palasonin, butrin, isobutrin, leucocyanidin.

Parts used: Seeds, gums, leaves, flowers and bark

Medicinal uses: Seeds are diuretic, purgative and anti-parasitic. Flowers and leaves are diuretic, astringent, aphrodisiac and anti-diarrhoeal. Tree bark gum is astringent and useful in haemorrhage. Roots are used to treat night blindness. Leaves are used to reduce blood sugar and diabetes.

Scientific name	:	*Canscora decussata* Schultes & Schuletes f.
Family	:	Gentianaceae
Name of Drug	:	Sankhapuspi

Botanical description: This is an erect, branching annual herb, up to 60 cm high. Stems are 4-winged. Leaves are 2.5-3.8 cm long, oblong-lanceolate, acute 3-nerved. Cymes are lax, the branches erect rigid. Flowers are white, those in the forks long-pedicelled; calyx 8-13 mm long, winged; corolla tube as long as calyx. Capsule are subcylindrical, 6-8 mm long, tipped by persistent style.

Distribution: India, Bangladesh

Active ingredients: Xanthones, amyrin, friedelin, gentianine, mangiferin

Parts used: Whole plant

Medicinal uses: The entire plant is useful in promoting digestive power, improving memory power and intellect. The drug is rejuvenating, astringent, aphrodisiac and digestive.

Scientific name	:	*Cassia occidentalis* Linn.
Family	:	Caesalpiniaceae
Name of Drug	:	Kasamarda

Botanical description: This is a smooth annual that can grow up to 2 m tall. The leaves are compound, leaflets, in 4-6 pairs, have a sharp tip. These leaflets are 2-9 cm long and 2-3 cm wide with a distinct gland 3-5 mm from the base of the stalk. Flowers occur in leaf axils. Sepals are green and 6-9 mm long. The petals are yellow and 1-2 cm long. The seed pods are dark brown, 8 to 12 cm long, 7-10 mm wide and curve slightly upward. The seeds are dull brown and flattened on both ends. The seeds can be roasted and made into a coffee-like drink.

Active ingredients: Emodin, anthraquinones, chrysarobin.

Parts used: Root, seed and leaf

Medicinal uses: It is used as an aphrodisiac. It improves digestion, cleanses throat and useful in cough. The seed is bitter and has purgative properties. It is also used as a diuretic, liver detoxifier, as a hepato-tonic (balances and strengthens the liver).

Scientific name	:	*Cassia tora* Linn.
Family	:	Caesalpiniaceae
Name of Drug	:	Chakramardah

Botanical description: It is an annual herb up to 1 m tall. It has pinnate leaves, which are about 10 cm long. Each leaf has three pairs of leaflets that are opposite, ovate, oblong and oblique at the base. Flowers are yellow. Pods are tapering.

Distribution: It is commonly found in rainy seasons as an annual weed in shady places in India.

Active ingredients: Anthraquinones

Parts used: Whole plants

Medicinal uses: The entire plant is laxative, vermifuge, carminative, acrid, sweet and bitter. Useful in treating skin diseases like ringworm and itching or body scratch and psoriasis. The drug is a remedy for itching dermatosis, eczema, respiratory diseases, constipation, obesity, helminthic, anorexia, dyspepsia and cough. It has anti-cancer potential, externally used for snake bite poisoning and reduces cholesterol.

Scientific name : *Caesalpinia bonduc* Linn.

Family : Caesalpiniaceae

Name of Drug : Kuberaksi

Botanical description: It is a vine-like shrub that reaches a length of 6 m) and scrambles over other vegetation. Stems are covered in curved spines. Its 2 cm grey seeds, known as nickernuts,

Distribution: Pantropical

Active ingredients: Seeds contain stearic, palmitic, octadeca -4, octadeca-2, ligoceric, linolenic and oleic acids. They also contain bonducellin, arginine, citrulline and aspartic acid.

Parts used: Roots, seeds, leaves and barks

Medicinal uses: The root bark is emmenagogue, febrigue expectorant, anthelmintic and stomachic. The seeds have a bitter taste, and are valued in indigenous medicine as a tonic. They are an ingredient of 'Ayush-64' an ayurvedic compound preparation used as an anti-malarial drug.

Scientific name : *Celosia argentea* Linn.

Family : Amaranthaceae

Name of Drug : Tandutiyah

Botanical description: A glabrous herb with erect , simple or ascending, grooved stems and linear acute, alternate leaves.

Distribution: It is found as weed in waste lands of India.

Active ingredients: Saponins, glycosides and tannins

Parts used: Whole plant

Medicinal uses: Stems and leaves applied as poultice, is used for treating of infected sores, wounds and skin eruptions. Whole plant is antidote for snake poison. Root used for colic, gonorrhoea and eczema. Flowers and seeds are used for bloody stools, leucorrhoea and diarrhoea.

Scientific name : *Cissampelos pareira* Linn.

Family : Menispermaceae

Name of Drug : Patha

Botanical description: It is a slender, tropical woody, climbing vine with orbicular shaped leaves, white-greenish flowers and inedible dark red, grape sized, berries.

Distribution: Southern India

Active ingredients: Oxo-aporphines, palmatine and isoquinolines

Parts used: Whole plant

Medicinal uses: The drug is used for the treatment of oedema, cystitis, kidney inflammation, intestinal worms and fever. Besides, It is also used as an infertility agent. It is helpful with symptoms associated with menstruation and balancing hormones in women.

Scientific name	:	*Clerodendrum serratum* Linn.
Family	:	Verbenaceae
Name of Drug	:	Bharangi

Botanical description: A shrub, 1-3 m tall with bluntly quadrangular stems. Young parts are usually glabrous. Leaves are usually three at the node, opposite, coarsely and sharply serrate, glabrous, base acute, stout petioles. Numerous showy flowers are borne in lax pubescent dichotomous cymes, with a pair of acute bracts at each branching and a flower in the fork, each in the axil of a large leafy bract and collectively forming a long lax terminal usually pyramidal erect panicle of 6-10 in long. Bluish corolla, glabrous outside, cylindric, hairy within the tube at the insertion of the stamens.

Distribution: Native of East India and Malasiya. Distributed throughout in forests of Srilanka and India.

Active ingredients: Sapogenin, Flavonoids and phenolic acids

Parts used: Barks, leaves

Medicinal uses: Root is pungent, bitter, acrid, dry, healing, anti-inflammatory, digestive, carminative, depurative, expectorant, antispasmodic, stimulant, appetizer and anthelmintic. It is used clinically in treatment of bronchitis, asthma, fevers, blood disease, tumours, inflammations, burning sensation, epilepsy, malaria, ulcer and wounds. Leaves are used in fever and hiccough.

Scientific name	:	*Centella asiatica* Urban.
Family	:	Apiaceae
Name of Drug	:	Mandukaparni

Botanical description: These are small trailing herbs with creeping stems. The stems bear roots on their nodes. The leaves are undivided, round or broad kidney shaped up to 4cm in diameter. The flowers are minute, pinkish red in colour, three to six in clusters, umbel. The fruits are laterally compressed, small with seven to nine ridges.

Distribution: Shady moist or wet locations in tropical and sub-tropical India

Active ingredients: Asiaticoside, brahmoside, asiatic acid, brahmic acid, centellose, centelloside

Parts used: Whole plant

Medicinal uses: The entire plant is cardiotonic, astringent and diuretic. It improves memory. The drug is employed in anaemia, diabetes, indigestion, nervous strengthening, insanity and cough.

Scientific name : *Citrullus colocynthis* Schrad.

Family : Cucurbitaceae

Name of Drug : Indravaruni

Botanical description: These are perennial herbs with vine like stems and palmate or angular leaves with three to seven undivided lobes. The flowers are yellow and solitary in the axes of leaves and are borne by yellow-greenish peduncles. The fruit is smooth, spheric with a 5– to 10-cm-diameter and extremely bitter taste.

Distribution: Southern India, tropical and subtropical North Africa and Asia.

Active ingredients: Cucurbitacin, fatty acids

Parts used: Roots and fruits

Medicinal uses: The fruit is purgative, bitter, hot, abortifacient and cathartic. The root is said to be effective in cough, asthma, abdominal enlargement, inflammation of breasts, ulcers, rheumatisms and urinary diseases.

Scientific name : *Coccinia grandis* Voigt.

Family : Cucurbitaceae

Name of Drug : Bimbi

Botanical description: *Coccinia grandis* is a fast-growing perennial vine that grows several meters long. Its leaves are arranged alternately along the stems; the shape of the leaves varies from heart to pentagon shaped. Flowers are large, white and star-shaped. The fruit is red, ovoid to elliptical, 25–60 mm long, 15– 35 mm in diameter, glabrous, hairless on stalks.

Distribution: India, tropical Asia, Africa and Australia.

Active ingredients: α- Amyrin Acetate, Lupeol, Cucurbitacin B, Taraxerone, Taraxerol, α-carotene, Lycopene, Cryptoxanthin, Xyloglucan, Carotenoids, α-sitosterol, Stigma-7-en-3-one

Parts used: Tender fruits, plant juice and leaves

Medicinal uses: The leaves are anti-diabetic, cooling and effective against eye, gonorrhoea, hypolipidemic, skin diseases, urinary tract infection. Fruits are hypoglycemic, analgesic, antipyretic, hepatoprotective, tuberculosis, eczema and anti-inflammatory. Stems are expectorant, antispasmodic, asthma, bronchitis, urinary tract infection, skin diseases. Roots are hypoglycemic, antidiabetic, skin diseases, removes pain in joint, urinary tract infection.

Scientific name : *Coleus amboinicus* Lour.

Family : Lamiaceae

Name of Drug : Karpuravalli

Botanical description: Itis small succulent shrub with tendency for climbing or creeping, it can reach over 1m hight and even more in width. It have strong stems with fleshy, hairy leaves and blue-violet or white-pink flowers in summer time.

Distribution: Southern India

Active ingredients: carvacrol, thymol, eugenol, quercetin, apigenin, luteolin, salvigenin, genkwanin caryophyllene, patchoulane and p-cymene.

Parts used: Leaves

Medicinal uses: This herb is antioxidant, antimicrobial, radioprotective, carminative, tonic, stimulant, emmenagogue, hepatoprotective, diaphoretic and antiepileptic.

Leaves are simply eaten fresh for coughs, colds, malarial fevers, asthma, bronchitis, mouth and nasal infections, diarrhoea, indigestion, flatulence, dyspepsia, epilepsy, rheumatism, kidney stones and helminthiasis. Fresh crushed leaves are applied on burns, sprains, skin infections, scorpion bites or on forehead to ease headache.

Scientific name	:	*Coleus zeylanicus* Cramer
Family	:	Lamiaceae
Name of Drug	:	Valakah

Botanical description: This is an erect herb with profuse branches, grown up to a height of 1.5 m. Flowers are borne in racemes.

Distribution: Kerala

Active ingredients: Abietane diterpenoids

Parts used: Whole plant

Medicinal uses: The drug is used against indigestion and urinal disorders. The whole plant is useful for the treatment of vomiting, diarrhoea, leucoderma, fever and chronic liver diseases.

Scientific name	:	*Convovulus pluricaulis* Chois.
Family	:	Convolvulaceae
Name of Drug	:	Sankhapuspi

Botanical description: It is a twining perennial herb that seems like morning glory. Branches are spread on the ground and more than 30 cm long. The flowers are blue in color (5 mm) and the leaves are elliptic in shape (2 mm), are located at alternate positions with branches or flowers. Known as 'Aloe weed' in English.

Distribution: Bihar and plains of West Bengal

Active ingredients: B. sitosterol glycoside, hydroxy cinnamic acid, octacosanol tetracosane, glucose

Parts used: Whole plant

Medicinal uses: The drug cures uterine disorders, epilepsy, dysuria, diabetes, coughs and intestinal worms. It is also tranquilising and anti-thyroid. The whole herb is used medicinally in the form of decoction with cumin and milk in fever, nervous debility, and loss of memory.

Scientific name	:	*Costus speciosus* Smith.
Family	:	Zingiberaceae
Name of Drug	:	Canda

Botanical description: It is an erect perennial herb. Leaves are almost without petiole, silky pubescent beneath, up to 30 cm long. Flowers are white with red bracts and tubular in shape. Fruits are red and seeds are black.

Distribution: Central and eastern Himalayas, Malaya Islands, Sri Lanka, Taiwan, Indo-China to New Guinea

Active ingredients: Diosgenin

Parts used: Roots and rhizomes

Medicinal uses: The rhizomes are bitter, astringent, acrid, cooling, aphrodisiac, purgative, anthelminthic, depurative, febrifuge, expectorant, tonic, improve digestion, and is a stimulant herb that clears toxins. It also has anti-fertility, anabolic properties. Rhizomes are also given in diseases as pneumonia, rheumatism, dropsy, urinary diseases, jaundice, and leaves are given in mental disorders.

Scientific name : *Crataeva magna* (Lour) DC

Family : Capparadaceae

Name of Drug : Varunah

Botanical description: It is tree, 10m tall with grey white smooth bark and trifoliate, alternate leaves. Flowers are bisexual and creamy white.

Distribution: Assam ,West Bengal, Kerala, MP

Active ingredients: Triterpenoid lupeol

Parts used: Leaves and barks

Medicinal uses: It is recommended for blood purification, urinary bladder problems, enlarged prostrate and calculi. This drug is effective in the treatment of kidney and urinary stones. The bark has appetizer, diuretic and litholytic activities. The tree is also used in the treatment of constipation and painful urination. It is extensively used in migraine and intestinal worms.

Scientific name : *Crotalaria pallida* Aiton.

Family : Fabaceae

Name of Drug : Sanapuspi

Botanical description: It is an erect, well-branched, sometimes robust perennial herb with woody stems. It can grow up to 2 metres tall.

Distribution: Asia, Southeastern USA

Active ingredients: Mucronatine, usaramine, nilgirine, mucronatinine, crotastriatine, luteolin, vitexin

Parts used: Roots and leaves

Medicinal uses: The plant is used in traditional medicine to treat urinary problems. The plant is also used to reduce fever. An infusion of the plant is used to bathe children in order to prevent skin infections and also to treat thrush. Boiled with salt, it is used to treat eczema and other skin problems. A poultice made of the roots is applied to painful swelling of joints. An extract of the leaves is taken as a vermifuge.

Scientific name : *Cucumis callosus* Cong.

Family : Cucurbitaceae

Name of Drug : Indravaruni

Botanical description: These are perennial climbers; stems 1.0-1.5 m, slender, angular, rough with short, rigid hairs. Leaves are alternate, 3-6 cm long, suborbicular, base cordate, palmately 5-7 lobed, lobes round or ovate-oblong, often narrowed at the base, apex round, dentate or lobulate; petioles 2-6 cm long, slender, hispid. Plants are monoecious. Male flowers often solitary; peduncles 0.5-1.0 cm long; calyx-tube narrow campanulate, ca. 3 mm long; lobes subulate, 1.5-2.0 mm long; corolla yellow, 6-7 mm long, lobes ovate, oblong, acute; staminal filaments short; anthers ca. 2 mm long, appendage of the connective shorter than anthers; pistillode ca. 1 mm long. Female flowers: peduncles 2-3 cm long; ovary densely hairy; style 1.5-2.0 mm long; stigmas converging, ca. 2.5 mm long. Fruits obovoid; seeds ca. 5 x 2 mm, 1 mm thick, oblong.

Distribution: Dry places of southern India

Active ingredients: Alkaloids, steroids, terpenes and triacylglycerols

Parts used: Fruits, root and seeds

Medicinal uses: The fruit is purgative, blood purifier and cathartic. The seed oil is used in epilepsy, for blackening of hair and bowel complaints. Roots are employed against rheumatism, cough, asthma and ulcers.

Scientific name : *Curculigo orchioides* Gaertn.

Family : Hypoxidacae

Name of Drug : Musali

Botanical description: It is a stemless herb with tuberous rootstock. The leaves are fleshy, elongate, lyrate or ensiform. The root is black.

Distribution: Peninsular India, Khasi hills and the subtropical Himalayas.

Active ingredients: Curculigoside, pcίysaccharides, yuccagenin, sapogenin, lycorin, hentriacontanol, sitosterol.

Parts used: Roots, flowers

Medicinal uses: Rootstocks are sweet, cooling, diuretic aphrodisiac, viriligenic and effective in skin diseases, asthma, bronchitis, jaundice, diarrhoea, dyspepsia, opthalmia and lumbago.

Scientific name : *Curcuma longa* Linn.

Family : Zingiberaceae

Name of Drug : Haridra

Botanical description: It is a tall herb with ovoid large rootstock, bearing sessile, orange coloured tubers. The leaves are oblong lanceolate, large and tapering to the base.

Distribution: Throughout tropical India

Active ingredients: Curcumin, sesquiterpenes

Parts used: Rhizome

Medicinal uses: The drug is used as antiseptic, aromatic, anti-rheumatic, anti-inflammatory, alterative, carminative, vermifuge, stimulant and tonic. It is used as a facial tonic and also skin diseases, eczema, itching, ringworm, chicken pox, small pox, cold and cough, conjunctivitis, liver affections and bronchitis.

Scientific name	:	*Curcuma zedoaria* Rosac.
Family	:	Zingiberaceae
Name of Drug	:	Kareurah

Botanical description: It is an herb grown up to 1.2 m in height. This plant has both vertical aerial stems and horizontal underground stems known as rhizomes, which allow the plant to spread and often grows in large clumps.The swollen underground stems are yellow or orange colored inside and are aromatic when crushed. The leaves are oblong, 81 cm long and 18 cm wide. This species is easily recognized by the presence of a purplish hue along the midvein of the leafblades. The clusters of flowers are produced in a dense aggregation on an above ground stalk that grows from a leafless underground stem.

Distribution: Native to Eastern Himalayas and Malay Islands, cultivated throughout India.

Active ingredients: Curcuminoid 1,7-bis(4-hydroxyphenyl)-1,4,6-heptatrien-3-one, and sesquiterpenes procurcumenol and epiprocurcumenol

Parts used: Rhizome

Medicinal uses: The drug is considered antiseptic, acrid, hot, bitter and aromatic. The rhizome is considered to aid digestion, to purify the blood, to provide relief for colic, and for the treatment of colds and infections. The essential oil is an active ingredient in antibacterial preparations.

Scientific name	:	*Cynodon dactylon* Pers.
Family	:	Poaceae
Name of Drug	:	Durva

Botanical description: It is a perennial grass having creeping and prostrate stems; with lanceolate or linear narrow, finely, acute leaves; and laterally compressed, sessile, imbricate spikelets.

Distribution: Throughout India

Active ingredients: Sterols and flavonoids

Parts used: Whole plant

Medicinal uses: The entire plant is pungent, bitter, fragrant, healing, appetizer, vulnerary, anthelmintic, antipyretic, alexiteric. It destroys foulness of breath, useful in leucoderma, bronchitis, piles, asthma, tumors, and enlargement of the spleen. Surprising effects of *Cynodon dactylon* are for nutrition, reducing menstrual problems, acidity curing , controlling sugar, immunity enhancement, constipation reduction, curing bleeding gums, skin rashes, nervous system, obesity control, wound healing, overcoming eye infections, curing piles, diarrhoea, blood purifiers etc.

Scientific name : *Cyperus rotundus* Linn.

Family : Cyperaceae

Name of Drug : Musta

Botanical description: These are glabrous perennial herbs with elongated, slender stolons and tunicated black tubers. Flowers are umbels, spikelets are loosely spicate of 3-8 spikelets.

Distribution: In warmer parts of India as weeds

Active ingredients: α-cyperone, β-selinene, cyperene, patchoulenone, sugeonol, kobusone, and isokobusone

Parts used: Tubers

Medicinal uses: The drug is used for treating fevers, indigestion, dysmenorrhea. This is anti-inflammatory, anti-arthritic, antipyretic, analgesic, anti-diabetic, anti-diarrhoeal, cyto-protective, anti-mutagenic, antioxidant and anti-microbial.

Scientific name : *Desmodium triflorum* DC.

Family : Papilionaceae

Name of Drug : Hamsapadi

Botanical description: A prostrate creeping herb with branched stems and forming dense mats. Leaflets broadly obovate, broad and emarginate at the apex, about 1 cm long, glabrous to finely appressed-pilose beneath. Flowers are pink or purplish, one to three in the leaf axils. Pods are about one-third indented on one side, approximately five-seeded, pubescent.

Distribution : Found as weeds in plains of India

Active ingredients: β-phenylamine (major), indole-3-acetic acid, tyramine, trigonelline, hypaphorine

Parts used: Whole plant

Medicinal uses: The entire plant is regarded as diuretic, carminative and tonic. It is also used in bilious complaints, skin diseases, cough, asthma, diarrhoea and convulsions. The plant juice is applied to wounds or abscesses for quick healing.

Scientific name : *Eclipta prostrata* Linn.

Family : Asteraceae

Name of Drug : Bhrngarajah

Botanical description: It is an annual erect or prostrate, strigose herb with sessile, oblong-lanceolate leaves and solitary-axillary heads.

Distribution: Wet and moist lands of India

Active ingredients: Wedelollavlone

Parts used: Whole plant

Medicinal uses: The leaf extract is considered liver tonic, rejuvenative and good for hair. The entire plant is used for headache, night blindness, asthma, jaundice, urinary infections and intestinal worms.

Scientific name	:	*Elephantopus scaber* Linn.
Family	:	Asteraceae
Name of Drug	:	Gojihva

Botanical description: It is an erect herb, up to 38 cm high; rootstock short, giving off many stout fibrous roots. Leaves are 12.5-20 cm long, mostly radical forming a spreading rosette on the ground, obovate-oblong, coarsely serrate-dentate, hairy. Heads are numerous, sessile, closely packed, forming a large flat-topped terminal inflorescence, nearly 2.5 cm across, and surrounded at the base by three large, broadly ovate, leafy bract. Flowers are small and violet in colour.

Distribution: Throughout India, Myanmar, Sri Lanka, Africa, Malaya Peninsula, Indo-China, tropical Asia and America

Active ingredients: Germancranolide, elephantopin and quinic acid esters

Parts used: Roots, flowers and leaves

Medicinal uses: The herb is diuretic, laxative, analgesic, alterative, ferbrifuge, cardiac and brain tonic; used in griping, inflammations and bronchitis. Root is used in fever and to arrest vomiting. Leaves are used in piles. Bruised leaves boiled in coconut oil are applied to ulcers and eczema. The flowers are aphrodisiac, tonic and expectorant; cures biliousness, liver troubles and cough. Decoction of the roots and leaves are emollient; given in dysuria, diarrhoea and dysentery. Root paste is applied in rheumatism, and with mustard oil given in amoebic dysentery.

Scientific name	:	*Embelia ribes* Burm. f.
Family	:	Myrsinaceae
Name of Drug	:	Vidangah

Botanical description: It is a climbing shrub. The leaves are leathery and the flowers are white to greenish white whereas the fruits are rounded berry shaped and possess only one seed each.

Distribution: South-East Asia, The Himalayas and the western ghats of Tamil Nadu, Kerala and Sri Lanka.

Active ingredients: Embelin, christembine

Parts used: Leaves, roots, fruits

Medicinal uses: The dried fruits or seeds are used for treating intestinal worms. The drug is also used for treating fever, paralysis, skin diseases and to purify blood.

Scientific name	:	*Erythrina variegata* Linn.
Family	:	Fabaceae
Name of Drug	:	Paribhadrah

Botanical description: Indian Coral Tree is a showy, spreading tree legume with brilliant red blossoms. It is a picturesque, broad and spreading, deciduous tree that can attain 15-20m tall. It has many stout branches that are armed with black tiger's claw spines. The leaves are compound, with three diamond shaped leaflets, each about 6 in long. Before the leaves come out in late winter or early spring, coral tree puts on a spectacular show with bright crimson flowers 2-3 in long borne in dense terminal clusters. The beanlike pods that follow the flowers are cylindrical, and constricted between the reddish brown seeds.

Distribution: India, Africa and Australia

Active ingredients: Pterocarpan, isoflavones, flavanones and chalcones

Parts used: Bark, roots and leaves

Medicinal uses: The bark and leaves are used to destroy pathogenic parasites and relieve joint pain; the juice from the leaves is mixed with honey and ingested to treat tapeworm, roundworm, and threadworm in India; women used to take this juice to stimulate lactation and menstruation; it is commonly mixed with castor oil to treat dysentery; a warm poultice of the leaves is applied externally to relieve rheumatic joints; and the bark is used as a laxative, diuretic, and expectorant.

Scientific name	:	*Euphorbia hirta* Linn.
Family	:	Euphorbiaceae
Name of Drug	:	Dugdhika

Botanical description: This is an erect herb upto 50cm height with greenish yellow or white flowers. Stems are covered with yellowish hair. The leaves are small in opposite pairs.

Distribution: It is found as weed in plains of India

Active ingredients: Euphorbianin, leucocyanidol, camphol, quercitrin , quercitol, gallic acid, myricitrin, euphorbins, β-amyrin

Parts used: Whole plant

Medicinal uses: The plant has been used for female disorders but is now more important in treating respiratory ailments, especially cough, coryza, bronchitis and asthma. In India, it is used to treat worm infestations in children and for dysentery, gonorrhoea, jaundice, pimples, digestive problems and tumour. The fresh milky latex is applied to wounds and warts and the root of the plant is used in sprains and inflammation, miscarriage, epilepsy, maggots in wounds and irregular growth of teeth.

Scientific name	:	*Euphorbia nivulia* Buch-Ham.
Family	:	Euphorbiaceae
Name of Drug	:	Snuhi

Botanical description: It is a tall shrub with cylindrical stem and branches. Spines are stipular, paired and blackish. Leaves are obovate-oblong, spathulate and borne at the end of branches. Flowers are borne in cymes.

Distribution: Northern and central India

Active ingredients: Cyclonivulinol, nivulian-II, nivulian-III.

Parts used: Leaves, root bark, stem, latex

Medicinal uses: Leaf juice is used as a purgative, diuretic. Latex is used for treating jaundice, dropsy, enlargement of liver and spleen and haemorrhoids. Leafy latex and roots are applied in skin disorders, ear disorders, swellings and worm infections. Root bark is used in dropsy.

Scientific name	:	*Euphorbia thymifolia* Burm.
Family	:	Euphorbiaceae
Name of Drug	:	Dugdhika

Botanical description: This is an annual herb, mainly found in waste lands, along roadsides and wall sides under humid conditions. Its stem is slender, smooth, and reddish in colour and profusely branched. Roots are fibrous, thin and delicate. The stem is 10 to 20 cm in length with a diameter from 1 to 3 mm. Leaves are opposite, elliptic, oblong or ovate. Flowering occurs from June to November. Inflorescence is solitary or severely clustered at axils of leaves; Fruits are cocci when mature and seeds are long, ovoid and tetragonal.

Distribution: Tropical Asia including India, Sri Lanka and all hot countries except Australia.

Active ingredients: Quercetin, sterol, epitaraxerol, euphorbol, methylene cycloartenol, quercretin galactoside, taraxerol, tirucallol.

Parts used: Whole plant

Medicinal uses: The fresh plant is considered vulnerary and galactagogue; used in ophthalmia and other eye troubles, sores, atrophy, dysentery and breast pain. It is an effective drug for bronchial asthma. Juice of the plant is used for ringworm, diarrhoea and dysentery; mixed with fresh goat milk is given to cure blood dysentery. Leaves and seeds are astringent, stimulant, anthelmintic and laxative; given to children in bowel complaints. Root is used in amenorrhoea.

Scientific name	:	*Evolvulus alsinoides* Linn.
Family	:	Convolvulaceae
Name of Drug	:	Visnukranta

Botanical description: It is a slender, more or less branched, spreading or ascending, hairy herb. The stems are 20 to 70 cm long. Leaves are densely clothed with white and silky hairs, lanceolate to ovate in shape. The flowers are pale blue.

Distribution: Found as wild in the plains of India, Deccan Peninsula and Tuticorin

Active ingredients: Evolvin, betaine, sterols, proteins

Parts used: Whole plant

Medicinal uses: It cures fevers, gastric and dueodinal ulcers, debility, dysentery and uterine bleeding. It is said to be powerful brain stimulant.

Scientific name	:	*Ficus microcarpa* Linn.f.
Family	:	Moraceae
Name of Drug	:	Plaksah

Botanical description: It is a tree species, 20m tall with aerial roots, smooth and grey bark, elliptical leaves, dense canopy and small green figs turning to yellow or red when ripe..

Distribution: Sri Lanka, India, Taiwan, Malaya Archipelago, Australia

Active ingredients: Sterols, terpenoids, glycoside, flavonoids, polyphenols, proteins and carbohydrates.

Parts used: Fruits, leaves, barks, roots, latex

Medicinal uses: Fruits are used a pain killer, leaves and buds to treat conjunctivitis, roots are used in swollen feet and rheumatism, roots and leaves are used in bruises and wounds, bark and leaves are used in headache and latex is used in colic and liver ailments.

Scientific name	:	*Flickingeria nodosa* Seiden f.
Family	:	Orchidaceae
Name of Drug	:	Jivanti

Botanical description: These are tufted epiphytes with creeping root stock and sympodially branched stems, 20-30 cm long, each branch ends in a fusiform pseudobulb. Leaves are one per pseudobulb, 7-10 x 2 cm, oblong-lanceolate, obtuse. Flowers are white, 1-3 from the tip of a pseudobulb; bracts are ovate, acuminate, 3-veined; dorsal sepals are oblong-lanceolate, acute, 5-veined; lateral sepals are obliquely oblong-lanceolate, acute, 7-veined; petals are lanceolate, acute, 3-veined; lip, 3-lobed; lateral lobes oblong, obtuse; midlobe sub quadrate, recurved, crenulate.

Distribution: Deccan plains

Active ingredients: Jibentine, Jivantic acids

Parts used: Roots, stems and pseudobulbs

Medicinal uses: The drug is considered aphrodisiac, sweet, cold, rejuvenative, digestive which promotes liveliness of vigour.

Scientific name	:	*Fumeria indica* Pugsley.
Family	:	Fumeriaceae
Name of Drug	:	Parpatakah

Botanical description: It is a delicate much branched annual herb with clusters of tiny pale pinkish to whitish flowers. Sepals are minute. Upper petals has short, down-curved sac like spur. Flower stalks are erect, shorter than bracts. Leaves are 2-3 times cut into narrow pointed segments. Stems are glaucous and leafy.

Distribution: Indo-Gangetic plains, lower Himalayas and Nilgiri Hills

Active ingredients: Fuyuziphine and alpha-hydrastine.

Parts used: Whole plant

Medicinal uses: The drug is used in aches and pains, diarrhoea, fever, influenza, constipation, dyspepsia and liver complaints. It is used as a blood purifier for skin diseases and applied externally in leucoderma and as a fomentation for swollen joints. The dried plant is also used as an anthelmintic, diuretic and diaphoretic.

Scientific name : *Garcinia pedunculata* Roxb.

Family : Clusiaceae

Name of Drug : Amlavetasah

Botanical description: The tree has a fluted trunk with short spreading branches. Leaves are lanceolate with prominent midribs. Male flowers are light green in sparsely flowered panicles. The female flowers are solitary. The roundish fruit with a diameter ranging between 8 and 12 cm. It has a juicy interior with edible aril.

Distribution: North-Eastern India

Active ingredients: Hydrocitric acid

Parts used: Fruits

Medicinal uses: It is an excellent cardiac stimulant, digestive and laxative. Used in Ayurved since the time of Charaka, it is also useful in cough, bloating, splenomegaly etc.

Scientific name : *Glomus oppositifolium* A.DC.

Family : Molluginaceae

Name of Drug : Konika

Distribution: Moist shady places as weed in different parts of India

Parts used: The whole plant

Medicinal uses: The entire plant is used in fever, inflammation of bladder and kidney, burning sensation of the body and thirst and eye diseases.

Scientific name : *Gloriosa superba* Linn.

Family : Liliaceae

Name of Drug : Langali

Botanical description: These are tuberous rooted perennial herbaceous climbers and plants may climb upto 3 metres. Stems are leafy. Leaves are sessile, ovate -lanceolate, tip ending in a tendril-like spiral. The flowers are orange, yellow or red.

Distribution: Tropical India in different parts. Also found inhabiting Malayasia, Indonesia, Madagascar and South Africa.

Active ingredients: Colchicines, sitosterol and cornigerine

Parts used: Tuber

Medicinal uses: The tuber is abortifacient, anthelmintic, germicidal, astringent, acrid and bitter. It cures piles, chronic ulcers, colic pain in the bladder, toxicosis, itching, leprosy. It is applied in the navel region of vagina for easy and quick expulsion of the placenta after delivery. The paste of root is applied in bites of poisonous insects, snake bites, scorpion sting etc.

Scientific name : *Hedyotis brachypoda* DC.

Family : Rubiaceae

Name of Drug : Parpatakah

Botanical description: These are slender herbs, annual, diffusely branched, to 50 cm tall; stems are terete to slightly flattened, glabrous. Leaves are sessile or subsessile;. Inflorescences are pseudoaxillary, 1-flowered (2-flowered and fasciculate), glabrous, sessile or with peduncle to 3 mm, ebracteate. Flowers are subsessile to shortly pedunculate, homostylous. Calyx glabrous; hypanthium portion globose, limb lobed essentially to base; lobes triangular. Corolla white, rotate, outside glabrous; tube, glabrous at throat; lobes triangular. Anthers are exerted. Stigma ca. 0.8 mm, exerted.

Distribution: Found as weed in low land in India. The plant is distributed in America and South East Asia.

Active ingredients: 2-Hydroxy-3-methylanthraquinone

Parts used: Whole plant

Medicinal uses: The species is most commonly used in traditional Chinese medicine for treating cancer. The drug is used as alterative, anodyne, antibacterial, anti-inflammatory, antitumor, cardiotonic, depurative, diuretic, febrifuge, and sedative.

Scientific name	:	*Hedyotis corymbosa* Wall.
Family	:	Rubiaceae
Name of Drug	:	Parpatakah

Botanical description: This is an erect ascending and spreading slender plant. The leaves are simple, opposite, sub-sessile, linear, narrow with recurved margins. The flowers are white in pairs or threes.

Distribution: Found as weed in cultivated low land fields.

Active ingredients: Hydroxy-geniposide, arabinose, fumaric acid, galactose, coumaric acid, rhamnose, rutin.

Parts used: Whole plants

Medicinal uses: It is diuretic, anthelmintic and digestive. Besides, the drug is used as hepatoprotective, anti-inflammatory, anti-malarial and anti-oxidant.

Scientific name	:	*Heliotropium indicum* Linn.
Family	:	Boraginaceae
Name of Drug	:	Vrscikali

Botanical description: This is an annual, erect, branched plant that can grow to a height of about 15–50 centimetres. Stems are hairy bearing alternating ovate to oblong-ovate leaves. It has small white flowers with a green calyx; five stamens borne on a corolla tube; a terminal style; and a four-lobed ovary.

Distribution: Throughout India, very common in moist and wet parts, Myanmar, Africa, Malayasia, China and America.

Active ingredients: Pyrrolizidine alkaloids, indicine (principal), echinitine, supinine, heleurine, heliotrine, lasiocarpine, its N-oxide, acetyl indicine, indicinine and anti-tumour alkaloid, indicine-N-oxide.

Parts used: Whole plant

Medicinal uses: Plants are astringent, emollient, vulnerary and diuretic. It is used as local application for ulcers, sores, wounds, gum boils, skin affections, stings of insects and rheumatism. Leaves are used for ringworm; juice is used in eye disease; decoction is used in fevers and urticaria. Roots are aphrodisiac and also used for the cure of night blindness. Decoction of the roots is used in coughs and fevers. Seeds are stomachic. The flowers are considered emmenagogue in small doses and abortifacient in large doses.

Scientific name : *Holarrhena anti-dysenterica* Linn.

Family : Apocynaceae

Name of Drug : Kutaja/Kurchi

Botanical description: A deciduous shrub or a small tree, growing up to 3 m in height. Short stem has pale bark and several branches. Oppositely arranges, ovate, obtusely acuminate leaves are 10-20 cm long. Leaf stalks are very short. White flowers appear in corymb-like cymes, 5-15 cm across, at the end of branches. Flowers have five white petals 2-3 cm long which turn creamish yellow as they age. The flowers are beautiful with oblong petals which are rounded at the tip, and remind one of frangipani.

Distribution: India and other Asian countries.

Active ingredients: Conessine, conessimine, kurchine, conamine, conimine, conessidine, conarrhimine, holarrhimine, holarrhine and kurchicine.

Parts used: Stem, root bark and seeds

Medicinal uses: Kutaja is primarily used for the treatment of dysentery but has several other therapeutic usages. It is particularly useful in bleeding disorders such as menorrhagia, haemorrhoids, diabetes and oedema and has been used for tumours, abscesses, aches and pains, bronchitis, colic, diarrhoea, splenitis and as a vermifuge, laxative and astringent. The bark is useful in treatment of piles, skin diseases and biliousness.

Scientific name : *Hemidesmus indicus* Linn.

Family : Asclepiadaceae

Name of Drug : Sariva

Botanical description: /t is a perennial, fast-growing thin creeper vine with tendrils at every node to cling to the surrounding vegetation for stability and support. The leaves are very slender, smooth and oval shaped, and shiny dark green colour throughout the year. The stems stiffens and become woody over time, and the bark varies in colour from dark red to rust to brown. The flowers are small, thin and elongated, and light green with a purple hue inside.

Distribution: South Asia

Active ingredients: Essential oil, Starch, Coumarin, Tannic acid, Triterpenoid saponins

Parts used: Roots

Medicinal uses: It is used as a healing herb as well as a magical-spiritual dream herb. It is used to treat stomach problems, cure rashes, ease the mind, quell the symptoms of syphilis, induce trance states and deep meditation, and to clarify and prepare the mind for the dream world. Ayurvedic healers also prescribed it to men suffering from low libido and sexual impotence, it is believed that one of active compounds produced by the roots improves male testosterone levels and therefore sexual desire, sperm count, and overall sexual performance.

Scientific name	:	*Homonoia riparia* Lour.
Family	:	Euphorbiaceae
Name of Drug	:	Pasanabhedah

Botanical description: It is a shrub grown up to 3 m tall. The leaves are linear-lanceshaped, 12-2- cm long. Upper surface of the leaves is green and shining and the lower surface brown and hairy. Flowers are reddish, 5-10 cm long, borne in spikes with obovate bracts.

Distribution: Rocky river banks, Sikkim Himalaya, Assam , Khasi Hills, Myanmar, Andaman Islands, South India, Sri Lanka and Java.

Active ingredients: Flavonoid glycoside

Parts used: Roots

Medicinal uses: The roots are diuretic, astringent, bitter and cool. The drug is employed in fever, cough, uterine disorders, ulcer, swellings, blood disorders and piles.

Scientific name	:	*Hybanthus enneaspermus* (L.f) Muell.
Family	:	Violaceae
Name of Drug	:	Padmacharni

Botanical description: It is a perennial herb or small shrub to 60 cm high, smooth or hairy. Leaves are linear to lance-like, 1-5 cm long, margins are recurved to revolute, occasionally flat; stipules are acuminate, 1-4 mm long. Flowers are pink-purple, spade-shaped and solitary.

Distribution: India, Sri Lanka, China, Africa, Madagascar

Active ingredients: Aurantiamide acetate, â-sitosterol and isoarborinol

Parts used: Whole plant

Medicinal uses: This herb is considered to be extremely beneficial to men, used as a diuretic, demulcent and tonic. The root is diuretic and is used in urinary affections and bowel complaints of children. Decoction of leaves and tender stalks is demulcent. The fruit is used to treat scorpion sting.

Scientific name	:	*Inula racemosa* HK.f.
Family	:	Asteraceae
Name of Drug	:	Puskaramulam

Botanical description: It is a tall stout herb up to 1.75m tall, with large short stalked flower heads, borne in a spike cluster at branch ends. Ray florets are slender, up to 2.5 cm long. Lower leaves are elliptic-lance-shaped, 18-45cm long, narrowed to a winged stalk. Upper leaves are lance shaped and half stem clasping.

Distribution: Jammu and Kashmir, Himachal Pradesh at higher elevations.

Active ingredients: Alantolactone, Inulin, Isoalantolactone, Inunolide, Germacronolide, Dihydroisoalantotolactone, D-Mannitol, Dihydroinunolide, Neo-A lantolactone, Inunolise, Alantodiene, Sesquiterpene, lactone- inunol

Parts used: Roots

Medicinal uses: Pushkarmool has path breaking results in respiratory discomfort, cough and chest pain. It also relieves stress and has anti allergic properties as well.

Scientific name	:	*Ipomoea digitata*Auct.
Family	:	Convolvulaceae
Name of Drug	:	Vidari

Botanical description: The plant is a climber with a large lobed leaves and attractive pink or purple flowers.

Distribution: Tropical and humid regions of India

Active ingredients: Pterocarpan tuberosin, pterocarpanone hydroxyl tuberosone

Parts used: Tuber

Medicinal uses: Raw tuber is taken for treating blood dysentery and used as astringent. Sun dried root powder, boiled in sugar and butter is recommended to promote weight gain, and to regulate menstrual discharge. Tubers are also used for the treatment of debility, spermatorrhoea, fever, bronchitis, scorpion stings and menorrhagia. The drug is also used for treatment of liver and spleen enlargement.

Scientific name	:	*Ipomoea marginata*
Family	:	Convolvulaceae
Name of Drug	:	Laxmana

Botanical description: It is a slender vine, 3-5 m long, commonly seen on bushes. Leaves are alternately arranged ovate-heart-shaped with purple botches in the center. Upper leaves are sometimes sagittate or hastate, smooth, base deeply heart-shaped or arrow-shaped, margin entire or minutely wavy. White to pink-purple funnel shaped flowers are borne in stalked cymes. Flowers are 2.5-4 cm long, with a narrow tube flaring open into a flat flower, and have a dark purple throat.

Distribution: Southern India

Medicinal uses: Used as a single drug for curing sterility in women and for promoting fertility and virility.

Scientific name	:	*Ipomoea reniformis* Choisy.
Family	:	Convolvulaceae
Name of Drug	:	Akhukarni

Botanical description: It is a slender, prostrate, creeping, smooth or hairy herb. The stems root at the node. The leaves are small, kidney shaped to heart shaped, often wider than long and irregularly toothed. One to three flowers occur on short stalks in the axils of leaves.

Distribution: Deccan Peninsula, Rajasthan, Bihar and tropical India, South Asia, Tropical Africa and Sri Lanka.

Active ingredients: Caffeic, p coumaric, ferulic and sinapic acid esters

Parts used: Whole plant

Medicinal uses: The entire plant is used in the treatment of cough, headache, neuralgia, rheumatism, diuretic, inflammation, troubles of nose, fever due to enlargement of liver and also in kidney diseases. Powder of leaves is used as a snuff during epileptic seizures, Juice acts as purgative and the root is having diuretic, laxative, and applied in the disease of the eyes and gums.

Scientific name : *Iris germanica* Linn.

Family : Iridaceae

Name of Drug : Puskaramulam

Botanical description: It is a tall bearded iris, perennial and 75cm tall. Iris have thick, fleshy, underground stems (called rhizomes) that store food produced by the sword-shaped, semi-evergreen leaves. Flowers are purple and lilac, fragrant.

Distribution: Himachal Pradesh and Eastern Himalaya

Active ingredients: Myristic acid and oleic acid

Parts used: Corms

Medicinal uses: The dried root is diuretic, expectorant and stomachic. It is taken internally in the treatment of coughs, catarrh and diarrhoea. Externally, it is applied to deep wounds. The juice of the fresh root is a strong purge of great efficiency in the treatment of dropsy.

Scientific name : *Jatropha glandulifera* Roxb.

Family : Euphorbiaceae

Name of Drug : Dravanti

Botanical description: It is an evergreen shrub, 3 - 5 m tall, often shrub-like, with sticky, transparent juice; trunk stout; hairy branches. Leaves are simple, alternate, 5-15 cm long, 2.5 - 5-5 cm wide, base rounded or flat, rarely shallowly heart-shaped, lower base with densely glandular hairs, tip caudate, entire or near base bilateral 2-lobed, lobes triangular, pointed, sometimes toothed, glandular-hairy above, hairless below, margins sparsely glandular-pilose; basal veins 5, lower base with densely glandular hairs; leaf-stalks 2 - 9 cm long, glandular-hairy. Flower are borne in cymes, greenish yellow. Fruit is a capsule, three-lobed.

Distribution: Peninsular India, Sri Lanka, Africa

Active ingredients: Alkannins, Jatropholone A, fraxetin and a coumarinolignan.

Parts used: Root and seeds

Medicinal uses: Root and oil from seed are purgative. Oil is used as antirheumatic, and antiparalytic. Used externally on ringworm and chronic ulcers. Roots are used for glandular swellings. Latex is applied to warts and tumour.

Scientific name : *Kaempferia galanga* Linn.

Family : Zingiberaceae

Name of Drug : Karcurah

Botanical description: It is a stem-less tuberous aromatic rootstock with horizontal spreading sub-orbicular and sub-sessile leaves, and central radical spikes.

Distribution: In shaded forest areas of Indian plains and Malaya Islands

Active ingredients: Borneol, 1,8-Cineole, Ethyl cinnamate, Ethyl p-methoxycinnamate, Gamma-car-3-ene, Pentadecane

Parts used: Rhizome

Medicinal uses: Rhizomes are used for the treatment of cold, cough, purify dirty blood, stomach inflammation and diarrhoea.

Scientific name : *Kaempferia rotunda* Linn.

Family : Zingiberaceae

Name of Drug : Hallakam

Botanical description: It is a stem-less herb with tuberous rootstock and oblong-lanceolate variegated leaves and radical spikes.

Distribution: In humid and shaded forest areas in southern India, Sri Lanka and Malay Islands

Active ingredients: Crotepoxide

Parts used: Rhizomes

Medicinal uses: The rhizomes are anti-inflammatory and stomachic. The ointment made from the plant is used in wounds or bruises to reduce swellings. It cures insomnia, mental disorder, burning sensation and purifies the blood.

Scientific name : *Kyllinga nemoralis* Hutch and Delz.

Family : Cyperaceae

Name of Drug : Musta

Botanical description: It is a perennial creeping sedge with erect 3-angled stems. Leaves are many, usually shorter than stems, linear with brown to purple brown sheath. Inflorescence is spherical terminal head. Spikelets in head are ovate to lanceolate, white and single flowered.

Distribution: Tropical and subtropical India

Active ingredients: Flavonoids, saponins, phenols, terpenes, lipids and glycosides

Parts used: Root tuber

Medicinal uses: Root tuber is employed in cough, bronchitis, fever, poisonous affections, indigestion, dyspepsia, vomiting and worm trouble. The paste is applied in eye and skin diseases.

Scientific name : *Lagenandra toxicaria* Dalz.

Family : Araceae

Name of Drug : Hallakam

Botanical description: Helophytes, rarely rheophytes, with thick creeping rhizome; leaf blade simple, ovate to almost linear, fine venation transverse-reticulate; spathe tube with connate margins; spadix entirely enclosed in spathe tube; flowers are unisexual, perigone absent.

Distribution: Andhra Pradesh, Goa, Karnataka, Kerala, Maharashtra, Tamil Nadu

Active ingredients: Diethyl phthalate, Oleic acid, Palmitic acid ethyl ester

Parts used: Whole plant

Medicinal uses: The plant contains an acrid juice and it is used in ointments for itch. Also used in the treatment of kidney disorders, heart diseases and swellings. It also has insecticidal properties. Infusion of tuberous rootstock is used in tuberculosis. Decoction of tuberous root is used to check growth of old tumour is any body part.

Scientific name	:	*Lagenaria siceraria* Standly
Family	:	Cucurbitaceae
Name of Drug	:	Iksvakuh

Botanical description: It is a vigorous annual herb. Stems are prostrate or climbing, angular, ribbed, thick, brittle, softly hairy, up to 5 m long, cut stems exude no sap. Leaves are simple, up to 400 mm long and 400 mm broad, shortly and softly hairy, broadly egg-, kidney- or heart-shaped in outline, undivided, angular or faintly 3-7-lobed, lobes rounded, margins shallowly toothed, crushed leaves non-aromatic. Leaf stalks up to 300 mm long, thick, often hollow, densely hairy, with two small, lateral glands inserted at the leaf base. Tendrils split in two. Flowers stalked (female flower stalks shorter than male), solitary, monoecious (male and female flowers on the same plant); petals 5, crisped, cream or white with darker veins, pale yellow at the base, obovate, up to 45 mm long, opening in the evenings, soon wilting. Fruit large, variable, subglobose to cylindrical, flask-shaped or globose with a constriction above the middle; fleshy, densely hairy to ultimately glabrous, indehiscent, green, maturing yellowish or pale brown, pulp drying out completely on ripening, leaving a thick, hard, hollow shell with almost nothing inside except the seeds. Seeds many, embedded in a spongy pulp, 7-20 mm long, compressed, with two flat facial ridges, in some variants rather irregular and rugose.

Distribution: It is cultivated as vegetable crop in hotter parts of India. The species inhabits tropical Asia and Africa.

Active ingredients: Cucurbitacins, fructose and glucose

Parts used: Whole plant

Medicinal uses: The fruits, leaves, stem, seeds and oils are traditionally used in the treatment of jaundice, diabetes, ulcer, piles, colitis, insanity, hypertension, congestive cardiac failure, and skin diseases. The fruit pulp is used as an emetic, sedative, purgative, cooling, diuretic, antibilious, and pectoral. The flowers are an antidote to poison. The stem bark and rind of the fruit are diuretic. The seed is vermifuge. Extracts of the plant have shown antibiotic activity. Leaf juice is widely used for baldness. The plant juice is an excellent remedy for heart problems, digestive and urinary disorders, and in diabetes. The juice prevents excessive loss of sodium, satiating thirst, and giving a cooling effect. Probably, the bitter principle found in the wild bottle gourds is responsible for the purgative property. Crushed leaves are used for baldness and applied on the head for the headache. Leaves are also used as alternative purgative. Fruit is rich source of water and minerals and are believed to possess vitamin A, C and B complex. Bottle gourd is believed to help the liver function in a balanced fashion. The juice from the leaves help cure jaundice and the juice from the gourd helps reduce graying of hair. The gourd juice helps treat burning sensation in the urinary passage if consumed with lime juice. It reduces fatigue and keeps you fresh, especially during summer. It helps fight constipation, as it is fibre rich. Because of its fibre and low fat content, Ayurveda highly recommends this food for diabetic patients and young children. Ayurveda also recommends the juice of this gourd in the treatment of acidity, indigestion and ulcers as it serves as an alkaline mixture. Indian traditional medicine claims that bottle gourd acts as a nerve tonic and can help improve obsessive compulsive disorder (OCD).

Scientific name : *Leptadenia reticulata* Wt. and Arn.

Family : Asclepiadaceae

Name of Drug : Jivanti

Botanical description: It is a twining shrub, with numerous branches. The stems have a cork-like, deeply cracked bark, hairless when young. Leaves are leathery, ovate, pointed, hairless above, finely velvety below. Flowers are greenish-yellow, in lateral or subaxillary cymes, often with small hairs.

Distribution: Found in the sub-Himalayan tracts of Punjab and Uttarakhand and throughout the Deccan peninsula up to an altitude of 900 m and found particularly in hedges. It is also distributed throughout Mauritius, Madagascar, Sri Lanka, the Himalayas and Burma.

Active ingredients: Stigmasterol, a-amyrin, tocopherols, hentriacontanol, leptadeno and sitosterol.

Parts used: Whole plant

Medicinal uses: It corrects the metabolism and enhance the nutritive status of the body and corrects the digestive system; it specially nourishes the eyes and the reproductive system. Jivanti is a cardio tonic and cures bleeding disorders. It is also used in bronchitis as an expectorant, an aphrodisiac and improves seminal quality through its specific action on the reproductive organs, used in spermatorrhoea, in dysuria, burning micturition and pyuria and it also benefits in fever, burning and also in defective eyesight.

Scientific name : *Leucas aspera* Spr.

Family : Lamiaceae

Name of Drug : Dronapuspi

Botanical description: This is an annual, branched, herb erecting to a height of 15-60 cm with stout and hispid acutely quadrangular stem and branches. Leaves are sub-sessile or shortly petiolate, linear or linearly lanceolate, obtuse, pubescent up to 8.0 cm long and 1.25 cm broad, with entire or crenate margin; petiole 2.5-6 mm long. Flowers are white, sessile small, in dense terminal or axillary whorls; bracts 6 mm long, linear, acute, bristle-tipped, ciliate with long slender hairs.

Distribution: Sikkim, Bihar, Punjab in India, Malaysia, Mauritius, Indo-China, Bangladesh, Java and Philippines

Active ingredients: Oleanolic acid, ursolic acid and 3-sitosterol.

Parts used: Whole plant

Medicinal uses: Medicinally, it has anti-microbial, insecticidal, fever reducing, larvicidal and inflammation reducing properties. It is useful in skin diseases. The juice removes toxins from the body. The leaf juice is recommended for cold and cough. The leaves are useful in chronic rheumatism.

Scientific name : *Leucas stricta* Benth.

Family : Lamiaceae

Name of Drug : Dronapuspi

Botanical description: This is an erect, annual herb, 15-25 cm high; stem and branches are 4-gonous, pubescent and sparsely hispidly hairy with spreading hairs. Leaves are oblong-lanceolate or linear, obtuse, entire or distantly serrulate, pubescent both sides and with scattered hairs on the veins, subsessile or shortly petioled. Flower-whorls upto 2 cm in diam., terminal, solitary, subtended by a whorl of floral leaves, rarely another whorl of flowers develops below the terminal one.

Distribution: It is widely distributed in sea coasts of southern India

Active ingredients: Laballenic acid, B- Sitosterol and its glycoside

oleanolic acid

Parts used: Whole plant

Medicinal uses: The herb is medicinal and regarded as antipyretic, antiseptic, anti-histaminic and wormifuge. The fresh juice extracted from leaves is used for jaundice and skin diseases.

Scientific name : *Luffa cylindrica* M. J. Roem.

Family : Cucurbitaceae

Name of Drug : Kosataki

Botanical description: It is a very fast growing climber. Its leaves are 7 - 20 cm across and have three lobes. Flowers are bright yellow. The fruits which grow to about 60 cm in length are oblong or cylindrical, smooth and contain many seeds. The fruit is brown when mature and dries on the vine to develop an inedible sponge-like structure.

Distribution: Coastal districts and tropical parts of India

Active ingredients: Triterpenoid saponins

Parts used: Leaves, flowers and fruits

Medicinal uses: Plant is bitter tonic, emetic, diuretic and purgative and useful in asthma, skin diseases and splenic enlargement. It is used internally for rheumatism, backache, internal haemorrhage, chest pains as well as haemorrhoids. Externally, it is used for shingles and boils. The dried fruit fibres are used as abrasive sponges in skin care, to remove dead skin and to stimulate the circulation. The fruits are anthelmintic, carminative, laxative, depurative, emollient, expectorant, tonic and galactagogue and are useful in fever, syphilis, tumours, bronchitis, splenopathy and leprosy. Fruit is intensely bitter and fibrous. It has purgative property and is used for dropsy, nephritis, chronic bronchitis and lung complaints. It is also applied to the body in putrid fevers and jaundice.

Scientific name : *Mallotus philippensis* Muell.-Arg.

Family : Euphorbiaceae

Name of Drug : Kampillakah

Botanical description: The tree can grow up to 10 m tall. Leaves are alternately arranged, ovate or rhombic ovate, rusty-velvety. Male and female flowers occur in different trees. Female flowers are borne in lax spike like racemes at the end of branches or in leaf axils. Male flowers occur three together in the axils of small bracts.

Distribution: Throughout India

Active ingredients: Phenols, diterpenoids, steroids, flavonoids, cardenolides, triterpenoids, coumarins, isocoumarin

Parts used: Leaves and fruits

Medicinal uses: The leaves are bitter, cooling and appetizer. Fruit is heating, purgative, anthelmintic, vulnerary, detergent, maturant, carminative, alexiteric and useful in treatment of bronchitis, abdominal diseases, spleen enlargement.

Scientific name : *Marsdenia tenacissima* Roxb.

Family : Asclepiadaceae

Name of Drug : Murva

Botanical description: It is a robust lianas, densely hairy to velvety throughout except for the upper surface of petals. Leaf-stalks are slender; leaves are ovate, base deeply heart-shaped with rounded sinus, tip long-pointed. Flower-clusters are much branched, broader than long and many flowered. Clusters are carried on stalks up to 2 cm. Flower-stalks are 6-8 mm. Sepals are elliptic-lanceolate with rounded tip. Flowers are greenish yellow, with a bell-shaped tube and spreading petals, very densely hairy on the underside.

Distribution: Himalayan region, Sri Lanka, West Bengal, China

Active ingredients: Cinamic acid, Steroidal compounds

Parts used: Roots

Medicinal uses: It is used for the treatment of cancer, asthma, blood purity and burns

Scientific name : *Marsilea minuta* Linn.

Family : Marsileaceae

Name of Drug : Sunisannah

Botanical description: It is a perennial fern with slender, rooted, creeping, branching rhizomes bearing erect leaves (sterile fronds) along their length. The leaves, which consist of four, 'clover-like' leaflets at the apex of a slender erect stalk, arise along the length of each rhizome. At the base of the petioles the sporocarps are formed on about 5 mm long stalks. The sporocarps are oblong with rounded ends (when seen laterally), with their long axis at right angles to the stalk.

Distribution: Naturally distributed in wet marshy lands in various parts of India

Active ingredients: Hydroxy-triacontan

Parts used: Whole plant

Medicinal uses: The entire plant is used to stop nose bleeding and in the treatment of diarrhoea, bronchitis, diabetes, epilepsy, hepatitis, kidney infection, as blood purifier, aphrodisiac and in the treatment of piles.

Scientific name	:	*Merremia tridentata* Hallier f.
Family	:	Convolvulaceae
Name of Drug	:	Prasarani

Botanical description: It is an annual or perennial herb or shrub, usually twining, but also prostrate and rooting at the nodes with tuberous roots. Leaves are alternate, variable in size and shape, entire, dentate, lobed or palmately or dentately partite or compound; petiole present; stipules absent. Inflorescence is axillary, few- to many-flowered, variously ramified, large to small; peduncle present to almost absent; bracts usually small. Flowers are bisexual, regular, small to large, white, yellow to orange in colour.

Active ingredients: Diosmetin, luteolin, diosmetin–7–o–β–d– glucoside, luteolin–7–o –β–d–glucoside.

Parts used: Whole plant

Medicinal uses: The plant is considered bitter, astringent, tonic and used in the treatment of rheumatism, swelling, urinary disorders and the decoction of the root is used for anti-diabetics and to improve hair growth.

Scientific name	:	*Mimosa pudica* Linn.
Family	:	Minosaceae
Name of Drug	:	Lajjalu

Botanical description: *M. pudica* is a low-growing, much-branched, prickly, perennial shrub. It grows 15-100 cm tall and may be erect, but more generally has a trailing, sprawling growth habit. The stems are woody at the base, stiff, cylindrical, reddish-brown or purple, pubescent and bear scattered prickles along the internodes. The prickles are compressed, slightly curved, hard and very sharp. The root is long and robust. The leaves are dark green, bi-pinnate and hairy. The pinnae are in two pairs (sometimes only one pair) arising close together from the tip of the petiole so the arrangement appears palmate. There are 12-50 leaflets, each 6-12 mm long, 1.5 mm wide, oblong-linear and pointed, glabrous above but with a hairy margin and lower surface. The stipules are linear-lanceolate.

Leaflets and leaves fold up rapidly when touched and also close at night.

Flowers are bright purplish-pink with four prominent stamens and occur in globular or ovoid heads about 9 mm in diameter.

Distribution: Naturally grown as weed in plains of India and found native in South America.

Active ingredients: Adrenalin, Crocetin dimethyl Easter, d-xylose and d-glucuronic acid, tubuline, turgorines The periodic leaf movement factors are reportedly the derivatives of 4-α-(b-D-glucopyranosyl-6-sulphate) gallic acid.

Parts used: Whole plant

Medicinal uses: Root is bitter, acrid, cooling, vulnerary, alexipharmic, and used in the treatment of leprosy, dysentery, vaginal and uterine complaints, inflammations, burning sensation, asthma, leucoderma, and fatigue and blood diseases. Decoction of root is used with water to gargle to reduce toothache. It is very useful in diarrhoea (athisaara), amoebic dysentery (raktaatisaara), bleeding piles and urinary infections. It arrests bleeding and fastens the wound healing process. It is mainly used in herbal preparations for gynecological disorders. It is also used to cure skin diseases. It is also used in conditions like bronchitis, general weakness and impotence. It is also used to treat neurological problems.

Scientific name	:	*Oxalis corniculata* Linn.
Family	:	Oxalidaceae
Name of Drug	:	Cangeri

Botanical description: This is a delicate perennial hairy herb, procumbent, with alternate, stalked, 3-lobed leaves. The flowers are pale pink to yellow, small in size and appear on long stalked inflorescence.

Distribution: In swampy and wet places throughout India, particularly in humid and tropical climate.

Active ingredients: Oxalic acid, potassium oxalate, cellulose, vitamin C

Parts used: Whole plant

Medicinal uses: The entire plant is considered digestive, stimulant and astringent. Leaves are used externally to remove corns, warts etc. Applied as poultice to inflamed parts for relieving pain. The leaves are used in the treatment of fevers, scurvy, stomach disorders, skin disorders, eye disorders, jaundice and diabetic symptoms.

Scientific name	:	*Melastoma malabathricum* Linn.
Family	:	Melastomataceae
Name of Drug	:	Tinisah

Botanical description: It is a shrub clothed with short acute pale scales, lanceolate, acute leaves narrowed at the base. Leaves are 3-5 nerved and with short petioled. The corymbs rae few flowered.

Distribution: South-East Asia

Active ingredients: Malabrathins B, C and D, nobotanin, B,G and H, kaempferol

Parts used: Leaves, flowers and roots

Medicinal uses: Leaves and flowers are used in diarrhoea, cholera and dysentery. Leaf is considered antiseptic and used in cuts and wounds. Leaves are also used in leucorrhoea and rejuvenation of wounds. The powdered leaves and roots are sprinkled on wounds. They are also used for haemorrhoids.

Scientific name	:	*Moringa oleifera* Linn.
Family	:	Moringaceae
Name of Drug	:	Sigru

Botanical description: *M. oleifera* is a fast-growing, deciduous tree, 10-12m tall. The bark is a whitish-grey colour and is surrounded by thick cork. Young shoots are purplish or greenish-white, hairy bark. The tree bears an open crown of drooping, fragile branches and the leaves build up a feathery foliage of tripinnate leaves.The flowers are fragrant and bisexual, surrounded by five unequal, thinly veined, yellowish-white petals. The fruit is a hanging, three-sided brown capsule of 20–45 cm size with dark brown, globular seeds. The seeds have three whitish papery wings and are dispersed by wind and water.

Distribution: Tropical and subtropical parts of India

Active ingredients: Zeatin, quercetin, beta-sitosterol, caffeoylquinic acid and kaempferol.

Parts used: Leaves, flowers, seeds, roots, barks and fruits

Medicinal uses: It acts as cardiac and circulatory stimulants, possess antitumor, antipyretic, antiepileptic, antiinflammatory, antiulcer, antispasmodic, diuretic, antihypertensive, cholesterol lowering, antioxidant, antidiabetic, hepatoprotective, antibacterial and antifungal activities, and are being employed for the treatment of different ailments in the indigenous system of medicine. Leaves are effective against headache, healing of wounds, fevers, bronchitis, ulcers, diarrhoea, skin diseases and anaemia. Flowers are useful for enhancement of breast milk, cold and urinary problems. Pods are used in de-worming, liver problems, joint pains and diarrhoea. Seeds are used for their antibiotic and anti-inflammatory properties to treat arthritis, rheumatism, gout, cramp, sexually transmitted diseases and boils. The roots and bark are used for cardiac and circulatory problems, as a tonic and for inflammation.

Scientific name	:	*Pavetta indica* Linn.
Family	:	Rubiaceae
Name of Drug	:	Paphanah

Botanical description: It is an erect, nearly smooth or somewhat hairy shrub, 2 to 4 m tall. The leaves are elliptic-oblong to elliptic-lanceolate, and pointed at both ends. The flowers are white, rather fragrant, and borne in considerable numbers in hairy terminal panicles. The sepals are very small, and toothed. The flower-tube is slender with obtuse petals about half the length of the tube.

Distribution: Tropical Asia and Africa

Active ingredients: D-mannitol. β-Sitosterol, α-amyrin, quercetin, caffeic acid, chlorogenic acid

Parts used: Bark, leaves and roots

Medicinal uses: The bark, in decoction, or pulverized, is administered, especially to children, to correct visceral obstructions. The decocted leaves are used externally to alleviate the pains caused by haemorrhoids. The root, pulverized and mixed with the ginger and rice-water, is given in dropsy. A local fomentation with the leaves is useful in relieving the pain of piles.

Scientific name	:	*Pavonia odorata* Willd.
Family	:	Malvaceae
Name of Drug	:	Valakah

Botanical description: This is an erect perennial herb, covered with sticky hairs. Leaves are heart-shaped-ovate, 3-5 angled or 3-5 lobed, 4-6 cm long, 5-7 cm broad. Flowers arise singly in leaf axils, or fascicled at the end of branches. Bracts are 10-12 in number, linear, and sepals are 5. Flowers are pink and twice longer than the sepal cup.

Distribution: North-West Provinces, Sindh and Banda, Western Peninsula, Myanmar, Sri Lanka, East tropical Africa.

Active ingredients: Alpha-terpinene, aromadendrene, azulene, isovaleraldehyde, isovaleric acid, pavonenol, pavonene

Parts used: Whole plant

Medicinal uses: Roots are cooling, demulcent, carminative, diaphoretic and diuretic; alleviates spasms and antiprotozoal (plant); reduces fever and astringent (root). Leaves and young shoots are used as an emollient. Used in dysentery, inflammation, intestinal bleeding, increasing the secretion of gastric juices, reducing fever and as a tonic and rheumatism.

Scientific name	:	*Pedalium murex* Linn.
Family	:	Pedaliaceae
Name of Drug	:	Goksurah

Botanical description: It is a shrubby, stiff-stemmed herb, native to India, grown for reputed medicinal and other uses. It is diffuse annual, much branched, spreading, succulent, glandular, up to 60 cm tall. Leaves are simple, opposite, ovate or oblong-obovate, irregularly and coarsely crenate-serrate. Flowers are yellow.

Distribution: Southern India, Pakistan, Sri Lanka and tropical Africa

Active ingredients: Caffeic acid, cumaric acid, ferulic acid, vanilic acid, ursolin acid and sitosterol

Parts used: Fruits and leaves

Medicinal uses: Leaves are used for ulcers. Fresh leaves and shoots are mixed with boiled milk and used as aphrodisiac. Fruit is demulcent, diuretic, anti-spasmodic and aphrodisiac. The juice of fruit is emmenagogue.

Scientific name	:	*Phaseolus delzellii* T.Cooke.
Family	:	Papilionaceae
Name of Drug	:	Masaparni

Botanical description: It is a perennial creeping herb or twine which form roots on nodes.

Distribution: Lower hills of Tamil Nadu

Parts used: Roots

Medicinal uses: It is used for curing indigestion, constipation, cough, pyrexia and haematemensis.

Scientific name	:	*Phaseolus mungo* Linn.
Family	:	Papilionaceae
Name of Drug	:	Masaparni

Botanical description: It is an erect, suberect or trailing, densely hairy, annual herb. The tap root produces a branched root system with smooth, rounded nodules. The pods are narrow, cylindrical and up to six cm long. The plant grows 30–100 cm with large hairy leaves and 4–6 cm seed pods.

Distribution: Throughout India

Active ingredients: Vitamins, minerals and potassium

Parts used: Roots and seeds

Medicinal uses: The drug promotes lymphatic circulation, increases body strength, colour of skin, hair growth and breast milk. It improves nervous system and increases sperm count and sperm motility. It is very effective in erectile dysfunction (impotence) and premature ejaculation of male reproductive system. This herb is effective in dysmenorrhea and primary amenorrhea.

Scientific name	:	*Phaseolus sublobatus* Wall.
Family	:	Papilionaceae
Name of Drug	:	Masaparni

Botanical description: It is an annual, herb or climber, 1 m tall. Flowers are white, yellow or purple. Stem are twining or prostrate. Leaflets are frequently lobed. Pods are cylindric, straight and densely hairy.

Distribution: Grown as weeds in different parts of India

Active ingredients: Vitamins, minerals

Parts used: Roots

Medicinal uses: Rejuvenating drug is prepared from the plant. It is considered aphrodisiac and roborant.

Scientific name	:	*Phyllanthus watsonii Airy-Shaw* Brunal and Roux.
Family	:	Euphorbiaceae
Name of Drug	:	Tamalaki

Botanical description: A semi-woody shrub, 1m tall. The flowers are in clusters, on slender racemes to 15 cm borne in groups on main branches behind the leafy twigs.

Distribution: Sri Lanka, Indonesia, West Indies and South India

Active ingredients: Terpenoids, alkaloids

Parts used: Whole plant

Medicinal uses: It is used for the treatment of diabetes, anaemia, hepatitis, bronchitis and jaundice.

Scientific name	:	*Phyllanthus amarus* Schum & Thonn.
Family	:	Euphorbiaceae
Name of Drug	:	Tamalaki

Botanical description: It is a branching annual glabrous herb, 30-60 cm tall with slender, leaf-bearing branchlets, distichous leaves which are subsessile elliptic-oblong, obtuse, rounded base. Flowers are yellowish, whitish or greenish, axillary, males flowers in groups of 1-3 whereas females are solitary.

Distribution: Grown as weed in south India and distributed in America.

Active ingredients: Alkaloids, flavonoids, tannins, lignins, polyphenolic compounds and tetracyclic triterpenoids.

Parts used: Whole plant

Medicinal uses: The drug is used for the treatment of diarrhoea, dysentery, dropsy, jaundice, intermittent fevers, urino-genital disorders, scabies and wounds. Further, these are used in the treatment of kidney problems, urinary bladder disturbances, pain, gonorrhoea, diabetes and chronic dysentery. Fresh leaf paste has wound healing capacity and used to cure white spots on skin & jaundice. The stem juice is also used as wound healers.

Scientific name	:	*Plumbago zeylanica* Linn.
Family	:	Plumbaginaceae
Name of Drug	:	Chitrakah

Botanical description: It is a much branched, evergreen shrub that reaches about 2 m in height. Dark green leaves are ovate to 6 inches long by half as wide. They are fast growing plants, but their size is easily controlled by pot size and pruning. The flowers are white in showy dense racemes and will flower all year long. Individual flowers are up to ½ inch (a bit more than 1 cm) across.

Distribution: South –East Asia

Active ingredients: Plumbagin, chitranone, zeylanone, zeylinone

Parts used: Bark

Medicinal uses: It is one of the powerful digestive and carminative herbs of Ayurveda. It is used in most of Ayurvedic medicines for indigestion.

Scientific name	:	*Portulaca oleracea* Linn.
Family	:	Portulacaceae
Name of Drug	:	Lonika

Botanical description: It is a succulent herb with smooth, reddish, mostly prostrate stems and alternate leaves clustered at stem joints and ends. The yellow flowers have five regular parts and are up to 6 millimetres wide.

Distribution: Found as weeds in different parts of India, particularly in moist and shady locations.

Active ingredients: Noradrenaline, calcium salts, dopamine, L-DOPA, malic acid, citric acid, glutamic acid, asparagic acid, nicotinic acid, alanine, glucose, fructose and sucrose.

Parts used: Whole plant

Medicinal uses: The drug is used to boost heart health, help in the proper development of children, treat certain gastrointestinal diseases, prevent certain cancers, protect the skin, boost vision, strengthen the immune system, build strong bones, and increase circulation.

Scientific name	:	*Psoralea corylifolia* Linn.
Family	:	Papilionaceae
Name of Drug	:	Bakuchi

Botanical description: It is a much branched annual herb, erect in habit and grows to a maximum height of 1.5 metres. The stems and branches are covered with conspicuous glands and white hair. The leaves are round, dotted with black glands on both surfaces. The flowers are axillary. Flowers are small, bluish purple in colour.

Distribution: It is found as weed during rainy season in different parts of India. The species is distributed in India, Sri Lanka, China, Myanmar, Pakistan and Arabia.

Active ingredients: Sugar, albumin, traces of manganese, limonene, corylin, psoralene.

Parts used: Seeds, fruits and roots

Medicinal uses: *Psoralea corylifolia* is valued in Chinese herbal medicine as a tonic remedy and is used to improve general vitality. It is also of value in the treatment of skin disorders, including vitiligo. The one-seeded fruits are highly regarded as an aphrodisiac and tonic to the genital organs. The seed is anthelmintic, antibacterial, aphrodisiac, astringent, cardiac, cytotoxic, deobstruent, diaphoretic, diuretic, stimulant, stomachic and tonic. It is used in the treatment of febrile diseases, premature ejaculation, impotence, lower back pains, frequent urination, incontinence, bed wetting etc. The root is used for treating dental caries. The plant yields a useful medicinal oleoresin which treats kidney disorders, impotence, lumbago. It is also used externally to treat various skin ailments including leprosy, leucoderma and hair loss. The antibacterial action of the fruit inhibits the growth of Mycobacterium tuberculosis.

Scientific name	:	*Rotula aquatica* Lour.
Family	:	Boraginaceae
Name of Drug	:	Pasanabhedah

Botanical description: It is a small branched shrub that grows characteristically among rocks and trailing over gravel in steam beds. This shrub or bush reaches a height of 2-3 m. The leaves are alternate, overlapping, ovate-oblong, and short-stalked. The flowers are small and crowded, and are borne in clusters at the end of branches. There are 5 narrow sepals which are about 4.5 mm long, with sharp points. The flowers are purple or pink, 5-6 mm long, with 5 petals.

Distribution: This is distributed in rocky river beds in different parts of India, South East Asia, Sri Lanka and Latin America.

Active ingredients: Baunerol, allantoin, sterol

Parts used: Root

Medicinal uses: It is used in traditional medicine for kidney and bladder stones. The root of the plant tubers are useful in treating coughs, fever, poisonings heart diseases, blood disorders, ulcers.

Scientific name : *Saccharum arundinaceum* Retz.

Family : Poaceae

Name of Drug : Sarah

Botanical description: It has long, grey-green leaves with a white midrib. This giant clumper makes a spectacular sight at 10' tall x 10' wide. In early to mid-October, the clumps are topped with narrow, purplish-pink, pampas-like plumes that extend another 2-3' above the foliage.

Distribution: Throughout the plains and low hills of India, Sri Lanka, Myanmar and China.

Parts used: Roots

Medicinal uses: The drug is aphrodisiac, galactogogue, cooing and sweet

Scientific name : *Saccharum officinarum* Linn.

Family : Poaceae

Name of Drug : Iksuh

Botanical description: A tall grass, which looks rather like a bamboo cane, and grows 3-6 m high with culms (stems) 20-45 mm in diameter. The thicker-stemmed forms are commonly known as 'thick' or 'noble' canes because of their tall, handsome, colourful stems. Leaves are broad (up to 6 cm wide), 70-150 cm long, borne alternately on the stem, with leaf base encircling the stem. Fruits are oblong caryopsis (small, dry, one-seeded fruit), 1.5 mm long.

Saccharum officinarum can be recognised by its hairless or short-haired panicle axis, and leaf-blades up to 6 cm wide.

Distribution: Cultivated in different parts of warmer regions of India

Active ingredients: Carbohydrates, vitamins, minerals

Parts used: Roots, stem juice

Medicinal uses: The fresh juice is considered sweet, cooling, laxative, aphrodisiac, restorative and diuretic. The drug is employed in jaundice, urinary diseases, general debility, impotence and haemophilic conditions. It increases libido, increases sperm count. It also helps in rectifying erectile dysfunction. It strengthens stomach, kidneys, heart, eyes, brain and sex organs.

Scientific name : *Saccharum spontaneum* Linn.

Family : Poaceae

Name of Drug : Kasah

Botanical.description: It is a perennial grass, growing up to three meters in height, with spreading rhizomatous roots. Leaves are harsh and linear, 0.5 to 1 meter long; 6 to 15 mm wide. Inflorescences are plumose panicles, which are white and erect, measuring 15-30 cm long, with slender and whorled branches, the joints covered with soft white hair.

Distribution: Throughout the warmers parts of India, Sri Lanka, South Europe and warmer regions of the old World.

Active ingredients: Quinone, alkaloids, carbohydrate, protein, coumarin

Parts used: Roots

Medicinal uses: Roots are sweet, astringent, emollient, refrigerant, diuretic, lithotriptic, purgative, tonic, aphrodisiac and useful in treatment of dyspepsia, burning sensation, piles, sexual weakness, gynoecological troubles, respiratory troubles.

Scientific name : *Saraca indica* Roxb.

Family : Caesalpiniaceae

Name of Drug : Asokah

Botanical description: It is an erect tree, small and evergreen, with a smooth, grey-brown bark. The crown is compact and shapely. Flowers are usually to be seen in the profusion of orange and scarlet clusters turns the tree into an object of startling beauty. Pinned closely on to every branch and twig, these clusters consist of numerous, small, long-tubed flowers which open out into four oval lobes. Yellow when young, they become orange then crimson with age and from the effect of the sun's rays. From a ring at the top of each tube spread several long, half-white, half-crimson, stamens which give a hairy appearance to the flower clusters. In strong contrast to these fiery blooms is the deep-green, shiny foliage. The foot-long leaves each have four, five or six pairs of long, wavy-edged, leaflets. Young leaves are soft, red and limp and remain pendent even after attaining full size.

Distribution: India, Burma and Malaya

Active ingredients: Epicatechin, catechin, leucocyanidin, sitosterol

Parts used: Bark, seeds

Medicinal uses: The bark of Ashoka Tree is used for its medicinal value and it is reported to have a stimulating effect on the endometrium and ovarian tissue. The bark is useful in all cases of uterine bleeding where ergot is indicated. It is also useful in menorrhagia due to uterine fibroids, in leucorrhoea and in internal bleeding.

Scientific name : *Sida cordata* Borss.

Family : Malvaceae

Name of Drug : Bala

Botanical description: It is a weed that grows wild in wastelands and along roadsides. They are found throughout the tropical and sub-tropical plains all over India. The herb grows to 30-80 cm in height. The plant though seasonal, is available throughout the year, and is especially luxuriant during the months of October to December. The leaves, 1-5.5 cm long, are heart shaped, one at each node. The flowers, 7-8 mm across, are produced at the leaf axils, and are yellow in colour.

Distribution: India, the Philippines, Malayasia

Active ingredients: Flavonoids

Parts used: Whole plant

Medicinal uses: The root is used against leucorrhoea, hemiplegia, sciatica, general debility, headache and fever. The juice of the plant is applied to boils and pimples. The juice of the leaves is used to treat cuts and wound.

Scientific name	:	*Sida cordifolia* Linn.
Family	:	Malvaceae
Name of Drug	:	Bala

Botanical description: It is an erect perennial that reaches 50 to 200 cm tall, with the entire plant covered with soft white felt-like hair. The stems are yellow-green, hairy, long, and slender. The yellow-green leaves are oblong-ovate, covered with hairs, and 3.5 to 7.5 cm) long by 2.5 to 6 cm. The flowers are dark yellow, sometimes with a darker orange center, with a hairy 5-lobed calyx and 5-lobed corolla.

Distribution: It is distributed as weed in West Bengal, Maharashtra, Kerala, Karnataka and also in Myanmar.

Active ingredients: β-phenethylamine,ephedrine, pseudo-ephedrine, S-(+)-N_b-methyltryptophan methyl ester, hypaphorine, vasicinone, vasicinol, choline, and betaine.

Parts used: Roots

Medicinal uses: It is administered in facial paralysis, hemiplegia, leuchorrhoea, diabetes, asthma, bronchitis, rejuvenating tonic, libido enhancing and anti-inflammatory.

Scientific name	:	*Sida rhombifolia* Linn.
Family	:	Malvaceae
Name of Drug	:	Bala

Botanical description: A summer annual with yellow flowers and very small spines at the base of each leaf and branch. This plant often occurs as a weed of peanuts, cotton, and soybeans. Leaves are arranged alternately along the stem. Leaves are widest at or above the middle and taper toward the leaf bases (rhombic). The upper ½ of the leaves have toothed or serrated margins while the remainder of the leaves are untoothed. Flowers occur singly on flower stalks (peduncles) that arise from the area between the stems and leaf petioles. Flowers consist of 5 yellow petals that are 4 to 8 mm long.

Distribution: West Bengal, Kerala, Karnataka, Maharashtra and in tropics of both the Hemispheres.

Active ingredients: Saponin, ephedrine

Parts used: Root

Medicinal uses: It is used in headache, opthalmia, dysuria, tuberculosis, fever and uterine disorders.

Scientific name	:	*Solanum indicum* Linn.
Family	:	Solanaceae
Name of Drug	:	Brhati

Botanical description: It grows upto 1-5 metres in height. The young branches are densely covered with minute star – shaped hairs. The prickles are yellow and shining, about 1-5 cm long. The leaves upto 10 cm long, their midribs and other nerves with sharp yellow prickles. The flowers are purple, about 2 cm long, in racemose extra – axillary cymes. The fruits round, 1.5 – 2 cm, dark yellow when ripe.

Distribution: Tropical parts of India

Active ingredients: Disogenin, lanosterol, sitosterol. Solasonnine, solamargine and solasidine

Parts used: Fruits and roots

Medicinal uses: It is an appetizer, digestant, diuretic and is beneficial to the heart. It alleviates cough and asthma. It is useful in the diseases like fever, cough, skin diseases, distaste and pain.

Scientific name	:	*Solanum nigrum* Linn.
Family	:	Solanaceae
Name of Drug	:	Kakamachi

Botanical description: It is an annual weed that grows up to 60cm tall, is branched and usually erect, growing wild in wastelands and crop fields. Alternate leaves are ovate deep green with an indented margin and acuminate at the tip. Flowers are white with yellow colored centre. The berries are green at early stage and turn to orange or black when ripened.

Distribution: It is found as weed in different parts of India and also found in Sri Lanka and all temperate and tropical parts of the world.

Active ingredients: Solasodine

Parts used: Fruit and whole plant

Medicinal uses: It is considered for good for cooling hot inflammation, ringworm, ulcer, testicular swellings, gout and ear pain. It is very good mouth sores and ulcers. Leaves and stems are diuretic. The juice of leaves is good for flatulence, peptic ulcers, painful periods, inflamed scotum, testicles and dysentery. The leaves and fruits are said to cure fever. The berries and flowers are good for cough and cold.

Scientific name	:	*Solanum virginianum* Linn.
Family	:	Solanaceae
Name of Drug	:	Kantakari

Botanical description: Thorny Nightshade is a herb which is erect or creeping, sometimes woody at base, 50-70 cm tall, copiously armed with sturdy, needlelike, broad-based prickles. Leaves are unequal paired; prickly; leaf blade ovate-oblong. Inflorescences are elongate racemes. Sepal tube is bell-shaped. Flowers are blue-purple. Berry re pale yellow, 1.3-2.2 cm in diameter.

Distribution: It is naturally distributed in different parts of India, Polynesia, Malayasia, Australia and South East Asia.

Active ingredients: Steroidal glycoalkaloids, solasonine, solamargine, solasurine, solanocarpine, solanine-S and alkaloidal bases, solanidine-S and solasodine.

Parts used: Whole plant

Medicinal uses: This plant is considered as aperients, pungent, bitter, digestive, and astringent. The extract of roots of this plant is an effective diuretic, expectorant and febrifuge. The extract of leaves of this plant if taken internally is an important agent for the treatment of tenacious phlegm. It is also useful in the treatment of influenza, enteric fever, cardiotonic, and laxative. It is also used against difficult urination.

Scientific name	:	*Solanum xanthocarpum* Schand Klendl.
Family	:	Solanaceae
Name of Drug	:	Kantakari

Botanical description: It is a very spiny diffused herb, with a height of up to 1.2 meters. The young branches are densely covered with minute star-shaped hair, while the mature branches are zigzag, covered with yellow, sharp shining prickles and spread close to the ground. The midribs and other nerves of the leaves have sharp yellow prickles and grow up to 10 cm in length. The purple flowers are borne in small bunches, sometimes opposite to the leaves. Kantkari plant bears glabrous, globular drooping berries as fruits, yellow or pale in color, with green veins.

Distribution: Throughout India

Active ingredients: Solanocarpine, carpesterol, solanocarpidine, potassium nitrate, fatty acid, diosgenin, sitosterol, isochlorogenic acid, neochronogenic acid, chronogenic acid, caffeic acid, solasodine, solasonine, solamargine, quercetin, apigenin, histamine, acetylcholine.

Parts used: Whole plant

Medicinal uses: It is useful in treating worms, cold, hoarseness of voice, fever, dysuria, enlargement of the liver, muscular pain, spleen and stone in the urinary bladder. The juice of the berries is used in curing sore throat. Fruits also facilitate seminal ejaculation, alleviate worms, itching, and fever and reduce fats. Seeds are helpful for treating irregular menstruation and dysmenorrheal in females.

Scientific name	:	*Sphaeranthus africanus* Linn.
Family	:	Asteraceae
Name of Drug	:	Hapusa

Botanical description: It is a herbaceous plant with 40-50 cm tall, usually robust stems, which are hairless or velvet-hairy, curved, ascending, branched, winged, wings entire. Leaves are obovate-oblong, velvety or becoming hairless, base narrowed, tip rounded. Inflorescence consists of clusters of white globose or ovoid flower-heads about 8 mm in diameter. Flower-heads are carried on hairless, winged stalks. Flower-head cup is bell-shaped.

Distribution: Coastal India in wet marshy places., the Philippines, Persia, Malayasia, China and Africa.

Active ingredients: Carvotanacetone, stigmasterol, spinasterol

Parts used: Whole plant

Medicinal uses: The plant pacifies vitiated vata, pitta epilepsy, migraine, jaundice, fever, cough, haemorrhoids, helminthiasis and skin diseases.

Scientific name	:	*Sphaeranthus indicus* Linn.
Family	:	Asteraceae
Name of Drug	:	Hapusa

Botanical description: East Indian globe thistle is a much branched, strongly-scented annual herb with winged stem and the wings toothed. Alternately arranged obovate-oblong leaves are narrowed at the base, dentate and serrate, 1-3 cm long. Flowers occur in purple spherical heads, 8-15 mm, consisting of numerous tiny flowers

Distribution: Wet lands of India, Myanmar, Malayasia, Australia and Sri Lanka

Active ingredients: Eudesmanolide, sterol glycosides, sesquiterpenlactones, flavonoids

Parts used: Whole plant

Medicinal uses: This herb is hot, laxative, digestible, tonic, fattening, alterative, anthelmintic and alexipharmic. It is used in insanity, tuberculosis, indigestion, bronchitis, spleen diseases, elephantiasis, anaemia, pain in uterus and vagina, piles, asthma, leucoderma, dysentery, vomiting and hemicranias.

Scientific name	:	*Spermacoce articularis* L.f.
Family	:	Rubiaceae
Name of Drug	:	Tukah

Botanical description: This is an annual herb, growing to 60 cm tall. The stems are arranged with few 4-angled branches near the base. Leaves are obovate to oblong-lanceolate, covered with fine hair. Flowers are borne in groups of few, and are stalkless. The flower-tube is narrow, white in color, and 3 mm long. The 4 oblong-ovate petals are tinged pale purple, and covered with fine hair at the back. The fruit is spherical-ellipsoid, 3 mm across.

Distribution: It is distributed in India, South West Asia and Tropical Africa.

Active ingredients: Alkaloids, glycosides, steroids, flavonoids and tannins

Parts used: Whole plant

Medicinal uses: Leaf extract of the plant is in against haemorrhoids, galls tones, jaundice and conjunctivitis. Roots are used to mouthwash to relieve toothache, decoction of the herb used to relieve headache, while seeds are demulcent in diarrhoea, dysentery and antimicrobial activity.

Scientific name : *Spermacoce hispida* Linn.

Family : Rubiaceae

Name of Drug : Tukah

Botanical description: It is a procumbent branched or hairy herb, 10-14cm long. The branches are greenish or purplish, stout, 4-angled. The leaves are ovate, speculate or elliptic and round or pointed at the tip. Flowers are borne in whorls in axil of leaves. The fruit is hairy capsule.

Distribution: In the coastal regions of Orissa, West Bengal, Western Himalaya, Assam, Sri Lanka, Singapore, Southern China and Malay Archipelago.

Active ingredients: Saponins, tannins, phenolics, steroids, essential oils, flavonoids and terpenoids

Parts used: Whole plant

Medicinal uses: It is administered to cure vesical calculus, dandruff, stomach ailments dysuria, anuria and restore morbid. Flowers are used for the treatment of boils, eruptions, swellings, cough and malaria.

Scientific name : *Stereospermum suaveolens* DC.

Family : Bignoniaceae

Name of Drug : Patala

Botanical description: It is a large deciduous tree with greyish or dark brown bark, exfoliating in scales. Leaves are imparipinnate; leaflets are elliptic. Flowers are purplish-yellow, fragrant, in large, lax panicles. Capsules are straight, cylindric, greyish with white dots.

Distribution: India, Bangladesh and Sri Lanka

Active ingredients: Naphthoquinone lapachol

Parts used: Barks and flowers

Medicinal uses: Root bark is an ingredient of Dasamula. Decoction of roots is used in intermittent and puerperal fevers and affections of the brain. Stem bark is diuretic and tonic. Flowers are given with honey to stop hiccup.

Scientific name : *Swertia chirayita* Karsten.

Family : Gentianaceae

Name of Drug : Kiratatiktah

Botanical description: It is an erect annual herb, 1.5m tall, much branched and robust. The leaves are opposite in pairs, stalkless and pointed at tips. Flowers are greenish yellow, tinged with purple. Fruits are ovoid.

Distribution: Temperate Himalayas from Kashmir to Bhutan and Khasi Hills

Active ingredients: Chiratamin, swertanone,

Parts used: Whole plant

Medicinal uses: An astringent stimulant prepared with the plant is an outstanding medication for weak stomach, particularly when it results in indigestion, bloating and nausea. In addition, this bitter tonic is also said to be effective in protecting the liver. Chirayata is a valuable bitter tonic. It is laxative and an appetizer. It also corrects the disordered process of nutrition and restores the normal function of the system. It is an effective herb for reducing all kinds of fevers. It is especially beneficial in the treatment of malarial fevers. It is also effective in hysteria and convulsions. It also shows hypoglycemic activity.

Scientific name	:	*Symplocos racemosa* Roxb.
Family	:	Symplocaceae
Name of Drug	:	Lodhrah

Botanical description: It is an evergreen medium size tree. Leaves are small stalked, circular to oval shaped and velvety to touch. Flowers are small, borne in clusters and cream coloured. Fruits are purple blackish and leathery.

Distribution: India, Burma

Active ingredients: Loturine, colloturine, loturidine

Parts used: Bark and roots

Medicinal uses: Bark has astringent properties which help in relieving diarrhoea and liver concerns. It is used as a tonic in disorders like conjunctivitis and ophthalmia. A decoction of the same is used to relieve bleeding gums, menorrhagia, and other uterine disorders.

Scientific name	:	*Syzygium cumini* Linn.
Family	:	Myrtaceae
Name of Drug	:	Jambuah

Botanical description: A slow growing species, 30 m tall with dense foliages. At the base of the tree, the bark is rough and dark grey, becoming lighter grey and smoother higher up. The leaves are pinkish when young, changing to a leathery, glossy dark green with a yellow midrib as they mature. *Syzygium cumini* trees start flowering from March to April. The flowers are fragrant and small, about 5 mm in diameter. The fruit is oblong, ovoid. Unripe fruit looks green. As it matures, its color changes to pink, then to shining crimson red and finally to black color. The fruit has a combination of sweet, mildly sour and astringent flavour and tends to colour the tongue purple.

Distribution: India, Bangladesh, Nepal , Pakistan, Malayasia, the Philippines, Indonesia

Active ingredients: Jambosine, jambolin, kaempferol, myrecetin

Parts used: Fruits, seeds

Medicinal uses: It is effective in the treatment of diabetes mellitus, inflammation, ulcers and diarrhoea. The leaves are antibacterial, and are used for strengthening the teeth and gums. The fruit and seeds are sweet, acrid, sour, tonic, and cooling, and are used in diabetes , diarrhoea and ringworm. The bark is astringent, sweet sour, diuretic, digestive and anthelmintic.

Scientific name : *Tephrosia purpurea* (Linn.) Pers.

Family : Fabaceae

Name of Drug : Sarapunkhah

Botanical description: A much branched, erect, perennial herb with hairy stems, lanceolate or linear leaves, reddish purple flowers and recurved glabrous pods.

Distribution: Sri Lanka and India

Active ingredients: Tephrosin, dengulin, quercetin, isotephrosin and rotenone.

Parts used: Whole plant

Medicinal uses: It is an important component of some herbal preparations such as Tephroli and Yakrifit used for liver disorders. In Ayurvedic system, this plant is used as remedy for impotency, asthma, diarrhoea, gonorrhoea, rheumatism, ulcer and urinary disorders. The plant is reported to cure diseases of kidney, liver spleen, heart and blood. The dried herb is effective as tonic laxative, diuretics and deobstruents. It is also used in the treatment of bronchitis, bilious febrile attack, boils, pimples and bleeding piles.

Scientific name : *Thespesia populnea* (Linn.) Sol. Ex. Corr.

Family : Malvaceae

Name of Drug : Parisah

Botanical description: It is a small to medium-sized tree, 10-12m tall. The trunk can be 2 feet in diameter at full maturity. The bark is corrugated, with scaly twigs. The branches are widely spread and usually horizontal. The glossy heart-shaped leaves are 3-5 inches across. Flowers are bell-shaped, pale yellow flowers with maroon or purple centers.

Distribution: India, Hawaii

Active ingredients: Gossypol, Populneol, Calycopterin, Thespesone, Thespone.

Parts used: Leaves, roots and barks

Medicinal uses: The ground up bark is used to treat skin diseases and in Mauritius dysentery and haemorrhoids. Leaves are applied to inflamed and swollen joints in South India. The young fruit secretes a yellow sticky sap used to treat ringworm and other skin diseases. Roots are used as a tonic. There is some recent investigation of the plant's effects on high blood pressure.

Scientific name : *Tinospora cordifolia* (Willd) Hook.f and Thoms.

Family : Menispermaceae

Name of Drug : Amrita

Botanical description: It is a large, deciduous extensively spreading climbing shrub with several elongated twining branches. Leaves are simple, alternate, exstipulate, long petioles up to 15 cm long, roundish, pulvinate, both at the base and apex with the basal one longer and twisted partially and half way around. Lamina broadly ovate or ovate cordate, 7 nerved and deeply cordate at base, membranous, pubescent above, whitish tomentose with a prominent reticulum beneath. Flowers are unisexual, small on separate plants and appearing when plant is leafless, greenish yellow on axillary and terminal racemes.

Distribution: India, Sri Lanka and Myanmar

Active ingredients: Columbin, tinosporaside, jatrorhizine, palmatine, berberine, tinocordifolioside, tinosporic acid, tinosporal, and tinosporon

Parts used: Whole plant

Medicinal uses: The drug is used for the treatment of fever, jaundice, diarrhoea, dysentery, general debility, leucorrhoea, cough, asthma, skin diseases, eye disorders etc.

Scientific name	:	*Tragia involucrata* Linn.
Family	:	Euphorbiaceae
Name of Drug	:	Duralabha

Botanical description: It is a perennial twinning herb, covered with stinging hairs. Alternately arranged leaves are oblong-lanceolate or ovate, toothed, base heart-shaped or rounded, tip long pointed. Flowers are borne in racemes in leaf axils. Female flowers are few, in lower part of inflorescence; male flowers are many in the upper part. Fruit is a 3-lobed capsule, containing 3 round smooth seeds.

Distribution: It is widely distributed in West Bengal, Assam, Deccan Peninsula, Kerala, Western Ghats and Punjab and also in Sri Lanka, Singapore and all tropical and sub-tropical countries.

Active ingredients: Linoleic acids, enzymes

Parts used: Roots

Medicinal uses: The drug is used to cure asthma, toxicosis, fevers, diarrhoea and purify blood. It is given in excessive urinary secretion, vomiting, dermatosis and thirst.

Scientific name	:	*Trianthema portulacastrum* Linn.
Family	:	Aizoaceae
Name of Drug	:	Punarnava

Botanical description: Stems are prostrate or rising, somewhat succulent, up to 50 cm long or more, smooth or sparsely velvety. Leaves are flat, elliptic to obovate or spade-shaped. Leaf stalks are expanded into a sheath joined with opposing leaf base to form a cup. Pink flowers are borne solitary, stalkless, largely hidden in leaf axils.

Distribution: Tropical and subtropical parts of India

Active ingredients: Ecdysterone, Punarnavine, trianthemine

Parts used: Whole plant

Medicinal uses: Punarnava is bitter, cooling, astringent to bowels, useful in biliousness, blood impurities, leucorrhoea, anaemia, inflammations, heart diseases, asthma, alternatives etc. The leaves are useful in dyspepsia, tumours, spleen enlargement and abdominal pains.

Scientific name	:	*Tribulus terrestris* Linn.
Family	:	Zygophyllaceae
Name of Drug	:	Goksurah

Botanical description: *Tribulus terrestris* is a taprooted herbaceous perennial plant. The stems radiate from the crown to a diameter of about 10 cm to over 1 m, often branching. They are usually prostrate, forming flat patches. Leaves are opposite and pinnately compound. The flowers are 4–10 mm wide, with 5 lemon-yellow petals, 5 sepals, and 10 stamens.

Distribution: Indifferent parts of India, western Tibbet, Sri Lanka and throughout the warm regions of the World.

Active ingredients: Harmane, norharmane

Parts used: Fruits

Medicinal uses: The drugs are considered tonic, demulcent, digestive, cooling and aphrodisiac. In renal calculus, chronic cystitis, difficult micturition etc. this drug is very effective. This also employed in impotence, diseases of heart, dropsy, burning sensation, rheumatism, piles, diabetes, cough and difficult breathing.

Scientific name	:	*Trichosanthes cucumerina* Linn.
Family	:	Cucurbitaceae
Name of Drug	:	Patolah

Botanical description: Snake Gourd is a tropical or subtropical vine, grown for its strikingly long fruit, used as a vegetable and for medicine. The narrow, soft-skinned fruits are 150 cm long. Leaf blade is kidney-shaped or broadly ovate, membranous, deeply 5-7-lobed. Lobes are triangular or rhombic. The white flower is beautiful and lacy, and open at night.

Distribution: South Asia and South-East Asia

Active ingredients: Cucurbitacin B, cucurbitacin E, isocucurbitacin B, 23,24-dihydroisocucurbitacin B, 23,24-dihydrocucurbitacin E, sterols 2 β-sitosterol, stigmasterol

Parts used: Whole plants

Medicinal uses: The drug is used to pacify vitiated pitta, constipation, skin diseases, burning sensation, diabetes, anorexia, flatulence, constipation, worm infestation, fever and general weakness.

Scientific name	:	*Trichosanthes tricuspidata* Lour.
Family	:	Cucurbitaceae
Name of Drug	:	Indravaruni

Botanical description: It is a large climber, 5-20 m long. It has strong, woody and grooved stem, with trailing branches. Tendrils are divided into 2-3 parts. Leaves are broadly heart-shaped, but palmately 3-5 lobed. The leaf margin is toothed. Flowers occur either singly or in 5-10 flowered racemes, in leaf axils. Flowers are 4-5 cm, white, with 5 wedge-shaped petals with frilly margins. Fruit is spherical, 4-5 cm across, red when ripe, streaked with 10 orange streaks.

Distribution: Australia, India, Japan, Malayasia and China

Active ingredients: Trichotetrol

Parts used: Fruits and roots

Medicinal uses: Fruits are useful in stomatitis, earache, asthma, worms, sciatica and fever. Roots are administered in cough, asthma and abdominal enlargement.

Scientific name	:	*Uraria picta* Desv.
Family	:	Fabaceae
Name of Drug	:	Saliparni

Botanical description: An erect, little-branched, annual undershrub, 0.9-1.8 m tall. Leaves are imparipinnate, 20-30 cm long. Leaflets on the upper part of the stem 5-7, (rarely 9), 10-20 cm long, linear oblong, acute, blotched with white, minutely pubescent beneath, base rounded. Flowers are borne in close fascicles along the rachis of spicate, cylindric racemes, 15-30 cm long; rachis and pedicels downy with hooked hairs.

Distribution: Found in dry grasslands and open forests in the sub-himalayan tract from Kashmir to West Bengal and Assam.

Active ingredients: Flavanoids, steroids, amino acids and fatty acids.

Parts used: Roots, pods and leaves

Medicinal uses: It is one of the components of Dasamoola. Roots are in nurture. They are mainly used for fracture healing properties. Its decoction is useful in cough, chills and fevers. Leaves are antiseptic in nature. They are mainly useful in gonorrhoea. Roots and pods used in the treatment of prolapse of anus in infants; pods also used in sore mouth.

Scientific name	:	*Valaris solanacea* (Roth) Kuntze.
Family	:	Apocynaceae
Name of Drug	:	Visanika

Botanical description: It is a hairy, climbing shrub. Stems are covered with a gray, spotted bark, emitting a white latex when wounded. Leaves are oppositely arranged and elliptic-oblong. Flowers are white, fragrant and bowl-shaped occur in 3-6 flowered cymes in leaf axils.

Distribution: South East Asia

Active ingredients: 2-acetyl-1-pyrroline, vallaroside, solanoside,vallarosolanoside,

Medicinal uses: Traditionally, the milky latex of V. solanacea canbe applied to treat ringworm and other skin infections, including sores, cuts and wounds. Leaves and barks of *V. solanacea* have been reported to possess anticancer, antimicrobial, analgesic, anti-in flammatory, anti-diarrhoeal and cardiotonic properties, and display toxicity to brine shrimp.

Scientific name	:	*Vernonia cinerea* Less.
Family	:	Asteraceae
Name of Drug	:	Sahadevi

Botanical description: It is an annual, erect, cylindrical herb. Leaves are petioled elliptical or lanceolate, obtuse and pubescent. The flowers are small heads borne in terminal corymb and pinkish violet in colour.

Distribution: It is a weed found distributed in different parts of India.

Active ingredient: Cosmosiin, Cynaroside, glucoside

Parts used: Whole plant

Medicinal uses: Plants are febrifuge, diaphoretic and alterative. Used as a specific herb for leucorrhoea, dysuria, spasm of bladder, strangury and for haematological disorders, as a blood purifier and styptic. Also used in asthma. Seeds are anthelmintic, antiflatulent, antispasmodic; used in dysuria, leucoderma, psoriasis and other skin diseases. Roots are anthelmintic; decoction used for colic.

Scientific name	:	*Vitex negundo* Linn.
Family	:	Verbenaceac
Name of Drug	:	Nirgundi

Botanical description: It is a woody, aromatic shrub growing to a small tree. It commonly bears tri- or penta-foliate leaves on quadrangular branches, which give rise to bluish-purple coloured flowers in branched tomentose cymes.

Distribution: Afghanistan, India, Pakistan, Sri Lanka, Thailand, Malaysia, eastern Africa and Madagascar

Active ingredients: Phenol, Dulcitol, Alkaloid-Vitricine, β-sitosterol, Camphene, α- And β- Pinenes, Angoside, Acunbin, Casticin, Artemetin, Orientin

Parts used: Leaf, roots and seeds

Medicinal uses: The drug is used as vermifuge. People used to keep leaves with pillow to dispel catarrh and headache and smoke the leaves for relief. Crushed leaf poultice is applied to cure headaches, neck gland sores, tubercular neck swellings and sinusitis. Essential oil of the leaves is also useful in treatment of venereal diseases and other syphilitic skin disorders. A leaf decoction with *Piper nigrum* is effective in catarrhal fever with heaviness of head and dull hearing. A tincture of the root-bark provides relief from irritability of bladder and rheumatism.

Scientific name	:	*Wedelia chinensis* Merr.
Family	:	Asteraceae
Name of Drug	:	Bhrngahajah

Botanical description: It is a tender, spreading, and hairy herb, with the branches usually less than 50 cm long. The leaves are oblong to oblong-lanceolate, and narrowed at both ends. The margins are entire or obscurely toothed; and both surfaces are covered with sharp-pointed, appressed, straight, and stiff hairs. The heads are stalked, and yellow. The involucral bracts are oblong-ovate. The ray flowers are 8-12, spreading, about equal to the bracts, and broad; the disk flowers number about 20, and are short, narrow, and pointed. The achenes are nearly cylindric, and hairy.

Distribution: India, China, Malayasia, tropical and subtropical regions of the world.

Active ingredients: Flavonoids, diterpenes, triterpene saponins and phytosteroids

Parts used: Whole plant

Medicinal uses: The leaves are used in dyeing grey hair and in promoting the growth of hair. They are considered tonic, alternative, and useful in coughs, cephalalgia, skin diseases, and alopecia. The juice of the leaves is much used as a snuff in cephalalgia. The seeds and flowers, as well as the leaves, are used in decoction, as a deobstruent. In decoction, the plant is used in uterine haemorrhage and menorrhagia.

Scientific name : *Zingiber officinale* Rose.

Family : Zingiberaceae

Name of Drug : Ardrakah

Botanical description: Ginger has a distinctive thickened, branched rhizome (underground stem) looks somewhat like a swollen hand. The rhizome has a brown corky outer layer and a pale yellow centre with a spicy lemon-like scent. Shoots (pseudostems), up to 1.2 m tall, arise annually from buds on the rhizome. These pseudostems are formed from a series of leaf bases (sheaths) wrapped tightly around one another with the long, narrow and mid-green leaf blades arranged alternately. The flowering heads, borne on separate shorter stems, are cone-shaped spikes and composed of a series of greenish to yellowish leaf-like bracts. The flowers are pale yellow in colour with a purplish lip that has yellowish dots and striations.

Distribution: Cultivated in different parts of India and tropical Asia

Active ingredients: Zingiberene, zingerone

Parts used: Rhizome

Medicinal uses: The drug is said to be anti-rheumatic, carminative, diurectic, aphrodisiac, cooling, cordial, digestive, acrid, hot, anodyne. Administered in colic, constipation, diarrhoea, cholera, diabetes, fever, cough, cold, dysuria, tongue and throat congestion, vomiting, cardiac disorder, anaemia, menstrual disorder and flatulence.

Table 1. Medicinal plants frequently used in Ayurvedic formulations

Sl.No.	Disease	Medicinal plants
1.	*Anaemia*	*Asparagus racemosus* (roots), *Withania somnifera* (roots), *Phyllanthus emblica* (fruits), *P. amarus* (leaves), *Tephrosia purpurea* (leaves), *Plumbago zeylanica* (roots), *Glycyrrhiza glabra* (roots) and *Piper longum* (fruits)
2.	Asthma bronchitis	*Solanum xanthocarpum* (whole plant), *Piper longum* (fruits), *Adhatoda vasica* (leaves), *Zingiber officinale* (roots), *Curcuma zedoaria* (roots), *Ocimum sanctum* (leaves) and *Phyllanthus emblica* (fruits)
3.	Arthritis	*Piper longum* (fruits), *S. xanthocarpum* (whole plant), *Withania somnifera* (roots), *Terminalia chebula* (fruits), *T. bellerica* (fruits), *Curcuma zedoaria* (roots), *Phyllanthus emblica* (fruits) and *Ricinus communis* (roots)

Contd...

Contd...

Sl.No.	Disease	Medicinal plants
4.	Blood circulation	*Zingiber officinale* (roots), *Piper longum* (roots), *Withania somnifera* (roots), *Phyllanthus emblica* (fruits), *Curcuma longa* (roots), *Terminalia bellerica (*fruits), *T. chebula* (fruits), *Ocimum sanctum* (leaves) and *Tephrosia purpurea* (leaves)
5.	Chronic constipation	*Holarrhena antidysenterica* (bark), *Plumbago ovata* (husk), *Terminalia bellerica* (fruits), *T. chebula* (fruits), *Phyllanthus emblica* (fruits), *Cassia angustifolia* (leaves) and *Glycyrrhiza glabra* (roots)
6.	Cancer	*Crataeva nurvala* (bark), *Terminalia chebula* (fruits), *T. bellerica* (fruits), *Holarrhena antidysenterica* (bark) and *Tinospora cordifolia* (stems)
7.	Chronic fever	*Tinospora cordifolia* (stems), *Ocimum sanctum* (leaves), *Adhatoda vasica* (leaves), *Azadirachta indica* (leaves), *Holarrhena antidysenterica* (bark), *Piper longum* (fruits), *Zingiber officinale* (roots) and *Terminalia bellerica* (fruits)
8.	Cough	*Phyllanthus emblica* (fruits), *Adhatoda vasica* (leaves), *Ocimum sanctum* (leaves), *Piper longum* (fruits), *Zingiber officinale* (roots), *Glycyrrhiza glabra* (roots) and *Solanum xanthocarpum* (whole plant)
9.	Cysts	*Terminalia chebula* (fruits), *Azadirachta indica* (bark), *Holarrhena antidysenterica* (bark), *Terminalia bellerica* (fruits), *Withania somnifera* (roots) and *Tinospora cordifolia* (stems)
10.	Dental diseases	*Azadirachta indica* (leaves, *A. arabia* (bark), *Areca catechu* (bark), *Achyranthus aspera* (leaves), *Ficus benghalensis* (bark), *Quercus infectoria* (fruits) and *Symplocos racemosa* (bark)
11.	Diarrhoea	*Holarrhena antidysenterica* (bark), *Aegle marmelos* (fruits), *Zingiber officinale* (roots), *Terminalia chebula* (fruits), *Cyperus rotundus* (roots), *Syzygium cumini* (seeds) and *Phyllanthus emblica* (fruits)
12.	Dislocation of bones	*Asparagus racemosus* (roots), *Withania somnifera* (roots), *Azadirachta arabica* (bark), *Terminalia arjuna* (bark), *T. chebula* (fruits), *T. bellerica* (fruits) and *Phyllanthus emblica* (fruits)
13.	Diabetes	*Gymnema sylvestre* (leaves, *Tinospora cordifolia* (stems), *Azadirachta indica* (leaves), *Phyllanthus emblica* (fruits), *Curcuma longa* (roots) and *Aegle marmelos* (leaves)
14.	Fistula	*Glycyrrhiza glabra* (roots), *Terminalia chebula* (fruits), *T. bellerica* (fruits), *Tinospora cordifolia* (stems), *Azadirachta indica* (leaves), and *Withania somnifera* (roots)
15.	Female sterility	*Asparagus racemosus* (roots), *Withania somnifera* (roots), *Glycyrrhiza glabra (roots), Phyllanthus emblica* (fruits), *Ficus glomerata* (bark) and *F. religiosa* (bark)

Contd...

Contd...

Sl.No.	Disease	Medicinal plants
16.	General health tonic	*Withania somnifera* (roots), *Asparagus racemosus* (roots), *Glycyrrhiza glabra* (roots), *Tribulus terrestris* (fruits), *Phyllanthus emblica* (fruits), *Terminalia arjuna* (bark) and *Centella asiatica* (leaves)
17.	Gastritis	*Zingiber officinale* (roots), *Piper longum* (fruits), *Mentha piperata* (leaves), *Terminalia chebula* (fruits), *T. bellerica* (fruits), *Phyllanthus emblica* (fruits), *Plumbago zeylanica* (roots) and *Tinospora cordifolia* (stems)
18.	Hair problems	*Eclipta alba* (leaves), *Centella asiatica* (leaves), *Terminalia chebula* (fruits), *T. bellerica* (fruits), *Phyllanthus emblica* (fruits), *Glycyrrhiza glabra* (roots), *Tinospora cordifolia* (stems) and *Tribulus terrestris* (fruits)
19.	High blood pressure	*Terminalia arjuna* (bark), *T. chebula* (fruits), *Asparagus racemosus (roots), Zingiber officinale* (roots) and *Withania somnifera* (roots)
20.	Heart tonic	*Withania somnifera* (roots), *Terminalia arjuna* (bark), *T. bellerica* (fruits), *T. chebula* (fruits), *Cyperus rotundus* (roots), *Phyllanthus emblica* (fruits) and *Ocimum sanctum* (leaves)
21.	Intestinal worms	*Holarrhena antidysenterica* (bark), *Mentha piperata* (leaves), *Tinospora cordifolia* (stems), *Butea monosperma* (seeds), *Azadirachta indica* (leaves), *Pyllanthus emblica* (fruits) and *Tribulus terrestris* (fruits)
22.	Epilepsy	*Centella asiatica* (leaves), *Withania somnifera* (roots), *Tribulus terrestris* (fruits), *Piper longum* (roots), *Acyranthus aspera* (leaves) and *Plumbago zeylanica* (roots)
23.	Leucorrhoea	*Symplocos racemosa* (bark), *Asparagus racemosus* (roots), *Adhatoda vasica* (leaves), *Aegle marmelos* (fruits), *Phyllanthus emblica* (fruits) and *Azadirachta indica* (bark)
24.	Leucoderma	*Psoralea cordifolia* (seeds), *Terrminalia chebula* (fruits), *Phyllanthus emblica* (fruits), *Azadirachta indica* (bark), *Areca catechu* (bark), *Tinospora cordifolia* (stems) and *Eclipta alba* (leaves)
25.	Liver tonic	*Holarrhena antidysenterica* (bark), *Eclipta alba* (leaves), *Tephrosia purpurea* (leaves), *Tinospora cordifolia* (stems) *Azadirachta indica* (bark), *Phyllanthus amarus* (whole plant) and *Plumbago zeylanica* (roots)
26.	Lack of appetite	*Zinziber officinale* (roots), *Piper longum* (fruits), *Pyllanthus emblica* (fruits), *Terminalia chebula* (fruits), *Tinospora cordifolia* (stems), *Cassia angustifolia* (leaves) and *Mentha piperata* (leaves)
27.	Male sterility	*Withania somnifera* (roots), *Mucuna pruriens (*seeds), *Tribulus terrestris* (fruits), *Glycyrrhiza glabra* (roots), *Terminalia arjuna*

Contd...

Contd...

Sl.No.	Disease	Medicinal plants
		(bark), *Phyllanthus emblica* (fruits), *Zingiber officinale* (roots) and *Piper longum* (fruits)
28.	Migraine	*Curcuma longa* (roots), *Glycyrrhiza glabra* (roots), *Azadirachta indica* (bark), *Tinosporacordifolia* (stems), *Terminalia chebula* (fruits), *Ocimum sanctum* (leaves) and *Eclipta alba* (leaves)
29.	Obesity	*Terminalia chebula* (fruits), *Terminalia bellerica* (fruits), *Phyllanthus emblica* (fruits), *Crateva nurvala* (bark), *Tribulus terrestris* (fruits) and *Zingiber officinale* (roots)
30.	Paralysis	*Curcuma zedoaria* (roots), *Withania somnifera* (roots), *Tribulus terrestris* (fruits), *Zingiber officinale* (roots), *Piper longum* (fruits), *Crataeva nurvala* (leaves) and *Plumbago zeylanica* (roots)
31.	Prostate enlargement	*Tinospora cordifolia* (stems), *Tribulus terrestris* (fruits), *Phyllanthus emblica* (fruits), *Zingiber officinale* (roots), *Butea monosperma* (seeds), *Adhatoda vasica* (leaves), *Terminalia chebula* (fruits), *T. bellerica* (fruits) and *Glycyrrhiza glabra* (roots)
32.	Piles	*Eclipta alba* (leaves), *Terminalia chebula* (fruits), *Terminalia bellerica* (fruits), *Phyllanthus emblica* (fruits), *Adhatoda vasica* (leaves), *Plumbago zeylanica* (roots), *Piper longum* (fruits) and *Aegle marmelos* (fruits)
33.	Sleeplessness	*Withania somnifera* (roots), *Centella asiatica* (leaves), *Piper longum* (roots), *Glycyrrhiza glabra* (roots) and *Terminalia bellerica* (fruits)
34.	Skin diseases	*Cyperus rotundus* (roots), *Tinospora cordifolia* (stems), *Azadirachta indica* (bark), *Terminalia chebula* (fruits), *T. bellerica* (fruits), *Curcuma longa* (roots), *Phyllanthus emblica* (fruits) and *Centella asiatica* (leaves)
35.	Sexual debility	*Withania somnifera* (roots), *Mucuna pruriens* (seeds), *Asparagus racemosus* (roots), *Sida cordifolia* (seeds), *Tribulus terrestris* (fruits) and *Glycyrrhiza glabra* (roots)
36.	Throat diseases	*Glycyrrhiza glabra* (roots), *Terminalia chebula* (fruits), *T. bellerica* (fruits), *Solanum xanthocarpum* (whole plant), *Piper longum* (fruits), *Sida cordifolia* (roots) and *Phyllanthus emblica* (fruits)
37.	Thyroid problems	*Crataeva nurvala* (bark), *Bauhinia variegata* (bark), *Sida cordifolia* (leaves), *Terminalia chebula* (fruits), *T. bellerica* (fruits), *Glycyrrhiza glabra* (roots) and *Zingiber officinale* (roots)
38.	Urinary tract	*Tribulus terrestris* (fruits), *Zingiber officinale* (roots), *Solanum xanthocarpum* (whole plant), *Crataeva nurvala* (bark), *Tinospora cordifolia* (stems), *Asparagus racemosus* (roots) and *Tephrosia purpurea* (leaves)

❑❑❑

6

ITK Used in Medicinal and Aromatic Plants

India is rich in its biological resources. The Eastern Himalayas, Western Ghats and Indo-Burma Region are the concentrated hotspots of India. Presently, it has rich vegetation of more than 45,000 plant species of which 15,000-20,000 plants have medicinal values. Out of these only 7500-8000 plants are used for medicinal purpose by traditional communities.

North-East India is comprised of eight states namely Arunachal Pradesh, Assam, Manipur, Meghalaya, Mizoram, Nagaland, Tripura and Sikkim and supports 50% of India's biodiversity. In addition, this region is also a home of 130 major tribal communities. Meghalaya, the abode of clouds, is one of eight sister states of NE India. It is bounded by Assam on north and north-east and Bangladesh on south and south-west. It has geographical area of 22,429 km^2 with 9,496 km^2 as recorded forest area.

Traditional Medicinal Plant Species of Meghalaya

Meghalaya is rich in its floral diversity and contributes about 18% of total flora of country. There are about 3128 species of flowering plants in the state of which 40% of total flora of state is endemic. Meghalaya is endowed with 850 medicinal plants, of which 377 species are used by majority of people for their primary health care needs. The indigenous people have vast knowledge of their plant resources as medicines and have been using over the years.

Table 1. Medicinal plant species used by tribal people of Meghalaya

Scientific name	Local name	Family	Parts used	Application
Aegle marmelos (Linn.) Corr	Sempri	Rutaceae	Leaves, Bark	Gastric problem, dysentery
Albizia odoratissima (Linn. f.) Benth	Siso	Fabaceae	Bark	Headache, dizziness

Contd...

Contd...

Scientific name	Local name	Family	Parts used	Application
Alstonia scholaris (Linn.) R. Br.	Sokson	Apocynaceae	Leaves, Bark	Fever, epilepsy andrespiratory diseases
Anacardium occidentale Linn.	Cashewnut	Anacardiaceae	Bark, Fruit and Leaves	Inflammation, urinary disorder
Annona squamosa L.	Ata bol	Anonaceae	Bark, Leaves, Roots, Unripe fruits and seed	Indigestion
Antidesma acidum Retz.	Arobakh	Euphorbiaceae	Leaves	Blood pressure
Aristolochia clematitis L.	Esamul	Aristolo-chiaceae	Leaves and roots	Snakebite
Artocarpus gomezianus Wall. Ex.Trecul	Armu	Moraceae	Bark	Headache, dizziness
Bauhinia variegata L.	Megong	Fabaceae	Leaves	Blood pressure
Bombax ceiba L.	Bolchu	Bombacaceae	Leaves	Diarrhoea
Bridelia retusa Spreng.	Khasi	Euphorbiaceae	Bark, fruit juice	Wound healing, earache
Calotropis procera (Linn.) R. Br.	Memangtebrong	Apocynaceae	Bark, Leaves, Roots, Latex andFlower	Paralysis, leprosy, skin diseases and cough
Carica papaya L.	Modipol	Caricaceae	Root, latex and seed	Dogbite, earache
Cassia fistula L.	Snaru	Fabaceae	Bark	Stomach pain
Cestrum nocturnum L.	Gamini	Solanaceae	Bark, leaves	Toothache, headache
Cinnamomum tamala Fr. Nees	Tejpata	Lauraceae	Bark	Urinary disorder
Cissus quadr angularis L.	Haljora	Vitaceae	Whole plant	Sprain and piles
Clerodendrum glandulosum Coleb.	Dongam	Verbenaceae	Leaves	Blood pressure
Crataeva nurvala Buch. Ham.	Jongchia	Oleaceae	Bark, leaves	Skin diseases, toothache
Cuscuta reflexa Roxb,	Dodimit	Convolvul-aceae	Leafless stem	Jaundice, headache

Contd...

Contd...

Scientific name	Local name	Family	Parts used	Application
Crypteronia paniculata Blume	Mosuginsep	Crypteroniaceae		Leaves Snake bite and woundHealing
Derris robusta (Roxb. Ex DC.) Benth.	Kakharu	Fabaceae	Leaves	Headache
Dillenia indica Linn.	Tedike	Dilleniaceae	Bark, leaves	Diarrhoea
DIllenia scabrella Roxb.	Agatchi	Dilleniaceae	Bark	Snakebite
Drimycarpus racemosus (Roxb.) Hk. f.	Babari	Anacardiaceae	Bark, leaves	Skin diseases
Duabanga grandiflora (Roxb. ex DC)Walp.	Bolchim	Lythraceae	Bark, leaves, root and flower	Headache, epilepsy
Erythrina stricta Linn.	Mandal	Fabaceae	Leaves, root	Asthma, epilepsy
Ficus hispida L	Thiwek	Moraceae	Root	Dysentery
Firminia colorata (Roxb.)R. Br.	Walgem	Sterculiaceae	Bark, leaves	Wound healing
*Gmelina arborea*Roxb.	Gambal	Verbanaceae	Bark, leaves	Cough, snakebite
Grevillea robusta A. Cunn,	Silver oak	Proteaceae	Bark, Leaves	Headache, dizziness
Hibiscus rosa-sinensis (Linn.)	Jova	Malvaceae	Flower	Headache, abortion
Jatropha curcas L.	Banglagash	Euphorbiaceae	Stem	Burn
Jatropha gossypiifolia L.	Banglagash	Euphorbiaceae	Latex	Dysentery
Justicia gendarussa Linn.	Dojagipe	Acanthaceae	Whole Plant	Cough,throat infection and sprain
Lagerstroemia parviflora Roxb.	Sidai	Lythraceae	Bark, Leaves	Skin diseases
Lagerstroemia speciosa L. Pers.	Asari	Lythraceae	Bark, root	Dysentery, jaundice
Litchi chinensis Sonn. T	Letchu	Sapindaceae	Bark, Root, Seed	Stomach and throat pain
Lygodium japonium Thunb. Ex Murr.	Ruatip	Lygodiaceae	Whole Plant	Wound healing

Contd...

Contd...

Scientific name	Local name	Family	Parts used	Application
Mangifera indica L. Tree	Tekatchu	Anacardiaceae	Bark, seed	Jaundice and itching
Macropanax undulatus (Wall ex D. Don)	Sambrong	Araliaceae	Leaves	Headache
Melia azadirachta L.	Neem	Meliaceae	Leaves	Fever and stomach pain
Mikania micrantha H. B. K	Sampangguri	Asteraceae	Leaves	Wound healing
*Mussaenda glabra*Vahl.	Sonarupa	Rubiaceae	Flower, Leaves,Root	Jaundice, leprosy and fever
Oryza sativa L.	Minil	Poaceae	Stem	Gall bladder stone and urinary disorder
Oroxylum indicum (Linn.) Vent.	Khiring	Bignoniaceae	Bark, Flower,Root, leaves and seed	Fever and jaundice
Phlogacanthus thyrsiflorus (Roxb.) Nees	Ellot	Acanthaceae	Bark and Leaves	Cough, fever, jaundice and gastric problem
Piper betel Linn.	Pan	Piperaceae	Leaves	Dysentery
*Piper longum*L.	Golmoris	Piperaceae	Fruit	Fever and cough
Plumbago indica L.	Achitragitchak	Plumbagin-aceae	Leaves, Root	Headache
Premna latifolia Roxb.	Dukhemi	Verbinaceae	Leaves, Bark	Diabetes
*Prunus persica*Linn.	Biispol	Rosaceae	Leaves	Skin diseases
Sabia lanceolata Colebr.	Madri	Sabiaceae	Bark, leaves, root	Epilepsy
Sida acuta Burm.	Chirotha	Malvaceae	Bark, Leaves	Malaria
Spilanthes acmella Murr.	Sonapul	Asteraceae	Stem, leaves, flower	Toothache
Spondias pinnata (L.F.) Kurz	Ambarerongtong	Anacardiaceae	Bark	Nasal bleeding
Sterculia villosa Roxb.	Olmak	Sterculiaceae	Bark	Throat pain
*Tamarindus indica*L.	Cheeng	Fabaceae	Leaves, Roots	Skin diseases

Contd...

Contd...

Scientific name	Local name	Family	Parts used	Application
Thevetia peruviana (Pers.) K. Schum.	Ceiling bol	Apocynaceae	Latex	Cough
Terminalia bellerica (Gaerth.) Roxb.	Chirore	Combretaceae	Bark, Fruit, Root, Seed	Headache, Jaundice and gastric problem
Terminalia chebula Retz.	Arithak	Combretaceae	Bark, Fruit	Dysentery
Terminalia myriocarpa Heurck. And Muell.	Rakseng	Combretaceae	Bark, Root	Urinary disorder, heartproblem
Tinospora cordifolia (Willd.) Miers	Dumandal	Menisperm-aceae	Stem	Urinary disorder
Vitex glabrata R. Br.	Bandi-kuri	Verbanaceae	Bark, Leaves	Inflammation
Ziziphus jujuba Lam.	Kangkil	Rhamanaceae	Bark	Cough, tuberculosis

Traditional Medicinal Plant Species of Assam

Assam, one of the biodiversity hotspots, occupies a special place in Northeastern India located between 24°44' N to 27°45'N latitude and 89°41'E to 96°02'E longitude, covering 2.4% of the geographical area of the country, i.e. 78,438 sq. km. The annual rainfall ranges between 305 cm. max. to 178 cm. min. with an average of 211.76 cm. The temperature recorded in summer is 37°C max. and 18°C min. and 26°C max. and 7°C min. in winter, with an average humidity of 83.00%. These types of climatic factors of Assam provide a congenial environment for growing diverse medicinal plants. In Assam, herbal plants are used at the household level in a self-help mode by the rural community. The rural people of Assam also consume many herbs as nutritional diet used in Indian system of medicine viz. *Bacopa monnieri* (L.) Penn., *Centella asiatica* (L), *Dioscorea bulbifera* Linn., *Emblica officinalis* Gaertn, *Eryngium foetidum* Linn., *Terminalia chebula* Retz., *Zanthoxylum alatum* Roxb., *Mentha spicata* Linn., *Ocimum sanctum* Linn., *Terminalia bellirica* (Gaertn.) Roxb.,*Paederia foetida* Linn., *Euryale ferox* Salisb., *Solanum nigrum* Linn., *Piper longum* Linn., *Garcinia cowa* Roxb. Ex D.C., *Garcinia Morella* (Gaertn.) Desr., *Garcinia pedunculata* Roxb., *Dillenia indica* Linn., *Calamus rotang* Linn., *Parkia roxburghii* G. Don., *Alpinia allughas* Rosc., *Clerodendrum glandulosum* Lindl., etc.

With its vast hills and forests, Assam is the home to a variety of medicinal herbs and plants such as eg. Sarpagandha (*Rauvolfia serpentina* (Benth) ex.Kurz.), Pippali (*Piper longum* Linn), Amlakhi (*Emblica officinalis* Gaertn), Hilikha (*Terminalia chebula* Retz.), Bhomora (*Terminalia belerica* (Gaertn.) Roxb.), Arjuna (*Terminalia arjuna*Wight & Arn.), Vaividang (*Embelia ribes* Burm.f.), Chaulmoodgra (*Hydnocarpus kurzii* King.), Mezankori (*Litsea citrate* Blume.), etc. About 900 species of medicinal herbs and plants are known to exist in abundance in the forest area of the state with the Brahmaputra valley itself having 50 species of herbs and plants of commercial value.

It has been found that the people apply their traditional knowledge of different medicinal plants to cure simple and common diseases as cold and fever but also deadly ones, as complicated as cancer. It is interesting to note that the roots and leaves of *Catharanthus roseus* are used as anticarcinogenic medicine. Other major ailments for which there are trade of medicinal plants include *leprosy, jaundice, dropsy, pneumonia, asthma, elephantiasis, piles, hysteria, malaria, calculi* both *gall-bladder* and *kidney, bronchitis,pharyngitis and rheumatism*. Highest numbers of plants are found to be used against snakebite. Several parts of a plant such as roots, leaves, flowers, seeds, tubers, stems, fruits are used as medicine. Different parts of plants used by the community according to their preference are being presented in the Table-2. Among the different parts of the documented plants they utilise the leaves mostly (43%), followed by roots (28%), sometimes the whole plant (17%), seeds (4%), flowers (4%), stems (2%), tubers (1%) and fruits (1%).

Table 2. Medicinal plant species used by rural people of Assam

Scientific name	Local name	Family	Parts used	Application
Adhatoda zeylanica Medic.	*Bahak tita*	Acanthaceae	Leaves	Cough, diarrhoea, dysentery and grandular tumours.
Ageratum conyzoides Linn.	*Gondhuabon*	Compositae	Leaves / Roots	Cut or injuries
*Alternanthera sessilis*R.Br.	*Matikaduri*	Amarantha ceae	Leaves	Growth of hair and stomach disorders, increase in milk flow
Amaranthus spinosus Linn.	*Hatikhutora*	Amaranthaceae	Roots / Stems	Snakebite, menorrhagia, gonorrhoea and to increase milk flow
Amaranthus viridis Linn.	*Khutora*	Amaranthaceae	Stem/ Leaves	Stem is used as antidote against snakebite. Leaves are good against scorpion sting.
Argemone maxicana Linn.	*Sialkatahi*	Papaverceae	Roots/ Seeds	Used in leprosy, scabies, and syphilis and gonorrhoea; used as antidote against snakebite.
Argyreia speciosa Sweet.	*Takoria alu*	Convolvulaceae	Leaves/ Tuber	Paste of the leaves, latex is used to

Contd...

Contd...

Scientific name	Local name	Family	Parts used	Application
				suppress small boils. The tuber is used as a contraceptive and also used in skin diseases.
Ardisia humilis Vahl	*Talotha poka*	Myrsinaceae	Leaves, fruits and roots	It is used as a stimulant and a carminative.
Artemisia vulgaris Linn.	*Chirota*	Compositae	Leaves	The juice is used as a blood purifier. Used against worm troubles, asthma, brain disorder and nervous problems
Asparagus racemosus Willd.	*Satamul*	Liliaceae	Roots	Roots are used as demulcents; diuretics, prepared medicated oil is good for rheumatic pain, nervous disorders. Also useful in dyspepsia, diarrhoea and dysentery
Bonnaya reptans Spreng	*Kasidoria*	Scrophulariaceae		Leaves Roasted leaves in banana leaves are used for cold bite and cut injuries.
Bryophyllum pinnatum Kuntz.	*Dupartenga*	Crassulaceae	Leaves	Leaves are useful in wounds, bruises, boils, jaundice, snakebite, dysentery, urinary trouble and for quick healing of wounds.
Calotropis gigantea (L)R.Br.	*Akon*	Asclepidaceae	Roots/ leaves /flower	Leaves are used in sprain and pain. Flowers are used

Contd...

Contd...

Scientific name	Local name	Family	Parts used	Application
				as digestive remedy and for cough, asthma, and cold. Root bark is used in dysentery; cough, skin disease, cutaneous disease, and elephantiasis.
Cannabis sativa Linn.	*Bhang*	Cannabinaceae	Leaves/ flower	Dried flower is used medicinally as sedative, analgesic, narcotic.
*Cardiospermum halicacabum*Linn.	*Lota kopalphuta*	Sapindaceae	Whole plant	Paste made with water is useful in pain, stiff-neck, rheumatism, fever, piles. Leaves mixed with castor oil are used in lumbago, nervous disorder.
Cassia occidentalis Linn.	*Charudoi ghigha/ Bonoria daldol*	Caesalpiniceae	Roots/ leaves /seeds	Used as remedy in skin disease, leprosy, psoriasis, ulcers, etc.
Cassia tora Linn.	*Bon medelua*	Caesalpinaceae	Roots/leaves /seeds	Leaf paste is used against skin diseases and leprosy. Roots are an antidote against snakebite, ring worm, psoriasis and eczema.
Catharanthus roseus(Linn.) G. Don.	*Nayantara*	Apocynaceae	Roots/ leaves	The parts of the plant are used as an anticarcinogenic agent.

Contd...

Contd...

Scientific name	Local name	Family	Parts used	Application
Centella asiatica (Linn.)Urbàn.	*Bor manimuni*	Apiaceae	Whole plant	Used in dysentery, liver trouble, nerve disorder, and stomach problems. It stimulates appetite, taken with milk to improve memory, good for skin disease and a blood purifier, given to women after childbirth.
Cissus quadr angularis L.	*Hasjora/ Harbhanga*	Vitaceae	Leaves/stems	It is used for muscular-skeletal disorders. It is considered as a tonic and analgesic, and is believed to help heal broken bones.
Citrus macroptora var *assamensis*D. et Bhat.	*Satkora*	*Rutaceae*	Fruits	For digestion, rheumatic pains, colic.
Clerodendrum colebrookianum Walp.	Nefafu	Verbenaceae	Bud / leaf	For the control of high blood pressure.
Clerodendrum infortunatum Gaertn.	*Dhopat tita*	Verbenaceae	Roots/ leaves	Leaves can be used against malarial fever. Roots are used externally against tumours and skin disease.
Clitoria ternatea Linn.	*Aparajita*	Fabaceae	Leaves	Juice of leaves mixed with salt applied around ears in earache and swelling of adjacent glands to

Contd...

Contd...

Scientific name	Local name	Family	Parts used	Application
				relive pain. The juice is also an antidote against snake poison.
Coccinia grandis (L.) J.Voigt	*Belipoka*	Cucurbitaceae	Fruits	It is useful for treating diabetes, leprosy, fever, asthma, bronchitis and jaundice.
Coffea bengalensis Roxb.exSchult.	*Kothona phool*	Rubiaceae	Leaves/Flowers	Young shoots and leaves used as medicine for cattle eye boil.
Colocasia esculenta (Linn.) Schott.	*Kola Kachu*	Araceae	Whole plant	Leaves used for blood coagulation in small injuries, roots used in pharyngitis.
*Commelina bengalensis*Linn.	*Kana simolu*	Commelinaceae	Branch/roots	Root juice is applied in eye-lid sore. Roots are useful in fever, bilious disease, snakebite and leprosy
Datura stramonium Linn.	*Boga dhotura*	Solanaceae	Whole plant	Smoke of leaves is used medicinally for asthma. A seed causes sleepiness. Poisonous and necrotic. Roots are good for toothache.
Diplazium esculentum (Retz.) Sw.	Dhekia	Athyriaceae	Leaves	Used in urinal complaints and to enhance sexual power.
Drymaria cordata Willd.	*Lai jabori*	Caryophyllaceae	Whole plant	Juice of the plant is laxative and antifebrile. The plant has cooling properties.

Contd...

Contd...

Scientific name	Local name	Family	Parts used	Application
Eclipta alba (Linn.) Hassk.	*Kehraj*	Compositae	Entire plant	Fresh leaves are used in elephantiasis, liver disease and dropsy. Juice is also used for jaundice and fever.
Enhydra fluctuans Lour.	Helosi sak	Asteraceae	Leaves	Leaves are useful for cough, skin diseases, nervous disease, gonorrhoea, etc.
Erechthites valerianaefolia DC.	*Bon kopah*	Compositae	Leaves	Juice of the leaves is used in cut wounds for quick healing.
Eryngium foetidum Linn.	*Ban dhania*	Apiaceae	Leaves/roots	An aromatic herb used as a condiments in soups, meat etc.
Eupatorium odoratum Linn.	*Naga bon*	Asteraceae	Leaves/flowers	Leaves and flower tops are used medicinally as emetic, cathartic, in cut wounds.
*Euphorbia hirta*Linn.	*Dudh bon /Gakhiroti bon*	Euphorbiaceae	Whole plant	The entire plant is considered as sedative, haemostatic, soporific, used medicinally in asthma, chronic bronchitis. The milky juice is useful in destroying warts.
Garcinia cowa Roxb. exD.C.	*Kau-thekera*	Clusiaceae	Fruits/ stems /resin / latex	Used in dysentery, diarrhoea and instomach pain.

Contd...

Contd...

Scientific name	Local name	Family	Parts used	Application
Garcinia laneaefolia	*Rupohi-thekera*	Clusiaceae	Leaves/fruits	Used in treatment of dysentery.
Garcinia pedunculata Roxb.	*Borthekera*	Clusiaceae	Fruits	Used in dysentery and diarrhoea.
Garcinia xanthochymus Hook. f.	*Tepor tenga*	Clusiaceae	Fruits, leaves	Used in dysentery, diarrhoea and, skin disease, diabetes etc.
Gnetum gnemon Linn.	*Mejherguti, Letera*	Gnetaceae	Leaves/fruits	Used to purify blood, stimulates urination
Gymnopetalum cochinchinensis (Lour.)Kurz.	*Kauri kerala*	Cucurbitaceae	Leaves/Roots	Roots made into paste and rubbed on body in body pain, pneumonia.
Heliotropium indicum Linn.	*Hatisuriya*	Boraginaceae	Whole plant	The juice of the plant is used in ulcers, wounds and local inflammation. Leaf paste is used against insect bite and boils.
Hibiscus subdariffa Linn.	*Mesta tenga*	Malvaceae	Leaves/fruits / Seeds	It is used for dysentery, fever, dyspepsia, general debility, etc.
Houttuynia cordata Thunb.	*Masandari*	Saururaceae	Whole plant	Used in stomach complaint and anaemia and in burn injuries.
*Hydrocotyle rotundifolia*Roxb.	*Soru manimuni*	Apiaceae	Whole plant	Used in dysentery, liver trouble, nerve disorder, and stomach problems. It induces appetite, taken with milk to improve memory, good for skin disease and is also a blood purifier.

Contd...

Contd...

Scientific name	Local name	Family	Parts used	Application
Impatiens roylei Walp.	*Bijolkoria*	Balsaminaceae	Leaves/ fruits	Leaf and fruit paste is used on the head as a remedy in high fever.
Ipomoea aquatica Forsk.	*Pani kolmou*	Convolvulaceae	Leaves	Leaf juice is used in jaundice and also in urinary trouble and nervous disorders.
Ipomoea quamoclit Linn.	*Kunjalata*	Covolvulaceae	Leaves	Pounded leaves are used in piles.
Justicia japonica Linn.	*Jooron*	Acanthaceae	Leaves	Leaves are used in ophthalmia.
Jussiaea suffruticosa Linn.	*Bon jolokia*	Onagraceae	Whole plant	Used in dysentery and fever.
Lawsonia inermis Linn.	*Jetuka*	Lythraceae	Leaves	It is astringent, sedative, detergent and is used as a prophylactic against skin diseases.
Leucas aspera Spreng.	*Duron bon*	Labiatae	Whole plant	Used against rabies; leaf juice mixed with garlic is good for the stomach, liver and spleen.
Melastoma malabathricum Linn.	*Phut kola*	Melastomaceae	Leaves	Leaf powder is used in wounds to prevent marks on the skin; wood tar is used in for blackening teeth.
Mikania micrantha Linn.	*Japanihabi*	Asteraceae	Leaves/ roots	Juice is useful in insect bite and scorpion sting
Mimosa pudica Linn.	*Lajuki lata*	Mimosaceae	Leaves	Leaf juice with milk is used as a good remedy for piles.
Mirabilis jalapa Linn.	*Godhuli gopal*	Nyctaginaceae	Roots	Roots are used in dropsy.

Contd...

Contd...

Scientific name	Local name	Family	Parts used	Application
Momordica charantia Linn.	Tita kerela	Cucurbitaceae	Leaves/fruits /seeds	Useful remedy for diabetes, rheumatism, night blindness, and dysmenorrhea.
Murraya koenigii (Linn.) Spreng.	*Narasingha*	Rutaceae	Leaves	Leaves are used in diarrhoea, dysentery, digestion problems, as memory enhancer and for curing vomiting.
Nasturtium indicum (Linn.) D.C.	*Bonhariah*	Brassicaceae	Seed	Seed juice is used in kidney trouble.
Ocimum canum Sims.	*Bon tulsi*	Labiatae	Leaves	Leaves made into paste applied to the fingers to relieve from fever. Seeds are used as remedy for dysentery. Leaves are used in skin disease.
Ocimum gratissimum Linn.	*Ram tulsi*	Labiatae	Whole plant	Plant juice is used as insect repellent. Seeds are used in headache. Leaves are used in for gonorrhoea, rheumatism and paralysis.
Oldenlandia corymbosa Linn.	*Bon jaluk*	Rubiaceae	Whole plant	Juice is applied in burning sensation of palms; it is also good for liver trouble, urinary disorder in children, jaundice, fever and bilious infection.
Oxalis corniculata Linn.	*Soru tengesi*	Oxalidaceae	Whole plant	Juice is used in stomach

Contd...

Contd...

Scientific name	Local name	Family	Parts used	Application
				problems, wine intoxication and dysentery.
Paederia foetida Linn.	*Bhebeli lata*	Rubiaceae	Leaves	The plant is very good in stomach, kidney and liver problems. Decoction of leaves increases appetite.
Peperomia pellucida (L) H.B. and K	*Ponownoa*	Piperaceae	Leaves/ roots	Plant paste is used on the head to reduce temperature.
Phlogacanthus thyrsiflorus Nees.	Tita phool	Acanthaceae	Flower/fruits	Useful in chronic bronchitis, asthma and phthisis, dysentery, neuralgia, scabies and malarial fever.
Phyllanthus niruri Linn.	*Bon amlokhi*	Euphorbiaceae	Whole plant	The whole plant is used in jaundice. Young leaves are good for dysentery. Root juice is used in urogenital troubles and gonorrhoea.
Polygonum microcephalum D.Don.	*Madhu holing*	Polygonaceae	Whole plant	It is useful in female weakness, bruises, piles and inflammation.
Pouzolzia indica Gaud.	*Dudhmor goch*	Urticaceae	Whole plant	Used against snakebite, syphilis, gonorrhoea. It is also used to heal children.
Rhynchotechum ellipticum A. DC.		Gesneriaceae	Leaves	Useful in cough.
Ricinus communis Linn.	*Era*	Euphorbiaceae	Leaves/ roots	Roots are used in urinary trouble;

Contd...

Contd...

Scientific name	Local name	Family	Parts used	Application
				juice with lime is used to suppress newly formed boils.
Siegesbekia orientalis Linn.	*Katampam*	Compositae	Whole plant	Used in healing gangrenous ulcers and skin disease. Useful against worms.
Solanum indicum Linn.	*Tita bhekuri*	Solanaceae	Roots/ leaves	Roots are used to cure toothache, asthma and in cough.
Solanum nigrum Linn.	*Pokmou*	Solanaceae	Roots	Root juice is used against asthma and whooping cough.
*Solanum spirale*Roxb.	*Tita kuchi*	Solanaceae	Whole plant	It is used against worms. The root isnarcotic and diuretic.
Solanum xanthocarpum Schrad. & Wendl.	*Kanta kori*	Solanaceae	Roots/ leaves	Roots are used in asthma and in chest pain. Leaf juice with black pepper is used in rheumatic pain.
Spilanthes paniculata Wall. Ex. D.C.	*Huhoni bon*	Asteraceae	Whole plant	Used in toothache, sore mouth
Urena lobata Linn.	*Bor sonborial*	Malvaceae	Roots/ leaves	Roots are diuretic, good in rheumatic pain.
Xanthium strumarium Linn.	*Ogaro*	Compositae	Roots/ leaves	Used against chronic malarial fever,urinary trouble.
*Zanthoxylum oxyphyllum*Edgew.	*Mejenga*	Rutaceae	Tender shoots	For deworming of tapeworms

Traditional Medicinal Plant Species of Arunachal Pradesh

Arunachal Pradesh, the treasure house of biological and socio-cultural diversity in the Eastern Himalayan Region of India. The state has 26 major tribes and over 110 sub-tribes who maintain a close relationship with the nature. The local inhabitants of the state have their own customs, tradition and medicinal system who mainly depended on forests and forest products for their day to day lives.

Table 3. Medicinal plant species used by tribal people of Arunachal Pradesh

Scientific name	Local name	Family	Parts used	Application
Acorus calamus Linn.	*Kiile tolyo*	Acoraceae	Rhizome	A paste of rhizome is applied to cure headache, joint pain, wounds and skin rashes. Juice is also taken for stomachache
Ageratum conyzoides Linn.	*Borbe tami*	Asteraceae	Leaves	Paste of leaves is applied to stop bleeding in minor cuts.
Allium cepa Linn.	*Byaku*	Amaryllidaceae	Leaves	Leaf juice is taken orally for cold and cough.
Allium hookeri Thwaites	*Lepi*	Amaryllidaceae	Leaves	Pounded leaves mixed with oil is warmed and massaged on body to get relief from body ache. It is also used against cold and cough.
Artemisia indica Willd. (Wild)	*Kukulyu*	Asteraceae	Leaves	Leaf decoction is used to getrelief from stomachache and loose motion. Paste is applied to stop bleeding in minor cuts. It also has anthelmintic property.
*Berberis wallichiana*DC.	*Tiipe tiire*	Berberidaceae	Leaves	Leaves are boiled in oil and oil isapplied against joint pain and swells
Cardamine hirsuta Linn.	*Padii hamang*	Brassicaceae	Whole plant	Whole plant is consumed raw

Contd...

Contd...

Scientific name	Local name	Family	Parts used	Application
				forindigestion, cold and cough.
Centella asiatica Linn.	*Ngiilyan akhoHamang*	Apiaceae	Leaves	Leaf juice is taken against gastric, jaundice and stomach ache. Raw leaves or juice is also taken for diarrhea and dysentery.
Clerodendrum colebrookianum Walp.	*Pato hamang*	Lamiaceae	Leaves	Boiled or steamed leaves are taken to control high blood pressure.
Crassocephalum crepidioides (Benth.) S. Moore	*Genda hamang*	Asteraceae	Leaves	Grinded leaves paste is applied on minor cuts to stop bleeding.
Eleusine coracana (Linn.) Gaertn.	*Sarse* (Cultivated)	Poaceae	Grains	Powdered grains are boiled andtaken orally to reduce cholesteroland raise blood percentage.
Gynostemma pedata Blume	*Riikoh* (Wild)	Cucurbitaceae	Stem and tuber	Grinded powder of stem or tubermixed with salt is used to get relief from throat pain and cough. Powder mixed with water is taken against dysentery, gastric, loose motion and stomachache.
Gynura cusimbua (D. Don) S. Moore	*Kochi hamang*	Asteraceae	Leaves	Raw juice is taken against stomachache and worms. It is also used as blood purifier and is

Contd...

Contd...

Scientific name	Local name	Family	Parts used	Application
				applied against allergy.
Houttuynia cordata Thunb.	*Siyan hamang*	Saururaceae	Tenderleaves and stem	Juice is extracted and taken to get relief from dysentery, indigestion, loose motion, and stomach ache. It is also known to provide sound sleep.
Hydrocotyle javanica Thunb.	*Subu tute* (Wild)	Apiaceae	Whole plant	Raw juice is taken for dysentery and stomachache.
Litsea cubeba (Lour.) Pers.	*Santero* (Wild)	Lauraceae	Tender leaves and fruits	Pounded tender leaves and fruitsmixed with water are taken for dysentery, indigestion, stomach ache, cold and cough. It also provides sound sleep.
*Mahonia napaulensis*DC.	*Taming*	Berberidaceae	Bark	Bark juice is applied against boils in lips, itching, wounds and skin rashes. Juice is also used for preparation of dye.
*Michelia champaca*Linn.	*Salyo sanii* (Wild)	Magnoliaceae	Pericarp and Seeds	Dried and grinded or boiled pericarps and seeds are taken forstomach ache and as an appetizer.
Mikania micrantha Kunth	*Mantami* (Wild)	Asteraceae	Tender leaves and stem	Tender leaves and stem are grinded into paste and applied on cuts and wounds to

Contd...

Contd...

Scientific name	Local name	Family	Parts used	Application
				stop bleeding. Juice is also taken orally to cure nose bleeding.
Molineria recurvata. (Dryand.) Herb	*Loli* (Wild)	Hypoxidaceae	Leaves androots	Grinded paste of leaves is appliedagainst body pain. Paste of roots isused against cuts and wounds forfast healing.
Oenanthe javanica (Blume) DC.	*Hugu-hamang* (Wild)	Apiaceae	Tenderleaves andstem	Raw tender leaves and stem is taken against stomach ache.
*Oxalis corniculata*Linn.	*O- khui hamang* (Wild)	Oxalidaceae	Leaves	Leaves are grinded into paste andare applied in cuts and wounds.Raw leaves are used as an appetizer. Leaves are also chewed for foul smell of mouth.
*Paederia foetida*Linn.	*Gandhali* (Wild)	Rubiaceae	Leaves and stem	Juice of leaves and stem is used tocure dysentery, diarrhea, gastric,indigestion and stomach ache.
*Plantago erosa*Wall. ex Roxb. Herb	*Mepi-hamang* (Wild)	Plantaginaceae	Leaves	Boiled leaves are taken to get relief from constipation and indigestion. Paste of raw leaves is applied on cuts to stop blood flow.
Plectranthus japonicus(Burm.f.) Koidz.	*Yode* (Wild)	Lamiaceae	Leaves	Leaves are grinded into paste andare applied on swells, wounds andon cuts to stop bleeding.

Contd...

Contd...

Scientific name	Local name	Family	Parts used	Application
Pteridium revolutum (Blume) Nakai	*Taree* (Wild)	Dennstae dtiaceae	Leaves	Juice of young or tender leaves isused to stop bleeding in minor cuts.
*Rhus javanica*Linn.	*Tamo sanii* (Wild)	Anacardiaceae	Fruits	Water of boiled or soaked fruits istaken to cure dysentery, gastric and stomach pain.
*Solanum nigrum*Linn.	*Hiiro hamang* (Cultivated)	Solanaceae	Leaves andtender shoots	Fresh leaves are chewed to cureboils in mouth and tongue. Boiledleaves and tender shoots are takento cure high blood pressure andstomachache.
Solanum xanthocarpum Schrad. & Wendl.	*Siitii byako* (Wild)	Solanaceae	Seeds	Dried seeds are wrapped with clean cloth, burned and fumes are allowed to enter into teeth cavity to get relief from toothache.
Spilanthespaniculata Wall. exDC.	*Yorkhun* (Cultivated)	Asteraceae	Flowers andleaves	Leaves are taken as raw for indigestion, stomach ache and throat pain. Flowers are chewed for toothache.
Swertia chirayita (Roxb. ex Fleming) Karsten	*Chirayata*	Gentianaceae	Whole plant	Decoction of the dried plant is taken in malarial fever and cold. It is also used as anthelmintic.
Taxus wallichiana Zucc.	*Talley Noori*	Taxaceae	Bark	Oil is extracted from bark and

Contd...

Contd...

Scientific name	Local name	Family	Parts used	Application
				isused in the treatment of cancer.
Valerianaj atamonsii Jones		Valerianaceae	Roots	Grinded paste of roots is applied for fast healing of boils and wounds.
Zingiber officinale Roscoe	*Taki*	Zingiberaceae	Rhizome	Juice mixed with honey is taken for cold and cough. It is also taken with hot water against asthma andindigestion. Raw rhizome is chewed to cure tooth ache.

Traditional Medicinal Knowledge of Underutilized Minor fruits as Medicine in Manipur

The state has a central valley (Imphal Valley) inhabited by the *Meitei* and *Meitei pangal* whereas the hilly areas are inhabited by 30 different tribes of *Naga and kuki tribes*. A number of underutilized minor edible fruits have been identified in the Imphal valley (excluding Jiribam sub-division, Imphal East) of Manipur. The therapeutic application of 39 plant representing 29 genera and 23 families used as medicine by the *Meitei* community. Tree species contributed in maximum having 76 % while shrubs 10% and herbs recorded 7 % of the total underutilized fruits as medicine. These plants are used for a wide range of common ailments like diuretic, laxative, jaundice, diabetes, diarrhoea, dysentery etc. Fruits and leaves are the major plant parts used for the preparation of medicine having 52 % and 22%.

Table 4. Underutilized minor fruits as medicine in Manipur

Scientific name	Common name	Local name	Family	Parts used	Application
Aegle marmelos	Bael	Heiri-khagok	Rutaceae	Fruit	Mature fruit as against diarrhea and dysentery and ripe fruit as tonic, laxative and good for heart
Annona reticulata	Bullock's heart	Ramphal	Annon -aceae	Fruit, leaves	Seed powdered paste are applied to the head to kill lice and leaves boil juice to reduce high blood pressure

Contd...

Contd...

Scientific name	Common name	Local name	Family	Parts used	Application
Aphanamixis polystachya	Pithraj tree	Hei-ranggoi	Meliaceae	Fruit	Liver constipation and leucorrhoea
Artocarpus lakoocha	Monkey jack	Heiri kothong	Moraceae	Fruit	Ripe fruit pulp is taken in constipation and in fever and also as anthelmintic
Artocarpus heterophyllus	Jackfruit	Theibong	Moraceae	Root, seed, ripe fruit	Treatment of diarrhoea from root and seed extract. Ripe fruit as laxative and gum exude from unripe fruit is burnt and the ash powder is applied in skin disease.
Averrhoea carambola	Carambola	Heinoujom	Oxalidaceae	Fruit, root	Ripe fruit pulp along with little common salt is eaten against jaundice, bleeding piles and for washing utensil. Besides, the crushed leaves for curing chicken pox, ring worm and scabies and its root extract is used as an antidote for poisoning.
Baccaurea ramiflora	Burmese grape	Moktok-hei	Euphorbiaceae	Fruit, bark	Fruit as a digestive and bark for skin disease
Calamus tenius	Rattans (Canes)	Heiri	Arecaceae	Fruit	Highly acidic fruit which use as digestive after mea; stem are used for handicraft and furniture work
Celtis timorensis	Stinkwood	Hei-kreng	Cannabinaceae	Leaf	Dysentery and jaundice
Citrus grandis	Pommelo	Nobab	Rutaceae	Fruit juice	Fruit juice as febrifuge, seed against dyspepsia and its dry pill to repel the mosquito
Citrus macroptera	Hatkora	Heiribob	Rutaceae	Fruit, peel	Fruit juice is used for stomach ailment, fruit pill as spice and in dyspepsia
Dillenia indica	Chulta / Elephant apple	Heigri	Dilleniaceae	Fruit	As a digestive and pickle preparation
Duchesnea indica	Mock strawberry	Heirong-kak	Rosaceae	Whole plant	Against stone formation in urinary tracts and kidney
Elaegnus umbellata	Japanese silvery	Heiyai	Eleaegnaceae	Fruit, seed	Fruit as a digestive and seed for curing cough

Contd...

Contd...

Scientific name	Common name	Local name	Family	Parts used	Application
Euphoria longan	Longan	Nongang hei	Sapindaceae	Fruit	Relaxation and stomachic from fresh fruit
Ficus auriculata	Elephant ear Fig	Heirit	Moraceae	Fruit and Bark	Used against dysentery, diabetes and lungs disease
Ficus palmata	Bedu	Heibala	Moraceae	Fruit	Prevent oxidative stress
Ficus glomerata	Cluster fig tree/ Gooler	Heibong	Moraceae	Fruit, root	Fruit are used against dysentery, diabetes and lung disease and bark are used on skin having boil or insect bite
Ficus hispida Linn.	Gobla	Asiheibong	Moraceae	Fruit, bark, leaves	Dysentery, ringworm and intestinal worm infection. Leaves are used in preparation of fermented soyabean locally called "hawaizar"
Flacourtia jangomas	Indian plum / Coffee plum	Heitroi	Flacourtia-ceae	Fruit	Used for bleeding gum and toothache and diabetes
Gardenia campanulata	Boilem	Lam-Heibi	Rubiaceae	Young leaves, fruit	Leave paste are applied on the skin of boil for removing pus and diabetes
Garcinia pedunculata	Sani	Heibung	Guttiferae	Fruit	Fruit as a digestive and stomach disorder
Glycosmis arborea	Chauldhua	Yong komla	Rutaceae	Fruit, leaves	Fever liver complaints, jaundice and hair lotion
Juglans regia	Walnut	Heijuga	Juglandaceae	Fruit, leaves	Fruit use for curing heart diseases and leaves are use as traditional medicine to reduce swell on joint, fever and antidairrhoeal
Litsea glutinosa	Medasaka	Thang-hidak	Lauraceae	Leaves and bark	Cut and injuries for early blood clotting and muscular sprain
Litsea monopetala	Meda	Tumid-la	Lauraceae	Leaves seed and bark	Diarrhoea and rheumatism of body pain

Contd...

Contd...

Scientific name	Common name	Local name	Family	Parts used	Application
Malus baccata	Crab apple	Heitup	Rosaceae	Fruit	As a digestive but not recommended to eat during cold fever
Meyna laxiflora	Moyna	Heibi	Rubiaceae	Young leaves, fruit	Intestinal worm and hoarseness
Musa paradisiaca	Hill banana	Ching laphu	Musaceae	Pseud-ostem, male flower	Pseudostem for better breast milk for newly born child mother and male flower for blood purification
Olea ferruginea	Indian olive	Chorphon	Oleaceae	Leaves, fruit	Leaves are used for pile treatment and fruit as a digestive
Phyllanthus acidulous	Star gooseberry	Kihori	Euphorbi-aceae	Fruit	Blood enhancer for the lungs and root as a purgative
Phyllanthus fraternus	Bhumy-amalki	Chakpa-heikru	Euphorbi-aceae	Whole plant	Leucoderma
Prunus domestica ssp. Syrica	Mirabelles plum	Kalen Heikha	Rosaceae	Fruit	Laxative
Prunus domestica ssp. Insititia	Damsons plum	Heikha	Rosaceae	Fruit	Laxative
Rhus chinensis	Nutgall tree	Heimang	Anacardi-aceae	Young shoot, fruit	Antiviral, antibacterial, anti-diarrhea, antioxidant activities and as a digestive
Rubus elliptiens	Yellow Raspberry	Heijampet	Rosaceae	Fruit, leaves & root	Fruit against diarrhea and root for dysentery and leaves for abortifacient
Rubus moluccanus	Ceylon blackberry	Heijampet amuba	Rosaceae	Fruit, leaves & root	Fruit against diarrhea and root for dysentery and leaves for abortifacient
Spondias pinnata	Indian hog plum	Heining	Anacardi-aceae	Fruit, leaves	Leaves are used as an ingredient for making herbal hair lotion "Chinghi", fruit are use against dysentery and dyspepsia
Terminalia chebula	Yellow myrobalan	Manahi	Combre-taceae	Fruit	Anti-inflammatory, cough and colds, pile, ulcer and mild purgative

Traditional Medicinal Plant Species of Sikkim

The unique geographical position and wide range of topography, high fertile soil , sufficient rainfall and presence of large number of perennial stream makes the state of Sikkim one of the treasure house of bio-diversity in the country. Sikkim has vast reserve of medicinal plants and rich culture of folk medicine. The rich flora of Sikkim has a number of raw drugs described in Ayurvedic texts. There are about 420 plants are used by the tribal people for various diseases in Sikkim Himalayas region, out of which few are in utilized on commercial basis.

Table 5. Medicinal Plant Species used by Indigenous People of Sikkim

Scientific name	Local name	Family	Parts used	Application
Abies forrestii C.C Rogers	Bobresall	Pinaceae	Leaf	Used for bronchitis, cough, hiccough, vomiting, dyspepsia, anorexia, indigestion etc.
Abies webiana Linn.F	Sala	Pinaceae	Leaf	Used for bronchitis, cough, hiccough, vomiting, dyspepsia, anorexia, indigestion etc.
Aconitum ferox wall.	Bikh, Bikhma	Ranuncula-ceae	Whole plant	It is used in Ayurveda in pain condition, fever, Inflammatory conditions ,diabetes, bronchitis and skin disorders
Aconitum hetrophyllum	Bikh, Paunkar	Ranunucula-ceae	Root	The root is bitter in taste and *Tridosagna* (maintain the body humors). It is the best drug for child's all types of problems like- Infant diarrhea, cough, cold, fever etc
Acorus calamus	Bajao	Acoraceae	Rhizome	The rhizomes give a bitter tasting tonic which is much used against fever, cough, pharingitis, bronchitis, rheumatism, diarrhea, dyspepsia and colic condition by the tribal.

Contd...

Contd...

Scientific name	Local name	Family	Parts used	Application
Aloe barbadensis Mill	Ghewkumari	Asphodel-aceae	Leaf	Fresh leaf juice is used as purgative, carminative, digestive, helpful in opthalmia, piles, muscular pain and inflammation. It is also used as a nourishment for growth of hair. It is also used in skin ailments, arthritis, burn, etc.
Angelica archangelica Linn	Khomog	Apiaceae	Root	Root is bitter in taste and exclusively used in Mental diseases particularly epilepsy and Insanity with emetic drugs. It is also used in itching, skin diseases, ulcer, infection and toxic condition.
Artemisia vulgaris L	Tetaypati	Asteraceae	Whole plant	Used in Ayurveda in skin diseases,irritable bowel syndrome, bleeding, various toxic condition and to maintain the body humors(Tridosaghna).
Asparagus recemosus Willd	Satmuli	Asparagaceae	Whole plant	It is also a Rasayana drug and good for pregnant women. It increase the semen , milk, memory and used in abdominal discomfort, dysentery, inflammation and all vata pitta condition.
Bergenia lingulata (Har.)	Pakhan bheda	Saxifragaceae	Root and rhizome	Local people used in only diarrhoea and dysentery but Ayurveda used in poor appetite, renal calculi, abdominal

Contd...

Contd...

Scientific name	Local name	Family	Parts used	Application
				abscess, infectious abscess, old goiter and gland, blood purifier and to maintain the body humors(Tridosaghna).
Callicarpa macrophylla Vahl.	Patharwar	Verbenaceae	Bark	It is bitter in taste and useful in blood dysentery, sweating, burning sensation and fever due to its cold potency. This is the best medicine for bleeding disorders and it reduces the bad smell from body.
Celastrus paniculatus Willd	Runglim	Celastraceae		The herb mainly used in vata-kapha disorders and it increase the *Medha*(memory), *buddhi* (judgment) and *pragyna* (Intelligence).
Cinnamomum tamala Nees & Eberm	Choti, Sinkoli	Lauraceae	Bark and leaves	It is a used in kapha pitta disorders like cough, rhinitis, piles, worm infestation of children. It is used in convulsion and vomiting.
Costus speciosus	Betlaure	Costaceae	Root	The root is pungent and bitter and used in cough, bronchitis, fever, rheumatism, urinary disorders, loss o appetite, loose motion and skin diseases. These underground tubers are considered now as source of Diosgenin, the precursor of steroid,

Contd...

Contd...

Scientific name	Local name	Family	Parts used	Application
				including sex hormones and oral contraceptives.
Dendrobrium nobile Lindl.	Sonahkiri	Orchidaceae	Leaf and pseudobulb	This herb juices is redness of eyes, for increase strength, vigor and vitality and *Rasayana*.
Eupatorium cannabinum L	Banmara	Asteraceae	Leaf and stem	It is important medicinal plant in Ayurveda to stop the bleeding and protect the wound from infection.
Hedychium spicatum Ham Ex. Smith	Kapur kacheri	Zingiberaceae	Root	Root is pungent and bitter in taste mostly used in tooth powder. It is used in cough, Bronchitis, ulcers , inflammation, pain etc.
Mallotus philippinensis Muel	Sinduria, Puroa, Safedmallata	Euphor biaceae	Fruit	Fruit *renu* is used as purgative and used kapha pitta disease like- worm constipation, infestation and abdominal diseases.
Nardostachya jatamansi DC	Japoy	Caprifoliaceae	Root	The dried root is used in epileptic bits, cholera, palpitation and also during hair loss.
Oroxylum indicum Vent.	Totola, Shivnak	Bignoniaceae	Bark	Bark is one of ingredient in *dosamula* and bitter in taste and used in *kaphapitta* and *ama* disorders. It is best medicine for urinary bladder problems and used in stones, diarrhea and anorexia

Contd...

Contd...

Scientific name	Local name	Family	Parts used	Application
Paederia foetida	Biri, Berihara	Rubiaceae	Leaves	In Ayurvedic preparation, it is considered alterative, anti-arthritic, antispasmodic cardiac, diaphoretic, expectorant and stomachic. It is also used in asthma, bowel complaints, diarrhoea, diabetes, rheumatism, seminal weakness etc.
Picrorhiza kurroa Royle ex. Benth	Kutki	Plantagin-aceae	Root	The roots of the plant is bitter in taste and the dried roots are used orally in malaria. The roots are used as bitter tonic, cathartic, stomachic, given in fever, dyspepsia, as strong purgative and also applied in scorpion and other insect bites.
Podophyllum hexandrum Royal	Panchpatey	Berberidaceae	Root	The roots are used as emetic, cholagogue,blood. purifier, purgative and alterative. It is considered as a cardiac tonic in small doses. It also finds use as a stimulant in peristalsis, allergy and skin inflammations.
Rubia cordifolia Linn	Manjitho	Rubiaceae	Fruits and roots	Root is bitter and used in Kaphapitta disorders. It has analgesic and inflammatory properties. It is used

Contd...

Contd...

Scientific name	Local name	Family	Parts used	Application
				in the diseases of the uterus, pains in the joint, rheumatic conditions, leucorrhoea, etc. Also used as febrifuge and against blood disorders in Ayurveda.
Swertia chirata Buch.-Ham.	Chirato	Gentianaceae	Whole plant	Due to its bitter taste, it is considered as tonic, astringent, stomachic, improves eye sight, pain in the joints, scabies
Saussuria lappa C.B.Cl	Asolow, Brahma kamal	Asteraceae	Fruits, roots and leaves	Ayurvedic physician used in cough, cold, bronchitis, hiccough, skin diseases and fever.
Smilax lanceifolia roxb.	Kukur, Ramdatun	Smilacaceae	Rhizome	Used in Ayurvedic medicine for sexual transmitted diseases, neurological problems and psychiatric problems. It is also used in epilepsy and purification of urine and stool.
Taxus baccata Linn.	Dhengresalla	Taxaceae	Bark and leaves	Taxol is isolated from the bark and leaf of the tree which is effective against certain forms of cancers, particularly breast, ovarian and lung cancer. young shoots has long been in use for the treatment of headache, giddiness, feeble and falling pulse, diarrhoea and severe biliousness.

Contd...

Contd...

Scientific name	Local name	Family	Parts used	Application
Valeriana hardwickii Wall.	Chammaha	Caprifoliaceae	Root	This plant is used in kapha vatta diseases and the drug of choice in neuro-psychiatric problems like- vertigo, drug dependence, epilepsy and other mental diseases.
Zanthoxylum alatum Roxb	Tumuru	Rutaceae	Fruits and barks	Used in kapha vata disorders, mostly in skin diseases, abdominal pain, anorexia, warm infestation and ataxia in Ayurvedic practice.

Traditional Medicinal Plant Species of Tripura

The tribal people of Tripura are accustomed to a wide variety of medicinal plants used in their herbal medicinal practices. A total of 25 of such plants were recorded from the medicine men and aged villagers of various villages. Some important plants include *Oroxylum indicum*, *Euphorbia nerifolia*, *Scoparia dulcis*, *Jatropha curcas* and *Kaempferia rotunda*.

Table 6. Medicinal plant species used by village people of Tripura

Scientific name	Local name	Family	Parts used	Application
Aegle marmelos Correa	Bael	Rutaceae	Leaves, gum from fruit, ripe fruit	Leaves pounded to paste with equal quantities of leaves of *Cajanus cajan* and a little water and 1 cupful extract taken in the morning in empty stomach in combination with molasses in jaundice. Fruit pulp as medicine and for prevention of dysentery.
Ageratum conyzoides Linn.	Shyamtulsi	Asteraceae	Leaves and twigs	Fresh leaf extract used as anti hemorrhagic. Juice used as an expectorant.

Contd...

Contd...

Scientific name	Local name	Family	Parts used	Application
Ananas comosus (L.) Merrill	Gihom	Bromeliaceae	Leaves	Leaf extract with milk and sugar candy in rheumatic swellings. Extract of leaf base is taken 1 teaspoon thrice daily in diarrhoea.
Azadirachta indica A. Juss.	Neem	Meliaceae	Leaves and stem bark	Leaves boiled in water to bathe patient with malaria and chicken pox. Smoke produced by burning leaves is used as mosquito repellent.
Cajanus cajan (L.) Millsp.	Arail	Fabaceae	Leaves and twigs	Mature seeds are cooked as a pulse and given in weakness. Leaf and twig paste is applied throughout the body during jaundice.
Carica papaya L.	Paypay	Caricaceae	Root and fruit	Unripe fruits is cooked as a vegetable, ripe fruit, eaten raw. Fruits are considered to be stomachic.
Centella asiatica (Linn.) Urban	Thunman-kuni	Apiaceae	Leaves and entire shoot	Eaten either as paste or cooked as a vegetable for dysentery and diarrhoea.
Clerodendrum viscosum Vent.	Bhati pataa	Verbenaceae	Leaves and root	Extract is used as expectorant. Decoction of the leaves is used to check high blood pressure. Root extract is as febrifuge.
Cynodon dactylon (L.) Persoon	Doob pataa	Poaceae	Shoot and roots	Shoot extract is anti haemorrhagic.
Euphorbia nerifolia Linn.	Sairapal	Euphorbi-aceae	Leaves	Vapor is inhaled as medicine during fever. Oven heated leaves

Contd...

Contd...

Scientific name	Local name	Family	Parts used	Application
				are placed on the chest to control cough.
Jatropha curcas Linn.	Girogaachh	Euphorbi-aceae	Branches	The sap is applied locally in gum infections. Raw leaves are used in high blood pressure.
Kaempferia rotunda Linn.	Bhojoraphul	Zingiberaceae	Flower and rhizome	Flower decoction is used to bathe patient with skin infections. Aqueous decoction of rhizome used in jaundice.
Kalanchoe pinnata Pers.	Khurajot	Crassulaceae	Leaves	Chewed raw with sugar to control dysentery and diarrhoea. Juice is administered in jaundice. Leaf paste is applied on skin infections and pimples.
Leucas aspera Spreng.	Doron pushpa	Laminaceae	Leaves and twigs	Extract is taken for cough and jaundice.
Marsilea quadrifolia Linn.	Susni	Marsileaceae	Leaves	Used in jaundice, brain tonic
Mimosa pudica Linn.	Dugjat lajari	Mimosaceae	Entire plant	Leaf paste is applied on the acne and pimples. Extract is used in equal proportion by weight along with bark extract of *Urena lobata* as a combination medicine against jaundice.
Momordica charantia Linn	Gangrauk	Cucurbitaceae	Fruits and twigs	Fruits is cooked as a vegetable and considered to be anthelmintic. Extract of twigs is used against dyspepsia.

Contd...

Contd...

Scientific name	Local name	Family	Parts used	Application
Musa paradisiaca Roxb	Mot	Musaceae	Flower and stolon	Juice of flowers is used in dysmenorrhoea and menorrhagia. Extract of the stolon is administered in dysentery and diarrhoea.
Phlogacanthus thyrsiflorus	Basokpata	Acanthaceae	Leaves	Leaf juice is used as an expectorant.
Psidium guajava L.	Sapri	Myrtaceae	Fruits and twigs	Young twigs is chewed in empty stomach every morning in dysentery and diarrhoea. Fruit is used in anaemia.
Scoparia dulcis L.	Naipun gchewk	Scrophulari-aceae	Leaves and twigs	Anthelmintic and used in jaundice.
Spilanthes paniculata Wallich ex. DC.	Ansha	Asteraceae	Leaves	Boiled leaves is used with rice in stomach troubles and liver problems.
Terminalia chebula Retzius	Bukhala buthai	Combretaceae	Dried fruit	Extract is used in stomachic and jaundice.
Oroxylum indicum (L.) Vent.	Taukharun	Bignoniaceae	Stem bark and immature fruits	Soup is used as stomachic and anthelmintic.
Ocimum sanctum L.	Tulsi	Laminaceae	Leaves	Extract is used as expectorant and jaundice.

Traditional Medicinal Plant Species of Uttrakhand

A total of 61 medicinal plants belonging to 28 families are used by local people to cure different diseases. Out of 61 medicinal plant species 13 plant species were having roots and rhizomes as medicinal parts, 7 species containing fruits and flowers as medicinal value, Whereas 28 plant species were having leaves as their medicinal plant parts. Almost all these medicinal plants identified were used to cure human diseases and in which 10 plant species were also used to cure diseases of cattle of the local communities.

Table 7. Medicinal plant species used by village people of Uttrakhand

Scientific name	Family	Parts used	Application
Agrimonia pilosa	Rosaceae	Roots	Cough, diarrhoea and suppressed urination.
Ajuga bracteosa	Lamiaceae	Leaf	Malarial fever, tonic, astringent and febrifuge.
Anaphalis contorta	Asteraceae	Leaves	Paste applied on cuts, wound, boils and insect repellent.
Artemisia nilagirica	Asteraceae	Leaf	Used against intestinal worms.
Artemisia roxburghiana	Asteraceae	Plant extract	Antipyretic, tonic and used in skin allergy.
Barleria cristata	Acanthaceae	Leaves, roots, seeds	Root decoction against bronchitis and pneumonia, leaves and root paste on woundswelling, seeds as antidote to snake bite.
Berberis aristata	Berberidaceae	Bark, root	Used in ophthalmic and fever.
Bergenia ciliata	Saxifragaceae	Rhizomes, leaves	Rhizomatous part used as tonic and febrifuge, dried leaves adulterated with tea.
Bidens bipinnata	Asteraceae	Leaves	Leprosy and juice in cuts.
Bidens pilosa	Asteraceae	Plant extract	With honey used in cough and bronchitis.
Canscora decussata	Gentianaceae	Whole plant	Plant extract as nerve tonic.
Cedrus deodara	Pinaceae	Bark, wood oil	Bowl complaints, applied in piles, lumbago, rheumatic arthritis and urticaria.
Clematis buchananiana	Ranunculaceae	Leaves	Poisonous to cattle and paste applied externally on skin ailments.
Clinopodium umbrosum	Lamiaceae	Plant extract, leaf	Astringent, carminative and blood purifier. leaf in gastric troubles.
Coleus forskohlii	Lamiaceae	Root, leaves	Against intestinal worms.
Cotoneaster microphyllus	Rosaceae	Leaves, fruits	Diarrhoea, cuts and wounds
Elephantopus scaber	Asteraceae	Roots, leaves	Fever, vomiting and blood diseases.
Euonymus tingens	Celastraceae	Bark	Used in eye diseases.
Ficus palmata	Moraceae	Fruits	Used for digestive disorders.
Fragaria indica	Rosaceae	Leaf	Diarrhoea and leucorrhoea.

Contd...

Contd...

Scientific name	Family	Parts used	Application
Geranium ocellatum	Geraniaceae	Roots	Antiseptic and given in liver troubles and fever.
Gnaphalium hypoleucum	Asteraceae	Plant extract and juice	Applied on cuts and wounds and juice is believed to increase in lactation.
Gonostegia hirta	Urticaceae	Roots	Used in fractured bones
Hedera nepalensis	Araliaceae	Leaf, fruit	Paste applied on ulcers, leaf juice given in dyspepsia
Hedychium spicatum	Zingiberaceae	Rhizome	Used in asthma and with saw dust of deodar used in tuberculosis.
Hedyotis corymbosa	Rubiaceae	Plant extract	Used in fever and liver diseases.
Hypericum oblongifolium	Hypericaceae	Leaves, stem	Decoction of leaves and stem given to facilitate delivery
Juglans regia	Juglandaceae	Bark, leaves, fruits	Leaves fungicides, insecticides, fruits used to intoxicate the fishes.
Kalanchoe integra	Convolvul-aceae	Bark, leaf	Leaf juice purgative and insecticidal, plat poisonous to cattle.
Leucas lanata	Lamiaceae	Young shoots	Whooping cough.
Lyonia ovalifolia	Ericaceae	Leaves, seeds	Young leaves poisonous to cattle, seed paste applied on wounds and boils.
Micromeria biflora	Lamiaceae	Leaves	Cold and sinusitis and in gastroenteritis.
Myrica esculenta	Myricaceae	Bark	Used to intoxicate the fishes.
Nepta ciliaris	Lamiaceae	Leaves	Taken in fever.
Origanum vulgare	Lamiaceae	Plant extract	Used in bronchitis, colic and diarrhoea.
Paspalum scrobiculatum	Juglandaceae	Root	Used in painful urination and in eye diseases.
Pimpinella diversifolia	Apiaceae	Plant extract	Given in digestive disorders and cold and cough.
Pinus roxburghii	Pinaceae	Saw dust	Asthma and bronchitis.
Pinus wallichiana	Pinaceae	Resin	Rheumatic pain.
Plectranthus rugosus	Lamiaceae	Leaves	Used in fever.
Polygala arvensis	Polygalaceae	Whole plant	Plant decoction given in paralysis.

Contd...

Contd...

Scientific name	Family	Parts used	Application
Potentilla fulgens	Rosaceae	Whole plant	Stomatitis and aphthae.
Potentilla sundaica	Rosaceae	Root, stem, leaf	Antidote to snake bite, applied on itches and abscesses.
Prinsepia utilis	Rosaceae	Seed, bark	Rheumatic pains and diarrhoea.
Pyrus pashia	Rosaceae	Fruits	Ripe fruits used in digestive disorders, used in pterygium disease to cure affected eyes of cattle.
Rosa brunonii	Rosaceae	Leaf, flower	Ophthalmia, diarrhoea and in wounds.
Rubia manjith	Rubiaceae	Root, stem, flower	Tonic, astringent and antidote to snake bite.
Rumex hastatus	Polygonaceae	Leaf	Applied on cuts and wounds to check bleeding.
Salvia lanata	Lamiaceae	Leaf, flower	Leaf extract given in colic and fever.
Sedum multicaule	Fabaceae	Leaves	Used as insecticides.
Senecio nudicaulis Herb	Asteraceae	Plant extract, leaves	Used in colic, fever and on some skin diseases.
Solidago virgaurea	Asteraceae	Whole plant	Used in kidney trouble, asthma and throat infection.
Sonchus asper	Asteraceae	Whole plant	Used in blood purification, hepatitis and on wounds.
Spiraea bella	Rosaceae	Seeds	Wash sores and wounds.
Taraxacum officinale	Asteraceae	Roots	Used in treatment of headache and hepatitis.
Urtica dioica	Urticaceae	Seeds, leaf	Leaf extract believed to stop baldness and roots in various skin ailments.
Vernonia cinerea	Asteraceae	Leaf, seeds	Used in dysentery and seeds in cough and cold.
Vervascum thapsus	Scrophul ariaceae	Plant extract, seeds	Plant extract in bronchitis and asthma, seeds used as narcotic.
Viola betonicifolia	Violaceae	Whole plant	Useful in skin and blood diseases, also used for fever and cough.

Traditional Medicinal Plant Species of Kerala

Kerala the Southernmost state of India is known for its biodiversity reserve, high cultural heterogeneity and high rate of literacy. There are five major ethnic groups namely Kurichia, Kuruma, Kattunaika, Adiyan and Paniya groups scattered along the western Ghats forest. These tribes use different medicinal plants for their primary healthcare and other purposes on the basis of their traditional knowledge. About 295 medicinal plant species belonging to 93 families have been recorded.

Table 8. Medicinal plant species used by tribal people of Kerala

Scientific name	Local name	Family	Parts used	Application
Acorus calamus L.	Vayambu	Araceae	Whole plant	The plant juice is administered orally to treat abdominal pain and diarrohea. Tuber paste is taken orally to control blood circulation.
Actinodaphne malabarica Balakr.	Vatt Kambilyvirinji.	Lauraceae	Bark	The bark paste of this plant with the bark paste of Pterospermum suberifolium mixed together and applied topically to bruised part of the body.
Aegle marmelos (L.) Correa.	Koovalam	Rutaceae	Root, Bark, Leaf, Fruit	Medicine for dysentery, fever, vomiting, body pain, pilesand jaun-dice. Promote digestion
Amorphophallus paeoniifolius(Dennst.)Nicols var.paeoniifolius	Kattuchena	Araceae	Corm	Edible. Medicine for piles.
Aristolochia tagala Cham.	Garudakkody	Aristoloch-iaceae	Whole plant	The plant Paste taken internally to cure abdominal pain.
Artocarpus hirsutus Lam.	Anjili	Moraceae	Leaf	Burn the leaves of *Artocarpus hirsutus,* the ash is taken internally to treat abdominal problems.
Baliospermum montanum (Willd.) Muell.-Arg.	Nagadenthi	Euphorbi-aceae	Root, Seed, Leaf	Root paste is applied externallyon the painfull swelling of piles.

Contd...

Contd...

Scientific name	Local name	Family	Parts used	Application
Calophyllum apetalum Willd.	Cherupunna	Clusiaceae	Seed, Flower	Seed powder istaken internally to cure menstrual disorders. The flower paste is applied on the body to get releaf from itching.
Celastrus paniculatus Willd.	Palulavam	Celasteraceae	Bark, Leaf, Seed	Brain tonic, Seed oil promotes intelligence
Chonemorpha fragrans (Moon) Alston	Perumku-rumba	Apocynaceae	Root	The rubbed root paste is applied on the skin cures skin diseases. The rubbed root paste is taken internally to purify blood.
Cinnamomum macrocarpum Hook.f.	Karuva	Lauraceae	Bark, Root, Leaf	The oil extracted from the root, bark and leaf is used to pre pare massaging oil for rheumatism. The powdered bark is taken inter-nally with honey to treat cough.
Cinnamomum sulphuratum Nees	Kaattu karuva	Lauraceae	Bark, Leaf	The leaf and bark paste is taken internally to treat cough and head ache.
Cinnamomum zeylanicum Blume	Karuva	Lauraceae	Bark	The bark powder mixed with honey is taken internally to treat cough and dysentery
Coscinium fenestratum (Gaertn Colebr.)	Maramanjal	Menisper-miaceae	Underground part Stem	By applying the grinded underground part cures skin diseases and reduces leucoderma. The grinded stem is used to treat bleeding piles.
Curcuma zedoaria (Christm.)Rosc	Kaattu manjal	Zingiberaceae	Tuber	Grinded tuber is a good medicine for psoriasis

Contd...

Contd...

Scientific name	Local name	Family	Parts used	Application
Cycas circinalis L.	Eanth	Cycadaceae	Bark, Seed	Eating the seed powder with ghee increases sperm production
Dipterocarpus bourdillonii Brandis	Karanjili Kalpayin	Dipterocar-paceae	Gum	The smoke coming from the burned resin Cures swelling due to rheumatism on the legs
Dysoxylum ma. labaricum Bedd.ex Hiern	Akil, Vellakil	Meliaceae	Bark	Anticancerous. Medicine for rheumatism. Kuruma Vulnerable
Embelia ribes Burm.f.	Vizhal	Myrsinaceae	Leaf	Grinded leaf is applied on the skin to cure skin problems.
Embelia tsjeriam-cottam (Roem. &Schult.)DC.	Kattu vizha	Myrsinaceae	Root, Leaf	Grinded root juice is a component of medicine for diarrhoea and pneumonia.
Garcinia gummi gutta(L.)Robs.var gummi -gutta	Kudampuli	Clusiaceae	Seed, Leaf, Fruit	The seed is fermented and the solution mixed with salt and garlic and applies 2 drops inside the throat. And the remaining smears outside the throat. It is used to cure tonsilites, ulcer, bleeding, piles, dysentery and diarrhoea.
Gloriosa superba L.	Menthonni	Amaryllid-aceae	Tuberous root	Bleeding piles,White discharge,skin diseases
Goniothalamus wynaadensis (Bedd.)Bedd.	Anappanal	Annonaceae	Bark	Medicinal
Hedychium coronarium Koening	Kalyanasou ganthi-kam	Zingiberaceae	Rhizome.	The rhizome paste is applied on the bruised part

Contd...

Contd...

Scientific name	Local name	Family	Parts used	Application
Holigarna beddomei Hook.f.	Malancheru	Anacardiaceae	Bark	Dried bark is used to cure dysentery
Hopea glabra Wight&Arn.	Irumbakam	Dipterocarp-aceae	Leaf	Antiviral
Humboltia brunonis. Wall	Malayasokam	Caesalpinia-ceae	Leaf, Bark	Medicinal
Hydnocarpus macrocarpa(Bedd.)	Malamarotty	Flacourtiaceae	Seed	The oil extracted from the seeds of this plant is used to treat skin diseases
Madhuka nerifolia	Aattu Elippa	Sapotaceae	Honey	Medicinal
Nothapodytes nimmoniana (Graham) Mabb.	Ulukkuvett	Icacinaceae	Leaf	Cut the leaves of Ulukkuvetty in to small pieces and fry it in a pan and take one egg and fry it in another pan and both mix together and fry again and inhale the smoke coming from it with eyes opened. It cures head ache after delivery. The leaves of ulukkuvetty and pachamanjal grind together and apply on the affected part and massage properly. It reduces swells.
Pseudarthria viscida (L.) Wight &Arn.	Moovila	Fabaceae	Root	Root juice is taken in to treat fever, diarrhoea, asthma, worms & piles.
Raphidophora pertusa (Roxb.)Schott.	Anachakkara	Araceae	Flower, Leaf	Mumps&tonsillitis
Rauvolfia serpentina (L.) Benth. ex Kurz	Sarpagandhi	Apocynaceae	Leaves	A leaf of sarpagandhi and nagam is a good medicine for snake poisons.
Saraca asoka (Roxb.)deWilde	Ashokam	Caesalpini-aceae	Bark	Used in the treatment of menorrhagia, uterine infection,

Contd...

Contd...

Scientific name	Local name	Family	Parts used	Application
				haemorrhagic dysentery, internal bleeding..
Smilax zeylanica L.	Kariyilanchi	Smilacaceae	Root	The root juice is used to cure rheumatism, skin troubles and blood dysentery.
Solenocarpus indicus Wight & Arn. Tree	Kattanbazham	Anacardiaceae	Fruit, Bark.	Medicinal
Symplocos cochinchinensis (Loureiro) S. Moore,	Paachotti	Symplocaceae	Leaf.	The leaf Paste is applied on the forehead to cure head ache. A mixture of coconut oil and the juice of pachotti leaves boiled and prepare the massaging oil applied on the affected part to cure Leucorrhea and Psoriasis.
Symplocos racemosa Roxb.	Valiya pachotty	Symplocaceae	Bark,	Leucorrhea, Liver diseases
Tinospora sinensis (Lour.)Merr	Kaattamruthu	Menispermaceae	Stem	The stem juice is taken internally to treat piles and ulcer

Table 9: Indigenous traditional medicinal plants of MP used in Ayurveda for the treatment of diabetes

Scientific name	Family	Local name	Chemical constituents	Uses
Abelmoschus esculentus Linn.	Malvaceae	Bhindi	9 flavonoid glycosides, quercetin-4-glucoside, quercetin-7-glucoside, quercetin-5-glucoside, quercetin-3 (o- glucosyl glucoside)-4-(o-glucosyl	Two fresh seedless tender fruits are given twice in a day till cure.

Contd...

Contd...

Scientific name	Family	Local name	Chemical constituents	Uses
			glucoside0 etc and 2 anthocyanins cyanidine 4 glucoside and cyanidine 3glucoside 4 glucoside	
Abutilon indicum (Linn.) Sweet.	Malvaceae	Kanghi	Alkaloids, steroids, glycosides, flavonoids, aminoacids, fatty acids	One table spoonful leaf powder is taken orally along with cow milk in the morning and evening. Seven fresh leaves are also given to patient for seven days.
Aegle marmelos (Linn.) Corr.	Rutaceae	Bael	Essential oil, aegelenine and aegeline, tannins, flavonoid glycosides, phlobotannins, umbeliferon, skimmianine, marmin, lupen, a-sitosterol, furocoumarins.	About 10 ml leaf juice is given with five black pepper seeds twice a day for about two months to control the blood sugar level. For the treatment of diabetes twenty leaves from tip to downwards are plugged and boiled in two cups of water till it become one cup. It is cooled and filtered and the decoction is given to the patient in morning. One table spoonful leaf powder of Aegle marmelos and Azadirachta indica and seeds of Syzygium cumini is also given to the patient along with water twice a day to cure diabetes.

Contd...

Contd...

Scientific name	Family	Local name	Chemical constituents	Uses
Annona squamosa Linn.	Annonaceae	Sharifa, Sitaphal		Ten leaves from tip to downwards are plugged, cleaned with a soft cloth and then made into small pieces which are soaked in a glass of water. It is kept over night and the following morning water is filtered which is given to the patient daily every morning
Azadirachta indica A. Juss.	Meliaceae	Neem	Leaves contain flavonone, meliacin - solannolide, solannin, triter-penoidnimocinol, azadirachtannin, tetranorterpenoid isoazaditolide, nimbocinolide, quercitin and b-siitosterol. Seeds contain nimbolin A and B, nimbenene tri terpinoids, 3-acetoxy-1-hydroxy-azadirachtol. Seed oil contains miliantriol. Bark contains deacetylnimbin.	The young leaves are chewed in the morning and it is also used in the form of paste. Leaf paste with black pepper is given once a day for one month. Roasted bark powder is taken with milk for thirty days. One teaspoonful oil is also given to the patient orally for one month in the morning in the empty stomach for the treatment of the disease.
Butea monosperma (Lamak,)Taub.	Papilion-aceae	Dhak, Palas,Tesu, Chheul	Alkaloids, flavonoids, B-sitosterol, fatty acids, enzymes,	The leaves and flowers juice (10 ml) is given to the patient to reduce the blood sugar and are also

Contd...

Contd...

Scientific name	Family	Local name	Chemical constituents	Uses
			acoumaranone glucoside-palsitrin from the plant and palasonin from seeds.	used in glycosuria. The leaf decoction is also given to the patient in the morning.
Cassia fistula Linn.	Caesalpini-aceae	Amaltas, Kirwar	Glycosides, Steroids, tannins, quinines, flavonoids, disac charides, phlo baphenes, oxyanthro quinone substances.	Fruit is eaten once a day for 30 days in the treatment of diabetes. One tea spoonful powder of dried fruits is given to the patient twice a day. It may also be given to the patient in the form of decoction to control the sugar level in blood. 5 ml juice of fresh root sheath along with 3 ml rhizome of Zingiber officinale and 2 black pepper seed powder is prescribed for 15 days.
Catharanthus roseus Linn.	Apocyn-aceae	Sadabahar	Leaves contain oleoresin, volatile oil. Flower contains diglucoside, pseudoindoxyl alkaloid-rosamine, b-carboline, bannucine and leurosinone. Roots contain ajmalacine. Fruits and seeds contain vindoline, dehydraouin	The leaves and flowers are effectively used in the treatment of diabetes. Its effect on reducing the blood sugar level has been observed by the authors in a few cases. For this purpose two to three leaves and flowers are chewed in the morning. Plants with white flowers have been found more effective. 25 ml leaf decoction is also given on empty stomach once a day

Contd...

Contd...

Scientific name	Family	Local name	Chemical constituents	Uses
			dolinine and corpmaridine.	for one and half month to cure the disease.
Clerodendrum multiflorum (Burm.f) O. Ktze	Verbenaceae	Aran	Leaves contain monoglucoside and Y-sitosterols, unidentified sterol, ceryl alcohol and palmitic and cerotic acids. Stem contains D-mannitol, B-sitosterol, B-D glucoside and ceryl alcohol.	The leaves of the plant are considered to be anti-diabetic. Aqueous extract / decoction of the leaves is given to the patient on empty stomach in the morning.
Convolvulus microphyllus Sieb. ex. Spreng.	Convolvulaceae	Shankha-holi		Whole plant is soaked in four cups of water over night and boiled in the same water until the water becomes one cup, it is cooled and filtered. Two spoonful of this filtrate is taken 3-4 times daily to manage the proper level of sugar in the blood.
Emblica officinalis Gaertn.	Euphorbiaceae	Amla, Aoula	Linolenic, linoleic, oleic, stearic, palmitic myristic acids, trigalloyglucose, terchebin, covilagin, ellagic and phyllemblic acids.	The fruit is effective in controlling diabetes. A tablespoon of Amla fruit juice mixed with fresh juice of karela (*Momordica charantia*) is administered daily for two month. This medicine stimulates the pancreas and enables it to secrete insulin, thus reducing

Contd...

Contd...

Scientific name	Family	Local name	Chemical constituents	Uses
				blood sugar level in diabetes. The fruit powder of *E. officinalis, Momordica charantia* and *Syzygium cumini* is also used to check diabetes. One tea spoonful powder of this mixture is given twice a day.
Ficus benghalensis Linn.	Moraceae	Bargad, Bar	Bark contains leuco-anthocyanins. Leaves contain quercitin-3-galactoside, rutin, b-sitosterol, taraxasterol, lupeol, B-amyrin along with psoralen bergapten. Ripe figs contain glycoside	The aqueous extract of aerial root and stem bark is used to reduce the sugar level in blood. The decoction of tender aerial roots is given to the patient in the morning. The bark is given in the form of powder twice a day. Root bark soaked in a cup of water over night and infusion administered daily morning for a month.
Gymnema sylvestre Gaertn. (Retz.) R.Br.	Asclepid-aceae	Gurmar, Gusmuru	Gymnemic acids I-VII, XV-XVIII, gymnamine, gymnimagenin, querci tol lupeol , s t igmas terol , gymnosaponin TN-2, quercitol, dammarone, gymnemic acids, saponin-gymnemosides, B-amyrin,	Root/leaf decoction is given to the patient till cure. Fresh leaves/dried powder is also taken twice a day till cure. Some diabetic persons claimed that powder of the entire plant of G. sylvestre used with that of tender leaves of *Aegle marmelos* and *Azadirachta indica* perfectly control diabetes. For this

Contd...

Contd...

Scientific name	Family	Local name	Chemical constituents	Uses
			gypenosides II, V, XLIII, LXXIV	purpose one tea spoonful powder of this mixture is taken along with water twice a day.
Helicteres isora Linn.	Sterculi-aceae	Maroorphal, Ainthi, Petmurri	Root contains cucurbitacin-B and Isocucurbitacin-B. Bark contains physterol, hydroxyl-carboxylic acid, saponins, phlobotannins and lignin	The roots in the form of powder and also as decoction mixed with honey is given to the patient. Seeds are also used in the form of powder. 20 gm root bark paste is also given daily once for a month.
Hibiscus rosa-sinensis Linn.	Malvaceae	Gurhal	Alkaloids, steroids, glycosides, flavones, organic acids, vitamins, anthocyanin pigment and cyanidin diglucoside.	One mature bud (unopened flower) is chewed and eaten in the morning in empty stomach till the desired level of sugar is obtained. This treatment is said to be good in managing the level of blood sugar. 5-10 ml infusion of petals is also given to patient.
Madhuca indica Gmel	Sapotaceae	Mahua, Mahuwa	Lupeol acetate, b-amyrin acetate, a-spinasterol, erythrodiol-monocaprylate, betulinic acid and oleanolic acid.	One cup decoction of the bark is given in the morning every day till the desired result is obtained.
Mangifera indica Linn.	Anacardi-aceae	Aam	Triterpinoids (mangiferin, taraerone,	The seed kernel is given to the patient in the form of powder

Contd...

Contd...

Scientific name	Family	Local name	Chemical constituents	Uses
			taraxerol, b-sitosterol), flavonoids, phenolic compounds and tannins.	(one teaspoonful twice a day). The young leaves are also chewed for lowering the blood sugar level. ½ teaspoonful shade dried leaves is also given to the patient twice a day till cure.
Momordica charantia Linn.	Cucurbitaceae	Karela	Plant contains Alkaloid momoridicine, saponin, carotene, glycoside and highly aromatic essential oils. Fruit contains acylglucosyl sterol, charantin, lipid, b-sistosterol glucoside, ascorbigen and stigmast-5, 25-dine-3,B-O-glucoside.	The juice of the green fruit is taken regularly in the morning to control the sugar level in blood of diabetic patient. One teaspoonful powder of dried fruit is also given to the patient twice a day. Decoction of the fruits is also used in the morning on empty stomach.
Murraya koenigii (Linn.) Spreng.	Rutaceae	Kathnim	Alkaloids cyclomahanimcine, mahanimcine, koenidien, mahanine, koenine, koenigine and scopoline, girinimbine, mahanimbine, isomahanimbine and koengicine.	7-10 leaves are consumed every morning for three months to check diabetes.
Phoenix sylvestris (Linn.) Roxb.	Arecaceae	Khajoor	Aminoacids, vitamins.	The seed powder is used in the treatment of diabetes.

Contd...

Contd...

Scientific name	Family	Local name	Chemical constituents	Uses
Syzygium cumini Linn.	Myrtaceae	Jamun	Leaves contain essential oil including a-pinene, b-pinene, limonene, cis-ocimene and trans-ocimine. Fruits contain anthocyanins delphinidin-3-gentiobioside, malvidine-3-laminaribioside. Seed oil contains oleic, myristic, linoleic, stearic, vernolic, lauric, sterculic and malvalic acid.	Leaf juice / stem bark juice is taken by the patient in case of diabetes. Fruits and seeds are also considered as a best remedy for the treatment of diabetes. One to two spoonful juice of ripe fruit is given to the patient in morning in empty stomach. One teaspoonful seeds powder is given to the patient twice a day along with water.
Swertia chirata Buch-Ham.	Gentian-aceae	Chirayta	Chiratin and ophelic acid, a phenolic compound, oleic, palmitic, stearic acids and a phytosterol.	Decoction of whole plant along with 2gm black cumin (*Nigella sativa*) is given once in a day during excessive urination. The plants are soaked in water over night and the filtrate is taken in empty stomach in the morning.
Tephrosia purpurea Linn.	Papilion-aceae	Sarphonka	Steroids, glycosides, flavones and fatty acids.	Decoction of the leaf is taken in the morning and evening for a period of 30-45 days. Root is also used in the treatment of diabetes.
Tinospora cordifolia (Willd.) Murs.	Menisperm-aceae	Gurch, Guduchi	Furanoid diterpene 10-hydroxycolumbin,	Oral administration of aqueous extract of root caused a

Contd...

Contd...

Scientific name	Family	Local name	Chemical constituents	Uses
			clerodane, clerodanef uranoditerpene I, II, III, IV, tinosporaside, tinosporidine, b-sistosterol.	significant reduction in blood glucose. The decoction of the stem with honey is taken in the treatment of diabetes. Leaf powder alone or mixed with milk is taken twice a day for one month.
Trigonella foenum-graecum Linn.	Papilion-aceae	Methi	2 flavonoid glycosides-quercetin and luteolin, Trigocoumarin, steroidal saponins from seeds, saponins yield diosgenin and gitogenin, alkaloid trigonellene.	One teaspoonful powder of the seeds is taken in the morning for 30 days.

❑❑❑

7

BREEDING STRATEGIES OF MEDICINAL AND AROMATIC PLANTS

The basis for use of medicinal plants is that these plants contain so called active principles that affect physiological processes of living organisms. Aromatic plants indicate plants having sweet smelling or fragrant aroma. In modern ages, Medicinal and Aromatic Plants (MAP's) are defined as fragrant ingredient containing group of medicinal plants. The basic taxonomical unit of MAP's is so called *species* (sp.) with related species constituting a *genus.*the categories *subspecies* (subsp.), *variety* (var.) and *form* (f.) are used to differentiate among varied populations of wild growing species. In fact, both natural and cultivated species are divided into *infra-specific* varieties.

Secondary metabolites usually derived from the precursors produced by the primary metabolic processes are found only in one plant species or related group of species. These metabolites protect plants against plants being eaten by herbivores or being infected by microbial pathogens. They serve as attractants for pollinators and as agents of plant-plant competition and plant microbe symbioses.

The rise of chemical taxa of MAP's can be considered as result of biochemical and metabolic processes mostly under genetic control. It is assumed that differentiation of cell is mostly manifested in chemo-differentiation at a molecular level and the fundamental differences of protein present in the organism, i.e. in differences between enzyme systems. The special chemical features of MAP's are attributed to dissimilarities in the metabolism. Besides, infra-specific chemical modifications can be caused by ecological and geographical conditions. These chemical differences can be called as polychemism. Polychemism is known to occur frequently in autogamous species.

The phenotypes of plants are influenced not only by environmental but also by genetic factors, which are inherited from parents by their progenies. The breeder creates the genetic variability and tries to change the genetically controlled reaction norm of a genotype towards the aspired direction – the improvement of the average performance and of the ecological stability. Breeding has become one of the most important key factors for improvement of the MAP branch because the genotypes can be adapted to the special demands of the players in the production chain and contribute to high quality, profitable and sustainable production. The exploitation of the genetic potential of MAP is still in its initial stage. Conventional breeding methods prevail due to the high natural variability available. Nevertheless, the use of biotechnological tools and research on genes controlling the secondary metabolite formation and their transmission are in progress. Success has been achieved in the development of expense saving selection procedures.

Common Breeding Objectives

Some general breeding aims for MAPs are:

- high and stable yield, in particular, of the used plant part (e.g, leaves)
- resistance to pests and diseases, as well as to abiotic stress (such as adapting to the climate of the cultivation region, e.g, frost and drought resistance)
- high functional value and safety
- suitability for technological processes in agriculture
- provision of low input, cost saving and sustainable production
- effective protection of plant breeders' rights (e.g, by hybrid varieties)

Specifics of MAP Breeding

Breeders of MAP cultivars confront with some special features. In case of agricultural crops, only a few results of breeding research are available - e.g. on the genetics of certain traits and on breeding methods, that MAP comprise a particularly great number of species, that often the breeding objectives differ on one and the same species depending on the field of usage, that the analyses of the important constituents are particularly costly, that only limited capacities for breeding research and breeding are available for these minor crops, and that refinancing of the breeders' expenses is insufficient due to the small saleable seed amount. But, the breeder can exploit high natural variability because breeding of MAP is in its initial stage only so that a high progress is achievable already after a few selection steps. In view of these specifics and to ensure an adequate cost-value ratio, breeders have to plan very carefully which crops, breeding objectives and breeding methods are to be chosen.

Utilization of Natural Diversity

Sources of natural variability are accessions collected in the wild, accessions from germplasm collections - e.g. from gene banks and botanical gardens -, furthermore old primitive cultivars and landraces, and contemporary used cultivars for last two decades. Introduction of a high performance population selected among a great collection of different accessions can already provide an enormous progress in adaptation of a MAP species to the demands, even without additional time-consuming and expensive breeding procedures.

Creation of New Variability

New variability should be created if the initial material does not provide the desired trait expression. Some common methods are described as follows:

Combination breeding by crossing: The aim of crossing is the combination of trait expressions of the parents (donors of certain characteristics) in common progenies. The pollen of the male parent is transferred on the stigma of an emasculated flower of the female parent, and the pollinated flower must be kept isolated, e.g. with a paper bag, during the period of stigma fertility. Elite plants with the desired trait combination are selected in the following segregating generations and bred to constancy. This method is most common in MAP breeding to create new variability. Recurrent back crossing is used if a valuable trait, e.g, of a wild accession, will be transferred to an established variety. The variety is repeatedly fertilised with the pollen of the donor accession and only plants combining the good characteristics of the variety and the valuable trait of the wild accession are selected in the progeny. This procedure is completed when the desired trait expression of the wild accession is thoroughly transferred to the variety.

Hybrid breeding: Hybrid breeding is very popular in important agricultural and horticultural crops but in MAPs it is still in the beginning stages. Hybrid varieties excel by high performance due to hybrid vigour and uniformity, and by the natural protection of plant breeders' rights because unlicensed seed propagation results in worthless segregating populations. Seeds from hybrid varieties are produced as F1-generation of selected parents with good combining ability. Good combining ability means that the performance of the F1-progeny of these special parents is particularly high. F1 seed production of certain parents needs controlled pollination. For hybrid breeding or outbreeding MAP, preferably lines with genic-cytoplasmatic pollen sterility should be used for this purpose to minimise the expense for emasculation. Hybrid varieties need the development of separate lines with the following characteristics; 1) male sterile; 2) male fertile with maintainer properties to maintain the male sterile line; 3) male fertile pollinator with good combining ability for F1 seed production; and 4) if generative organs are harvested and the pollinator parent has no restorer properties, a male fertile restorer has to be added to the F1-seeds to ensure a sufficient seed set in the production field. Advances in the development of hybrid cultivar systems are reported.

Synthetic cultivars: Synthetic varieties based on several – more than two as known for hybrid varieties – well combining parental populations. The seeds are produced from advanced generations. Their potential performance can also reach the level of hybrid varieties, but the uniformity will be lower in general. The combining ability of parental populations (inbred lines or clones) is assessed by combining ability tests. Lines with the best general combining ability are selected as parental components of the synthetic variety and are maintained by the breeder. For seed production, the selected components are cultivated together. The seeds of this panmicticly reproduced mixture are sold as seeds for production only in the second or third generation because the heterozygosis as basis for high performance can largely survive in the following generations.

Induced mutation: The spontaneous mutation rate of plants can be increased by the use of e.g, radiation or mutagenic chemicals. The change of the genome by a mutation inducing agent results in a different reaction norm of the mutated plants compared with non- mutated original plants. A disadvantage of this method is that the emerging mutations and the changes of the related trait expressions occur by chance and may be erratic. Induced mutation results in new variability of the mutants themselves or in segregation of the progeny of mutants crossed with existing genotypes. Examples of successful breeding by mutation induction are the high yielding peppermint-variety 'Multimentha', which has a high essential oil content and resistance against *Puccinia menthae*. Further examples of successful mutation breeding are *Chamomilla recutita*('Zloty lan', 'Bodegold'), *Origanum majorana* ('Miraz', 'Tetrata'), *Verbascum phlomoides* ('Polyverb') and *Lavandula vera*.

Somaclonal variation: Also the *in vitro* culture stress in cell and tissue cultures coupled with processes of de- and redifferentiation acts mutagenically. Selection is accomplished on the cell level *in vitro* or on the level of fully functional plants regenerated from tissues or cells. Several authors have studied somaclonal variation on *Datura* sp. and*Viscum album*.

Somatic hybridisation (protoplast fusion): Protoplast fusion is indicated if there exist natural barriers(incompatibility) to cross different donors of desired trait expressions. In this method, the protoplasts of different genotypes are fused without any meiotic recombination, the two heterozygous genomes are combined and the fused protoplasts are regenerated to functional plants. Protoplasts of e.g, *Catharanthus roseus/Vinca minor*, *Rauwolfia serpentina/ Vinca minor* and *Rauwolfia serpentina/Catharantus roseus* were successfully fused.

Molecular gene transfer: Molecular gene transfer shortens the breeding procedures at considerable rate. A lot of on-going research also with MAP is concerned with the clarification of the biosynthetic pathway of important secondary metabolites, the identification of related enzymes, the isolation of encoding genes, and the establishment of protocols for the gene transfer. These investigations are very expensive and till date successful breeding of transgenic cultivars with economic importance is scarcely known.

Clone breeding: Clone breeding means the vegetative propagation of selected high performance individual plants. Vegetative propagation involves identical reproduction of the clone variety –particularly important in the case of allogamous species – without the time consuming procedures for breeding to consistency. Sterile plants, e.g, *Mentha x piperita*, can be propagated only vegetatively. Clone breeding remunerates only for high priced MAPs like *Baptisia tinctoria* (L.) R. Br due to the high costs of vegetative propagation – mostly carried out by micro propagation *in vitro* – in comparison to generative reproduction. The generation of artificial seeds is an other method of vegetative propagation as mentioned with *Cymbopogon martini* and *Echinacea angustifolia*.

SELECTION

If a population with an appropriate natural or artificially generated variability is available, the breeders initiate the improvement of the population by selection. Selection accumulates the genes in a population which control the desired trait expressions. It retains fitness while inbreeding increases uniformity. Selection in wild populations or landraces is the most common way in medicinal plant breeding, because most of the species are yet in the stage of wild plants with high genetic variability. Only a few selection steps provide already satisfying results.

Selection methods: Common selection methods in MAP breeding are mass selection, recurrent selection, pedigree breeding (selection of elite plants and testing their progenies), selection breeding in apomicts, and clonal breeding. Clonal breeding has a particularly high importance for MAP because this method shortens the breeding procedures considerably. Due to the high costs of vegetative propagation mostly done by micropropagation *in vitro*, clonal breeding is appropriate only for high priced MAP.

Acceleration of generation sequence, vernalization and photoperiodic reaction: Further tools for reducing the breeding procedure are: to accelerate the sequence of generations by eg, additional generations during winter in the greenhouse with *Carum carvi* ; flower induction by vernalisation of *Salvia officinalis* with *Thymus vulgaris*, regarding the typical photoperiodic reaction of a species. Early selection saves expenses by selecting and removing low performance individuals in an early developmental stage.

Doubled haploids: The time consuming breeding for homozygosis can be saved by generating doubled haploids of *Allium cepa* and several other MAPs.

Selection in apomicts: Selection in apomictic plants with asexual seed formation without fusion of two reduced gametes is particularly effective because the progeny has the same genetic constitution as the mother plant, and it is already true breeding without further selection steps. Apomixis is also be found in some MAPs, for example in *Coriandrum sativum* and *Ocimum basilicum*. Recent experiences were also gained with the apomictic *Hypericum perforatum*.

Rapid analytical methods: One of the most important breeding objectives is the improvement of the content of essential constituents. But, the huge expenses for chemical analysis are a limiting factor. Therefore, the development of special analytical methods for MAP breeding is an indispensable prerequisite to provide a sufficient selection intensity by analysing an appropriate amount of individual plants. Analytical methods must fulfill the following demands: time and cost saving to allow the evaluation of as many individual plants as possible, non-destructive for keeping the investigated plant material alive for following breeding procedures, small sample quantities because single plants or parts of them are to be evaluated, direct analysis without preceding sample preparation is preferred, the precision must be reliable. A lot of expenses can be saved considering the fact that the accuracy must not comply with the high level of standard methods. For example, the rapid and non-destructive near infrared spectroscopy and the solid phase micro-extraction proved well in MAP breeding.

Chemo-variation

Differences of flower colours and scents as well as fruit tastes from similar looking plants have been known from the everyday experience of mankind since ancient ages. Infra-specific chemical diversities (IsCh) were observed based on chemical, biochemical and biosynthetic results since the beginning of the 20th century. Worldwide results provided evidence that direct procedures of clearing genetic similarities on chemistry of plants were efficient methods. The new research trend provides a new opportunity for disclosing real DNA and biosynthetic causes of chemo-differentiation, opening great perspectives for the breeding of new, high-powered chemo-cultivars of medicinal and aromatic taxa.

Molecular Markers in the Improvement of the Medicinal Plants

With the advancement of DNA based different molecular markers and molecular breeding strategies provides many new strategies to plant breeders and geneticists to resolve many of decades old problems faced during conventional breeding of medicinal plants. Molecular markers are unique and related directly to the plant's genotype.

Genetic diversity among individuals or populations can be estimated using morphological, biochemical and DNA based molecular markers. Till date, however, there have been relatively few reports of molecular marker-based approaches to medicinal plant improvement, and not even the most skeletal of genetic maps is available for any of the important industrially important species except for *Artemisia annua*. Selection assisted by genetic markers is an extension of traditional crop breeding, which has been used extensively in food crop improvement. Again, it is a way to recognize desirable genotypes at an early stage to speed up the selection process. Identifying a novel functional gene and useful marker sequences linked to them in a plant species is a lengthy and costly technical affair. Plant genome sequencing has progressed rapidly since the first genome (*Arabidopsis thaliana*) was completed in 2000 followed by the 389-Mb rice genome was completed in 2004. There is a high degree of match in the DNA sequences of functional genes between different plant species; therefore, DNA probes from one species can often be used to identify homologous sequences in another closely related species. The key developmental challenges for molecular markers now depends on developing new breeding strategies where the objectives will be increasing the germplasm base and increasing the number of traits that can be effectively selected simultaneously. The new marker technologies that offer greatly reduced costs in marker screening and high multiplexing capabilities will be central to these developments. Essentially the whole genome-based selection strategies are adopted where specific recombinational events are sought and changes will be assessed on a genome-wide scale. In this way, one can look for better management of chromosome regions that may come from wild relatives or land races, track several traits at once and keep the population size as small as possible. It is believed that despite the relatively small adoption of markers in different medicinal plant breeding to date, there will be a greater level of adoption in the next decade and beyond.

Factors that should enforce to a greater adoption of MAS in medicinal plants include:

- Establishment of facilities for marker genotyping and staff training within many medicinal plant breeding institutes in different parts of the world.
- Currently available data on genes/QTLs controlling traits and the identification of tightly-linked markers in different plants
- development of effective strategies for using markers in breeding
- establishment and creation of public databases for QTL/marker data especially for different medicinal plants
- Available resource for generating new markers from DNA sequence data arising in parts from different medicinal plant genome sequencing and research in functional genomics.

It is also critical that future endeavors in marker assisted selection are based upon lessons that have been learnt from past successes and failures in using MAS. Further optimization of marker genotyping methods in terms of cost-effectiveness and a greater level of integration between molecular and conventional breeding represent the main challenges for the greater adoption and impact of MAS on different medicinal plants in the near future. The choice of the most appropriate marker system, however, needs to be decided on a case-by-case basis and will depend on many issues including the availability of technology platforms, costs for marker development, species transferability, information content and ease of documentation.

Breeding Objectives Relevant to Postharvest Processing

Homogeneity: Homogeneity of the plant material facilitates all mechanised and postharvest procedures, and provides the prerequisites for the production of uniform qualities.

Constituents: The value of herbal drugs is determined by the biologically active constituents, while the extraction yield depends on their concentration in the crude drug.

Therefore, MAP breeding is mainly focused on essential constituents: There should be:

- High content of desired compounds to economise extraction and allow microbiological decontamination procedures, which often lead to losses of active constituents to a certain extent
- Absence of harmful substances in order to avoid expenses for eliminating them after harvest
- High dry matter content to lower the energy consumption for drying
- High antimicrobial properties of the constituents to inhibit decay of the herbal material

Morphology

The morphology of the plant influences the feasibility of mechanised harvest and therefore the suitability of the harvested crops for postharvest processing. Morphological considerations include:

- thick roots with limited ramification (eg, roots of *Valeriana officinalis*) to ease root washing after harvest and industrial processing
- upright growth type to facilitate mechanised harvest of herbal drugs (eg, *Origanum majorana*, *Thymus vulgaris*, *Digitalis lanata* and *Plantago lanceolata*)
- narrow flower horizon (eg, of *Chamomilla recutita* and *Hypericum perforatum*) stands to enrich the flower portion in the harvested cropsresistance of *Chamomilla recutita* flower heads to disaggregation after drying
- small grained *Foeniculum vulgare* fruits for good functioning of machines for filling tea bags
- resistance to disintegration of both achenes of *Coriandrum sativum* fruits
- high portion of leaves and low portion of stems in the herb (eg, *Mentha* x *piperita, Melisse officinalis* and *Origanum majorana*) to improve the yield of leaves after mechanical separation
- high ratio of ray florets of *Calendula officinalis*
- low share of stems adhering to the fruits of *Carum carvi* to facilitate cleaning procedures

- resistance to grain shattering during the maturity period (eg, of *Foeniculum vulgare* and *Carum carvi*)
- oil glands (eg, of *Mentha x piperita*, *Origanum majorana* and *Salvia officinalis*) should be resistant to damage from mechanical stress caused by crushing and milling.

Table 1. Improved varieties of medicinal and aromatic plants

Common name	Botanical name	Improved varieties
Isabgol	*Plantago ovata*	Gujarat Isabgol-I, Gujarat Isabgol-2, Haryana Isabgol-5, Jawahar Isabgol-4, Niharika
Opium poppy	*Papavar somniferum*	Jawahar Aphium-16 (JA-16), Jawahar Opium 539, Jawahar Opium 540, Trishna(I.C. 402), Shaktiman, Kirtiman, Chetak=Aphium, E.C. 179777, EC. 196430 (Finland), E.C. 196433 (Hungary), Sujata, TOPI, Mandakini, Subhra
Senna	*Cassia angustifolia*	Anand Late Selection, Sona, Tinnevelly Senna, KKM-1
Periwinkle	*Catharanthus roseus*	E.C. 120837 (Russia), I.C. 49581, Nirmal
Liquorice	*Glycyrrhiza glabra*	E.C. 114304 (Russia), E.C. 41911E.C. 128587 (Pakistan), Haryana Mulhatti-1
Ashwagandha	*Withania somnifera*	R.S.1. Jawahar Asgand-20, NMITLI-118, Jawahar Asgand-134,Rakshita, Posita, Manasa Local, Kukedesher
Jasmine	*Jasminum grandiflorum*	Pitchi, Arka Surabhi
Palmarosa	*Cymbopogon martinivar. motia*	I.W. 31245, "CIMAP HARSH', Rosha Grass-49, CI-80-68, Vaishnavi Trishna, Tripta ;PRC-1
Vetiver/ khus	*Vetiveria zizanioides*	K.S.I., Sugandha, Pusa Hybrid-7, 8, Kesari, Gulabi, Dharini
East Indian Lemon grass	*Cymbopogon flexuosus*	Sugandhi, , Pragati, Nima, Cauvery, Krishna, NLG-84, O.D. 410
Lemon grass	*Cymbopogon pendulus*	Praman Hybrid of *C. khasianus* × *C. pendulus* CKP-25
Chamomile	*Matricaria chamomilla*	CIMAP "SAMMOHAK"
Lemon grass	*Cymbopogon khasianus*	"CIMAP SUWARNA"

contd....

Common name	Botanical name	Improved varieties
Mint	*Mentha arvensis*	"CIMAP SARYU", Hybrid 77
Mint	*Mentha piperita*	Siwalik
Mint	*Mentha spicata*	Punjab Spearmint-1
Sarpagandha	*Rauwolfia serpentina*	R.S.1
Khasi kateli	*Solanum viarum*	Arka Sanjeevani, Arka Mahima
Steroidal yarns	*Dioscorea floribunda*	PB(c) I, Arka Upkar
Foxglove	*Digitalis lanata*	D 76
Long Pepper	*Piper longum*	Viswam
Mushakbala	*Valeriana jatamansi*	Dalhousi Clone
Java citronella	*Cymbopogon nardus var. mahapengiri*	'Jorlab C2', 'RRL JOR-3-1970', 'CIMAP/Bio-13' ,CIMAP/73-1', Manjari , Jal Pallavi, Medini , Manjusha, Mandakani
Patchouli	*Pogostemon patchouli*	'Johore', 'Singapore', 'Indonesia
Tuberose	*Polyanthes tuberosa*	Calcutta single', 'Mexican Single', 'Rejat Rekha', 'Suvarna Rekha
Aloe	*Aloe barbadensis*	IC-111271, IC 111280, IC 111269 and IC 111273
Guggal	*Commiphora wightii*	Marusudha, GAU-1
Safed Musli	*Chlorophytum borivilianum*	Jawahar Safed Musli 405
	Coleus forskholii	Maimul, Garmai
Black Henbane	*Hyoscyamus niger*	IC 6, Aela
Indian Basil	*Ocimum basilicum*	Clo-Ocimum

Breeding Approaches of Important Medicinal and Aromatic Plants

Aloe vera

Description: Aloe L. belongs to the family Liliaceae , which represent perennial succulent plants, bearing rosettes of leaves at the end of juicy green branches. *Aloe vera* is native to Africa and introduced to India. The species is found in the Arabian Peninsula, through North Africa as well as Sudan and neighbouring countries. *Aloe vera* grows in arid climates and is widely distributed in Africa, India, and other dry areas. In India, it is commonly observed in Rajasthan, Andhra Padesh, Gujarat, Maharashtra and Tamil Nadu.

Reproductive biology: The flowers are orange or red, bisexual, containing both female and male parts arranged on spike in pendulum manner. The first sign of anthesis is indicated by appearance of longitudinal crack at the apex of corolla and widening at the middle of the bud and slowly one after another or simultaneously the petals of the bud separated and six stamens and stigma became visible. This process completed within 30 to 50 min. Anthesis started in the morning hours from 7.00 am and continued up to 3.00 pm. The head shaped stigma remains receptive for 48 hours after the opening of flower. The receptivity is found higher on the next day of the anthesis. The bursting of pollen from anther i.e. dehiscence is observed maximum between 10.00 am to 12.00 am. Dehisced pollen grain appears as smooth, creamish yellow mass and remained accumulated on the surface of the two lobed anthers. Freshly dehisced pollen grains examined under the microscope are yellow coloured spheroidal oval in shape. Pollen viability is recorded 80 to 90 per cent. Pollen remains viable for 6 to 8 hours. There is fruit setting in the bagged raceme, which corroborate the self- pollination and it is due to hermaphrodite nature flower. Since there are honeybees observed on flower indicates some extent of cross pollination. It is reported that bees and sunbird are responsible for pollination.

Vegetative propagation: Vegetatively aloes are cultivated through suckers. About 3 to 4 months old suckers are used for planting. The suckers are planted in 15cm deep pits at 60 x60 cm spacing. About 28,000-34,000 suckers are needed for one hectare planting. Under rainfed conditions, the suckers are planted in July-August.

Micro-propagation: Shoot bud explants raised *in vitro* is used to produce multiple shoots after surface sterilization. As many as 28 shoot buds can be obtained from a single shoot bud after 2 weeks of cultures on ½ MS basal medium supplemented with BA, IAA and sucrose. After multiplication, micro-shoots (1-2cm) are cultured on half strength MS basal salts supplemented with IBA (0.50 mg/l), NAA (0.5 mg/l) and IAA (0.5 mg/l) for rooting. Roots emerge within 8 days. After 15 days of rooting the plantlets should be transferred into pots containing sterilized soil: sand: FYM (1:1:1) and kept in a room where temperature, RH and Light were 23-27°C, 60-70% and ca.4500 lux, respectively and covered with perforated polybags. After 4-5 days, the pots are shifted to a net house having 75% light intensity. After 15 days about 98% plants survived when planted outside. Micro-propagation through organogenesis is also reported. For callus induction, small leaf bases should be inoculated on ½ MS basal medium supplemented with BA, 2,4-D and ascorbic acid and incubated at 23-27°C with 4500 lux illumination from cool, white influorescent lamps having 16h photoperiod. Callus formation starts at leaf base within 16-17 days. After 3-4 weeks the calli should be sub-cultured in ½ MS media containing BA and IAA for shoot bud regeneration. The calli differentiate into green nodular structures all over the callus mass after 3-4 weeks that further develop into dark green shoot buds. After regeneration, micro-shoots (1-2cm) should be cultured on ½ MS basal salts supplemented with IAA and sucrose for induction of rooting.

Artemisia annua

Description and floral biology: *A. annua* is a large shrub, more than 2.0 m tall, usually single-stemmed with alternate branches. The aromatic leaves are deeply dissected and 2.5 to 5 cm in length. Leaves and flowers contain both 10-celled biseriate trichomes and 5 cell filamentous (T) trichomes. The nodding flowers (capitula), only 2 to 3 mm in diameter, are greenish-yellow enclosed by numerous, imbricated bracts. Capitula are displayed in loose panicles containing numerous central bisexual florets and marginal pistillate florets, the latter extruding stigmas prior to the central flowers. Both flowers have synpetalous tubular corolla with the top split into five lobes in the hermaphroditic florets and 2–3 lobes in the pistillate florets. The receptacle is glabrous, not chaffy, and triangular in shape. Both florets and receptacle bear abundant 10-celled biseriate trichomes; T-trichomes (filamentous) occur at the pedicel and bracts. The biseriate glandular trichomes sequester artemisinin as well as highly aromatic volatile oils (essential oils).

***In vitro* propagation:** *A. annua* is easily propagated *in vitro* by standard protocols. Cytokinins increase shoot proliferation but decrease rooting and increase vitrification. *A. annua*is grown and propagated by microcuttings in a hormone-free medium. Artemisinin is produced in shoots *in vitro* and is enhanced by the presence of roots. None or trace levels of artemisinin are found in roots, callus, cells, or cell free medium. There is no evidence that *in vitro* production of artemisinin in tissue culture will ever be commercially feasible. However, experimental evidences have shown success in producing artemisinic acid (a precursor) in yeast cultures and artemisinin production, through its precursor, in bioreactors might be a future avenue to explore.

Breeding: The Centre for Novel Agricultural Products (CNAP) at the University of York is applying fast-track breeding technologies to Artemisia with the aim of creating new, non-genetically modified (GM) varieties with increased artemisinin yields. Many thousands of plants are being screened at both the phenotypic (trait) and genetic levels. High yielding varieties will secure the supply of artemisinin – remove shortage-driven price inflation reduce the cost of production of artemisinin. The main objectives of the breeding are increasing the leaves on a plant, increasing the trichomes on a leaf, increasing the artemisinin per trichome, increasing the genetic diversity, identifying individuals with high artemisinin, tilling for changes in target genes, genes controlling metabolisms for supply of precursors, competing pathways, sink/source relationships, transporters etc., genes controlling development for trichome density, leaf, flowering etc. and developing robust new varieties by back crossing into wild type parent, confirming heritability of phenotype, further back crossing, field trials and user-tests for new hybrid and roll out of new variety.

Cassia aungustifolia

Description: Senna (*Cassia aungustifolia*) is a native of Yemen and Hadramaunt province of Saudi Arabia. Commonly available species are *C. obovata*, *C. italica*, *C. tora* and *C. acutifolia*. In India, it is commonly called as senna, sona, sanai etc. This is a small erect perennial shrub, 100-120cm tall with pinnate, lanceolate leaves. The flowers are bright yellow and borne in racemose inflorescence The plants are rich in sennosides, aloe-amine, kaempferein and iso-rhein. Senna is laxative or purgative and used in the treatment of skin problems. This crop is commercially cultivated In Tamil Nadu, Andhra Pradesh, Rajasthan, Gujarat, Maharashtra, Karnataka, West Bengal and Tripura.

Breeding: Genetic variability is recorded high for pod and seed yields but moderate for leaf yield. Broad sense heritability is also high for all traits except number of branches. The three important traits. viz., dry leaf yield, dry pod yield and seed yield manifested high his coupled with high GA and the latter two traits are favourably correlated as well and therefore most amenable for straight selection.

Propagation: The crop is propagated through seeds. About 6 kg of seeds are required for line sowing in one hectare. In case of broadcasting, 20-25 kg of seeds per hectare is recommended. The optimum time for sowing of seeds is western India is June-July and in South India is September-October. Seed production is done in drier areas of Rajasthan, Gujarat and Haryana. This crop is usually grown as rainfed crop. An average rainfall of 25-40 cm from June-October is sufficient to get good harvest. Two to three sprayings of Dithane M-45 at 0.15% at weekly interval is effective to control of leaf diseases. Seed treatment with Thiram at 3g/kg of seed is useful to reduce the damping off disease.

Catharanthus roseus

Periwinkle (Catharanthus roseus (L) G. Don) has become one of the very important medicinal plants after the discovery of two powerful anti-cancer alkaloids, vinblastine and vincristine, in its leaves more than 50 years ago. These alkaloidal drugs are still in clinical use for cancer treatment. In addition, periwinkle is still the only source of these alkaloids and their precursors, catharanthine and vindoline. Catharanthus roseus is an evergreen subherb or herbaceous plant growing to 1 m. tall. The leaves are oval to oblong and broad, glossy green hairless with a pale midrib and a short petiole about 1-1.8 cm. long and they are arranged in the opposite pairs. The flowers are white to dark pink with a dark red centre.

Breeding: The commonly observed flower colors in periwinkle were earlier explained by a three gene (*R, W,* and *I*) model.According to their model, the *R* allele produces three anthocyanidins—A_1, B_1, and C_1 (located in the center of the corolla)—and is epistatic to the *W* and *I* alleles, which function only in its presence. The *W* allele produces pigments A_2, B_2, and C_2. The *I* allele also produces the same pigments as the *W* allele, but in smaller quantities and mainly in the center of the corolla. The *r, w,* and *i* alleles do not produce any pigments.

The commonly found flower colors in periwinkle (*Catharanthus roseus*)—pink, white, red-eyed, and pale pink center—are reported to be governed by the epistatic interaction between four genes—*A, R, W,* and *I*. The phenotype of the F_1 plants and segregation data of F_2 and backcross generations suggested the involvement of two more interacting and independently inherited genes, one (proposed symbol *E*) determining the presence or absence of red eye and another (proposed symbol *O*) determining orange-red corolla.

Catharanthus species are predominantly self-pollinated and need emasculation of the flower buds 2-3 days before anthesis by cutting the corolla tip between stigma and anthers. Natural interspecific hybridizations of periwinkles have been observed when the different species are grown together in botanical gardens or when the geographical distribution overlaps and most of these hybridizations concerned *C. longifolizis* x *C. roseus*.

Although *Catharanthus* species are predominantly autogamous, low rates of outcrossing in *C. roseus* have been measured using markers involved in seedling pigmentation and flower colour.

Mutation breeding is another approach for creation of genetic variability, which is also applied in improvement of *C. roseus* (L.) G. Don. Chatterjee *et al.* (1978)attempted induced mutagenesis in *C. roseus* for upgradation of leaf alkaloid content and isolated several morpho- chemotypes. Kulakarni *et al*(1999) reported three mutants, one dwarf, one semidwarf with high alkaloid content in the roots and one mutant with wavy leaf margin and high alkaloid content in its leaves, were isolated after chemical mutagenesis with EMS and N-nitroso-N ethylurea in the variety 'Nirmal'. On the basis of inheritance studies on morphological traits, they found that the mode of inheritance of all three mutants was controlled by monogenic recessive genes. Kulakarni *et al*(1999b) reported a mutant with indehiscent anther (functional male sterility), following an induced mutagenesis programme, which was taken up with an objective of inducing mutants with high alkaloid content in variety 'Nirmal'. The mutant had smaller anther, with lesser number of pollen grains. Rai (2000) reported EMS induced and Kumar mutant in which embryo had only one cotyledon but upon supplementation with the natural cytokinin, kinetin it developed two cotyledons. Mishra and Kumar(2003)reported recessive salt resistant mutant with monocotyledonary and hsf mutant with herkogamous flower. Kulkarni *et al.* reported a periwinkle variety'Dhawal'(NCIMB Accession No. 41147), a high alkaloid producing variety produced by chemical mutagen treatment of seeds followed by rigorous selection in widely cultivated variety 'Nirmal'. Maithy *et al.*(2004)reported six morphological mutants formed at different doses of gamma rays and EMS with their inheritance. Anjalika *et al.* (2005)carried out a mutagenesis experiment using gamma rays and EMS as mutagens and reported six morphological mutants viz. very tall (vt), dwarf with obvate leaves (dwob), medium tall with small leaf area (mtsl), non trichomous (nt), upright oriented elliptical leaf (Upef) and spoon shaped leaf (Spl) and described dwob mutant as most promising mutant for cultivation in India. Kumar *et al* (2007)reported a unique inflorescence bearing mutant (lli / lli) which is a monogenic recessive having leafless inflorescence architecture with increased flower frequency. This distinct plant of *C. roseus* was developed through chemical mutagenesis followed by salt tolerance selection (US Patent PP18315). Kulkarni and Baskaran (2008)reported two mutants, one with small pollen and pollenless anthers (OR-EA) and another with 'Pin' flower (EMS-13-2), in contrast to "thrum' flower found in normal periwinkle plants, after induced mutagenesis in strain 'OR" and cultivars 'Dhawal' respectively. Baskaran *et al*(2013)reported two EMS induced macro-mutants ('necrotic leaf' and 'nerium leaf) of periwinkle with enhanced contents of total root and leaf alkaloids and anticancer leaf alkaloids, vincristine and vinblastine than the parental variety.

Chlorophytum spp.

Safed musli (*Chlorophytum borivilianum*) is widely distributed in Asia, Tropical Africa, America and Australia. Tubers are of great medicinal value containing the steriod sapoginine (1-2%), proteins (10-12%) and calcium to some extent with some extent of water-soluble minerals. Tubers are useful on *pitta* and *vata*. It is effective on *fatigueness, daaha* and in blood purification. It is useful in certain diseases like renal calculus, dhupani, sangrahani, leucorrhoea and diabetes. It is basic ingredient in *chyanprash*. It is lactating, energetic to heart and aphrodisiac. Daily use of about 3g of musli powder with milk is helpful in providing healthiness. This crop is grown in Southern Rajasthan, North Gujarat and western Madhya Pradesh.

Safed Musli is a herb with sub-erect leaves and tuberous root system. It can grow upto a maximum height of 45-50cm. Tubers can grow upto a depth of 25cm. The plant sports a rosette of long, slender leaves, growing from a thick, fleshy rhizome. The flowers are small, white, produced on sparse panicles. Leaves are slightly yellowish and white flowers with 6 petals are arranged on the flowering stalk which emerge from the centre of the plant. About 20-25 flowers on the flowering stalk appear in July.

Propagation: Usually safed musli is propagated by tubers. However, the propagation of musli can be done by seed also. But the germination percentage by seed is very less and it takes about 18 months for harvesting tubers. Hence the sprouted tubers are useful for planting. It takes about 6 months for maturity in the rainy (kharif) season. Previous year's tubers are used for cultivation. The bunch of the tubers should be removed from soil in May and tubers are separated from the bunch in such a way that some portion of crown should remain attached to each tuber. Tubers separated should be stored in gunny bags filled with mixture of fine sand and silt with moist condition under shade. In June, tubers get sprouted naturally and these are used for planting.

In-vitro clonal multiplication using leaf base, stem disc as explants has been standardized. Multiplication rate can be increased four fold within three weeks. Protocol for rapid multiplication through organogenesis and somatic embryogenesis have also been standardized. Germination of somatic embryos is found successful on MS medium supplemented with 17.6 BAP. Plantlets produced from *in-vitro* embryo germination gave 90% survival in field on transfer. An *in-vitro* propagation technique for C. *borivilianum* using encapsulated shoot buds has also been developed. 4mm long shoot buds encapsulated in 3 % sodium alginate matrix polymerised by 100 mM solution of $CaCl_2.2H_2O$ yielded best results.

Breeding: Although the species is cross-pollinated in nature, self-pollination is also feasible artificially. Successful use of synthetic seed has also been reported.

A variability and association study of *Chlorophytum borivilianum* on yield and yield-contributing morphometrical traits revealed the scope for genetic improvement in major yield attributes with some limitations for root length and diameter. An interrelationship among the yield and yield attributes indicated that improvement in yield can be attained through improvement in root numbers, root length, and diameter; however, more emphasis must be paid on root diameter for organic conditions. The cause and effect study on the interrelationship exhibited the importance of productive root yield as a causation to improvement in yield.

Conservation of identified germplasm is an important component for efficient and effective management of plant genetic resources. Since Chlorophytum species are important medicinal plants, studies were carried out for identification and establish genetic relationships in three species of Chlorophytum and two high yielding genotypes of *Chlorophtum borivilianum* using RAPD markers. Out of one hundred primers tested, 47 decamers amplified a total of 454 distinct bands ranging from 0.25–3.0 kbp to identify and to evaluate genetic relationships between and among three species of Chlorophytum and two genotypes of *Chlorophtum borivilianum.* The cluster analysis indicated that three species of Chlorophytum and two genotypes (NRCCB-1 and NRCCB-2) of *C. borivilianum* formed two major clusters. The first major cluster constituted *C. arundinaceum* and *C. tuberosum*, and the second major cluster composed of two subclusters; the first subcluster represented NRCB-1 and NRCB-2 where as the second subcluster represented *C. borivilianum*. Thus, the RAPD markers have the potential for identification and characterization of genetic relatedness among the species and genotypes. *C. borivilianum* along with two genotypes also showed similar banding patterns which could be chosen as candidate markers for differentiating the other two species such as *C. arundinaceum* and *C. tuberosum*. This would helpful for breeding programmes and provides an important input in conservation biology.

Citronella

Citronella is a tufted perennial grass, the leaves of which on distillation yield an yellowish-brown essential oil with *citrus odour*. Two types of citronella are identified, viz, *Ceylon and Java citronella*, the former gives inferior quality oil while the latter yields a superior oil of commercial importance. Java oil serves as a starting material for the extraction of geraniol and citronellal which can be converted into aroma chemicals such as citronellol, hydroxy citronellol, synthetic menthol and esters of geraniol. These are extensively used in soap, perfumery, cosmetic and flavouring industries throughout the world. Soaps, soap flakes, detergents, household cleansers, insecticides and other technical products are often perfumed exclusively with this oil. Citronella grass belongs to the family Graminae (Poaceae) and the species *Cymbopogon nardus* (L.) Rendle with 2n=20.

Both Ceylon and Java citronella have probably originated from *Mana grass* of Sri Lanka which according to Finnemore occurs in two wild forms, viz, *Cymbopogon nardus var. linnaei (typicus)* and *C. nardus var. confertiflorus*. The Java citronella which is called *Mahapengiri* in Sri Lanka is the result of a selection from the Ceylon citronella and the name *Cymbopogon winterianus* is given to this selected variety to commemorate Mr. Winter, a prominent citronella oil distiller of Sri Lanka who first cultivated and distilled *Mahapengiri* type of citronella grass in Sri Lanka.

Cymbopogon nardus var. lenabatu: It is known as *Ceylon citronella*. It is the older form and less important than Java type. Ceylon type is cultivated in Sri Lanka. It is hardier. The leaves which are smaller than that of Java type yield an inferior oil with a lower citronellal content. The plant is a robust, stoloniferous perennial grass, 0.5-1.0 m tall, leaves are broad, panicles are large and mostly compound, spikelets awnless, lanceolate and flattened on back.

Cymbopogon nardus var. mahapengiri: This is sometimes recognized as a distinct species, *C. winterianus* Jowitt. It is known as *Java citronella*. Java citronella is a tufted aromatic perennial grass with fibrous roots. Plants are more than 2 m in height. Clumps are stout, erect, terete, leafy and glabrous at the nodes. Leaf blades are linear, with long acuminate tip and serrate margins. Ligules are scarious. Inflorescence is a very large compound panicle. Racemes 20 mm long, one sessile and the other pedicelled with the two lower spikelets, homogeneous, male or neuter, the remaining pairs in both racemes heterogeneous. Ovary is superior with one carpel, unilocular. Stigma is bifid, lateral and leathery. Ovule is single and erect.

Propagation: Java citronella flowers copiously in South India and at higher attitudes in the hills of North Eastern India. Due to irregularities in meiosis and chromosome polyploidy viable seeds, however, are not formed and therefore, the species can be propagated only vegetatively. The entire clump of grass is splitted into a number of slips and each slip contains 1-3 tillers. Each slip is a unit of propagation and on planting establishes itself as plants or bushes. Before planting, leaves should be trimmed off from the slips. Monsoon is the best time for plantation of Java citronella and can be initiated any time during the year.

Breeding: Genetic variability among citronella clones has been made through phenotypic recurrent selection for different traits. Heritability estimates revealed highest values in no. of tillers, linear growth, and herb dry yield and viability percentage. Coefficient of variation (C.V. %) varied in all generations and all characters. One the other hand, herb dry yield was high positive correlation with linear growth and no. of tillers. Highly significant differences were shown among generations in three characters only (number of tillers, linear growth and viability percentage). This finding revealed that clones carried genes with different additive effects.

Cymbopogon species display wide variation in morphological attributes and essential oil composition at inter- and intraspecific levels.Although, morphological traits can be used to assess genetic diversity they are strongly influenced by environment conditions and show little variation at the intra-specific level. A number of breeding techniques like clonal selection, induced mutagenesis and *in –vitro* somaclonal selection are used for varietal development.

Therefore, molecular markers are often used to obtain more reliable estimates of genetic diversity. Amongst them, RAPDs have been widely employed for the identification of cultivars, ancestors and discerning the genomic diversity in *Cymbopogons*. Simple sequence repeats (SSR) markers have features that make them particularly suitable for assessing the genetic diversity among cultivated lines or germplasm including that they are co-dominant, multiallelic, locus specific and highly reproducible. Comparative genetic analyses have shown that different plant species share orthologous genes for similar function and have highly conserved gene content and gene order. Therefore, EST-SSRs possess considerable potential not only for genome mapping within species but also for comparative studies between species. Moreover, EST-SSR markers are known to provide better estimates of functional diversity existent in a gene pool.

Coleus forskholii

C. forskohlii is a perennial plant that grows to a height of about 45 - 60 cm. It has four angled stems that are branched and nodes are often hairy. Leaves are usually pubescent, narrowed into petioles. Inflorescence is raceme, 15 - 30 cm in length; flowers are stout, 2 to 2.5 cm in size, usually perfect and calyx hairy inside. Upper lip of calyx is broadly ovate. The blue or lilac corolla is bilabiate. Lower lobes are elongated and concave so that they enclose the essential organs. The ovary is four parted and stigma is two lobed and the flower is cross-pollinated by wind or insects. This is the only source of the diterpenoid forskolin. Forskolin is used for the treatment of eczema, asthma, psoriasis, cardiovascular disorders and hypertension.

Propagation: It is propagated by seeds as well as vegetatively by terminal stem cuttings. Seed propagation is difficult and slow whereas propagation by terminal stem cutting is easy and economical. 10 to 12 cm long terminal cuttings with 3 to 4 pairs of leaves are planted in nursery beds to induce rooting. *In vitro* propagation is useful for mass multiplication and germplasm conservation of any plant species. *C. forskohlii* being succulent in nature responds well to *in vitro* propagation and various explants viz., nodal segments, shoot tip, leaf etc., are effectively used. Complete plantlets of *C. forskohlii* were developed within 35-40 days by culturing shoot tip explants in MS medium containing 0.57 M IAA and 0.46 M kinetin through direct multiplication at the rate of 12.5 shoots per explant.

Breeding: *C. forskohlii* is a diploid species with a basic chromosome number varies from 14 to 17. Most of the traits studied in coleus are simply inherited and controlled by a single gene.

The following is observations made on different populations and habitats of *C. forskohlii.*

- *C. forskohlii* is a subtropical and warm temperate species naturally growing at 600-1800 m elevation
- The species grows on sun-exposed hill slopes and plateaus in arid and semi-arid climatic zones
- The species inhabits loamy or sandy-loam soil with 6.4 to 7.9 pH
- The species is herbaceous with annual stems and perennial rootstock
- Populations from different ecogeographic areas vary greatly in their morphology
- Growth habit is strikingly variable being erect, procumbent or decumbent
- Shoot height varies from 15.0 to 120 cm
- Lamina length varies from 1.5 to 15.5 cm^2
- Inflorescence length ranges from 3 to 40 cm
- Root morphology in different populations is fascinatingly diverse, being tuberous, semi-tuberous or fibrous
- Fresh root yield in different populations ranges from 1 to 500 g per plant
- Forskolin content in roots varied from 0.07 to 0.58% of dry matter within and between the populations

As the crop is vegetatively propagated, success obtained in induced colchiploidy and inter-varietal hybrids involving both tuberous and non-tuberous accessions could be exploited to advantage as donor parents in inter-varietal hybridisation. Induced autotetraploids can be produced using colchicine in different accessions also in inter-varietal hybrids. Significant increase in forskolin content is recorded both in induced autotetraploids and hybrids. Attempt can also be made for improvement of biomass yield and forskolin content through mutation. Attempts to obtain high yielding plants of *C. forskohlii* by somaclonal variation or UV mutation can also be made.

Dioscorea spp.

Yam (*Dioscorea* spp.) is a multi-species, polyploid and vegetatively propagated tuber crop that is cultivated widely in the tropics and subtropics. Over 90% of world yam production occurs in the yam belt of West and Central Africa with Nigeria alone accounting for about 68 percent of the world's total. Of the 600 known species of yams, only six are edible or food yams. *Dioscorea rotundata* and *D. cayenensis* (both known as Guinea yam) are the most popular and economically important yams in West and Central Africa where they are indigenous, while *D. alata* (referred to as water or greater yam) is the most widely distributed species globally. Guinea yams were domesticated in West Africa, while water yam is believed to be a true cultigen that might have been domesticated in Indo-China from *Dioscorea hamiltoni* and *D. persimilis*. In West and Central Africa where Guinea yams were domesticated about 7000 years ago, farmers have selected genotypes that best suit their needs, and thus have generated a large number of traditional cultivars.

Breeding: International and national yam breeding programs and activities have been set up to develop high-yielding improved varieties. The yam breeding program at IITA has evolved from being yield focused to one that considers pest and disease resistance, techno-functional characteristics, nutritional value, secondary metabolites, and to lesser extent, market demands. The approach focuses on the selection of germplasm-targeted agroecologies to develop varieties for broad adaptability; the breeding scheme has mainly been a mass phenotypic recurrent selection, using conventional propagation system (tuber set) and breeders' evaluations to recommend germplasm to national programs. The breeding scheme consisted of hybridization, seedling nursery, first clonal evaluation, preliminary yield trial, advanced yield trial, and uniformity yield trial as key phases to recommend varieties to the national programs lasting between 9 -11 years.

In spite of recent changes introduced to the breeding scheme, the breeding cycle still takes five years for the two species.

Year 1. Synchronization of male and female flowering period for inter- and intraspecific hybridization using planting dates. Monoecious and sexed diallel crosses to determine GCA and SCA; Open pollination and biparental populations to generate multi-trait populations

Year 2. Using one-node vine cuttings, in two months old seedlings, the time to start preliminary trials has been reduced from 4 to 1 year.

Year 3. Preliminary trials using mini-tubers as seed set generated from the F1C1 multiplication in vertical sacks.

Year 4. Family replicated trials in three environments, using full and half sib families, each family / 3 groups – 3 block, data recorded for each clone; in these trials, selection can be done within each block, reps allow for a statistical phenotypic selection, replicated information for each family is obtained and selection index (SI) is applied, using the relative values of the parents that generated the families, as an estimation of breeding value (additive genetic effect).

Year 5. Advanced Trials for the target traits, where data validate results of FRT, and progenies are selected and recommended to national programs.

New technologies to advance yam breeding: The information along with phenotyping and genotyping data of the breeding collections is being integrated to assess genetic diversity and identification of marker(s) associated with targeted traits for marker-assisted selections or genomic selection. The next-generation sequencing-based genotyping is used to understand the genetic diversity of *D. rotundata* and its wild relatives. The advent of molecular markers, genome studies, and plant genetic transformation has opened avenues for circumventing breeding obstacles in long growth cycle and heterozygous crops like yam. Flow cytometry is increasingly used to determine or screen the ploidy level in this genus. Diploid, tetraploid and octoploids individuals have been identified in genus.

Eucalyptus spp.

The eucalyptus tree is originated from Tasmania, Australia and other Indo-Malaysian islands.There are approximately 700 species of eucalyptus, all with great environmental value; 37 of these species are of interest for the forest industry while only 15 are used for commercial purpose.The most important of these species are *Eucalyptus grandis*, *E. urophylla*,*E. pellita*, *E. globulus*, *E. nitens*, *E. dunni*, *E. camaldulensis*, *E. tereticornis* and *E. saligna*, and the hybrids *E. urophylla x E. grandis*, *E. camaldulensis x E. grandis* and *E. globulus x E. nitens*. Some species of Eucalyptus produce a single stem from the time of germination, while others are characterised in their juvenile stage by several horizontal and/or oblique shoots, one of which will later become the major vertical stem. In most species, the leaves of adult trees hang vertically, while those of juvenile trees are near horizontal. Morphological characters that are used to differentiate Eucalyptus species, and the subspecies within a species, include the structure of the flower (number of capsules per umbel, size of the capsule, number of ribs per capsule), and the size and shapes of leaves. The inflorescences of Eucalyptus species are formed in the axils of leaves. Most species have a simple inflorescence, but some possess a compound inflorescence which may be either lateral or terminal. The flowers of most Eucalyptus species have functional male and female structures. Eucalyptus species are usually capable of self-fertilisation, but in most cases breed by a combination of self-pollination and outcrossing, with the latter being more common. Outbreeding rates are relatively high and are maintained by protandry.

Propagation: Most eucalypt species can be artificially propagated by rooted cuttings provided the cuttings are taken from young seedlings. However, natural asexual means of reproduction such by the use of rhizomes and stolons, and the ability of tissues to give rise to small plantlets (natural vegetative propagation), is extremely rare. Only in isolated instances, amongst tropical woody species such as *E. porrecta*, *E. ptychocarpa*, and *E. jacobsiana*, as well as *E. moluccana* and *E. stellulata* which are found in more temperate climates, have rhizomes and/or stolons been observed.

Breeding: Breeding programmes in Eucalyptus have focused on the crossing of relevant elite lines containing desirable traits. The open pollination of flowers (i.e. pollination through natural mechanisms) is often used to produce hybrid seed, but controlled (hand) pollination is also practised. Open pollination is obviously easy, but suffers from the disadvantage that it will usually result in the presence of undesired self-pollinated individuals. Controlled pollinations have traditionally involved multiple visits to flowers, emasculation, and wounding or cutting of the stigma or style to enhance its receptivity to germination of pollen and the formation of pollen tubes. Such techniques are both time-consuming and costly, but so-called one-stop pollination, involving a single visit to a flower, has been developed for some species. At least for *E. globulus*, it is possible to obtain successful fertilisation by the pollination of immature styles prior to flower dehiscence, and also by cutting the style but not concurrently utilising emasculation. This technique has now been optimised for several eucalypt species and given the term artificially induced protogyny; it has led to a significant advance in the ability to generate large quantities of seeds from controlled crosses.

The development of techniques for the clonal (asexual) propagation of plants has meant that plantations in countries such as Brazil and India now consist largely of clonal plantations, the sowing of seeds having been largely abandoned. Clonal propagation can be conducted by the use of rooted stem cuttings, but this technique has proven unsuitable for a number of commercially important species, either because of the difficulty in obtaining roots or the large number of rooted plants which have developmental problems. However, cuttings taken from cotyledons, shoot apices and axillary sprouts are much easier to manage. The mini-cutting technique, now widely adopted in Brazil and some other countries, currently represents the most efficient way to clonally propagate eucalypts. Experiments in Africa with terminal shoots of *E. urophylla* x *E. grandis* have demonstrated that root development is superior if they come from juvenile trees (at least during the dry season), but shoots derived from the re-growth of coppiced trees root equally well regardless of the age of the felled tree.

Clones of Eucalyptus can also be generated by somatic embryogenesis, a technique which allows the formation of embryos from somatic or haploid cells, avoiding the need for gamete fusion. Somatic embryogenesis has been used to propagate *E. citriodora*, *E. dunnii*, *E. grandis*, *E. nitens* and *E. globulus* at the experimental level. For *E. globulus*, a protocol for somatic embryogenesis has been developed that may enable the industrial production of such tissue although no operationally viable protocol exists. In that plant it is apparent that induction of somatic embryogenesis is under additive genetic effects, in particular the so-called general combining ability effect. Importantly, micro-propagated tissues and embryos are amenable to protocols for the transformation of genes, such as those using *Agrobacterium tumefaciens*.

In Brazil, trees with desirable traits, such as above average rates of growth, are selected in screenings of individuals, and cloned by using one of these techniques. However, at least in some cases, further enhancement of populations through the selection of individuals from already improved trees is proving difficult, and techniques such as artificial hybridisation may be useful. This latter technique involves the controlled crossing of individuals and the field evaluation of the progeny.

Modern molecular techniques, such as measuring the expression levels of genes known to influence a trait, may prove useful in selecting the individual trees for clonal propagation. Even for trees which are the product of natural or traditional breeding, such techniques can be employed to measure the success and/or levels of outcrossing. In order to compare the expression of genes between species, especially when employing the polymerase chain reaction (PCR), a reference gene is needed for normalisation. Genes suitable for this role have been identified for *E. globulus*, and should prove useful in other species.

Genetic linkage maps, based on RFLP, RAPD and AFLP molecular markers, have been constructed for species such as *E. grandis*, *E. urophylla* and *E. globulus*. A microsatellite map, covering at least 90% of the genome of Eucalyptus and containing over 230 mapped loci, has also been published which was recently significantly expanded by using much higher throughput marker technologies. Microsatellite markers have been widely used for genotyping species of Eucalyptus, especially those in the subgenus Symphyomyrtus. These latter markers were based on the sequences of existing ESTs, and it is expected that they will be useful in differentiating individuals and become part of work which necessitates clone fingerprinting and the testing of parentage. Future marker assisted breeding programmes will likely use a range of different molecular markers, including those arising from high throughput techniques such as Diversity Arrays Technology or genotyping by sequencing.

Constitutive overexpression and controlled stress induction of C-repeat binding factor (CBF) genes in *E. grandisx E. urophylla* has resulted in the isolation of "freeze tolerant" plants. In a different approach, cold stress tolerance in *E. saligna* has been addressed by the transformation into plants of the "1 -pyrroline-5-carboxylate synthetase gene (P5CS, coding for a key enzyme in proline biosynthesis) from *Vigna aconitifolia*. Transformation of Eucalyptus has been achieved both through biolistics and the use of *Agrobacterium tumefaciens*. However, Agrobacterium mediated transformation was used to generate all of the genetically modified plants. Other reports pertaining more generally to Agrobacterium transformation relate to *E. camaldulensis*, *E. globulus* ,*E. grandis* x *E. urophylla, E. occidentalis* and *E. tereticornis*. Usually explants of shoots, leaves and cotyledons are used for Agrobacterium transformation.

Geranium

Common geraniums are actually members of the genus *Pelargonium,* while members of the genus *Geranium* include native wildflowers and herbaceous perennials.

Common Garden Geraniums or Zonal Geraniums : *Pelargonium* x *hortorum* is the common geranium. They are compact in habit and often have fancy leaves marked by distinct bands or zones of darker pigments, tricolored leaves or leaves with silver or white markings. Flowers may be single or double, are clustered into heads, and range from red, pink, salmon and white.

Martha Washington Geranium : *Pelargonium* x *domesticum* is grown marketed by florists during the winter as a flowering pot plant. It is not heat tolerant, and will not perform as well outdoors as common geranium.

Ivy-Leaved Geraniums : *Pelargonium peltatum* is the ivy-leaved geranium. It has a vine-like growth habit with smooth, leathery leaves and flowers with narrower petals and less dense flower heads. It is commonly seen in Europe used in window boxes, and is attractive in hanging baskets where it may grow to 3 feet or more in length.

Scented-Leaved Geraniums : *Pelargonium graveolens* is the Rose scented geranium; *P. crispum* is the Lemon-scented geranium; *P.* x *fragrans* is the Nutmeg geranium; *P. odoratissimum* is the Apple-scented geranium; and *P.tomentosum* is the Peppermint geranium. Plants have a wide range of foliage types and habits; and are used for making potpourris, sachets, and to flavour teas. They make excellent houseplants, and many have soft, finely textured foliage which is pleasing even though the flowers are small and not attractive.

Mosquito Geraniums : A new plant developed by implanting genes of one species into another is the "Mosquito Plant," a scented *Pelargonium* species with introduced genes which code for production of oil of citronella. The citronella fragrance is released when the leaves are rubbed or crushed. The "Mosquito Geranium" is known as a natural mosquito repellent.

Varieties: 'Dynamo' Series, 'Elite' Series, 'Multibloom' Series, Orbit' Series, 'Pinto' Series, 'Ringo 2000' Series, 'Maverick Star', Orange Appeal, 'Tetra Scarlet', 'Freckles',

Evening Glow', 'Julia', 'Lucille', and 'Medallion Dark Red', : 'Angel', 'Cotton', Candy', 'Gypsy', 'Lollipop', 'Melody', 'Morning', Mist', 'North Star', 'Peaches', and 'Sassy Dark Red'; 'CIM Pawan'.

Reproductive biology: The zygormorphic flowers are borne in a compound umbel. Flowers, measuring 3 to 6 cm across, are perfect and complete. The flower bud is usually enclosed by five sepals. The androecium of regals consists of five to seven stamens with filaments united at the base. Both genotype and environmental conditions can affect pollen fertility. Some genotypes consistently produce sterile stamens, while others may produce sterile stamens when temperatures are too warm or irradiance is very low.

Breeding: The hooded-leaf pelargonium hybridizes freely in the wild with the birch-leaf pelargonium (*P.betulinum*) and the large-flowered pelargonium (*P. saniculaefolium*). The latter resembles another large-flowered pelargonium (*P.grandiflorum*), and they may be synonymous. It is reported that the large-flowered pelargonium was used in hybridization with the hooded-leaf pelargonium early in the 19th century. The large-flowered pelargonium was introduced to England in 1794, nearly a century after the hooded-leaf pelargonium had been introduced. Other species that may have contributed to the genetic composition of regal cultivars are the rose-scented pelargonium (*P. capitatum*), the heart leaved pelargonium (*P. cordifolium*), and the celandine-leaved pelargonium ; the apricot geranium (*P. scabrum*), the rose geranium (*P.graveolens*) and *P. angulosum*). *Pelargonium angulosum*, sometimes referred to as the maple-leaf geranium (*P. acerifolium*) has more recently been assigned as *P. cucullatum* subspecies *strigifolium*.Orange geranium (*P. citrosum*), the lemon geranium (*P. crispum*) and the oak-leaved geranium (*P. quercifolium*) were used in development of the Romany, Royal and Tudor regals in the late 1900s in England. Classification (1993), the section *Pelargonium*, includes the birch-leaf pelargonium, the rose-scented geranium, the lemon-scented geranium, the heartleaved pelargonium, the hooded-leaf geranium, and the rose geranium. The large-flowered geranium (syn. *P. saniculaefolium*) belongs to the section Glaucophyllum and the celandine leaved geranium belongs to the section *Polyactium*. The designated lectotype for the genus *Pelargonium* is the hoodedleaf pelargonium. In a review of interspecific crossability in *Pelargonium*, species in the section *Pelargonium* were crossed with species in *Polyactium*, *Hoarea*, and *Glaucophyllum* resulting in viable seeds. While the basic chromosome number for the section *Pelargonium* is x = 11, variable ploidy levels for species contributing to the development of regals. Specifically, It is reported 2n = 36 for the apricot geranium, 2n = 22 for the celandine-leaved pelargonium, 2n = 72 for the rose geranium, and 2n = 54 for the rose scented pelargonium. Another report says 2n = 22 (apricot geranium), 88(rose geranium), and 66 (rose-scented geranium), 2n = 22 for birchleaved pelargonium, heart-leaved pelargonium, and *P. cucullatum* subsp. *strigifolium* and 2n = 22 for the large-flowered pelargonium. Given the above chromosome counts, it is concluded that the basic chromosome number in the section *Pelargonium* is x = 11, designating plants having 2n = 22 as diploids, those having 2n = 66, hexaploids, and those having 2n = 88, octoploids.

The interspecific hybridization between species in the development of regals has contributed to the complex cytogenetics of regal cultivars. Regals are believed to be polyploids of complex parentage. Reported chromosome numbers for regal cultivars vary from 2n = 43 and 44 , 45 , to 17, 18, and 35 and 36, 42, and 45. Thus, aneuploidy is not uncommon among regal cultivars.

Early hybridization efforts: German hybridizers modified the existing small-flowered type regals into the large flowered types which also had an extended flowering period. Regal cultivars developed in Germany were introduced to California in the 1920s. Many of these cultivars performed well and were leading cultivars of the time, including 'Grossmama Fischer' (translation: 'Grandma Fischer'), 'Fruhlingzauber' (translation: 'Spring Magic'), and 'Ostergruss' (translation: 'Easter Greeting'). Sales of Faiss' 'Marie Vogel' exceeded other regal cultivars, in all countries up to the early 1950s. From the cross between 'Marie Vogel' and 'Beverly Fabretti' the cultivar 'Grand Slam', was developed from which the sport 'Lavender Grand Slam' originated.

Current hybridization efforts: One major production limitation is the cool temperature requirement for flower induction and development. Other limitations that must be overcome for successful greenhouse production include ease of rooting, predictable flowering response, whitefly resistance, shipping ability, and postproduction quality.

Currently two hybridization programs exist in the United States that are addressing these problems: Oglevee, Ltd. of Connellsville, Pa., and Pennsylvania State (Penn State) University pelargonium breeding program at University Park. Efforts to improve horticultural qualities of regals began at Penn State under the direction of Richard Craig in 1977. The first patented cultivars released from this program belong to the Elegance series: 'Allure', 'Crystal', 'Fantasy', 'Flair', 'Majestic', and 'Splendor'. The most important criteria in the development of these cultivars were predictable flowering response and improved postproduction quality. Temperatures for floral initiation were maintained at a minimum of 14 °C to achieve the desired flowering response in the development of new genotypes. New additions to the Elegance series have been developed both at the Penn State and at Oglevee, Ltd.: 'Tiara', 'Brilliance', 'Dandy', 'Symphony', 'Fascination', 'Enchantment', 'Sandra', 'Erin', 'Lois', 'Debutante', and 'Rapture'. David Lemon, currently with Oglevee, Ltd., first began working on regals in 1986. Many of the cultivars he developed are included in the Maiden series: 'Maiden Lilac', 'Maiden Orange', 'Maiden Rose Pink', and 'Maiden Petticoat' are grown and marketed by Oglevee, Ltd. and were released beginning in 1996. Two additional cultivars in the Maiden series were originally sold as Vavra cultivars: 'Maiden Sunrise' and 'Maiden Red'. The Vavra cultivars were originally developed by Wolfgang Kirmann in Austria and included the cultivars 'Josy', 'Dolly', 'Mary', 'Peggy', 'Macy', 'Honey', and 'Shirley'. In general, Vavra cultivars are difficult to root, require night temperatures around 16°C for floral initiation and are less heat tolerant than other modern cultivars. For these reasons many Vavra cultivars have been discontinued for sale in the United States, but are still available in Europe. Oglevee, Ltd. also markets the Royalty series that were originally developed by Ernest Walters, a hobbyist regal breeder in England. Walters' primary goal was to develop regals with extended flowering periods. He accomplished that, along with dwarf habit, natural free-branching habit, and inflorescences held high above the foliage. While many of the original Royalty cultivars have been superseded in the United States by newer cultivars, the traits they exhibited have been integrated into newer cultivars. Of the original Royalty series, only the cultivar

Baroness is still available through Oglevee, Ltd. David Lemon has added 'Imperial', 'Excalibur', 'Emperor', and 'Monarch' cultivars to the Royalty series, and the former Vavra cultivar 'Empress' is also now included in the Royalty series. 'Camelot' is a new addition to the Royalty series that was developed at Pennsylvania State University. Both Richard Craig at Penn State and David Lemon at Oglevee, Ltd. continue to develop new cultivars of regals. Objectives for genetic improvement include resistance to petal abscission, elimination of cool temperature and high irradiance requirements for flowering, and greenhouse whitefly (*Trialeurodes vaporariorum*) resistance. Other goals include improved performance under adverse environmental conditions including heat and humidity tolerance, and improved petal tolerance to rain. Root establishment problems can be circumvented with somatic embryogenesis. Efforts to develop this alternate propagation method with regals have been advanced by the research. Micropropagation techniques have been and propagation by callus-derived protoplasts was studied andin *vitro* culture was investigated. One or several of these techniques may replace cutting propagation as the preferred method of production in the future. Ovule culture of *P.×domesticum* hybrids have been reported.

It has been reported on successful *Agrobacterium*-mediated transformation of the regal cultivar 'Dubonnet'. The goal of their study was to transfer a range of flavonoid and plant morphology genes. Transgenic plants of rose-scented geranium (*Pelargonium graveolens* cv. Hemanti) have been produced from *Agrobacterium rhizogenes* (strains A4 and LBA9402) mediated hairy root cultures. Amongst the explants tested, leaves were most responsive followed by the petioles and internodal segments, respectively.

Gloriosa superba

Gloriosa superba L. (Colchicaceae) also known as Malabar glory lily or "kembang telang" (Java, Indonesia) is a perennial tuberous climbing herb, extensively scattered in the tropical and sub-tropical parts of the India, including the foothills of Himalayas. It is called as 'Mauve beauty', 'Purple Prince', 'Modest', 'Orange Gem', 'Salman glow' and 'Orange glow'. There are several associated species of *Gloriosa* including *G. superba*, *G. simplex*, *G. grandiflora*, *G. lutea*, *G. plantii*, *G. lipolidii*, *G. longifolia*, *G. rothschildiana*, *G. virescens*, *G. sudanica*, etc. These species are distributed mainly in Africa. Besides many hybrids popular garden cultivars like the purple prince, African chief, orange gem, orange gift, lavender lady, and orange ball are grown for ample range of flowers and color combination of the perianth, which bloom for long during the autumn. *Gloriosa superba* L. is an important medicinal plant of Asia and Africa. It is used in diseases, like cancer, gout, scrofula and act as antipyretic, antihelmintic, purgative and antiabortive. *Gloriosa superba* usually multiply by corm and seeds but due to low germination capability it restricts for the regeneration. Therefore, in order to safeguard and preserve this important plant biotechnological approaches would be very useful. It is reported that 92% of the cultures of apical and axillary buds of young sprout from naturally grown *G. superba* plants regenerate four shoots per culture in MS basal medium fortified with 1.5 mg/L BA + 0.5 mg/L NAA.

Breeding: The genetic variability also is low owing to the continued vegetative propagation through tubers which has reduced the vigour, tolerance to biotic and abiotic stress causing low yields. The growing demand for the seeds of *G. superba* in the international market and the wider popularity it has gained among the farmers necessitates attempts to induce new variability with high yield, high colchicine content, dwarf stature and leaf blight resistant of the plant as well. Traditional or conventional breeding has not been attempted so far as there is only one ecotype under cultivation and genetic wealth is limited. Introduction of new variability is the only option for the breeders at present to create new variability for selection of high yielding cultivars. The genetic architecture of *G. superba* was 2n=22 and that of *G. rothschildiana* was 2n=66.The failure in the interspecific crosses may be due to the failure of fertilized ovules to develop into mature seeds. The causative factors include the presence of lethal genes and failure or early breakdown of the endosperm. Arrest of normal development of embryo and endosperm may also result from unfavorable interaction between the embryo and surrounding ovular tissues. Often, embryo breakdown is associated with proliferation of the nucellar or integumentary cells. Failure of endosperm leading to death of potentially viable hybrid embryos is the common result of causes between closely related species. This may also be due to post-fertilization barriers, in which hybrid embryos obtained after fertilization degenerate during their development, can sometimes be overcome by embryo rescue techniques such as embryo or ovule culture.

Mutation breeding is envisaged as a complementary to conventional breeding due to the existence of polyploidy, crossing barriers, poor seed set and germination. Hence induced mutagenesis may be one of the accepted tools to improve the economic traits of the crop plants especially in vegetatively propagated crops like *G. superba*.

An investigation was carried out to induce mutants in *Gloriosa superba* L. possessing high value alkaloid *viz*., colchicine and colchicoside. VM2 generation was raised from EMS, DES and gamma ray treated VM1 tubers of Glory lily. The mutants had apparent morphological changes in plant height, number of leaves and flowers as well yield characters. Five high yielding mutants were subjected to colchicine estimation using High Performance Liquid Chromatography. These promising mutants seemed to possess high colchicine content, than the control. The maximum colchicine content of 0.707 per cent was observed in 2.00 per cent EMS, followed by 0.702 per cent in 1.00 per cent DES.

Lemon grass

Lemongrass is an aromatic tropical grass cultivated primarily for culinary use. There are two of the 55 species in the genus *Cymbopogon* that are used almost interchangeably as lemongrass. The so called East-Indian lemongrass (*C. flexuosus*) (also known as cochin or Malabar grass) is native to India, Sri Lanka, Burma and Thailand, while West-Indian lemongrass (*C. citratus*), is of a Malaysian origin and is more typically used for cooking.

This perennial herb is commercially grown in the tropical and subtropical regions, but can also be grown as an annual in cooler climates. The plants grow in dense clumps up to 1.8 to 2m in diameter, with leaves up to 90 cm to 1m long. They require warm, humid conditions in full sun to thrive. The coarse, strap-like leaves are 1.25 to 2.5cm wide, finely serrated on the edges (rough enough to cut skin), tapered at the end and have gracefully drooping tips. Plants will flower, with a branched cluster of stalked flowers. They are easily propagated by roots or plant division. Cultivated varieties available are Sugandhi, Pragati, Praman, RRL-16, CKP-25, RRL-39, Kaveri, Krishna, SD-68, GRL-1.

Breeding: Although breeding work has been mainly restricted to selections from natural populations, gains through selection have been substantial and a few number of 6 varieties have been released. The common breeding methods are phenotypic recurrent selection, clonal selection and hybridization.

The variability and association of genetic parameters of lemongrass (*Cymbopogon flexuosus*) studies indicate that the genetic differences are highly significant for oil, citral, geraniol and geranyl acetate contents and herb and oil yield. High heritability estimates in broad sense are observed for all the characters. High heritability with low genetic advance recorded for oil and geranyl acetate contents indicated that these characters are under non-additive gene action. Oil yield was positively and significantly associated with herb yield and oil content. Maximum direct effect of herb yield on oil content and oil yield was also observed.

Cymbopogon species display wide variation in morphological attributes and essential oil composition at inter- and intraspecific levels. Although, morphological traits can be used to assess genetic diversity they are strongly affected by environment conditions and show little variation at the intra-specific level. Therefore, molecular markers are often used to obtain more reliable estimates of genetic diversity. Amongst them, RAPDs have been widely employed for the identification of cultivars, ancestors and discerning the genomic diversity in *Cymbopogons*. However, RAPDs have some limitations as molecular markers, the most important being poor reproducibility of the profiles across the laboratories. Simple sequence repeats (SSR) markers have features that make them particularly suitable for assessing the genetic diversity among cultivated lines or germplasm including that they are co-dominant, multiallelic, locus specific and highly reproducible. Comparative genetic analyses have shown that different plant species share orthologous genes for similar function and have highly conserved gene content and gene order. Since expressed sequence tag-simple sequence repeats (EST-SSRs) are derived from the transcribed portion of the genome, they are expected to be more conserved compared to anonymous/random sequences . Therefore, EST-SSRs possess considerable potential not only for genome mapping within species but also for comparative studies between species. Moreover, EST-SSR markers are known to provide better estimates of functional diversity existent in a gene pool. Hence, it would be of interest to explore the feasibility of using EST-SSRs from related model grass species like rice for the genomic analyses of less studied grass species such as *Cymbopogon* aromatic grasses.

Mint

Mints are aromatic perennial herbs with quadrangular stem and bearing leaves with essential oil present in glands located in the sub-cuticular region. Mints are a group of plants belonging to the family Labiatae (Lamiaceae) and the genus *Mentha*. Four major mints are recognized in this genus.

1. *Mentha arvensis* L. Japanese mint
2. *M. piperita* L. Pepper mint
3. *M. spicata* L. Spear mint
4. *M. citrata* Ehrh. Bergamot mint or Lemon mint

Spearmint is derived from a range of species, but the two most important ones are native spearmint (*Mentha spicata* L.) and Scotch spearmint (*M.* x *gracilis* Sole). *M.*

spicata is an origin species. *Mentha* x gracilis is a hybrid derived from *M. arvensis* and *M. spicata*. The main components of spearmint are carvone and limonene. Peppermint, *M.* x *piperita* L., which is tetraploid (2n = 72), is a sterile natural hybrid of *M. aquatica* L. (2n = 96) and *M. spicata* (2n = 48) and produces typical peppermint cyclic monoterpenes, menthol and menthone. Spearmint, *M. spicata*, is either tetraploid (2n = 48) or triploid (2n = 36) and produces monoterpene carvone as the major oil component.

Breeding: the genus *Mentha* centers around two primary constituents and four species, all of hybrid, polyploidy origin. Menthol is the primary constituent of the essential oil of peppermint (*Mentha* × *piperita* L. 'Mitcham' and derived cultivars) and Chinese corn mint or Japanese peppermint (*Mentha canadensis* L.). Carvone is the primary constituent of the essential oil of Scotch spearmint (*Mentha* ×*gracilis*Sole 'Scotch') and Native or "American" spearmint (*Mentha* × *villosonervata* Opiz/*M. spicata* L.). 'Mitcham' peppermint arose in England as a hybrid of *M. aquatica* L. × *M. spicata*, prior to the 18th century. The plant has 2*n*=72 and is extremely sterile. Chinese cornmint is a naturally occurring hybrid of *M. arvensis* L. × *M. longifolia* (L.) L. It has 2*n*=96 and is fertile, but gynodioecious.

Scotch spearmint arose in Scotland as a hybrid of *M. arvensis* × *M.spicata* prior to the 18th century, has 2*n*=84, and is almost completely sterile. Native Spearmint arose as a hybrid of *M. spicata* × *M. longifolia* or as a self of *M. spicata* prior to the 18th century, has 2*n*=36, and is almost completely sterile.

In *Mentha*, an example of transgressive segregation is the origin of 'Mitcham' peppermint. *M.spicata*: The phenomenon of cytomixis is observed in leptotene to pachytene states of the first meiotic prophase. The migration of nuclear material involved all of the chromosomes or part of the chromosomes of the donor cell. The occurrence of PMCs [pollen mother cells] with chromosome numbers deviating from the tetraploid number (*n*=48), derived from the chromosome numbers deviating from the tetraploid number (*n*=48), derived from the process of cytomixis indicated the possibility of aneuploid and polyploidy gamete production.

Controlled hybridization in the field resulted in the release of *M. canadensis* 'Himalaya' and 'Kosi'. The natural causes of cytomixis are postulated to be: (1) genes, especially male-sterile genes, altered by environmental factors (pollution, fungal infection, and othes); (2) abnormal formation of the cell wall during premeiotic division; and/or (3) the microenvironment of the anthers. However, cytomixis can also be artificially induced by: (1) colchicine; (2) MMS (methylmethane sulfonate), EMS (ethyl methane sulfonate), rotenone, sodium azide, Trifluralin, and others, and/or (3) γ- irradiation.

Plant tissue culture has great potential to introduce genetic variability in peppermint genotypes through somaclonal variants, somatic hybrids or transgenic plants. However, a prerequisite to applied plant biotechnology is the development of suitable and reproducible plant regeneration system. Due to the sterility of many mint species, conventional breeding methods are often difficult. Thus, *in vitro* techniques such as protoplast culture or protoplast fusion are expected to provide a new possibility for increasing genetic variability and a means to transfer desirable traits to plants. Several hybridizations by protoplast fusion were carried out between two species of mints. Diseases resistance hybrid plants are one example of *in vitro* techniques used for mint. Somatic hybrid plants of 'Black Mitcham' and *M. spicata* cv. native spearmint were obtained after protoplast fusion. Hybrid plants exhibited an intermediate level of *Verticillium* resistance but oil profile resembles that of spearmint, a high level of carvone and almost no menthone or menthol.

Another example of hybridization in mint plants is protoplasts isolated from leaf mesophyll cells of *M. piperita* 'Blackmint' which were fused with those of a garden variety mint called 'gingermint' (*M. gentilis* L. cv. variegata) by electrofusion method. Genetic transformation of *Mentha* has been achieved using *Agrobacterium tumefaciens*-meditated DNA delivery. The stable integration of GUS (β-glucuronidase) and NPTII (antibiotic resistance) genes in *M. arvensis*and *M. spicata* has been achieved. It is reported on transformation of the limonene synthase gene into peppermint. Essential oil profiles of transgenic plants showed a high menthone, low menthol, high menthofuran and pulegone content in comparison to typical mid-west peppermint. An efficient protocol for plant regeneration from protoplast of peppermint 'Mitcham Digne 38', 'Mitcham Ribecourt 19' and 'Todd's' was developed by stepwise optimization of first cell division, microcalli formation and shoot differentiation.

In mutation breeding, γ- irradiation resulted in in the formal release of two verticillium-wilt resistant clones, 'Todd Mitcham' and 'Murray Mitcham.' Additional clones currently recognized by the Mint Industry Research Council (MIRC) include M-83-7, B-90-9, and 'Roberts Mitcham,' of which all were essentially derived from 'Mitcham' through mutation breeding.

***Ocimum* spp.**

Ocimum L., is a versatile aromatic genus (family- Lamiaceae) well known for medicinal properties and also for economically important essential oils. The genus is very variable and possesses wide range of intra- and inter-specific genetic diversity. Species like *O. sanctum* L. (tulsi), *O. gratissimum* (Ram Tulsi), *O. canum* (Dulal Tulsi), *O.basilicum* (Ban Tulsi), *O. kilimandscharicum* Guerke (Kapoor Tulsi), *O. camphora* and *O. micranthum* are examples of known important species of the genus which grow in different parts of the world and are known to have medicinal properties. The plants are generally herbs, shrubs or under importance shrubs, annuals or perennials in habit. They possess glandular hairs or sessile glands secreting strongly scented volatile oils. Flowers appear to be uniform in the appearance throughout the group but they are of great taxonomic importance for the demarcation of species. The seeds contain edible oils and a drying oil similar to linseed oil.

Clocimum is a new hybrid strain of *Ocimum gratissimum* which is a source of clove type oil, rich in eugenol. Improved strain of clocimum had 70-75 % eugenol compared to 45-60 % in the original races. *Ocimum* is highly cross pollinated. Therefore, a certain amount of heterozygosity is essential for vigour, high oil yield and high eugenol content. These characters appear to be controlled by polygenes due to additive action.

***Ocimum sanctum* L. (syn. *O. tenuiflorum* L.):** This is a perennial with woody root-stock; branchlets purplish. Leaves are ovate or ovate-oblong, elliptic-oblong, distantly serrate, crenate or entire, acute, pubescent on both sides; minutely gland dotted. Flowers are 4.5 to 5cm. long, purplish or crimson in racemed. Popular varieties are Sacred basil, Purple sacred basil.

***Ocimum canum* Sims. (syn. *O. americanum* L.):** It is an erect very pubescent undeshrub. Leaves are elliptic or elliptic-lanceolate, glandular, glabrate on maturity. Flowers are white or pale purple. Popular varieties are Lime basil, Lemon basil.

***Ocimum gratissimum* L.:** This is a shrub, 1.2 to 2.5m tall, much branched, woody below. Leaves are 5 to 10cm. Racemes are strict, slender; whorls rather close-set; pedicels shorter and bracts longer than the calyx. Popular varieties are East Indian basil.

***Ocimum basilicum* L.:** The plant is erect, herbaceous, glabrous, pubescent. Leaves are ovate toothed or entire. Bracts are petiolate. Flowers are white pinkish or purplish. Popular varieties are Asian basil, Bush Green basil, Spicy Globe basil, Green Globe basil, Green Gate basil, Mammoth lettuce leaf, Purple delight basil, Sweet basil, Sweet Salad basil, Blue spice basil, Thai basil.

Breeding: The genus *Ocimum* is characterized by great variability among its constituent species, including morphology, growth habit, the color of flowers, leaves stems and chemical composition.In modern field of crop improvement programme, genetic studies have attained significant importance in evolving varieties studied to various environmental conditions. In such studies, heritability estimates of quantitative characters play an important role in expressing the reliability of variance value as a selection guideline to the plant breeder during the succeeding generations.

The large variability occurred for various morphological characters are leaf area, number of racemes /plant, number of flowers/raceme, plant height, 1000 seeds weight and days to maturity and natural select plays an important role in biological diversity in *Ocimum sanctum.*

A long term cross breeding programme is required to combine both downy mildew resistance and basil of the "Genovese" type which is needed by the consumers. Selection tests are conducted in the greenhouse on infested soil in order to improve resistance up to five generations. This work has resulted in a registered variety resistant to Fusarium wilt ('Nufar'). All 'Nufar' individuals and F_1 individuals obtained from the cross of 'Nufar' × susceptible source plants were resistant to Fusarium wilt. In further studies it is found that this resistance is a monogenic dominant trait. At present, fresh market varieties resistant to Fusarium, with good yields and performance, as well as acceptable leaf-size, shape and shelf life, are being developed and tested for their aroma.

An ever green, cold tolerant with spreading habit recombinant developed between *Ocimum basilicum* and *O. kilimandscharicum* – G 34; one high linalool content in oil G 39 and two high oil (1.2%), 105.92g/plot, 0.95% oil content and 95.20g/plot oil yield with methyl chavicol chemotypes – G 14 (methyl chavicol=55.00%), G 25 (methyl chavicol=89.75%). Based on the AMMI model, these chemotypes showed the widest adaptability and stability due to its ability to tolerate wide environmental conditions over years. DNA markers have been demonstrated to serve as efficient tools in identifying basil species due to their characteristics which include abundance, highly polymorphism and freedom from environmental influence. Various types of DNA markers are applied in basil genetic diversity studies including AFLP, inter-simple sequence repeat (ISSR) markers.Recently, sequence-related amplified polymorphism(SRAP)markers have been demonstrated to be reliable, effective and cheap in several studies.

Palmarosa (*Cymbopogon Martinii* Wats)

Cymbopogon martinii var motia yields an essential oil with high geraniol content (60-90%) which is also called as East Indian Geranium. It has dark green leaves, which are leathery, prominently mid ribbed, roundish at the base and form an obtuse to right angle with the stem. It is a diploid (2n=20) and the oil is rich in geraniol, geranyl acetate and linalool. Breeding behaviour of palmarosa studied in self, cross and open pollinations. The hermaphrodite flowers constituted only 25 to 35% of the florets in an inflorescence and the rest are staminate florets. Pollen fertility varied from 84 to 96% in staminate and hermaphrodite florets. The seed set is 5% in self, 50% in cross, and 64% in open pollinations. Few pollen tubes are noticed to traverse inside stigma and penetrate through the stylar region under self-pollination. Low seed set on setting may be due to some self-incompatibility system in palmarosa.

Breeding: Production of quality planting material for obtaining quality products enriched with main chemical compounds of health care and industrial values is priority of crop improvement using existing medicinal and aromatic plants with minimum input to reduce the cost of the produce and more acceptability and safety of the products. Palmarosa is one of the essential oil bearing hardy, rainfed and drought tolerate aromatic grass which can be grown in marginal and agricultural lands with varied types of soil including saline soil and wide climatic conditions, requires less irrigation and capital input. It has also been reported to have no or minimum incidence of insect-pests and diseases during the cultivation in field. This study revealed that the amount of the essential oil and geraniol in palmarosa increases with maturity of the plants especially in inflorescence, while that of geranyl acetate decreases. The amount of the geraniol was found to be higher in the leaves at all the developing stages compare with the inflorescence but the essential oil yield was higher in inflorescence than leaves. Fully matured spikelets with brown inflorescence are found to be the best stage for harvesting to obtain maximum essential oil with geraniol. Some of the promising clone (DCM2, DCM4, DCM9, DCM13 and DCM14), performed better in terms of essential oil and geraniol contents compare with check, could be useful for the commercial cultivation. Further the new clones obtained will be evaluated in different soil conditions and agro-ecological zones of India to assess their performance and suitability in regions. It will aid to identify some new commercial varieties for production of palmarosa oil with immense commercial values. These superior clones will be further utilized to develop smart palmarosa varieties resilient to stress through genetic improvement.

Conventional breeding techniques include selection in open pollinated seedling plants, synthetic population breeding, mass selection and composite population breeding and popular varieties evolved through these breeding methods are PRC-1, Trishna, Tripta, Vaishnavi, Cim-Harsha. Besides, polycross breeding and mutation, as well as the modern biotechnological approaches of plant tissue culture, somatic embryo regeneration, cell suspension culture, and genetic engineering. Recent advances in tissue culture propagation, cytogenetic stability, stabilization of useful somaclonal variants as well as induction of tolerance to adverse soil conditions (e.g. salt tolerance) show a lot of promise for improvement in palmarosa crop.

Molecular markers provide extensive polymorphism at DNA level and can be used for differentiating closely related genotypes and also to find out the extent of genetic diversity. DNA based markers such as Random Amplified Polymorphic DNA (RAPD) and Amplified Fragment Length Polymorphism (AFLP) have been widely used in a number of plant groups for variety of purposes such as cultivar identification, diversity studies, parentage determination, developing breeding programmes and conservation strategies.

Papaver somniferum

Opium poppy belongs to the family papaveraceae and opioids are an important class of compounds produced by opium poppy which are used in medicine as a pain reliever. These opioids interact with the opioid receptor present in the central nervous system and gastro-intestinal tract. However, several of these medicinal compounds can be made synthetically but alkaloids belonging to various groups viz., Phenanthredene (Morphine, Codeine, Thebaine), the true Benzylisoquinilone (Papaverine) and Phthalideisoquinilone (Narcotine) are only obtained from opium which place opium poppy at the highest place among the diverse array of medicinal plants. The most important and potent alkaloid is morphine which can be used for both short term as well as long term pain control, is widely used in many prescriptions of pain medications.

Breeding: Generation mean analysis of cross NB-5x58/1 and its reciprocal cross was carried out to understand the nature of gene action in opium poppy. The significance of A, B, C and D scaling tests indicated presence of non-allelic interaction in the inheritance of traits except capsule size and husk yield/plant for reciprocal cross. Additive as well as dominance components of gene action were found in both the crosses. Most of the traits had greater non fixable dominance 'h' and dominance x dominance effects 'l' than fixable additive (d) and additive x additive effects (i) except leaves/plant, branches/plant, capsules/plant, stem diameter, capsule weight/plant, husk yield/plant, opium yield/plant, codeine and narcotine content which showed greater importance of additive (d) and additive x additive effects (i) effects. Inter-mating of the best parents, diallel selective mating or biparental mating in early segregating generations followed by recurrent selections were suggested for genetic improvement of opium poppy. Based on numerical and graphic analyses of 6 x 6 diallel cross progenies (n (2) = 36) over two environments, dominance was found to be most prevalent in parents for latex yield, and morphine, codeine, thebaine, and narcotine contents. However, the additive component was also significant for the former three traits. Dominant alleles mostly with positive effects were, by and large, asymmetrically distributed in parents except for narcotine content. Heritability estimates were moderate and corresponded with low genetic advance estimates for morphine, codeine, and thebaine contents. Ample genetic diversity among parents was reflected for all traits except narcotine content. On an overall basis, heterosis breeding is suggested to be a rewarding proposition for genetic improvement of the opium poppy.

Breeding Objectives

- Breeding for modified opium yield, seed yield and specific alkaloid variety
- Breeding for disease resistant variety, causal organism and their management
- Breeding for growing conditions

The scientists are now trying to develop varieties with specific alkaloid in opium latex through conventional and molecular techniques. In recent days, scientists are working to develop low morphine or morphine less varieties to check its illegal cultivation.

The development of low morphine or morphine less varieties can also help Narcotics Department, as it will not require issuing license for growing opium poppy to the cultivators. Both conventional and molecular approaches are being applied aiming at this target, fortunately a variety "Sujata" has been developed by Central Institute of Medicinal and Aromatic Plants, Lucknow.

The development of multiple disease resistant varieties is in requirement from very long time in opium poppy. A number of diseases occur which ruins the entire crop and ultimately the opium products. Several researchers especially plant breeders have faced many challenges during specific breeding objectives due to severe disease in opium poppy. Scientists have put their best possible efforts and continuously trying to develop such varieties resistant to major diseases through molecular and conventional tools.

One of the foremost steps in the genetical improvement of any crop through conventional breeding program is to study the genetic diversity available in the introduced plant/crop material. To conduct any breeding program judiciously, diversity analysis based on morphological and biochemical traits is prerequisite. In opium poppy, several of the exotic collections at different research institutes have been evaluated for genetic diversity.

A breeding programme focused to develop improved varieties requires knowledge about the genetic variability that exists for the concern trait. It is documented that sufficient variation for composition and content of secondary metabolites occurs in a number of medicinal plant. Several studies have been carried out in opium poppy to study the existing variability in different set of materials which showed varying results for composition of secondary metabolites and other chemical compounds along with morphological variations.

Experimental evidences have shown that F8 genotypes obtained through interspecific cross between *Papaver somniferum* and *Papaver setigerum* had higher oil (>40%) and fatty acid concentration than respective parental species. They also obtained varying results for linoleic (68%-74.4%) and oleic acid (13.6%-20.3%) content in F8 genotypes. High oleic desaturation ratio and C18 polyunsaturated fatty acid with very low linolenic (18:3) acid (0.37%) indicated the possibility of using poppy oil for edible purposes. However, oleic (18:1) acid was not correlated with other fatty acids, except for significant negative correlation with linoleic (C18:2) acid. Correlation and path coefficient analysis for qualitative and quantitative traits in four poppy cultivars indicate positive and significant correlation of morphine yield with morphine content, seed yield, capsule yield, oil yield; capsule yield with oil yield; seed yield with capsule yield, oil yield were noticed. Through path analysis, it was found that morphine content, capsule yield, seed yield and oil yield had positive direct effect on morphine yield.

Mutation breeding is carried out using physical and chemical mutagens to develop non-narcotic opium poppy from narcotic crop. Two families containing twenty latex less/opium-less and twelve partial latex bearing plants in M1 generation which gave similar observations in M2 generations also isolated. The best mutant genotype, LL-34 of family C1-Comb-113-2 with 5.66 g seeds/capsule had 52.6% oil was designated as cv. 'Sujata'. This is the world's first opiumless and alkaloid free seed poppy cultivar, offers a cheap and permanent (fundamental) solution to the global problem of opium-linked social abuse. Simultaneously, it serves as a food grade crop with proteinacious seeds along with healthy unsaturated seed oil. Experiments conducted on induced mutation through gamma rays, EMS and their combined doses in two varieties of opium poppy (NBRI-1 and NBRI-5) to create new genetic variability for isolation of high yielding genotypes along with specific alkaloids.

It has also been found a variant plant of opium poppy (*Papaver somniferum* L.) having high thebaine content. Besides, it was also tried to broaden the genetic variability and to evaluate the advance generations for different agronomic and chemotypic traits in the experimental high yielding varieties i.e. NBRI-1 and NBRI-5 through induced mutations. A mutant variety known as 'TOP 1' ('thebaine oripavine poppy 1') in opium poppy (*Papaversomniferum*) was developed by Tasmania Company. In this mutant, the morphinan pathway is blocked at thebaine results in absence of codeine and morphine.

In Opium poppy, tetraploidy is confirmed with chromosomal and stomatal studies. Differential gene expression of the diploids and autotetraploids which led to the elucidation of dosage regulated gene expression leading significant enhancement in morphine content in tetraploid plants. Their study in auto-tetraploids opens avenues towards the development of hexaploids and amphidiploids which can give multifold increase in specific alkaloids. This study also opens a new vista towards understanding of ploidy level changes in term of phenotypic, genetic and genomic and a better understanding of the complex mechanism involved in polyploidization.

Varieties Developed by National Botanical Research Institute, Lucknow

BROP-1—In this variety ,the plants are medium sized having 3-4 capsules/plant and capsules are of three types viz. oily, parrot coloured and black peduncle. Flowers are white. It is a synthetic variety stabilized after hybridization/intermating between three high yielding cultivars viz. kali dandi (black peduncle), suga pankhi (parrot color) and sufaid dandi (white peduncle) followed by selection. The average opium yield, seed yield and morphine content are up to 54kg/ha, 1000-1200kg/ha and 13% respectively. Geographically ,it can be cultivated mainly in Northern Indian plains.

NBRI-1—This variety is developed through selection. The plants are medium tall having large fringed leaves and white flowers. The average opium yield, seed yield and morphine content are up to 52kg/ha, 1000kg/ha and 12-13% respectively. Geographically it can be cultivated mainly in Northern Central India.

NBRI-2—This variety has intermediate tall plants, thick stem, broad leaves, long peduncle with big capsules and flowers are white. This variety is also developed through selection amongst local collection having above characters. Average opium yield, seed yield and morphine content are up to 52 kg/ha, 1200kg/ha and 15%respectively. Northern Central part of India is recommended for its cultivation.

NBRI-6—In this variety, plants are medium tall with narrow leaves and white flowers. It is developed by hybridization between two germplasm lines BR007 and BR008 (BR007 x BR008) followed by rigorous selection generation after generation up to eight generations. Average opium yield, seed yield and morphine content are up to 55kg/ha, 1200kg/ha and 13-14% respectively. Geographically, Northern Indian plains are mainly recommended for its cultivation.

NBRI-9—In this variety, plants are intermediate sized with white flowers and large capsules. The variety is high yielding (seed yield) and is developed by hybridization between germplasm lines S-10 x S-18 followed by rigorous selection until the variety is stabilized (up to eight generations). It's average opium yield, seed yield and morphine content is upto 52kg/ha, 1400kg/ha and 12% respectively. For the cultivation of the variety, Northern Indian Plains are recommended as most suitable.

NBRI-10—The plants in this variety are medium tall, having dark green leaves and white flowers. The development of variety was done through hybridization germplasm lines (IC-30 x S-10) followed by rigorous selection up to eight generations. Average opium yield, seed yield and morphine content are up to 50kg/ha, 1200kg/ha and 12% respectively. Geographically, Northern Central plains are recommended for its cultivation.

Madakini—It is a high yielding variety for opium poppy, have multiple disease resistance and is granted US patent no.7,442,854B2 in 2009. The variety is developed by hybridization germplasm lines (BR007 x BR008) followed by rigorous selection. Plants of the variety are vigorous having dark green leaves, white flowers with blackish flowering stalk at the bottom of capsule at maturity. Average opium yield, seed yield and morphine content are up to 64kg/ha, 1200kg/ha and 15% respectively. Northern Central India is recommended for its cultivation.

Varieties Developed by Central Institute of Medicinal and Aromatic Plants, Lucknow

Rakshit—It is a disease resistant and morphine rich variety in CPS (concentrated poppy straw). The plants are 106-112 cm tall with 20-26 cm long green peduncle and oblong capsules with waxy surface. The variety is developed by hybridization and selection generation after generation up to eight generations. Average seed yield and straw yield of the variety are up to 1200-1400kg/ha and 900-1100kg/ha respectively.

Sanchita—In this high yielding variety, plants are 107 cm tall and have 2-3 capsules/plant. The average seed yield and straw yield are 840kg/ha and 640kg/ha respectively. Morphine content of this variety is very low in CPS (approx. 0.74%).

Vivek—The plants of this variety are 112 cm tall with 2-3 capsules/plant. It is also a high yielding variety having seed yield and straw yield up to 840kg/ha and 760kg/ha respectively. Morphine content of this variety is also very low in husk (approx. 0.73%).

Sweta—It is high yielding variety with 66.5kg/ha opium yield and about 18% morphine content in latex.

Subhra—In this variety plants are medium sized having 3-4 capsules/plant. The average seed yield and husk yield of the variety are approx. 910kg/ha and 790kg/ha respectively. The morphine percentage in husk is approx. 0.77%.

Shyama—In this variety plants are 105 cm tall with black peduncle and also has erect incised leaves. It is a high yielding variety having seed yield, husk yield and morphine content up to 720kg/ha, 650kg/ha and 0.75% ,respectively.

Sujata—In this variety plants are 80-100 cm tall having 3-4 flat glabrous capsules with 18-20 cm long erratic black peduncle. It is an opium less, alkaloid less and non - narcotic variety.

Narendra Dev University of Agriculture and Technology, Faizabad has developed a downy mildew resistant variety by selection and named ;Kirtiman' (NOP-4). The plants of this variety are quite tall having white flowers and 1-2 oval capsules. Opium yield, seed yield and morphine content of the variety ranges 35-46kg/ha, 900-1100kg/ha and up to 12% respectively. Eastern U.P. region is best suited for its cultivation.

National Bureau of Plant Genetic Resources, New Delhi developed a variety 'Trishna' (IC 42) for resistance to frost, root rot and downy mildew through inbreeding and selection. The plants of the variety are tall with 5-7 capsules/plant and pink flowers. Opium yield, seed yield and morphine content ranges up to 49-53kg/ha, 1000kg/ha and 12-14.78%, respectively.

Rajasthan Agricultural University, Udaipur has developed a resistant variety to disease and lodging and named it as **'Chetak' (UO 285)**. The plants of this variety are average tall with big capsules. Flowers are white with smooth petals. Opium yield, seed yield and morphine content ranges up to 54kg/ha, 1000-1200kg/ha and 12% respectively. Geographically Rajasthan is most suitable for its cultivation.

Jawaharlal Nehru Krishi Vishwavidyalaya, College of Agriculture, Jabalpur (M.P.) has also succeeded in developing a downy mildew resistant variety by pure line selection and named as 'Jawahar Aphim 16' (JA-16). Plants of this variety are tall having white flowers and 1-3 big capsules/plant. The variety has opium yield, seed yield and morphine content up to 45-54kg/ha, 900-1000kg/ha and 12% respectively. Madhya Pradesh is geographically recommended for its cultivation.

Pogostemon patchouli

Pogostemon cablin Benth. syn. *Pogostemon patchouli* Pellet., a native of South East Asia is a hardy perennial herb adapted to hot humid climatic conditions. The plant is erect, branched and pubescent. The leaves are long petioled, ovate to obtuse, acute to acuminate, crenate, simply and doubly toothed to incised and the base of the leaf is cuneate. The commercially important patchouli oil, whose major constituents are sesquiterpenes is carried in the external glandular trichomes and specialized internal cells of the leaf. The flowers are borne on slender pubescent to tomentose panicled spikes. The floral whorls when present are sub-globose and many, densely flowered and placed distinctly on the infloresence. The flowers are sessile, complete, zygomorphic, bisexual and hypogynous.

The height of the plant, inter-node length, petiole length and number of leaves per plant are the parameters considered for the morphological characterization of patchouli. The Chlorophyll Stability Index (CSI) is the measure of a plant's ability to survive stress and it is considered a study parameter in the present investigation. The lower the CSI, the higher is the plant's ability to tolerate stress.

Breeding: A systematic study of genetic diversity and improvement in propagation methods can help substantially in reconstruction of the genetic base of patchouli. Somaclones can be selected randomly from a recommended variety and multiplied through stem cuttings and evaluated in SC_2 and SC_3 generations to study the extent of somaclonal variation generated for plant height, herb yield, essential oil content, essential oil yield, and seven constituents of the essential oil. Significant or highly significant somaclonal variation is found for plant height, herb yield, essential oil content, essential oil yield, and contents of patchouli alcohol, α-guaiene, α, δ-patchoulene, and α-bulnesene in the essential oil. A high positive correlation is observed between essential oil yield and herb yield suggesting that selection for herb yield would be effective in improving essential oil yield. Patchouli alcohol content in the essential oil is negatively correlated with all the studied traits. Somaclonal variation, heritabilities of traits, and inter-trait correlations are reported in patchouli.Common molecular markers used for detection of genetic diversity are RAPD (Random Amplified Polymorphism DNA) and ISSR (Inter Simple Sequence Repeat). The genetic diversity of patchouli is estimated by using PCR-RAPD and PCR-ISSR. Patchouli does not flower , thus it is propagated using stem cutting or *in vitro* multiplication. The diversity may have resulted from a long adaptation time to climatic condition and cultivation system. Patchouli plant with good adaptation to environment can be used as genetic source in patchouli breeding. Genetic improvement of patchouli requires wide variation of germplasm which can be obtained through induced mutation or somaclonal variation.

Phyllanthus amarus

Phyllanthus is one of the very important and ancient medicinal plants cultivated for its highly priced fruits and other parts. The genus *Phyllanthus* L. (Phyllanthaceae, formerly Euphorbiaceae) consists of about 800 species which are distributed in a wide range of habitats. *Phyllanthus* are considered to be very effective and rich in biochemical compounds used in health care, food and cosmetic industry. According to unani system of medicine, fruits are useful for tubercular ulcers, wounds, sores, scabies and ringworms and are anti-inflammatory. *P. amarus* has been extensively used in pharmacological research due to its anti-Hepatitis-B and antioxidant properties.

Breeding: One of the main objectives of *Phyllanthus* species identification is to use its specific genetic material for breeding programmes. The genetic distance between varieties can be accessed through morphological characterization and genetic markers. The characterized material then helps the plants breeders to select the genotypes to be utilized in hybridization programme.

The extent of genetic diversity has been investigated in *P. amarus*, *P. debilis,* and *P. virgatus* using RAPD (random amplified polymorphic DNA) and ISSR (Intersimple sequence repeats) markers, and an average polymorphism of 68.2% and 69.7%, respectively, is observed. Genetic diversity analysis using RAPD markers, within *P. amarus* collected from different geographical locations in India, showed that the accessions from the southern part of India have high intrapopulation variation. This variation may be attributed to the cross-pollination mechanisms in populations and also because they grow as weeds without much anthropogenic intervention. Isozymes also have been used to assess the genetic variability in south Indian populations of *P. amarus* to identify superior genotypes for improving drug quality and for formulating strategies for *in situ* conservation and sustainable utilization.

Species-specific SCAR (sequence characterized amplified regions) markers have been developed for identification of *Phyllanthus* species (*P. amarus, P. fraternus, P. debilis,* and *P. urinaria*) used in dry leaf bulk trade. Amongst RAPD, SCAR and AFLP, AFLP is a better polymorphic marker.

The development of a somoclonal variant and isozyme marker for *Phyllanthus amarus* Schum & Thonn can be done using inter-nodal segments and the enzyme peroxidase. Maximum callus proliferation is obtained on Murashige and Skoog's medium supplemented (MS) with 1.0 mg/l of 2,4-Dichlorophenoxyacetic acid. Three weeks-old pale yellowish white semi-friable callus is used for organogenesis; the maximum percentage of multiple shoot formation (85%) is achieved after 4 weeks when callus is cultured on Murashige and Skoog's medium fortified with 1.0 mg/l o 6-Benzylaminopurine. The multiple shootlet formations are also achieved in the presence of the same concentration. The maximum formation of rootlets is observed on MS medium augmented with 1.0 mg/l of Kinetin and 0.5 mg/l of Naphthalene acetic acid. The banding pattern and phytochemical constituent differences are observed between mother plants, directly regenerated via nodal segments, calli, and calli mediated plants. The calli mediated somoclonal variation is confirmed through isozyme (peroxidase) and phytochemical analysis. The isoperoxidase banding profile shows a difference in calli and calli mediated plants. The phytochemical study confirms the presence of more alkaloids, saponins, tannins and others from calli and calli mediated shoots and roots. Hence the isozyme banding patterns can be used as molecular markers in future plant breedings or genetic improvement programmes.

Phyllanthus amarus Schum & Thonn. is rich in various pharmacologically active compounds such as phyllanthin, hypophyllanthin, gallic acid, catechin, and nirurin, a flavone glycoside. A genetic transformation method using *Agrobacterium tumefaciens* is developed for this plant species for the first time and transgenic plants can be produced at 50% success rate. Shoot tips of full grown plants are used as explants for *Agrobacterium*-mediated transformation. Transgenic plants can be obtained by co-cultivation of shoot tips explants and *A. tumefaciens* strain LBA4404 containing the pCAMBIA 2301 plasmid harboring neomycin phosphotransferase II (NPT II) and α-glucuronidase encoding (GUS) genes in the T-DNA region in the presence of 200 ìM acetosyringone. Integration of the NPT II gene into the genome of transgenic plants is verified by PCR and Southern blot analyses. Expression of the NPT II gene is confirmed by RT-PCR analysis.

Varietal development: The present invention relates to a cultivar of *Phyllanthus amarus* 'CIM-Jeevan', producing high amount of herb, phyllanthin and hypophyllanthin, wherein above cultivar is developed through γ-irradiation of superior gemplasm, the said plant produces high amount of herbage yield ranging between 1.0-1.15 kg per sqm fresh total plant herb, possesses high vegetative erect growth with a height ranging between 60 to 65 cm, produces phyllanthin ranging between 0.70-0.77% in dry herb, produces hypophyllanthin ranging between 0.32-0.37% in dry herb, and shows seed germination of about 90%.

Plantago ovata (Isabgol)

The genus Plantago of family Plantaginaceae includes about 275 species. They are small, annual or perennial herbs with various medicinal properties. The word Plantago is taken from the "planta" which means "sole of the foot". *Plantago ovata* is the only cultivated and economically important species of the monotypic genus Plantago and yields Isabgol- a very effective laxative and its export earns India foreign exchange worth INR 2.5 billion.The seeds are rich in secondary metabolites and contain mucilage, fatty oil, large quantities of albuminous matter, the pharmacologically inactive glucoside *viz.* aucubin and a plantiose sugar. The husk constitutes 25%–30% (by weight) of the total seed yield and is the most economical part.

Breeding : *Plantago ovata* Forsk. ($2n = 2x = 8$), commonly known as Isabgol and blond psyllium, is an important medicinal plant of the family, *Plantaginaceae*.

P. ovata is generally an in-breeder and is characterized with narrow genetic base, because of low chromosome number, chromosome size, chiasmata frequency, recombination index and abundance of heterochromatin in chromatin material. Various plant breeding methodologies like selection, hybridization, induced mutations, polyploidy and tissue culture have been used for genetic amelioration of this crop. The mitosis studied from feulgen stained young plant root tips and/or germinated seeds in four species of *Plantago, P. ovata Forsk., Plantago lagopus L., P. lanceolata L. and P. major L.*, revealed normal karyotypes, which were then sub-jected to idiogram preparation and morphological characterization of chromosomes. Out of 200 wild species of *Plantago ovata* some are medicinally important. These wild species are a reservoir of important genes, which if introgressed to cultivated species through marker-assisted breeding could revolutionize the production of isabgol. There are few reports on exploitation of randomly amplified polymorphic DNA (RAPD) and combined RAPD and inter-simple sequence repeat (ISSR) markers in assessing genetic diversity in isabgol. Equally, a very small number of microsatellite markers are available for other species of Plantago *viz.* *P. major*, *P. lanceolata*, *P. coronopus* and *P. intermedia.* However, no simple sequence repeat (SSR) markers are available for *P. ovata*.

The genetics and inheritance of economic characters, such as number of panicles, panicle length, seed yield, and swelling factor, are measured by diallel analysis of F_1 progenies of parents. The additive component of genetic variance is observed significant for days to flowering, plant height, branches/plant, peduncle length, panicle length, days to maturity, and swelling factor, whereas the dominance component of genetic variance is found significant for all the characters except panicles/plant. Additivity is not significant for all the characters, indicating absence of non-allelic interactions (epistasis) in controlling gene expression. Heritability in the narrow sense is recorded very high for panicle length, days to flowering, and plant height, and moderate for branches/plant, panicles/plant, days to maturity, seed yield, husk yield, and swelling factor. However, the degree of genetic improvement is only high for panicle length, seed yield and husk yield. It can be concluded that hybridizations, isolation of superior genotypes by sib selection and recurrent selection, and exploitation of hybrid vigour in specific parental-cross combinations are good strategies for isabgol crop improvement. In *Plantago ovata*, mutation breeding has been attempted on account of the narrow genetic base of the crop. Treatment with two mutagens, gamma rays and ethylmethane sulphonate, revealed that 20 kR dose of gamma rays can boost seed yield by more than 200%.

The technique for callus initiation, growth and plant regeneration from cultured hypocotyl explants of *Plantago ovata* cv. GI-2 shows best initiation and growth of callus on Murashige & Skoog's medium containing 2,4-dichlorophenoxyacetic acid (1.0 mg/l^1) and kinetin (1.0 mg/l). The callus showed maximum shoot differentiation on medium containing kinetin (4.0 mgl^{-1}) and α-naphthaleneacetic acid (0.01 mg/l). Root formation of shoots was found best on half-strength medium supplemented with 3-indolebutyric acid. The regenerated plants were successfully transferred into pots.

Varieties

GI_3: High yielding, susceptible to downy mildew, aphids, seed shattering

GI_2: Susceptible to downy mildew, aphids, seed shattering

DPO 13: Early maturing mutant

DPO14: Early maturing mutant with desirable traits, such as high seed yield, early and uniform seed maturation and high harvest index.

EC 124345: Resistant to downy mildew; line from Pakistan

***Rauwolfia serpentina* (Sarpagandha)**

Rauwolfia serpentina (L.) Benth. ex. Kurz. (2n=22), an important medicinal plant under shrub, belongs to *Apocynaceae* family. It is indigenous to India and other tropical countries of Asia and is pantropic in distribution. There are approximately 85 species in the genus *Rauwolfia* mainly found in tropical regions. *Rauwolfia serpentina*, *R. caffra*, *R. canescens*, *R. micrantha*, *R. tetraphylla* and *R.vomitoria* are some of the most important species of the genus. Reserpine content in the root of *Rauwolfia serpentina* has attracted worldwide attention for drug development. More than 50 alkaloids have been extracted from*Rauwolfia serpentina* including the therapeutically important reserpine, rescinnamine, deserpidine, ajamalacine, ajmaline, neoajmalin, serpentine, and yohimbine.

Roots are main source of drug. *Rauwolfia serpentina* is a rich source of indole alkaloids of medicinal value such as reserpine, ajmaline, ajmalicine and serpentine which are used in the treatment of circulatory disorders. It helps to reduce blood pressure by dilating blood-vessels, depresses activity of central nervous system and acts as a hypnotic. In Ayurveda, its roots and whole plants are used for the treatment of cardiovascular disorder, snake bite, rheumatism, hypertension, insanity, epilepsy and leaves are used in removal of opacities of cornea.

Seedlings of 40-50 days which have 4-6 leaves are ready for transplanting. Vegetative propagation by root or shoot cutting is advocated for raising plantations to collect the seeds as well as to quickly multiply the genetically superior clones. Large tap root s with few lateral secondary rootlets are used. Cuttings of 2.5-5.0cm length are planted in holes at the beginning of the monsoon and are completely covered with earth. About 100 kg of root cuttings are required to plant one hectare. Propagation by root stumps is done by using about 5 cm of root with a portion of stem above the collar. This method gives about 90-95% success. Tissue culture techniques for the mass propagation are highly desirable. Endophytic micro flora is a major cause of contamination for *in vitro* culture of *R.serpentina*. Hence, a proper sterilization is needed for successful culture. *In vitro* propagation has become the boon to satisfy the commercial demand for production of *Sarpagandha*. It uses shoots and leaves as explants and cultured on MS media supplemented with phytohormones 2, 4-D, IBA, BAP. It includes callus induction from leaf and stem tissues and direct regeneration from apical and nodal tissues of field grown *R. serpentina.*

Breeding: In *Rauwolfia serpentina* (L.) Benth. ex Kurz (2n = 22) two different sizes of pollen grains (82.5 ± 5 and 52.3 ± 5 μm) with viability up to 85–95 % are observed. The receptivity of stigma is found to be one day prior to anthesis between 08.45 and 09.15 a.m. Study reveals that this species is preferentially cross pollinated and is well adopted to achieve effective pollination. Reproductive success in terms of fruit set is poor due to limitation of compatible pollens. Variability in the population is the prerequisite for any successful breeding programme. The genetic diversity of *R. serpentina* could be explained as a result of abiotic (geographical) and biotic (pollination, seed dispersal etc,) factors, sampling frequency, pollen and gene flow, genetic drift, local selection pressures, obligate outcrossing, inbreeding depression and self-incompatibility.

The variety "CIM Sheel" of Sarpgandha is developed by using directed breeding efforts (chemically induced mutagenesis) having high root and alkaloid yields with defined reserpine content. Considering all the available genotypes in terms of their yield potential as well as reserpine content the proposed variety as evaluated is the most superior genotype. This variety will also provide a defined planting material to the growers and ultimately will ensure a supply of defined quality material to the industry.

Rosemary(*Rosmarinus officinalis*)

Rosemary belonging to family Lamiaceae is an evergreen shrub that has been cultivated throughout time for its lovely fragrance. It is a native of southern Europe. Rosmarinic acid (RA) is a very strong antioxidant produced in rosemary leaves and used to protect plant tissues against free radicals and lipid peroxidation caused by stress. Also it is used in medicine, food preservatives and cosmetics. Increasing use of this species is mainly due to the commercialisation of its fresh shoots as condiment and has stimulated its cultivation.

Breeding : Indiscriminate supplying of biomass from wild populations of rosemary has resulted disrupted the ecological balance of the species. The intense, recurrent and selective biomass removal may results in a regression of the species diffusion and a reduction of the biodiversity of rosemary.

It is reported on substantial genetic diversity among three Tunisian Lamiaceae species, *Rosmarinus officinalis*, *Lavandula multifida* and *Thymus algeriensis*, using alloenzyme electrophoresis. Only *R. officinalis* showed a significant relationship between population structure and geographic distance. Intraspecific populations' genetic similarity was high. A high genetic diversity due to the species mating system and the large size of populations before fragmentation within population was revealed by isozymes and essential oil variations from 14 natural populations of *Rosmarinus officinalis* L. using isozymes and terpenoids. Essential oil composition varied among populations. Genetic and chemical data were highly correlated. Shoots obtained from mother plants showed considerable phenotypical variability for leaf number, size and density.

Varieties

'Cala Gonone': This is developed by clonal selection. The plant is a shrub with a spreading habit and medium vigour. This cultivar shows a good vigour and an abundant flowering. The essential oil is particularly rich in 1,8 cineole, borneol and bornyl acetate.

'Sette Fratelli': This is developed by clonal selection. The plant is a small and scarcely vigorous shrub with a prostrate habit. Cultivar with a prostrate shape and low vigour of the plants. Essential oil is particularly rich in verbenone.

'Gerrei': This is developed byclonal selection. The plant shows a medium vigour and an upright habit. Cultivar with a very upright habit, shoots with close internodes and a great number of leaves. The composition of essential oil is particularly rich in verbenone.

'Sant.Antioco': This is developed by clonal selection. Plants show an high vigour, with spread vegetative habit. Spring shoot length was 34.9 cm and internode length was 1.51 cm. Cultivar with a very high vigour. The essential oil resulted particularly rich in camphor and it has 1,8 cineole.

'Vignola': This is developed by clonal selection. The plant is a shrub of medium vigour and an intermediate-upright habit. Cultivar characterized by shoots with short internodes and abundant leaves.

'Costa Paradiso': This is developed by clonal selection. The plant has a compact habit and a medium vigour. Essential oil is characterized by high percentages of borneol and bornyl acetate.

Stevia rebaudiana

Stevia rebaudiana (Bertoni) is a herbaceous perennial plant (2n=22) of genus Stevia Cav., which consists of approximately 230 species of herbaceous, shrub and sub-shrub plants. Leaves of stevia produce diterpene glycosides (stevioside and rebaudiosides), non-nutritive, non-toxic, high-potency sweeteners and may substitute sucrose as well as other synthetic sweetners, being 300 times sweeter than sucrose. In addition to its sweetening property, it has medicinal values and uses. Stevia is self-incompatible plant and the pollination behaviour is entomophilous. Rebaudioside-A is of particular interest among the glycosides produced in the leaves of stevia because of the most desirable flavour profile, while, stevioside is responsible for aftertaste bitterness.

Breeding: Stevia is self-incompatible and probably insect pollinated. It has been reported that the amount of selfing ranged from 0 to 0.5%, while outcrossing ranged from 0.7 to 68.7%, indicating that the self-incompatibility system is operating. Since stevia is self-incompatible, seed collected from an individual plant would represent a half-sib family.

Breeding objectives include higher leaf yield per unit area of farm, Higher leaf : stem ratio, Higher steviol glycoside content in the leaves, Higher content of specific glycosides, Rapid Growth Rate, Better photosynthetic activity, Resistance to pest and diseases, Resistance to environmental stresses and Less frequency of flowering. Development of new varieties of *S. rebaudiana* with a higher content of rebaudioside-A and a reduced content of stevioside is the primary aim of plant breeders concerned with the improvement and utilization of this source of natural sweeteners. The proportions of rebaudioside-A and -C are controlled by a single additive gene known to be co-segregating suggesting synthesis by the same enzyme. Stevioside and rebaudioside-A are negatively correlated, while rebaudioside-A and -C are positively correlated. Conventional plant breeding approaches such as mass selection, pure line and intercrossing among various desirable genotypes is the best method for improving quality traits in a highly cross-pollinated crop like stevia. Various plant types with larger amounts of specific glycoside have already been patented, such as RSIT 94-1306, RSIT 94-75, RSIT 95-166-1 through selection and inter-crossing. Other selections are Madhuguna, Yungri, SM 4. In *S. rebaudiana*, where self-incompatibility ensures a high degree of heterogeneity, recurrent selection is the most effective method of increasing foliage yield as well as total glycoside content. Composites and synthetics can be used to capture part of the available heterosis because of the high degree of natural out-crossing and the absence of an efficient system of pollination control. Synthetics and composites like "AC Black Bird" and "PTA-444" have already been developed. Polyploidy results in better adaptability of individuals and increased organ and cell sizes. Triploid plants have higher Reb. A content and larger leaves than normal diploid plants. Tetraploids have larger leaf size, thickness and have potential use in increasing biomass and yield in comparison with diploid strains. Characters of interest with low variability in the population may be improved through mutation breeding. Use of biotechnological approaches, such as tissue culture for the mass propagation of elite genotypes, anther culture for development of pure homozygous doubled haploid and molecular marker technology for identification of marker loci linked to rebaudioside-A trait, can create new opportunities for plant breeders. Understanding the mechanism and pathway of biosynthesis of steviol glycosides can help to improve the glycoside profile by up-regulation and down-regulation of genes. Plant regeneration from *in vitro* culture can be achieved by embryogenesis or organogenesis. In stevia, regeneration has been achieved by organogenesis from different explants, such as leaves, axillary shoots, stem tips, suspension cultures and anthers. Somatic embryogenesis has previously been reported from leaves and stems. Somatic embryogenesis can also be obtained from floret explants of *S. rebaudiana* cultured on MS medium supplemented with 2,4-D (9.05 and 18.10 mM) and kinetin (0 to 9.29 mM).

Vetiveria zizanioides

Vetiver (*Vetiveria zizanoides* (Linn) Nash.) or *Khus* of family Poaceae, is a perennial clumped grass, 1 to 2 m tall with erect stems and rigid leaves. The panicles are 15-30 cm long, brownish-purple in colour with whorled,long branches. The spikelets are borne in pairs. The root system of vetiver is finely structured and very strong. It has no stolons or rhizomes. Vetiver's fibrous roots grow downward, 2-4 m in depth, and are strongly scented. Vetiver is mainly cultivated for the fragrant essential oil distilled from its roots. The main chemical components of the oil are benzoic acid, vetiverol, furfurol, vetivone and vetivene. Due to its excellent fixative properties, it is used widely in perfumes. Dry roots are also used for making mats, fans, screens, pillows, baskets, incense sticks and sachet bags.

Two distinct morphological complexes of vetiver are observed to inhabit spatially separated geographic regions in India: one in the north along the Indo-gangetic plains and adjoining areas mainly in the states of Rajasthan, Madhya Pradesh, Uttar Pradesh and Bihar, and the other in the south along the east and west coasts of Indian peninsula in the states of Andhra Pradesh, Karnataka, Tamil Nadu and Kerala. The two races are distinctly different. The north Indian wild types represented by "Bharatpur type" are profuse flowering, high seed-setting having narrowleaves with vigorous roots producing low concentration superior quality laevorotatory root oil (ruh-khus or khus oil) and the south Indian "cultivated type" that are late and low flowering with high pollen sterility and non-seed-setting with widerleaves producing low quality dextrorotatory root oil (vetiver oil) resembling Java vetiver.

Breeding: Higher order of genetic diversity with respect to ecological / geographic adaptation, morphometric traits, reproductive behaviour and essential oil concentration and composition is reported in the Indian subcontinent, followed by Indonesia. In addition to Indian subcontinent, vetiver grows wild in subtropical and tropical areas of world. However, a primitive type of clone that had only male florets compared to the normal occurrence of hermaphrodite flowers has been reported from wild population in south India. Occurrence of tremendous diversity especially for reproductive features ranging from nonflowering to late-flowering and range of pollen sterility point to the possibility of India being the center of origin.Extensive studies have been done on karyomorphological and nuclear DNA estimation analysis on diverse clones of vetiver. Based on geographical distribution patterns and detailed chromosomal evolutionary parameters applied, it was concluded that south Indian peninsula is the area of primary center of origin of vetiver from where it has diverged in two directions: (i) towards the north in the warm and dry northwest and the warm and humid east-central Indian plains, and (ii) towards south-east Asia and other parts of the world. Vetiver cultivars found outside south-east Asia are supposedly of south Indian origin having non / low seed-setting characteristics.

Extensive collections made from natural habitats in north India have been subjected to genetic analysis and performance evaluation of their clonal and seed progenies and hybridization with south Indian types carried out at the Indian Agricultural. Research Institute, New Delhi, National Bureau of Plant Genetic Resources, New Delhi, Central Institute of Medicinal and Aromatic Plants, Lucknow, Indore Agriculture College, Indore, Regional Research Laboratory, Bhubaneswar, and CS Azad University of Agriculture and Technology, Kanpur, Harcourt Butler Technical Institute, Kanpur and Kerala Agricultural University's research station at Oddakali. This has led to isolation of several superior clones, north-south hybrids, artificial polyploids for high productivity of essential oil and high value perfumery notes ranging from earthy-to-rosy-to- saffron odour genotypes suitable for commercial cultivation.

Varieties

KS-1: Essential oil yield 17.8 kg/ha

Sugandha: Yields 21.2 q fresh roots/ha, 1.4% oil content and 19.7 kg ha oil yield / ha

Hybrid 8: Root yield 12-15 q/ha, 1% essential oil content and possess 70-85% vetiverol

Keshari: Saffron flavoured, oil yield 30 kg/ha

Gulabi: Has a rosy odour, tolerant to sodic soil, yields 2.8 t dry roots and 25-30 kg/ha of essential oil, can be cultivated in marginal soils and waste lands

Dharini: Possesses longer, thicker and dense roots, tolerant to sodic soil, is a good soil binder and useful for soil and water conservation, oil yield 39 kg/ha

South Indian types: Pusa Hybrid-7, Hybrid-8, CIMAPKS- 2, Sugandha, KH-8, KH-40, ODV-3

***Withania somnifera* (Ashwagandha)**

Withania somnifera belongs to the family Solanaceae, Ashwagantha is one of the important medicinal plants in India. There is an evidence of considerable morphological variation in the species and need for systematic work on the various morphological and geographical forms. The plants are generally erect, branching shrubs, up to one meter in height. All the over ground parts, especially the stem, the vein, and the calyx, are covered with a sparsely hairy tomentum. The leaves are simple, entire acute and glabrous. The leaves on vegetative shoots are alternate and large while those on floral branches are opposite, flowers are short pedicellate, gamosepalous sepals 5- parted persistent with acute linear lobes.

The plant has immense therapeutic uses in traditional and also in modern system of medicine. Ashwagantha roots are compared with ginseng roots for their restorative properties of vigour and vitality, hence the name "Indian ginseng". The whole plant, specially the leaves and root bark are abortifacient, adaptogen, antibiotic, aphrodisiac, deobstruent, diuretic, narcotic, strongly sedative and tonic. The fruit and seed are diuretic while latter are hypotonic. The roots of ashwagandha contain bioactive compounds with sedative and anti-inflammatory, anti-tumor, anti-bacterial and anti-spasmodic properties. Ashwagandha acts mainly on the reproductive and nervous system having a rejuvenative effect on the body and is used to improve vitality and aid recovery after chronic illness. The medicinal properties of the species are due to its chemical constituents like alkoloides and withanolides (steroidal lactones) primarily in roots. Witheferin A is present in the leaf. Withaferin A and Withanolide A are the key components for therapeutic uses.

Breeding: Floral structure of cultivated plants favours self-pollination because of short stigma covered with anther lobes whereas in wild type having long projected stigma inviting cross-pollination. Both the types have same type of ploidy levels. A variety of breeding systems ranging from obligate selfing to obligate out-crossing operate in different species of medicinal plants including *Withania*.

Knowledge of reproductive biology of medicinal plants is essential for improvement, effective conservation and management plans to evolve genetically superior varieties. Flowering (peak) of *Withania* takes place during April–July and anthesis occurs between 08:00 and 11:00 h. The period of stigma receptivity coincides with anther dehiscence. Fruit set on pollination treatments ranges from 90.8% (passive autogamy), 72% (assisted autogamy), 30.30% (xenogamy), and 56.50% (geitonogamy) through 50.40% (open pollination). Xenogamy brings about very low fruit set, seed-set and seed germination percentages. It is inferred that *W. ashwagandha* is predominantly an autogamous and self-compatible species. Self-compatibility is mainly accomplished due to close proximity of stigma and anthers. The reported chromosome number of *Withania somnifera* is 24, 48 and 72. The gamma rays affect the normal cytological behavior of ashwagandha species.

The effect of gamma irradiation on the seed germination, seedling survival, days to first flower, plant height, number of branches per plant, number of leaves per plant, number of berries per plant, seed yield per plant (gm) and root yield per plant (fresh weight (gm)) are gradually decreased from lower to higher doses of gamma rays treatment. Then particular dose is gradually decreased especially 20, 25 and 30KR of gamma rays. The highest germination percentage and seedling survival are observed at lower dose of gamma rays when compared to other doses. The LD50 value for 50 per cent reduction of seed germination, (LD50 lethal dosage), seedling survival is observed in 25KR of gamma rays treatments.

The chromosomal aberrations increase with increase in the doses of gamma rays to optimum level of 30KR, because it causes changes in the chromosome structure, cellular structure and metabolism of plants. The chromosome aberrations include Sticky metaphase, Precocious moment chromosome, Fragments, Anaphasic bridge, Anaphasic laggard, Telophasic laggard A dose dependent increase in the mitotic aberrations is observed in root tips of ashwagandha. It is clear from the observation that gamma rays caused more chromosomal damage in Ashwagandha and also a dose dependent reduction is observed in seed germination, seedling survival and plant height was also observed. Although gamma rays showed inhibitory effect on seed germination, plant height and seedling survival but the gamma rays have enough potential for induction of genetic variability and to exhibit quantitative and qualitative variations which can be favorably exploited by cytogenetic and plant breeders in improving the genotype of *Withania somnifera* (L.) and to improve the yield characters by cause of gene mutation.

□□□

8

CULTIVATION OF IMPORTANT MEDICINAL TREES

Trees are largest life form of the plants. There is a rich diversity of the plants over the world out of them trees registered for their maximum diversity in variable environmental condition. These are erect, branched, long lived plants that are important not only for soil conservation, environmental purification but also useful for their Medicinal and aromatic value for human beings. Tree species of Medicinal and aromatic plants are also important for providing shade to the small herbaceous, shrub nature plants during high temperature range in summer season.

Scientific name	:	*Acacia catechu* Willd.
Family	:	Mimosaceae
Common name	:	Khair

Description: It is a moderate sized deciduous tree, 9-12m tall with dark greyish or brown rough bark and hooked short spines. Leaves are bipinnately compound, leaflets 30-50 pairs, main rachis pubescent with a large conspicuous gland near the middle of the rachis. Flowers are pale yellow, sessile in peduncled axiallary spikes. Fruits are flat brown pods, shiny and with a triangular beak at the apex and narrowed at the base. Pods are 3 to 10 seeded. The gummy extract of the wood is commercially known as '*Kath*' or '*Cutch*.

Cultivation: Catechu is well adapted to hilly areas and rocky places. The plant is propagated by seeds. Seeds are soaked in water for 6 hours and sown in seedbeds. Seeds germinate within a month. At four-leaf stage, seedlings are planted in polybags. Two months old seedlings from the polybags are used for transplanting. Pits of size 50cm x 50cm x 50cm are taken at a distance of 4-5m between plants and filled with topsoil, sand and dried cowdung in 1:1:1 ratio. Seedlings are planted in these pits. Application of organic manure every year during the rainy season is beneficial. Regular weeding is required. Pruning of branches and tender shoots developing from the base of the plant is done from second year onwards. Tree are trained to as single stemmed one. Flowering and fruiting commences from fourth year onwards. At the end of tenth year, the tree are cut and heartwood collected.

Medicinal uses: Heartwood contains kaempferol, dihydro kaempferol, taxifolin, iso rhamnetin(+)- afzelchin, α dimeric procyanidin, quercetin, (-)epi-catechin, (-)catechin, fisetin, quercetagetin and (+)-cyanidanol. The main constituent of heartwood is catechin and catechu tannic acid. The tree is commonly used as a blood purifier and for leoprosy and leucoderma. Catechu or Cutch tree bark is useful in melancholia, conjunctivitis and haemoptysis. It is useful in vitiated conditions of *kapha* and *pitta*, catarrh, cough, pruritus, leprosy, leucoderma, skin diseases, helminthiasis, anorexia, diarrhoea, dysentery, foul ulcers and wounds, haemoptysis, haematemesis, haemorrhages, intermittent fever, inflammations, odontopathy, anaemia, diabetes, splenomegaly and pharyngodyna. The gummy extract of the wood (*kath*) is used in laryngopathy, flatulence, anorexia, ulcers, wounds, helminthiasis, leucoderma, leoprosy, skin diseases, urorrhea, colporrhagia, erysipelas and odontopathy. For leprosy, root, leaf, flower, bark and fruits are made into a decoction which is given orally as well as for external dressing.

Scientific name : *Acacia nilotica* Linn.

Family : Mimosaceae

Common name : Babul

Description: It is an evergreen, usually moderate-sized tree, 2.5- 25 m tall tree with a short, thick and cylindrical trunk. Leaves are alternate, bi-pinnately compound, 5-15 cm long. Flowers many, scented, yellow, borne in heads, crowded, stalkless, 6-8 mm long, composed of 5-toothed corolla 3 mm long; many yellow, threadlike stamens. Pods long, narrow, flattened, straight, mostly narrowed.

Cultivation: This is drought resistant and occurs in plain, flat or gently undulating ground and ravines. Trees grow well on alluvial soils in ravine areas subject to periodic inundation. Otherwise it can be grown on a wide variety of soils, seemingly thriving on alluvial soils, black cotton soils, heavy clay soils, and can tolerate poorer soils. It can withstand extremes of temperature (> 50°C) and conditions of drought however; adequate moisture is required for full growth and development. It is susceptible to frost when young and trees of all age classes are adversely affected by conditions of severe frost. It is fire tender and both seedlings and saplings are aversely affected by fire. The average annual rainfall varies from 250-1500 mm. Trees are propagated by seeds.

Medicinal uses: Trees are rich in phenolics, consisting of condensed tannin and phlobatannin, gallic acid, protocatechuic acid, pyrocatechol, (+) -catechin, (-) epi-gallocatechin-7-gallate and (-) epigallocatechin-5, 7-digallate.

The Zulu of South Africa use the bark for cough. The Maasai of East Africa take a bark decoction as a nerve stimulant, and the root is said to cure impotence. An astringent from the bark is useful in diarrhoea, dysentery and leprosy. Bruised leaves are poulticed onto ulcers. The gum or bark is used for cancers and tumours of the ear, eye or testicles and indurations of the liver and spleen, condylomas and excess flesh. Other ailments treated by these products include colds, congestion, fever, gallbladder, haemorrhage, haemorrhoids, leucorrhoea, ophthalmia, sclerosis and smallpox. Bark, gum, leaves and pods are used medicinally in West Africa. Sap or bark, leaves and young pods are strongly astringent because of the tannin the possess, and in Senegal are chewed as an antiscorbutic, and in Ethiopia as a lactogogue. A bark decoction is effective for intestinal pains. Other preparations are used for, gargle, toothache, ophthalmia and syphilitic ulcers. In Tonga, the root is used in tuberculosis. In Lebanon, the resin is mixed with an orange-flower infusion for typhoid convalescence. In Somalia, the wood is useful in smallpox.

Scientific name	:	*Aegle marmelos* Linn.
Family	:	Rutaceae
Common name	:	Bael

Description: It is a medium sized armed deciduous tree, 8m tall with straight sharp axillary thorns and yellowish brown shallowly furrowed corky bark. Leaves are alternate, trifoliate and aromatic; leaflets ovate or ovate-lanceolate, crenate, pellucid-punctate, the laterals sub-sessile and the terminal long petioled. Flowers are greenish-white, sweet scented, borne on axillary panicles. Fruit is globose, woody berry with golden yellow rind when ripe. Seeds are many, oblong, compressed and embedded in the orange brown sweet gummy pulp.

Cultivation: Common varieties are 'Mirzapuri', 'Kagzi Gonkda', 'Kagzi Baranasi', 'Ayodhya', 'Dhar Road' 'KB- 11', 'KB- 1', 'Narendra Bael (NB) – I'(oblong), 'Narendra Bael (NB)- 2' (round), 'NB-5', 'NB-6' (round), 'Pant Shivani' (ovoid), 'Pant Aparna' (globose). It is a hardy fruit crop and can be grown in arid, semi-arid or sub-tropical hot and humid climate upto an altitude of 1200m. It is not damaged by the temperature as low as – 8°C. It can thrive well in a wide range of soil and climate, can even tolerate strong alkaline swampy soils. Bael can be grown in well-drained sandy and clay acid or alkaline soil with a pH rage of 5-10. Hot and dry summer, mild winter and well drained sandy soil are ideal for this crop. Bael is usually propagated by seeds. Recalcitrant fresh seeds are sown in the nursery in June and seedlings became ready for transplanting 7 weeks after sowing. Patch or shield budding on seedling rootstock between June – July give better success. Air layering is successful under humid tropical climate. Budded plants are planted at a spacing of 8m x 10m whereas seedlings are at 10 m x10m during Feb to March or July to August. Bael trees are trained in modified central leader system. Pruning is done twice in a year once in May and other in August. In May, dead, diseased, weak and criss cross branches are pruned off. In August, healthy leaves are pruned for sale. A dose of 15 to 20kg FYM, 500gN, 750g P_2O_5 and 500g K_2O is to be applied per plant per year. Leguminous vegetables or forage crops. Common pests are *Phyllocnistis citrella,Aonidiella aurantii* and *Papilio demoleus*, and these are controlled by spraying with Endosulphan (0.05%). Bacetial shot hole on leaves, fruit cancer and gummosis are the serious diseases. Mature green fruits are harvested for preservation. Fruits take about 11 months to ripen. These are harvested with fruit stalk when colour changes to yellowish green. Budded plants start bearing 5 years after planting and trees yield about 100-150 fruits/ tree or 18-20 tones/ha at the age of 10years. Seedling trees require 8 years to bear fruits giving around 400-1000 fruits/ tree depending upon the cultivars. Fruits harvested at full mature stage can be stored up to 15 days.

Medicinal uses: Bael contains a number of coumarins, alkaloids, sterols and essential oils. Roots and fruits contain coumarins such as scoparone, scopoletin, umbelliferone, marmesin and skimmin. Fruits, in addition, contain xanthotoxol, imperatorin and alloimperatorin and alkaloids like aegeline and marmeline identified as N-2-hydroxy-2-[4 -(3',3'-dimethyl allyloxy) phenyl] ethyl cinnamide. β- sitosterol and its glycoside are also present in the fruits. Roots and stem barks contain a coumarin - aegelinol. Roots also contain psoralen, xanthotoxin, 6,7-dimethoxy coumarin, tembamide, mermin and skimmianine.

Every part of the tree is medicinal. The roots are used in many Ayurvedic medicines for curing diabetes and leprosy. It is an ingredient of the '*dasamoola*'. The bark is used to treat intestinal disorders. Leaves contain an alkaloid rutacin which is hypoglycaemic. '*Two leaves before breakfast*' is said to keep diabetes under control. Leaves and fruits are used in controlling diarrhoea and dysentery. Fruit pulp is used as '*shampoo*' and cooling agent. It is also a rich source of carbohydrate, protein, fat, fibre, minerals and vitamin B and C. Fruit pulp is useful in curing mouth ulcers as it is the richest natural source of riboflavin (1191 units/ 100 g). '*Bael sharbat*' is prepared by mixing the fruit pulp with sugar, water and tamarind juice, which is very useful for stomach and intestinal disorders. The rind of the fruit is used for dyeing and tanning.

Scientific name : *Alstonia scholaris* R. Br.

Family : Apocynacae

Common name : Saptaparna

Description: It is a medium-sized evergreen tree, usually 12–18 m tall, with close-set canopy. Leaves are four to seven in a whorl, and are thick, oblong, with a blunt tip. Flowers are fragrant, greenish-white or greyish-yellow in umbrella-shaped cymes. Fruits are follicles are narrowly cylindrical, fascicled, with seeds possessing brown hair.

Cultivation: The species is adapted to a variety of climatic conditions in India, ranging from dry tropical to sub-temperate. However, it grows well in areas where annual rainfall is about 100–150 cm, as it prefers a fairly moist habitat. The species prefer the red alluvial soil having proper aeration. Seeds are the best planting material for raising the crop. About 25 g of clean seeds are required/m^2 of nursery beds. When the seedlings in mother beds are about 5–10 cm high, they are picked and transferred to polybags, so that they attain a height of about 30–45 cm in about two months. Approximately, 250 g seeds in a bed of 10 m × 1 m size are required to raise seedlings for one hectare plantation. About 400 saplings are accommodated in one hectare of land at a spacing of 5 m × 5 m. Pits of size 45 cm × 45 cm × 45 cm are dug at a spacing of 5 m × 5 m or 5 m × 6 m. The pits are then refilled with equal amounts of FYM, sand, and soil, and watered so that the soil settles. The polybag seedlings are transplanted in prepared pits at a spacing of 5 m × 5 m or 5 m × 6 m in the rainy season in June–July. Pit irrigation is required in summer season at weekly intervals when planted without intercrop. In case of intercropping, flood method of irrigation is beneficial. Light irrigation at an interval of 15–20 days in winter season is sufficient. About 15–20 litres of water per pit per week is needed for proper growth of the plant. It takes about 8–10 years for the tree to reach maturity, and thereafter bark or other parts are harvested. Peeling of bark from the trees is normally done for trees that have attained a girth of 50–62 cm, in the intermittent and strip manner. It is recommended that partial debarking should be done from each tree by removing bark in strips of 15 cm × 15 cm, 30 cm apart. October–January is the best season for collecting of stem bark. It has been estimated that for getting a sustainable yield without damaging the tree, a maximum of 2.5 kg of fresh bark can be extracted per tree from 10-year-old plantations, which on drying, weighs about 625 g. Therefore, one hectare of plantation yields approximately 250 kg dry stem bark every year.

Medicinal uses: Mature bark is rich in 0.16%–0.27% total alkaloids containing mainly echitamine, ditamine, and ditaine. It is a bitter tonic, febrifuge, diuretic, anthelmintic, stimulant, carminative, stomachic, aphrodisiac, galactagogue, and haemostatic. It is used as a substitute for cinchona and quinine for the treatment of intermittent periodic fever. An infusion of bark is used in fever, dyspepsia, skin diseases, liver complaints, chronic diarrhoea, and dysentery.

Scientific name	:	*Averrhoea carambola* Linn.
Family	:	Oxalidaceae
Common name	:	Kamrakh

Description: *Averrhoa carambola* is an attractive, small, slow-growing evergreen tree with a short-trunk. The branches are drooping and the wood is white and turns reddish. It has a bushy shape with many branches producing a broad, rounded crown. The compound leaves are soft, medium-green, spirally arranged around the branches in an alternate fashion. The pinnate leaves have a single terminal leaflet and 5 to 11 nearly opposite leaflets, each leaf is 15–20 cm long, and the 3.8–9 cm long leaflets are ovate or ovate-oblong in shape. The lilac or purple-streaked, downy, flowers are produced in the axils of leaves at the end of twigs. The fruits are showy with an oblong shape, longitudinally 5- to 6-angled and 6.35–15 cm long and up to 9 cm wide. The fruits have a thin, waxy skin that is orange-yellow colored. The juicy fruits are yellow inside when ripe and have a crisp texture and when cut in cross-section are star shaped.

Cultivation: Carambola requires warm moist climate and can be grown in India on the hills up to 1,200m. A well distributed rainfall encourages normal growth and cropping. It can be grown on any type of soil with good drainage, but deep rich soil supports better plant growth. It prefers acid soil. Trees growing on calcareous soils sometimes require zinc sprays. Carambola is often propagated through seeds, but budding, grafting and layering are recommended to produce true to type plants. The seeds have low viability and hence should be sown fresh. After cleaning and drying the seeds, they should be sown in plastic bags and allowed to harden. They are then transplanted to the field. When seedlings are to serve as rootstock, they should be at least one year old, on which budding (shield) or grafting (veneer or approach) can be taken up. The planting of carambola trees on commercial scale is seldom done. Planting is done by providing 8m x 8m spacing when one plans to raise a regular orchard. It is advised to manure the trees once in a year (40-50kg of farmyard manure/tree) and also apply some nitrogenous and phosphatic fertilizers, the doses depending upon the age and size of the tree and the fertility of the soil. or better orchard establishment, operations like watering of young plants as and when required, shaking, basin cleaning and protection from cold are important and should be carried out as per need. Irrigation is given when there is a long dry spell. This helps to secure a good crop. Seedlings take about 4 years to come to bearing but grafts produce crop within 1-2 years. Carambola produces flowers and fruits even on the trunk also. The fruits are borne year round under Indian conditions, but peak yields are obtained during January–February and

September-October. Fruit yield varies according to the age, variety and plant health. About 80 kg fruits can be harvested from a full grown tree.

Medicinal uses: The fruit is found to contain proanthocyanidins, epicatechin, gallic acid in gallotannin form and L-ascorbic acid. In India, the ripe fruits or its juice are used as anti-pyretic, laxative, appetite stimulant, sialogogue, astringent and antiscorbutic. In Brazil, the fruit is recommended as diuretic in kidney and bladder related problems. In Chinese Materia Medica, it is used to quench thirst and to increase the secretion of saliva. In Ayurveda, the ripe fruit is considered as digestive, tonic and causes biliousness. the fruits are also used to treat throat inflammation, mouth ulcer, toothache, cough, asthma, hiccups, indigestion, food poisoning, colic, diarrhea, jaundice, malarial splenomegaly, hemorrhoids, skin rashes, pruritis, sunstroke and some eye related problems. They are used as aphrodisiac for both men and women. In women, the fruits can be used to increase lactation and in large doses, they can act as an emmenagogue. The crushed leaves or shoots are externally applied in the treatment of chicken-pox, ringworm and headache. A decoction of boiled leaves is used to relieve aphthous stomatitis and angina. The boiled flowers are used as vermifuge, in fever and malaria. The roots of Averrhoa carambola are used to treat arthralgia, chronic headache, epitaxis and spermatorrhea. The roots with sugar are considered as an antidote for poison. A decoction of the crushed seeds acts as a galactagogue and emmenagogue. The powdered seeds are used for asthma and colic.

Scientific name : *Azadirachta indica* A. Juss.

Family : Meliaceae

Common name : Neem

Description: Neem is a hardy medium to large, mostly evergreen tree, 20m tall and 2.5m girth. It has a short bole with wide spreading branches and glabrous twigs forming a round to oval crown. The bark is thick, dark-grey with numerous longitudinal furrows and transverse cracks. Leaves are imparipinnately compound, alternate, exstipulate and 20-38cm long. Inflorescence is long, slender, axillary or terminal panicle. Flowers are white or pale yellow, small, bisexual, pentamerous and bracteate. Fruit is one seeded drupe with woody endocarp, greenish yellow when ripe. Seeds are ellipsoid, cotyledons thick fleshy and oily.

Cultivation: Neem grows well in tropical arid regions with high temperatures, altitudes between 50m and 1000m, as little rainfall as 130mm/year and long stretches of drought. Well drained sunny hill places are ideal. It can grow on most kinds of soils including dry, stony, shallow, nutrient deficient soils with scanty vegetation, moderately saline and alkali soils, black cotton, compact clays and laterite crusts. However, silty flats, clayey depressions and land prone to inundation are not congenial for its growth. Soil pH of 5.0 to 10.0 is best. It brings surface soil to neutral pH by its leaf litter. It has extensive and deeply penetrating root system capable of extracting moisture and nutrients even from highly leached poor sandy soils. Neem is propagated easily by seed without any pre-treatment, though it can be regenerated by vegetative means like root and shoot cuttings. Seeds are collected from June to August. These remain viable for 3-5 weeks only which necessitates sowing within this short time. Seeds are depulped and soaked in water for 6 hours before sowing. Seeds are sown on nursery beds at 15x5cm spacing, covered with rotten straw and irrigated. Germination takes 15-30 days. Seedlings are transplanted after two months of growth onwards either to polybags or to mainfield. Neem is grown along with agricultural crops like groundnut, bean, millets, sorghum and wheat. It is also suitable for planting in roadsides, for afforestation of wastelands and under agroforestry system. For field planting, pits of size 50-75 cm cube are dug 5-6m apart, filled with top soil and well rotten manure, formed into a heap, and seedling is planted at the centre of the heap. FYM is applied at 10-20 kg/plant every year. Chemical fertilizers are not generally applied. Both irrigation and weeding are required during the first year for quick establishment. More than 38 insect pests are reported on Neem which may become serious at times. The important ones are seed and flower insect (*Scirtothrips dorsalis* Hood), defoliators (*Boarmia variegata* Moore and *Eurema* sp.), sap suckers (*Helopeltes antonii* Signoret and *Pulvinaria maxima* Green) , root feeders (*Hototrichia consanguinea* Blanchard), mealy bug (*Pseudococus gilbertensis*), scale insect (*Parlatoria orientalis*) and a leaf webber (*Loboschiza Koenigiana*). They can be managed by the application of 0.01-0.02% monocrotophos or dimethoate. No serious diseases are reported in Neem. A full grown tree produces about 350 kg dry leaves and 40-50 kg berries per annum. Fresh fruits give 60% dry fruits which yield 10% kernel which contains 45% fixed oil, on an average. After 10 years of growth the wood is cut and used as timber.

Medicinal uses: Leaves contain the flavanoid quercetin, nimbosterol (β-sitosterol), kaempferol and myricetin. Seed and oil contains desacetylnimbin, azadirachtin (C35H44O16), nimbidol, meliantriol, tannic acid, and amino acids. Trunk barks are rich in nimbin, nimbinin, nimbidin, nimbosterol, essential oil, tannins, margosine and desacetylnimbin. Every part of the tree, namely root, bark, wood, twig, leaf, flower, fruit, seed, kernel and oil has been in use from time immemorial in the Ayurvedic and Unani systems of medicine. *Nimbarishta, nimbadi churna* and *nimbharidra khand* are well known preparations. It is valuable as an antiseptic, used in the treatment of small pox. Small twigs are used as tooth brushes and as a prophylactic for mouth and teeth complaints. Extract from the leaves are useful for sores, eczema and skin diseases. Boiled and smashed leaves serve as excellent antiseptic. Decoction of leaves is used for purifying blood. Neem oil is used in soaps, toothpaste and as a hair tonic to kill lice. Seed is used in snake bite. Extracts of neem seed oil and bark check the activity of male reproductive cells and prevents sperm production. Neem seed oil is more effective than the bark for birth control. Neem based commercial products are also available for diabetes treatment (Nimbola, JK-22), contraceptive effect (Sensal, Nim-76) and mosquito/ insect repelling.

Scientific name : *Bauhinia variegata* Linn.

Family : *Caesalpiniaceae*

Common name : Rakta Kanchan

Description: It is a moderate sized deciduous tree with vertically cracked grey bark, wood moderately hard, greyish brown with irregular darker patches. Leaves are of two leaflets, connate for about two-thirds up. Leaflets are ovate with rounded apex, 10-15cm long, pubescent beneath when young and coriaceous. Flowers are white or pink, the uppermost petal darker and variegated usually appearing before the leaves borne in short axillary or terminal racemes. Fruits are flat dehiscent pods with 10-15 seeds.

Cultivation: Well drained hilly areas are suited for the cultivation of Bauhinia. The plant is propagated by seeds. Seeds are formed in February-March. Seeds are to be collected from the dried pods, soaked in water for 12 hours before sowing in seedbeds. At four-leaved stage they are to be transferred to polybags. Two month old seedlings from polybags are used for field planting. Pits of size 60cm x 60cm x 60 are to be taken and filled with 10kg dried cowdung mixed with topsoil and formed into a mound. On these seedlings are to be planted at a distance of 6- 7.5m. Irrigation is to be given in the first year. Two weedings and application of organic manure once is required in a year. The plant is free from attack by any serious pests and diseases. The plant flowers on the third year. At the end of tenth year the trees are cut and wood used for medicinal purposes.

Medicinal uses: Flowers are rich in flavanoids-kaempferol-3-galactoside and kaempferol-3-rhamnoglucoside. Stem bark yields hentriacontane, octacosanol and stigmasterol. Stem yieldsβ-sitosterol, lupiol and a flavanone glycoside-5, 7-dimethoxy flavanone 4-O-α-Lrhamnopyranoside- β-D-glucopyranoside. Seeds have human blood agglutinating activity. Stem bark is hypothermic, CNS active and depressant. Bud, flower, leaf and stem bark are antibacterial. Stem possesses juvenoid activity. Bark is alterative, tonic,. antileprotic and anti-rheumatic. Bud is antidysenteric. Root is carminative and antidote for snakebite. Bark, flower and root promote suppuration. Bark and bud are astringent and vermifuge.

Scientific name	:	*Butea monosperma* (Lamk.) Taub.
Family	:	Papilionaceae
Common name	:	Flame of the Forest

Description: It is a medium sized tree with compound leaves, each leaf with three leaflets; leaflets are silky tomentous and strongly veined on lower surface. Flowers are scarlet to orange colour, borne in dense, small clusters on leafless branches. The fruit is a flat pod and single seeded.

Cultivation: The tree grows well in open plains in natural habitat and is highly drought resistant. The trees are well adapted to a sunny situation on moderately moist soil. The substrate should be sandy-loamy or gritty-loamy soil. They are propagated by seeds and suckers. The natural reproduction period in plant starts early in the rainy season. Plentiful and sustained moisture is required for germination of the seed. The plant can be raised by line sowings in irrigated or unirrigated land. For cultivation, the pods may be planted 25-30 cm apart along the lines and lightly covered with soil. Alternatively, the plants are raised at 10 x 10 cm distance in the nursery by sowing the ripe seeds before the rainy season. Regular watering and weeding is carried out and the seedlings are transplanted during the rainy season after the stem is pruned to 5 cm. Clonal propagation by air layering helps in the quick establishment of plantation of this tree. It is also found that the growth of seedlings of palas is better in pure black soil. The trees pollard and coppice well and produce root suckers freely. They can also withstand heavy annual lopping. Well suited for silvopasture at wide spacing (10-15 m) in extensive tracts in central and western India, where it is managed by repeated coppicing on a roughly 5-year rotation. Coppice shoots are also cropped in intermediate years for the larger leaves.

Medicinal uses: Flowers contain flavanoids like Butein, Isobutrin, Butin and other flavonoids and their glucosides like Coreospin, Isocoreospin, Sulphuretin, Monospermoside and Isospermoside, Dihydromonospermoside, chalcones,aurones, isobutyine. Palsitrin, Histidine, aspartic acid alanine of phenyl-1-alanine, Myricyl alcohol,stearic, palmitic,arachidic and lignoceric acids, fructose,glucose. Seeds and pods contain oil, proteolytic and lipolytic enzymes, plant proteionase and polypeptidase, palsonin. α-amyrin, β-sitosterol, β-sitosterol-β-D-glucopyranoside and sucroseheneicosanoic acid δ –lactone, Palasonin, Monospermin. Bark contains Kino-tannic acid, Gallic acid, pyrocatechin, palsitrin.

Bark has anti-diabetic, antiobesity, antidiarrhoel and wound healing property, The bark is reported to possess astringent, bitter, pungent, aphrodisiac and anthemintic properties. It is useful in tumors, bleeding, piles and ulcers.

Leaves have anti-diabetic, astringent, diuretic, tonic and aphrodisiac properties. They were also used to cure boils, pimples and timorous hemorrhoids and are internally given in flatulent, colic worms and piles. Flowers have antihepatotoxic, antihyperglycemic, anti-inflammatory, antioxidant, chemopreventive and anti-cancer properties. The flowers are reported to possess astringent, sweet cooling, constipating, aphrodiasiac, haemostatic, diuretic, febrifuge and tonic. They are useful in vitiated conditions of pitta and kapha, diarrhoea, haemorrhoids, menorrhagia, fever, leprosy, skin diseases, swelling, hyperdispesia, arthritis, burning sensation, bone fracture. Seeds have antihyperglycemic and antihyperlipaemic effects, anthelmintic and anti-microbial activity. The roots are useful in elephantiasis and in curing night blindness and other defects of sight. The root bark is used as an aphrodisiac and as an analgesic and antihelmintic. It is also applied in sprue, piles, ulcers, tumors and leprosy.

Scientific name : *Cassia fistula* Linn.

Family : Caesalpiniaceae

Common name : Amaltas

Description: It is a medium sized tree, 10-20m tall with compound leaves and large, shining, dark green leaflets. Flowers are bright yellow, borne in very large, hanging bunches. Fruits are pod, 50-60cm long, black or shining dark brown, almost cylindric.

Cultivation: It grows well in tropical and subtropical climate in clay, loam, sandy, acidic and well drained soil. The plants are moderately tolerant to drought and salt. Trees will require occasionally pruning when they are young to control shape and develop a uniform crown. The tree coppices vigorously and produces root suckers freely.

Medicinal uses: The root bark contains tannin, phlobaphanes and oxy - anthraquinone. The pulp contains rhein, a volatile oil, a resinous substance as well as three wax like substances. The flesh of the fruit is used as a laxative, while the bark can be used to treat skin infections. It is used to relieve constipation, common cold, fever and intestinal disorders.

Scientific name : *Cinchona officinalis* Linn.

Family : Rubiaceae

Common name : The Quinine Plant

Description: It is an evergreen tree, 10-15m tall. Leaves are opposite, elliptical, ovate- lanceolate, entire and glabrous. Flowers are reddish-brown in short cymbiform, compound cymes, terminal and axillary. Fruit is capsule ovoid-oblong; seeds are elliptic, winged margin octraceous, crinulate-dentate.

Cultivation: The plant grows well in tropical regions having an average minimum temperature of 14°C. Mountain slopes in the humid tropical areas with well distributed annual rainfall of 1500-1950mm are suited for its cultivation. Well drained virgin and fertile forest soils with pH 4.5-6.5 are ideal for its growth. It does not tolerate waterlogging. Cinchona is propagated through seeds and vegetative means. Most of the commercial plantations are raised by seeds. Vegetative techniques such as grafting, budding and softwood cuttings are practised in countries like India, Sri Lanka, Java and Guatemala. *Cinchona succirubra* is commonly used as root stock in the case of grafting and budding. Hormonal treatment induces better rooting. Seedlings are first raised in nursery under shade. Raised seedbeds of convenient size are prepared, well decomposed compost or manure is applied, seeds are broadcasted uniformly at 2g/m^2, covered with a thin layer of sand and irrigated. Seeds germinate in 10-20 days. Seedlings are transplanted into polythene bags after 3 months. These are transplanted into the field after 1 year at 1-2m spacing. Trees are thinned after third year for extracting bark , leaving 50% of the trees at the end of the fifth year. The crop is damaged by a number of fungal diseases like damping off caused by *Rhizoctoria solani*, tip blight by *Phytophthora parasatica*, collar rot by *Sclerotiun rolfsii,* root rot by *Phytophthora cinnamomi*, *Armillaria mellea* and *Pythium vexans*. Field sanitation, seed treatment with organo-mercurial fungicide, burning of infected plant parts and spraying 1% Bordeaux mixture are recommended for the control of the diseases. Harvesting is done in one or two phases. In one case, the complete tree is uprooted, after 8-10 years when the alkaloid yield is maximum. In another case, the tree is cut about 30cm from the ground for bark after 6-7 years so that fresh sprouts come up from the stem to yield a second crop which is harvested with the under ground roots after 6-7 years. Both the stem and root are cut into convenient pieces, bark is separated, dried in shade, graded, packed and traded. Bark yield is about 9000-16000kg/ha.

Medicinal uses: Over 35 alkaloids have been isolated from the plant; the most important among them being quinine, quinidine, cinchonine and cinchonidine. These alkaloids exist mainly as salts of quinic, quinovic and cinchotannic acids. The cultivated bark contains 7-10% total alkaloids of which about 70% is quinine. Similarly, 60% of the total alkaloids of root bark is quinine. Quinine is isolated from the total alkaloids of the bark as quinine sulphate. Commercial preparations contain cinchonidine and dihydroquinine. They are effective for the treatment of malarial fever, pneumonia, influenza, cold, whooping couphs, septicaemia, typhoid, amoebic dysentery, pin worms, lumbago, sciatica, intercostal neuralgia, bronchial neuritis and internal hemorrhoids. They are also used as anesthetic and contraceptive.

Scientific name	:	*Cinnamomum zeylanicum* Bl.
Family	:	Lauraceae
Common name	:	*Cinnamon*

Description: This is an evergreen tree, 8-17 m tall in wild state, but in cultivation it is coppiced to develop as a bush. Leaf is stiff; petiole 1-2 cm long, grooved on upper surface; lamina ovate or elliptic, strongly 3-veined from the base. Leaf is reddish when young. The anthocyanin pigments present in the flushes are reported to be cyanidin glucoside, cyanidin xyloside and cyanidin galactoside. Inflorescence is axillary or terminal panicle borne at the ends of twigs with creamy white peduncle. Flowers very small, 3 mm in diameter, with foetid smell, each subtended by a small ovate hairy bract. Fruit is fleshy berry, black, 1-seeded, and ovoid.

Cultivation: Wild cinnamon trees are adapted to tropical evergreen rain forests upto 1800 m from MSL. Sandy loam soils with admixture of humus or vegetative mould is ideal for sweet and fragrant bark. Proximity to sea, humid conditions and saltish water are good for the crop. It is propagated mainly by seed and rarely by cuttings of young 3-leaved shoots, layering of shoots and by the division of old rootstocks. Seeds lose their viability very soon and therefore, should be sown fresh after the removal of the pulp. Germination takes 2-3 weeks time. Seeds are sown thickly in nurseries in May-June. When 4 months old, seedlings are transplanted into polybags or baskets. After a further 10-12 months they are planted in the main field at 2-3 m spacing after sufficient hardening.

Cattle manure or compost at 20 kg/tree/year is applied. Inorganic fertilizers are applied at 20:20:25 g N, P_2O_5 and K_2O/seedling in the first year which is gradually increased to 200:180:200 g/tree/year for grown up plants of 10 years or more. Regular weeding is practised in the early stages of growth and the seedlings are irrigated till they are established, if there is long drought period. Plants are pruned when they are 2-3 years old at a height of 15 cm above ground level. Side shoots growing from the base are cut to encourage growth of more side shoots till the whole plant assumes the shape of a bush. Leaf spot and die back diseases caused by *Colletotrichum gloeosporioides*, grey blight caused by *Pestalotiopsis palmarum* and sooty mould caused by *Phragmocapinus betle* are the common diseases of cinnamon. For the preparation of *quills*, the plants are harvested 3 years after planting when the shoots have grown 2-2.5 cm in diameter and 1.5-2 m in length. Harvesting is done in May or November. The correct time for cutting the shoots for peeling is determined by noting the sap circulation between the wood and the corky layer. Peelers can judge this by making a test cut on the stem with a sharp knife. If the bark separates readily the cutting is taken immediately in the early morning with sharp knife to prevent breaking and splitting of cut ends.

Medicinal uses: Bark oil contains cinnamic aldehyde (60-75%), eugenol (10%), benzaldehyde, methyl amyl ketone, phellandrene, pinene, cymene, nonylaldehyde, linalool, cumin aldehyde, caryophyllene and esters of butyric acid. Cinnamon is a coagulant and prevents bleeding. Cinnamon also increases the blood circulation in the uterus and advances tissue regeneration.In addition to being an antioxidant, anti-inflammatory, antidiabetic, antimicrobial, anticancer, lipid-lowering, and cardiovascular-disease-lowering compound, cinnamon is reported to have activities against neurological disorders, such as Parkinson's and Alzheimer's diseases.

Scientific name	:	*Citrus medica* Linn.
Family	:	Rutaceae
Common name	:	Citron

Description: Citron is a slow-growing dwarf tree, 2 to 5 m tall with irregular straggling branches and stiff twigs and long spines at the leaf axils. The evergreen leaves are green and lemon-scented with slightly serrate edges, ovate-lanceolate or ovate elliptic. Petioles are usually wingless or with minor wings. The clustered flowers of the acidic varieties are purplish tinted from outside, but the sweet ones are white-yellowish.The citron fruit is usually ovate or oblong, narrowing towards the stylar end.

Cultivation: In tropical regions, citron grows well at an elevation of below 1300m in full sun. The optimal temperature for growth and cultivation range between 15-27°C with a minimal temperature of 15°C in winter. Generally, growth ceases below 13°C and above 35°C. It is very sensitive to frost, intense drought and heat. The plant prefers a deep, well drained, moist, loamy soil with a pH range of 5.5 to 8.0. They are propagated by seeds or leafy cuttings from 2 to 4 years old branches or layering. A mature tree can yield 40 kg fruits per year.

Medicinal uses: The citron is used mainly for medical purposes: to combat seasickness, pulmonary troubles,intestinal ailments, scurvy and other disorders. The essential oil of the flavedo (the outermost, pigmented layer of rind) is also regarded as an antibiotic. Citron juice with wine is considered an effective antidote to poison. In the Ayurvedic system of medicine, the juice is still used in nausea, vomiting, and excessive thirst. The juice of the citron is rich in Vitamin C content and used medicinally as an anthelmintic, appetizer, tonic, in cough, rheumatism, vomiting, flatulence, haemorrhoids, skin diseases and weak eyesight.

Scientific name	:	*Diospyros melanoxylon* Roxb.
Family	:	Ebenaceae
Common name	:	Kendu

Description: Trees are 25m tall with opposite or alternate leaves and coriaceus. Flowers are mauve in colour. Fruits are olive green, ovoid or globose with soft and sweet yellow pulp. Seeds are compressed and shiny.

Cultivation: It grows well on poor denuded soils, hot and dry hill slopes, stony soils with quartzite, shale and sandstone, and also heavy clays. It however, attains best growth and development on loose, porous soils in cool and moist sheltered valleys where it tends to be gregarious. It demands light; drought and frost hardy but sensitive to waterlogging. Trees are multiplied through seedling, coppice and root suckers.

Medicinal uses: Fruits and powdered seeds are eaten. Seeds are used in the treatment of mental disorders, nervous break downs and palpitations of the heart. Fruits have a cooling and astringent effect. Dried flowers are effective in urinary, skin and blood diseases. The bark is astringent and used against diarrhoea.

Scientific name	:	*Embelia tsjeriam-cottam A. DC.*
Family	:	Myrsinaceae
Common name	:	Bhingi/ Babrang

Description: It is a small tree arranged with brownish hairy young branches and glabrous mature branches. Branches are specked with white glands. Leaves are 12cm long, narrowed at both ends, dotted all over the glands, margins simple or toothed. Flowers are minute, pale green, borne in short axillary, densely hairy clusters. Fruits are small, globose, pointed , red when ripe.

Cultivation: It grows well upto an altitude of 1750 msl in tropical humid regions. A well drained light cotton soil or sandy loam soil is ideal for for its cultivation. Plants are propagated by seeds or stem cuttings. 50 kg seeds are required for raising of 10,000 plants. Polythene bags of 12cm x 18 cm filled with sand, soil and FYM (1:1:1) is ideal for its transplanting. In main fields, pits of 30cm x 30xm x 30cm are dug out at a spacing of 2-5m, filled with FYM, sand and forest soils (1:1:1) and saplings are transplanted during June-July.

Medicinal uses: The plant contains embelin 1, α benzoquinone, and gallic acid 2, α polyphenol. Both the compounds are reported to possess antioxidant and anticancer properties. The plant is highly esteemed in Ayurvedic medicine as a powerful anthelmintic, and is also an important ingredient of a number of formulations. The fruits are anthelmintic, antispasmodic and carminative. The bark of the root is reportedly used to treat toothache.

Scientific name : *Emblica officinalis*Gaertn.
Family : Euphorbiaceae
Common name : Amla

Description: Trees are small to medium sized, 8 to 18m tall with crooked trunk and spreading branches and simple subsessile leaves. Flowers are greenish yellow. Fruits are nearly spherical, light greenish yellow in colour, smooth and hard on appearance with 6 vertical strips or furrows.

Cultivation: Aonla prefers a subtropical dry climate and can be grown upto an elevation of 1800m above sea level. These are hardy and drought resistant tree. It can tolerate a wide range of temperature starting from freezing to high temperature even upto 46°C. It can withstand water stagnation. This is an ideal plant for arid and semi-arid regions. Aonla can be cultivated in marginal soils slightly acidic to saline/acidic soil. A well drained loamy soil is ideal for amla cultivation. The pH range from 6.5 to 8.5 is suitable to the plants.

The plants are propagated from seeds. The seeds should be treated with H_2SO_4 for 3 minutes, washed in water and then soaked in 500ppm GA_3 for 24 hours for easy germination. The seeds are hard, takes longer time for germination if not treated. They are easily propagated through patch or modified ring budding in North India during mid May to September. Veneer grafting is also successful. Six months to one year old seedling raised from Desi varieties are used as rootstocks for budding or grafting programmes.Grafted or budded plants are planted 7 to 10m apart during July-August or February. Young plants need protection from severe drought during summer and water stagnation in rains. A pit mixture of 10kg FYM + 1kg neem cake + topsoil is ideal. Planting is done in square or hedge row; two varieties in alternate rows. The best combination is 'Amrit' and 'Neelum' or 'Kanchan'. Plants are allowed to develop a medium headed tree leaving 0.75cm to 1.0m above ground level and trained into a modified central leader system.

In early years, 2 to 4 branches with wide crotch angle appearing in the opposite directions should be encourged. An one year old plant removes 10kg FYM, 100g N, 50g P and 100g K annually. In sodic soils, 100-500g of B and zinc sulphate is incorporated along with fertilizer. Fertilizer application is done in two split doses equally divided, once at fruit set during April-May and another dose in September-October. The plants are irrigated at 10-15 days intervals during summer season. Basin system of irrigation is ideal for Aonla. Alternate drip irrigation with 60 CPE is appropriate with a water saving upto 40-45%. Mulching with organic materials like paddy straw, sugarcane trash and FYM is effective in salt affected soils. Being deep rooted and deciduous, aonla is suited for intercropping with fruits, vegetables and ornamentals. Following model intercropping may be followed:

(i) Aonla + Ber (ii) Aonla + guava (iii) Aonla + Ber + Phalsa (iv) Aonla + Dhaincha + Wheat and (v) Aonla + Dhaincha + Onion/ garlic/brinjal. Fruits are ready for harvest in November to December when the mature fruits turn light green to dull, yellow with red blush. Budded or grafted aonla tree start fruiting from third year onwards, whereas seedling trees take 6-8 years to start bearing. A full grown amla tree produces 1-3q fruits per tree weighing 15 to 20 tonnes/ ha.

Medicinal uses: Amla is nutritious and richest source of vitamin C, minerals and amino acids. The edible part contains protein concentration three fold and ascorbic acid concentration 160 fold as compared to that of apple. Major active components recorded are emblicanin A & B, puniglucanin, pedunculagin, ellagic acid, hexahydroxy-diphenic acid and conjugates. The fruit is pickled with salts, oil and spices and used as a primary ingredient in the Ayurvedic tonic (Chyawanprash). Fruits are used for making beverages, squash, jam, jelly and wine. It is useful in anaemia, jaundice, dyspepsia, haemorrhage, diabetes, asthma, bronchitis and insomnia. Amla increases red blood cell counts and regulates blood sugar, and it is a heart tonic. It cleanses mouth, strengthens teeth, aids eye sight and reduces worms, acidity, eye and lung inflammations, ulcerations, painful urination and internal bleeding.

Scientific name	:	*Eugenia jambolana* Lam.
Family	:	Myrtaceae
Common name	:	Jamun

Description: It is a large evergreen tree with opposite leathery, smooth and shining leaves. Flowers are small, dull white and borne in large bunches. Fruits are 1.5-4.0cm long, ovoid, purplish when young, almost black when ripe, juicy and single seeded.

Cultivation: Java Plumgrows well in warm and dry climate. It generally grows up to 1200 meter above sea level at temperature of 20 - 40°C. All kinds of soils are ideal. Well drained fertile soil with good organic matter content are best. However, *Eugenia jambolana* perform well in moist soil having good drainage. A soil pH of 6 -7.5 is best. Grafted plants or plants from large fruited varieties are planted. In light to medium soil the planting is done at 10 x 10 meter. In good fertile soil the planting is done at 12 x 12 meter. Planting is done in July- August. Pits of 0.9 meter width x 0.9 meter length x 0.9 meter deep should be taken top soil from these pits should be mixed with 50 kg farm yard manure / well decomposed cow dung manure, 1 kg single super phosphate, to ensure proper distribution of the manure. Fertilizer and pesticide soil from bottom of the pit can be used for making a bund / ring around the pit can for keeping in the water. Leaves and grass can be placed these pits as a mulch. Watering is done every 10 to 15 days intervals in winter and at weekly intervals during summer. Drip irrigation is recommended. In pre- bearing period, 20 to 25 kg well rotten farmyard manure or compost / plant / year should be applied. The grown up trees should be applied 500 g nitrogen, 600 g phosphorus, and 300 g potassium / plant / year.

Medicinal uses: The plants are rich in phenols, tannins, jambosine, triterpenoids, volatile oils. The bark, fruits and seeds are medicinal. The bark is astringent, and is useful in sore throats, bronchitis, asthma, ulcers and dysentery. It is also given for purifying blood and as a gargle. The seeds are useful in diabetes. Fruits are used in diabetes.

Scientific name	:	*Ficus spp.*
Family	:	Moraceae
Common name	:	Nalpamaram

Description

Ficus racemosa **Linn. (Cluster Fig)**

Description: It is a moderate to large-sized spreading laticiferous, deciduous tree without many prominent aerial roots. Leaves are dark green and ovate or elliptic. Fruit receptacles are 2-5cm in diameter, subglobose or pyriform arranged in large clusters on short leafless branches arising from main trunk or large branches. Figs are smooth or rarely covered with minute soft hairs. When ripe, they are orange, dull reddish or dark crimson. They have a pleasant smell resembling that of cedar apples. The bark is rusty brown with a fairly smooth and soft surface. Surface is with minute separating flakes of white tissue. Texture is homogeneously leathery.

Medicinal uses: Stem-bark gives gluanol acetate, β-sitosterol, leucocyanidin-3-O-β-D-glucopyrancoside, leucopelargonidin-3-O-?-D-glucopyranoside, leucopelargonidin - 3-O-α-L-rhamnopyranoside, lupeol, ceryl behenate, lupeol acetate and α-amyrin acetate. Roots are useful in treating dysentery. The bark is used as a wash for wounds, highly efficacious in threatened abortions and recommended in uropathy. Powdered leaves mixed with honey are given in vitiated condition of *pitta*. A decoction of the leaves is a good wash for wounds and ulcers. Tender fruits (figs) are used in vitiated conditions of *pitta*, diarrhoea, dyspepsia and haemorrhages. The latex is administered in haemorrhoids and diarrhoea. The ripe fruits are sweet, cooling and are used in haemoptysis, thirst and vomiting.

Ficus microcarpa **Linn. (Plaksah)**

Description: This is a large glabrous evergreen tree with few aerial roots. Leaves are short petioled, 5-10cm long, and apex shortly and bluntly apiculate or slightly emarginate. Main lateral nerves are not very prominent and stipules are lanceolate. Fruit receptacles are sessile and globose occurring in axillary pairs. It is yellowish when ripe without any characteristic smell. Bark is dark grey or brown with a smooth surface except for the lenticels. Outer bark is corky and crustaceous thin and firmly adherent to inner tissue. Inner bark is light and flesh coloured with firbrous texture.

Medicinal uses: The bark contains tannin, wax and saponin. *Plaksah* is considered as cooling, astringent, and curative of *raktapitta doshas*, ulcers, skin diseases, burning sensation, inflammation and oedema. It has good healing property and is used in preparation of oils and ointments for external application in the treatment of ulcers. The bark and leaves are useful in wounds, ulcers, bruises, flatulent colic, hepatopathy, diarrhoea, dysentery, diabetes, hyperdipsia, burning sensation, haemaorrhages, erysipelas, dropsy, ulcerative stomatitis, haemoptysis, psychopathy, leucorrhoea and coporrhagia

***Ficus benghalensis* Linn. (Banyan Tree)**

Description: It is a very large tree, 30m tall with widely spreading branches bearing many aerial roots functioning as prop roots. Bark is greenish white. Leaves are simple, alternate, arranged often in clusters at the ends of branches. They are stipulate, 10-20cm long and 5-12.5cm broad, broadly elliptic to ovate, entire, coriaceous, strongly 3-7 ribbed from the base. The fruit receptacles are axillary, sessile, seen in pairs globose, brick red when ripe and enclosing male, female and gall flowers. Fruits are small, crustaceous, achenes, enclosed in the common fleshy receptacles. The young bark is somewhat smooth with longitudinal and transverse row of lenticels. In older bark, the lenticels are numerous and closely spaced; outer bark easily flakes off.

Medicinal uses: The bark contains flavanoid compounds A, B and C; A and C are identified as different forms of a leucoanthocyanidin and compound B a leucoanthocyanin. All the 3 are effective as hypoglycaemic agents. Leaves give friedelin, β-sitosterol, flavonoids- quercetin-3-galactoside and rutin. Heart wood gives tiglic acid ester of α taraxasterol. It is widely used in treatment of skin diseases with pitta and rakta predominance. Stem-bark, root-bark, aerial roots, leaves, vegetative buds and milky exudate are useful in medicine. It improves complexion, cures erysepelas, burning sensation and vaginal disorders, while an infusion of the bark cures dysentery, diarrhoea, leucorrhoea, menorrhagia, nervous disorders and reduces blood sugar in diabetes. A decoction of the vegetative buds in milk is beneficial in haemorrhages. A paste of the leaves is applied externally to abcesses and wounds to promote suppuration, while that of young aerial roots cure pimples. Young twigs when used as a tooth brush strengthen gum and teeth. The aerial roots are used in obstinate vomiting and leucorrhoea and are used in osteomalacia of the limbs. The buds are useful in diarrhoea and dysentery. The latex is useful in neuralgia, rheumatism, lumbago, bruises, nasitis, ulorrhagia, ulitis, odontopathy, haemorrhoids, gonorrhoea, inflammations, cracks of the sole and skin diseases.

***Ficus religiosa* Linn. (Peepal Tree)**

Description: It is a large deciduous tree with few or no aerial roots. It is often epiphytic with the drooping branches bearing long petioled, ovate, cordate shiny leaves. Leaves are bright green, the apex produced into a linear-lanceolate tail about half as long as the main portion of the blade. The receptacles occurring in pairs and are axillary, depressed globose, smooth and purplish when ripe. The bark is grey or ash coloured with thin or membranous flakes and is often covered with crustose lichen patches.

Medicinal uses: Bark yields β-sitosterol and its glucoside. An aqueous extract of the bark has an antibacterial activity against *Staphylococcus aureus* and *Escherichia coli*. It is useful in gonorrhoea, diarrhoea, dysentery, haemorrhoids and gastrohelcosis. A paste of the powdered bark is a good absorbent for inflammatory swellings. It is also good for burns. Leaves and tender shoots have purgative properties and are also recommended for wounds and skin diseases. Fruits are laxative and digestive. The dried fruit pulverized and taken in water cures asthma. Seeds are refrigerant and laxative. The latex is used for the treatment of neuralgia, inflammations and haemorrhages. Decoction of the bark if taken in honey subdues *vatarakta*.

Cultivation: *Ficus* species is cultivated in rocky areas, unused lands, or other wastelands of the farmyard. The plant is vegetatively propagated by stem cuttings. A few species are also propagated by seeds. Stem cuttings of pencil thickness taken from the branches are to be kept for rooting. Rooted cuttings are transplanted to prepared pits. No regular manuring is required. The plant is not attacked by any serious pests or diseases. Bark can be collected after 15 years. *Ficus* species generally has an economic life span of more than hundred years. Hence, bark can be regularly collected from the tree. Root, bark, leaves, fruits and latex form the economic parts.

Scientific name : *Hydnocarpus kurzii* (King) Warb.

Family : Flacourtiaceae

Common name : Chalmogra

Description: This is a tree of 15m tall with a tall trunk and a crown of drooping branches. Leaves are 20cm long, pointed at tip and leathery. Flowers are small, yellow, borne in small axillary clusters. Fruits are round, brown and many seeded.

Cultivation: The tree grows well in wet tropical climate where it can be found at elevations up to 1,800 metres. The plant grows best in areas where annual daytime annual temperatures fall within the range 28 - 34°C, though it is tolerant of 7 - 42°C. It is tolerant of temperatures to below 2.5°C. It prefers a mean annual rainfall in the range 3,500 - 6,500mm, but can tolerate 2,000 - 8,000mm. It thrives well in full sun and light shade. The tree grows best on well-drained, sandy, alluvial flats and floodplains along rivers, Propagated by seeds.

Medicinal uses: Chaulmoogra oil contains palmitic acid, hydnocarpic acid and oleic acid. The bark of the chaulmoogra tree yields tannins that are useful for reducing fever. In addition, the oil extracted from the seeds of chaulmoogra tree is used to cure a number of skin complaints, including different chronic conditions and scaly eruptions. A liniment or ointment is prepared by blending equal parts of lime water and chaulmoogra oil and applied topically to alleviate rheumatic pains, cure leprous ulcerations, burnt heads as well as scruff on the head. The crushed seeds are applied in the form of a paste to heal skin diseases like eczema, scabies and ringworm, as well as to cure wounds.

Scientific name : *Lagerstroemia speciosa* (Linn.) Pers.

Family : Lythraceae

Common name : Putranjeeva

Description: This is a small to medium sized tree, grows upto 40m with bushy and spreading crown. Leaves are opposite, simple and entire. Flowers are pink, white or purple borne in a large axillary or terminal panicle. Fruits are large, woody, capsule with a persistent calyx. Seeds are winged.

Cultivation: It is distributed at low to medium altitudes in comparatively open habitats, in disturbed or secondary forest, grassland, and along rivers. The habitat may vary from well drained to occasionally flooded but not peat soil. It is resistant to fire. Mean annual rainfall ranges from 1200- to 2400 mm and mean annual temperature ranges from 25 to 28°C. The tree requires fairly fertile soils. Due to poor natural pruning a spacing of 3m x 1 m is recommended. At 15.5 years the clear bole yield in the trial was 47-55 m cu/ha. Propagated by seeds.

Medicinal uses: A decoction of the bark is useful in diarrhoea and abdominal pains. A leaf poultice is used to relief malarial fever and is applied on cracked feet. A preparation from dried leaves, known as banaba, is widely used in the Philippines to treat diabetes and urinary problems.

Scientific name	:	*Madhuca indica* Gmel.
Family	:	Sapotaceae
Common name	:	Mahua

Description: It is a large deciduous tree, with shorter bole but larger crown. Leaves are 12-25cm long, thick, leathery, pointed at the tip. Flowers are small, fleshy, pale or dull white, borne in cluster near the end of branches, stalks of flowers bent downwards, covered with brownish hair. Flowers are strongly musk scented. Fruits are berry, fleshy, and greenish white and changes into orange brown when ripe..

Cultivation: It is a tree of dry tropical and subtropical climate. The tree grows on a wide variety of soils but prefers sandy soil. It also performs in alluvial soil. Trees are propagated through seeds, and transplanted seedlings. One month old seedlings are transplanted in plastic container of 15 x 25cm. Seedlings of 6 to 12 months old seedling are used for are used for planting in the main field. Besides, cuttings may be used as propagation materials. Trees become reproductive from 10th year onwards and produce around 20-200kg seeds annually. Trees are tolerant to varied weather conditions and insect pests and diseases.

Medicinal uses: The bark, leaves, flowers and seeds of the tree constitute the drug. Decoction of the bark is applied in curing itch, bleeding gums and ulcers. It is given internally in diabetes. Leaves are astringent. Flowers are used in cough and bronchitis. Seeds increases formation and flow of milk. Seed oil is laxative and on skin diseases.

Scientific name	:	*Mallotus philippensis* (Lamk.) Muell. Arg.
Family	:	Euphorbiaceae
Common name	:	Kamla tree

Description: The trees are 10 m tall, alternately arranged with ovate or rhombic ovate leaves which are rusty-velvety. Male and female flowers occur in different trees. Female flowers are borne in lax spike like racemes at the end of branches or in leaf axils. Male flowers are yellow, occur three together in the axils of small bracts. Capsule is trigonous-globular, covered with a bright crimson layer of minute, easily detachable reddish powder.

Cultivation: This tree in naturally distributed from the western Himalayas, through India, Sri Lanka, to southern China, and throughout Malesia to Australia and Melanesia. The tree is available usually mixed with other species, both in forests and in open scrubland. Kamala tree is common in evergreen forest, especially in secondary forest, and sometimes even dominant in the undergrowth. It can withstand considerable shade, frost-hardy and resistant to drought. It can be grown up to an altitude of 1600 msl with mean annual temperature of 16-28°C and mean annual rainfall of 800-2000 mm. Kamla tree tolerates a wide range of soli types, including infertile soils, limestone, acid and rocky land. The tree is light-demanding. It will tolerate light shade when young, but full overhead light is necessary for full vigour. Propagated by seeds. The tree is usually planted in rows with field crops. The species coppices well. Loosening of soil and regular weeding are necessary for at least 2 years after sowing.

Medicinal uses: All parts of the tree are applied externally to treat parasitic infections of the skin. The fruits and bark are reported to be used medicinally to treat stomach ulcers and tapeworm. According to Ayurveda, leaves are bitter, cooling and appetizer. Fruit is heating, purgative, anthelmintic, vulnerary, detergent, maturant, carminative, alexiteric and useful in treatment of bronchitis, abdominal diseases, spleen enlargement etc.

Scientific name : *Messua nagassarium* (Burm.f.) Kosterm.
Family : Clusiaceae
Common name : Nagkesar

Description: It is a medium sized to large evergreen tree, 18-30m tall with reddish brown bark which peels off in thin flakes. Leaves are simple, opposite, thick, lanceolate, coriaceous, covered with waxy bloom underneath, and red when young, acute or acuminate and with inconspicuous nerves. Flowers are white, very fragrant, axillary or terminal, solitary or in pairs. Stamens are numerous, golden yellow, much shorter than the petals. Fruits are ovoid with a conical point surrounded by the enlarged sepals. Seeds are 1-4 in number, angular, dark brown and smooth.

Cultivation: The plant grows well in plains, riverbanks or places which do not experiences moisture stress for its luxuriant growth. Silty loam soil is ideal for its cultivation. The plant is propagated by seeds. Seed formation occurs in November-March. Seeds are to be collected and sown in seedbeds or polybags. 3-4 months old seedlings are used for transplanting. Pits of 45cm x 45cm x 45 cm are to be taken at a distance of 3-3.5m and filled with a mixture of 10kg FYM, sand and top soil and made into a mound. Seedlings are transplanted into small handpits taken on these mounds. FYM is to be applied twice a year. Regular irrigation and weeding are to be done. The tree flowers in the fourth year.

Medicinal uses: Seed oil gives 4-phenyl coumarin analogues-mesuol, mammeigin, mesuagin, mammeisin and mesuone. Bark gives ferruols A and B. Heartwood gives xanthoneseuxanthone, mesuaxanthones A and B and a tetroxygenated xanthone named ferraxanthone. The essential oil from the stamens is antibacterial, antifungal, anthelmintic and that from fruit is antifungal. An Ayurvedic preparations containing *M. ferrea* has haemostatic and astringent properties and is particularly useful in uterine bleeding. Aerial part is CVS active, spasmolytic and diuretic. Phenol containing fraction of seed oil is antiasthmatic and antianaphylaxis. Bark is used as tonic after childbirth. Bark and unripe fruit is sudorific. Leaf and flower is an antidote for snakebite and scorpion sting. Flower bud is antidysenteric. Flower is stomachic and expectorant. Seed oil is antirheumatic. Unripe fruit and flower is astringent.

Scientific name : *Michelia champaca* Linn.
Family : Magnoliaceae
Common name : Nagkesar

Description: It is a graceful tree of 10 to 12 m tall. Leaves are ovate-oblong, acuminated, acute at the base with the ribs beneath up to 25 cm long and 7 cm broad.

Flowers are sweet scented, lemon yellow in colour, peduncles are short axillary, one flowered. This plant produces flowers throughout the year in a suitable climate. This species is cultivated in ReunionIsland and in Nossi-Be (Madagascar) for extraction of oil from flowers. There are also different types of flower colours in this species like white, sovereign red and creamy light yellow with large number of petals. White flowered types are more sweetly scented.

Cultivation: The tree grows in tropical and subtropical climate up to an altitude of 2000 m with mean annual temperature of 7-38 °C. It prefers moist, deep and fertile soils. Plants are propagated by grafting and from seeds. Trees propagated from seed take 8-10 years to flower whereas asexually propagated trees flower in 2-3 years. It is planted at a distance of 3 m x 3 m. The tree is a light demander and is susceptible to fire. It coppices well. Flowers appear in well grown plants during April to May and also in September to October.

Medicinal uses: The flowers yield champaka perfume of commerce. The oil contains linalool, methyl ester and a phenol. The constituents of flower concrete are phenylethyl alcohol, ionones, di-hidroionones, dihydro-beta-ionol, indole and methyl anthranilate, methyl palmitate, methyl linoleate and oximes of ionones. Root bark contains sesquiterpene lactones costunolide. parthenolide, dihydroparthenolide, micheliolide. Bark contains tannin. Bark is useful in fevers; flowers are stimulant, good for eyes and relives burning , and in skin diseases. Root bark is purgative and used in amenorrhoea. Flowers are tonic, stomachic, carminative, used in dyspepsia, nausea, fever, also useful as a diuretic in renal diseases. Flower oils are used in cephalgia. Bark is stimulant, diuretic and febrifuge. Dried root and root-bark are used as a purgative and emmenagogue.

Scientific name	:	*Mimusops elengi* Linn.
Family	:	Sapotaceae
Common name	:	Bakul/Maulsiri

Description: It is an evergreen tree with dark grey fissured bark and densely spreading crown. Leaves are oblong, glabrous and leathery with wavy margins. Flowers are white, fragrant, axillary, solitary or fascicled. Fruits are ovoid or ellipsoid berries. Seeds are 1-2 per fruit, ovoid, compressed, greyish brown and shiny.

Cultivation: It prefers moist soil rich in organic matter for good growth. The plant is propagated by seeds. Fruits are formed in October-November. Seeds are collected and dried. Seeds are to soaked in water for 12 hours without much delay and sown on seedbeds. Viability of seeds is poor. After germination they are to be transferred to polybags. Pits of 45cm x 45cm x 45cm are taken and filled with 5kg dried cowdung and top soil. To these pits, about 4 months old seedlings from the polybags are to be transplanted with the onset of monsoon. Addition of 10kg FYM every year is beneficial. Any serious pests or diseases do not attack the plant. Flowering commences from fourth year onwards. Bark, flowers, fruit and seeds are the economic parts.

Medicinal uses: Bark, flowers, fruit and seeds are used in medicine.β-sitosterol and its glucoside, α-spina-sterol, quercitol, taraxerol and lupeol and its acetate are present in the aerial parts as well as the roots and seeds. The aerial parts in addition gave quercetin, dihydroquercetin, myricetin, glycosides, hederagenin, ursolic acid, hentriacontane and β-carotene. Saponins from seed are spermicidal and spasmolytic. The aerial part is diuretic. Pulp of ripe fruit is antidysenteric. Seed is purgative. Bark is used as a gargle for odontopathy, ulitis and ulemorrhagia. Tender stems are used as tooth brushes. It is also useful in urethrorrhoea, cystorrhoea, diarrhoea and dysentery. Flowers are used for preparing a lotion for wounds and ulcers. Powder of dried flowers is a brain tonic and is useful as a snuff to relieve cephalgia.

Scientific name : *Moringa oleifera* Linn.

Family : Moringaceae

Common name : Drumstick Tree

Description: This is a small tree, upto 12 m tall and fast growing. The leaves are bi- or tripinnate, alternate and spirally arranged on twigs. Flowers are bisexual, yellowish white borne on slender spreading or drooping axillary clusters. The fruits are pendulous, linear, three sided pods with nine longitudinal ridges.

Cultivation: It is mainly grown in semi-arid, tropical and sub-tropical climate. It prefers sandy red soils, sandy loam or loam or black soils of pH 6.0-7.5. It is a drought tolerant plant. In south India, Nov-Dec and in Nortern India. June-July is the best time for planting. On an average, 550g seed is required for one hectare of land. Pits of 45cm x 45cm x 45cm size are dug out at a spacing of 2.5m x 2.5m and filled with 5kg well rotten FYM and soil mixture. Usually, two seeds are planted in each pit. Perennial types are propagated vegetatively. One year old stem cuttings of 1.0-1.5m long and 14-16cm in girth is used. Stem cuttings are planted 3-5m apart. On an average, an individual plant responds 90gN, 16g P_2O_5 and 30g K_2O and 10-15kg rotten manure annually for better growth and flowering.

Medicinal uses: Leaves are rich in beta-carotene, vitamin C, protein and iron. The leaves are cooked and eaten with spinach. Seeds contain 38-40% oil. Drumstick seeds are aphrodisiac. The leaves are reported to have a stabilizing effect on blood pressure and to control glucose levels. They are also used to treat anxiety, diarrhoea and inflammation of the colon, skin infections, scurvy, intestinal parasites and many other ailments.

Scientific name : *Murraya koengii* Linn.

Family : Rutaceae

Common name : Curry Leaf

Description: Itis a small aromatic tree with dark grey bark and closely crowded spreading dark green foliage. Leaves are imparipinnate and alternate. Leaflets are alternate, obliquely ovate or somewhat rhomboid, gland dotted and strongly aromatic. Flowers are white, borne in much branched terminal corymbose cymes and fragrant. Fruits are subglobose or ellipsoid berries, purplish black when ripe and 2-seeded.

Cultivation: Curry leaf tree does not choose a specific climate and can grow even in dry climate too. In places where minimum temperature goes below 13°C, the growth of the shoot will be slightly affected. It comes up very well in light textured red soils. *DWD-1* and *DWD-2* are two improved varieties released from UAS, Dharwad. Curry leaf is propagated by seed. A spacing of 90-120cm is followed on either side. Pits of size 30cm x 30cm x 30 cm are dug out one month before planting and filled with top soil mixed with well decomposed FYM at the time of planting. Healthy seedlings are planted in the centre of the pits. Then long furrows are formed connecting all the pits to facilitate easy irrigation. The seedlings are irrigated once in 5-7 days upto 3 years and once in 15 days afterwards. The field should be kept free from weeds. Plants may be trained and pruned to maintain a bush of one metre in height. For better growth and yield, each plant is fertilised with 20kg of FYM besides 150:25:50g of N, P_2O_5, K_2O/ year. Attack of aphids in the vegetative stage can be controlled by spraying of dimethoate at 2ml/litre of water. Leaves from such sprayed plants should be harvested only after 10 days. Spraying carbendazim at 1g/ litre can take care of leaf spot diseases. The crop becomes ready to first harvest at the end of first year. The yield of leaves account to 400kg/ha at the end of first year, 2000 to 2200kg/ha in the second and third year harvested at an interval of four months and 2500kg/ha in the fourth year harvested at three months interval. From fifth year onwards it is harvested at 2.5-3 months interval giving a yield of 3500-5000kg/ha.

Medicinal uses: All parts of the plant, especially the leaves are rich in carbazole alkaloids. The plant is highly esteemed for its leaves which promote appetite and digestion and destroy pathogenic organisms. It is reported to be useful in emaciation, skin diseases, homeopathy, worm troubles, neurosis and poisons. They are useful in vitiated conditions of *kapha* and *pitta*, hyperdipna, colic, flatulence, diarrhoea, dysentery, vomiting, inflammations and foul ulcers. "*Kaidaryah*" drug is prepared using this plant which improves voice, stimulates digestion and destroys concocted poisons in the system.

Scientific name	:	*Nyctanthes arbortristis* Linn.
Family	:	Oleaceae
Common name	:	Night Jasmine

Description: It is a small tree upto 10m height with rough, scaly and grey bark. It has rough, tetragonal and scabrous spreading branches. Leaves are shortly petiolate, decussately opposite and ovate which are cuneate to subcordate at base and acute or acuminate at apex. The leaf margin is entire, scabrous above with bulbous based hairs and pubescent beneath. The Flowers are fragnant having 5-15mm long white lobes, axillary or terminal placentation with imbricate aestivation, a bracteate cyme consisting of 2-7 flowered corymbs with quadrangular and slender peduncle and is sessile.

Cultivation: This iswidely cultivated in tropical and subtropical regions all over the world. The plant is tolerant to moderate shade and can grow on rocky ground in dry hillshades, dry deciduous forests or at sea-level upto 1500 m altitude with a wide range of rainfall patterns. The tree prefers a loamy soil. It is often cultivated in gardens due to its most pleasant and peculiar fragrance. Though the tree can be propagated by seeds as well as by cuttings, the seeds have been found to exhibit a poor germination rate because of phenolic compounds leaching out of the imbibed seeds. These inhibitory phenolic compounds are stored in the pericarp assisted by the seed coat. The rate of germination is improved by either removing both the coverings or treating seeds with a solution of antioxidants like polyvinylpyrrolidine (PVP) and polyvinylpolypyrrolidine (PVPP) prior to germination.

Medicinal uses: It is used to provoke menstruation, for treatment of scabies and other skin infections, as hair tonic, chalogogue, laxative, diaphoretic, diuretic, in treatment of arthritis, malaria, bronchitis and as anthelmintic.

Scientific name : *Olea europaea* Linn.
Family : Oleaceae
Common name : Eurpean Olive

Description: The olive tree, is an evergreen tree, 8–15 m tall. The leaves are silvery green, oblong, measuring 4–10 cm long and 1–3 cm wide. The trunk is typically gnarled and twisted. The flowers are small, white, feathery with ten-cleft calyx and corolla, two stamens and bifid stigma, are borne generally on the previous year's wood, in racemes springing from the axils of the leaves. The fruit is a small drupe 1–2.5 cm long, thinner-fleshed and smaller in wild plants than in orchard cultivars. Olives are harvested in the green to purple stage.

Cultivation: Olives like hot weather and sunny positions without any shade while temperatures below "10°C may injure even a mature tree. They prefer coastal climate. They perform better in well drained light soils. Olives are propagated by various methods. The preferred ways are cuttings and layers; the tree roots easily in favourable soil and throws up suckers from the stump when cut down. However, yields from trees grown from suckers or seeds are poor; they must be budded or grafted onto other specimens to do well. The olive can also sometimes grown from seeds. To facilitate germination, the oily pericarp is first softened by slight rotting, or soaked in hot water or in an alkaline solution. The olive tree, is very hardy: drought-, disease- and fire-resistant and it can live to a great age due its robust root system and good regeneration capacity. Most olives today are harvested by shaking the boughs or the whole tree.

Medicinal uses: Olive fruits are rich in several types of polyphenols, mainly tyrosols, phenolic acids, flavonols and flavones, and for black olives, anthocyanins. The main bitter flavor of olives before curing results from oleuropein and its aglycone content. Barks, fruits, leaves, wood, seeds, and oil are used in different forms, alone or sometimes in combination with other herbs. Oil of seeds is taken orally as a laxative and also applied externally as a balm for inflammation. Decoctions of dried leaves and fruit are used orally to treat diarrhoea, respiratory and urinary tract infections, stomach and intestinal diseases, and as mouth cleanser. Continuous application of olive oil is also useful to prevent hair loss. It is applied externally to treat burns, ores, and rheumatism and to promote circulation. Infusion of fresh leaves is also taken as anti-inflammatory. Tincture of olive leaves is taken in Italy as a febrifuge and applied externally as an emolument for ingrown nails and a restorer of epithelium. Boiled extract of fresh or dried leaves is taken orally to reduce asthma and hypertension and to induce diuresis.

Scientific name : *Oroxylum indicum* Vent.
Family : Bignoniaceae
Common name : Sonapatha

Description: This iss a medium-sized, soft-wooded tree, 10–16 m tall. Stem bark is dull brown in colour. Leaves are broad, 60–120 cm in length and pinnately compound. Leaflets are ovate, wavy, and acuminate. The tree is recognized by ternately bipinnate leaves. Flowers are large, fleshy, violet coloured, and foul smelling with 2.5-cm-long pedicle. Fruit is a follicle, 30–90 cm long and 5–10 cm broad, strap/sword shaped, compressed, and two-valved. Seeds are flattened and winged.

Cultivation: The plant grows well in tropical areas that receive well-distributed rainfall, between 85 cm and 130 cm per annum. Sandy-loam fertile soil is ideal for good growth and development of the species; but it can also be grown well in medium to deep black soils to sandy loam soils. Seed is the best propagation material and should be collected before splitting of pods during February–March. Seed germination percentage is 80%–90% without any pre-treatment. In all, 250 g of seeds are required to raise stock for plantation one hectare of land. Pits of appropriate size, preferably 60 cm × 60 cm × 60 cm, are dug in a square planting geometry (2 m × 2 m). Approximately, 10 kg FYM, 150 g of nitrogen, and 250 g of single super phosphate are mixed in the topsoil, and the pit is loosely filled back before planting. Transplanting in the main field is done in the first week of July (with the onset of monsoon in central India). A spacing of 2 m × 2 m is considered best. An optimum crop stand of 2500 plants/hectare is achieved with this spacing. About 20 kg of well-decomposed FYM, 150 g of nitrogen, and 250 g of super phosphate are required per plant in three split doses at intervals of six months for two years after transplanting. Potash is required only in potassium-deficient soils. Normally, two weedings alongwith hoeing are considered sufficient. In the first year, irrigation should be done at least six to eight times but more frequent irrigation during the summer months (at an interval of 7–10 days) is desirable for maximum productivity. The check basin method is found good for irrigation. The tree starts flowering and fruiting after three years, and forms viable seeds even in first fruiting. It takes about three to five months for fruits to get maturity. The plants are uprooted for harvesting of root bark only after six to eight years. A six-year-old tree yields only one kg of dried root bark. Therefore, the estimated yield of root bark is about 4.0 quintals/hectare per year from sixth year onwards.

Medicinal uses: The root bark is rich in chrysin, baicalein, dehydrobaicalein, and orozylin. Stem bark contains flavonoids such as oroxylin, baicalein, scutelarin and 7-rutinoside, chrysin, and p-coumaric acid. Heartwood yields β-sitosterol and isoflavone-prunetin. Root bark of sonapatha is an astringent, tonic, anti-diarrhoeal, diuretic, anodyne, and is useful in curing dropsy. It is an ingredient of 'dashamoolarishta' of Ayurvedic medicine. Stem bark is anti-rheumatic. An infusion of bark powder is diaphoretic. Tender fruits have spasmolytic, carminative, and stomachic properties, while seeds are purgative.

Scientific name	:	*Phyllanthus acidus*Vent.
Family	:	Phyllanthaceae
Common name	:	Star Gooseberry

Description: The plant is a small tree, reaching 2 to 9 m high.The crown of the tree is dense and bushy, composed of thickish, tough main branches, at the end of which are clusters of deciduous, greenish, 15-to-30-cm long branchlets. The branchlets are arranged with alternate leaves that are ovate or lanceolate in form, with short petioles and pointed ends. The leaves are 2-7.5 cm long and thin, green and smooth on the upperside and blue-green on the underside. In general, the Star gooseberry tree very much looks like the bilimbi tree.The flowers can be male, female or hermaphrodite. They are small and pinkish and borne clusters in 5-to-12.5-cm long panicles. Flowers are formed at leafless parts of the main branches, at the upper part of the tree. The fruits are numerous, oblate, with 6 to 8 ribs, and densely clustered. They are pale yellow or white, waxy, crisp and juicy, and very sour. Four to six seeds are contained in a stone at the center of each fruit.

Cultivation: Gooseberry tree thrives well in the tropics at low and medium altitudes in places with a short or prolonged dry season. It usually prefers hot, humid tropical lowlands. It flourishes from sea-level up to an altitude of 1,500 - 1,800 metres. The temperature range for growth is reported to be 14 - 35°C with the optimum between 20 - 29°C. The annual rainfall range for growth is reported to be 700 - 4,200mm with the optimum between 1,500 - 2,500mm. The tree requires a position in full sun or part day shade. It grows well in dry area. It prefers growing in a moist site. Seed is the most common way of propagating the plant. Other methods are green wood cuttings, budding and air layering. The saplings are planted at a distance of 8 m x 8m.

Medicinal uses: The root is an active purgative. An infusion of the root is used to alleviate asthma. An extract from the root is used in curing skin diseases especially to bring relief from itching. A leaf decoction is applied to urticarial. An infusion of the leaves is used as a dieting aid for people who wish to remain slim. A decoction of the bark is useful in bronchial catarrh. The fruit is used as a laxative. They are also taken as a liver tonic to enrich the blood. The juice of the fruit is applied on eyes to treat eye-troubles.

Scientific name	:	*Pongamia pinnata*Linn.
Family	:	Papilionaceae
Common name	:	Indian Beech

Description: It is a moderate sized, semi-evergreen tree, 18m or more high, with a short bole, spreading crown and greyish green or brown bark. Leaves are imparipinnate, alternate, leaflets 5-7, ovate and opposite. Flowers are lilac or pinkish white and fragrant, borne in axillary recemes. Pods are compressed, woody, indehiscent, yellowish grey when ripe varying in size and shape, elliptic to obliquely oblong, 4.0-7.5cm long and 1.7-3.2cm broad with a short curved beak. Seeds are usually one, elliptic or reniform, wrinkled with reddish brown, leathery testa.

Cultivation: The plant grows well in tropical areas with warm humid climate and well distributed rainfall. Though it grows in almost all types of soils, silty soils on river banks are most suited for its cultivation. It is tolerant to drought and salinity. The tree is used for afforestation, especially in watersheds in the drier parts of the country. It is propagated by seeds and vegetatively by rootsuckers. Seed setting is usually in November. Seeds are soaked in water for few hours before sowing. Raised seed beds of convenient size are prepared, well rotten cattle manure is applied at 1kg/m^2 and seeds are uniformly broadcasted. The seeds are covered with a thin layer of sand and irrigated. One month old seedlings are transplanted into polybags, which after one month can be planted in the field. Pits of size 50cm x 50cm x 50cm are dug at a spacing of 4- 5m, filled with top soil and manure and planted. Organic manure are applied annually. Regular weeding and irrigation are required for initial establishment. The trees flower and set fruits in 5 years. The harvest season extends from November- June. Pods are collected and seeds are removed by hand.

Medicinal uses: Seed, leaves, bark and root are used for medicinal purposes. Seeds and seed oil, flowers and stem bark contain karanjin, pongapin, pongaglabrone, kanugin, desmethoxykanugin and pinnatin. Seed and its oil also contain kanjone, isolonchocarpin, karanjachromene, isopongachromene, glabrin, glabrachalcone, glabrachromene, isopongaflavone, pongol, 2'-methoxy-furano[2",3":7,8]-flavone and phospholipids. Stem-bark contains pongachromene, pongaflavone, tetra-O-methylfisetin, glabra I and II, lanceolatin B, gamatin, 5-methoxyfurano [2",3":7,8]-flavone, 5-methoxy-3',4'-methelenedioxyfurano[2",3":7,8]-flavone and α-sitosterol. Heartwood yields chromenochalcones and flavones. Flowers are reported to contain kanjone, gamatin, glabra saponin, kaempferol, g-sitosterol, quercetin glycocides, pongaglabol, isopongaglabol, 6-methoxy isopongaglabol, lanceolatin B, 5-methoxy-3',4'-methelenedioxyfurano[8,7:4",5"]-flavone, fisetin tetramethyl ether, isolonchocarpin, ovalichromene B, pongamol, ovalitenon, two triterpenes- cycloart-23-ene,3b,25 diol andfriedelin and a dipeptide aurantinamide acetate. The whole plant and the seed oil are used in ayurvedic formulations as effective remedy for all skin diseases like scabies, eczema, leprosy and ulcers. The roots are used for cleaning teeth, strengthening gums and in gonorrhoea and scrofulous enlargement. The bark is used in haemorhoids, beriberi, ophthalmopathy and vaginopathy. Leaves are good for flatulence, dyspepsia, diarrhoea, leprosy, gonorrhoea, cough, rheumatalgia, piles and oedema. Flowers are given in diabetes. Fruits overcome urinary disease and piles. The seeds are useful in inflammations, otalgia, lumbago, pectoral diseases, chronic fevers, hydrocele, haemorrhoids and anaemia. The seed oil is recommended for ophthalmia, haemorrhoids, herpes and lumbago. The seed oil is also valued for its industrial uses.

Scientific name	:	*Pterocarpus marsupium* Roxb.
Family	:	Fabaceae
Common name	:	Vijaysar

Description: The tree are 30 m tall, up to 2.5 m in girth with straight and clear bole. Bark is scaly, rough, and longitudinally fissured. Leaflets are generally five to seven in number, 8–13 cm long, oblong or elliptic, or rotund, with 15–20 pairs of lateral veins. Oleo-resin obtained from tree trunk is called kino-gum, which is fragrant, brittle, almost black in colour, angular and glistering, and occurs in small flakes. Flowers are fragrant, yellow occur in about 1–5 cm long large panicles. Pods are reported to be flat, orbicular, winged, and up to 5 cm in diameter. Seeds are one to three in number, bony and convex in shape.

Cultivation: The tree thrives well in open sun in tropics under moderate rainfall of 80–200 cm. It prefers fertile, deep clayey loam soil with good drainage. It can tolerate excessive temperatures in summer. Freshly collected seeds are used for raising the plantations. About 100 g of viable seeds are required to raise seedlings for one hectare of land for planting at 8 m × 8 m spacing. Pits of appropriate size (50 cm × 50 cm x 50cm) are dug at a spacing of 8 m × 8 m to accommodate 160 plants per ha. About 25 kg FYM (farmyard manure), along with 200 g of nitrogen and 150 g of phosphorus, is mixed with soil of each pit as basal dose. FYM @ 25 kg per plant, nitrogen @ 200 g/plant, and phosphorus @ 150 g/plant are applied every year for the first three years. The fertilizer is applied in two split doses, the first in September and the second in January. Two manual weedings, the first one in August and the second in November, are recommended. Irrigation should be given six times in the first year (preferably once a month) through check basin system or filling the basin of the pit with water. The tree is harvested after 10–15 years for production of heartwood. Kino gum is collected through incision in the bark before logging of tree, and dried well in shade. Each mature tree yields approximately 500 kg of dry heartwood after 10–15 years. On an average, an estimated yield of 750–800 quintals/hectare is obtained.

Medicinal uses: Isoflavanoids, terpenoids, and tannis are reported from heartwood. Roots yield liquid-ritigenin, garbanzol, pterosupin, pseudo-dobatigenin, and 5-deoxy-kaempferol. Kino gum contains kitannic acid. Heartwood is antibiotic and hypoglycaemic, and is used to control blood sugar. Kino gum, obtained from incisions in bark, has astringent, anti-diarrhoeal, and anti-haemorrhagic properties. Leaves are used externally to treat boils, sores, and other skin diseases, while flowers are febrifuge.

Scientific name : *Pterocarpus santalinus* Linn. f.

Family : Fabaceae

Common name : Rakta Chandan

Description: This is a moderate-sized deciduous tree, 10 m tall and 0.9-1.5m girth with clear trunk and dense rounded crown. The bark is blackish-brown, is fissured and resembles crocodile skin. The inner bark, when injured or cut, oozes red coloured 'santolin' dye. The wood is extremely hard and dark red in colour. The leaves are imparipinnate, petiolate, and alternate, simple at seedling stage but trifoliate or rarely pentafoliate later. The flowers are yellow, densely arranged in simple or sparingly branched racemes. Pods are oblilquely orbicular, 3–4 cm in diameter, including the wing, and gradually narrow down into a short stipe. There is only one seed per pod, 1–1.5 cm in length and reddish brown in colour.

Cultivation: Well-drained red soils with gravelled loam are ideal for the cultivation of Lal Chandan species. It survives well in dry hot climate and requires rainfall ranging from 800 mm to 1000 mm annually for good growth. In nature, two types of rakta chandan trees are observed—wavy grained and straight. The seeds retain their viability for about one year under normal storage conditions and are the best propagation material. Fruits are collected in March for the extraction of seeds. Seedlings or stumps prepared from two-year-old nursery seedlings are used for planting in the field. About 1 kg seeds are required to raise seedlings for one hectare of plantation. Seeds show only about 50%–60% germination and about 40% plant survival. Seed treatment with GA_3 (gibberellic acid)+BA (benzyladenenine) @ 250 ppm helps in early and optimum germination. Pits of size 45 cm × 45 cm × 45 cm are dug at a spacing of 4 m × 4 m to accommodate 600 plants per ha. The pits are filled with topsoil mixed thoroughly with 10–15 kg FYM (farmyard manure) and 10 g lindane dust to protect the planting stock from soil-borne fungi. The best time for planting the crop in the field is end of May to June, i.e. onset of rainy season. About 10–15 kg FYM per plant per year and 150:100:100 g NPK (nitrogen, phosphorus, and potassium) per plant per year are required for at least five years. Fertilizer should be applied in 15–20-cm-deep circular trenches dug around the plant at a distance of 60 cm. Full dose of P and K along with one-third N should be applied at the end of February. The remaining N should be applied bark in two split doses during June–July and October–November. Application of inorganic fertilizers should always be followed by irrigation. Gap filling is done one month after planting. Weeding is done manually as and when necessary and particularly just before manuring. The soil around the basin is loosened frequently by hoeing. The plants are irrigated immediately after transplantation. Thereafter, irrigation is done on alternate days up to 15 days. After the seedlings get established, irrigation may be done at an interval of 10–15 days, depending on the weather condition. The fruits require almost 11 months to mature. Twenty- to twenty-five-year-old trees are used for the extraction of bark and wood. Pod yield from 15-year-old trees is 30 tonnes per hectare. Heartwood yield per tree after felling is 250 kg. Therefore, 15 000 tonnes per hectare of heartwood is expected after 15–20 years.

Medicinal uses: Bark, heartwood, and sapwood contain santalin A and B, isopterocarpolone, pterrocarptriol, isipterocarpene, pterocarpdiolone, pterocarpol, acyl olealonic aldehyde, and acetyloleanolic acid. The heartwood of red sanders is used as an astringent and tonic, and is bitter, sweet, cooling, analgesic, anti-inflammatory, and febrifuge. Its decoction is given in chronic dysentery. It is also used in vitiated conditions of pitta, burning sensation, vomiting, skin diseases, leprosy, ulcers, fistula, and haemorrhages.

Scientific name	:	*Saraca asoca*(Roxb.) de Wilde.
Family	:	Caesalpiniaceae
Common name	:	Asoka

Description: This is a medium sized evergreen tree, 9m tall with numerous spreading and drooping glabrous branches. Leaves are pinnate, 30-60cm long having 2-3 pairs of lanceolate leaflets. Flowers are orange or orange yellow, arranged in dense corymbs and very fragrant. Fruits are flat black pods, leathery and compressed with 4-8 seeds/pod. Seeds are ellipsoid oblong and compressed. The bark is dark brown to grey or black with a warty surface.

Cultivation: Asoka thrives well in areas with well distributed rainfall and in slightly shady areas. It requires soil rich in organic matter and moisture. The best season of planting is June-July. It is also grown in summer, if irrigation facilities are available. The plant is propagated through seeds. Seeds are formed usually during February-April. Seeds are collected when they are ripen and fall down and are sown after soaking in water for 12 hours on the prepared beds. Seeds germinate within 20 days. The seeds are then planted in polybags. 2-month-old seedlings from the polybags are used for transplanting. Square shaped pits of 60cm depth are taken at 3m spacing and filled with topsoil, sand and dried cowdung. On this the seedlings are planted. Application of FYM at 10kg/tree/year is highly beneficial.

Chemical fertilisers are not usually applied. Irrigation during summer months is essential. No serious pests or diseases are generally noted in this crop. If properly cultivated, *Asoka* can be cut after 20 years and the bark collected. It is cut at a height of 15cm from the soil level. If given irrigation and fertilisers, the cut wood will sprout again and harvested again after 5 years. This can be continued.

Medicinal uses: Flowers contain β-sitosterol, flavonoids and flavone glycosides-quercetin, kaempferol-3-O-β-D-glucoside, quercetin-3-O-β-D-glucoside. The anthocyanins present are pelargonidin-3, 5-diglucoside and cyanadin-3, 5-diglucoside. Bark gives catechol and sterols—24-methyl cholest-5-en-3β-ol, 24-ethylcholesta-5, 22-dien-3 β-ol and -24-ethyl cholest-5-en-3β-ol, a wax containing n-alkanes, esters and free primary alcohols.

The bark is used in dyspepsia, fever, dipsia, burning sensation, visceromegaly, colic, ulcers, menorrhagia, metropathy, leucorrhoea and pimples. The leaf juice mixed with cumin seeds is useful in stomachalagia. The floweres are reported to be uterine tonic and are used in vitiated conditions of *pitta*, syphilis, cervical adinitis, hyperdipsia, burning sensation, haemorrhoids, dysentery, scabies in children and inflammation. The well-known Ayurvedic preparations are "*Ashokarishta*" and "*Ashokaghrita*". *Ashokarishta* is prescribed in leucorrhoea, haematuria, menorrhagia and other diseases of genitourinary system of females.

Scientific name : *Sterculia urens* Roxb.

Family : Sterculiaceae

Common name : Karaya

Description: This is a medium sized deciduous tree with white or greenish white, shining, smooth papery bark. Leaves crowded at the ends of branches, large, 20-40cm in diameter, 5 lobed, very densely hairy on the under surface. Flowers are yellowish brown, small, borne in hairy erect branches. Fruits of 4-5 large, thick, red carpels, densely covered with stinging hairs.

Cultivation: In India the tree is distributed in the outer Himalayas and sub Himalayan tract from the Ganges to east words, Gujarat, Rajasthan, Central India, Deccan, and on the West Coast in Konkan, and North Kanara; it is also been recorded from the Lushai hills in Mizoram. The species *Sterculia urens* is commonly occurred in the dry deciduous forests of Andhra Pradesh. In Andhra Pradesh State, the species growth is found to be common in the 400 mm - 1300 mm rainfall areas of the forests. In Andhra Pradesh, the habitat of the trees is in rocky soils, plain rocks, clay lands. It is observed that the trees are growing vigorously in complete rocky lands and clay soils. Otherwise, it grows well in the arid to moist tropics, at elevations from 300 - 750 metre. It grows best in areas where annual daytime temperatures are within the range 30 - 42°C, though it can tolerate a temperature range of 7 - 47°C. The tree requires a sunny position, though seedlings are shade tolerant.

Medicinal uses: The gum exudate from the stem of the tree is used in medicine. The gum is used as a mechanical laxative. The gum is also reported to have aphrodisiacal properties.

Scientific name : *Symplocos racemosa* Roxb.

Family : Symplocaceae

Common name : Lodh

Description: This is a small tree, 6m tall. Leaves are dark green, leathery, pointed at the tip, margin entire or toothed. Leafstalks are small. Flowers are small, white or pale yellow in colour, borne in small axillary branches. Fruits are purplish black.

Cultivation: The tree occurs in plains and lower hills of eastern and central India. The tree grows in full to partial sun in moist, fertile, well-drained soil.

Medicinal uses: Bark contains triterpenoids, alkaloids loturine, loturidine and colloturine, glycosides, reducing sugars, phytosterols. The drug is used in digestive disorders, eye diseases and ulcers. A decoction of the bark is useful in a gargle in bleeding gums. It is used in plasters and applied on wound for promoting maturation of wounds. It is astringent and is useful in excessive bleeding during menstruation, other uterine disorders. The astringent properties are used in diarrhoea, dysentery, liver complaints and dropsy; also used in opthalmia and conjunctivitis.

Scientific name	:	*Strychnos nux vomica* Linn.
Family	:	Loganiaceae
Common name	:	Kuchla

Description: It is a medium-sized, deciduous tree, with fairly straight and cylindrical bole and dark-grey or yellowish-grey bark with minute tubercles. Leaves are simple, opposite, orbicular to ovate, 6–12 cm long and 6–10 cm broad, coriaceous, glabrous, and five-nerved. Flowers are white or greenish white and fragrant. They appear in many flowered terminal cymes. Fruit is an indehiscent berry, 5–6 cm in diameter, thick shelled, orange-red when ripe with fleshy pulp. Seeds are discoid, compressed, and coin like, concave on one side and convex on the other, and covered with fine grey silky hairs.

Cultivation: The plant grows well in dry or humid tropical areas of the country. It thrives over laterite, sandy, and alluvial soils. Seeds are the best material for propagation of kuchila tree. The collected seeds are dried in the sun after removing the pulp. Preferably, fresh seeds are used. The plant can also be propagated through cuttings. About 1 kg seeds are required to raise one hectare of plantation. Germination can be enhanced by treating the seeds with hot water (50°C) for 6–12 hours prior to sowing. Pits of size 45 cm × 45 cm × 45 cm are dug at a spacing of 5 m × 5 m and refilled with a mixture of soil and manure in 1:1 ratio. Appropriate quantities of sand may be added if the soil is heavy. About 10 kg of well-rotten FYM is applied in each pit at the time of its refilling. In South India, the seedlings are transplanted with the onset of South-west monsoon in May or early June. An optimum spacing of 5 m × 5 m is recommended, which gives a stand of 400 trees per hectare. An additional 10 kg manure is again applied to the soil around the plants during October–November at the time of weeding. A total of 20 kg FYM/plant/year in subsequent years results in best growth of plants. This is to be applied in two split doses in June–July and September– October. Supplementary doses of inorganic fertilizers like nitrogen, phosphorus, and potassium do not lead to any significant additional growth in the young plantations. The area around the basin of the plant should be kept weed-free by frequent weeding. The tree has a long span of life, about 50–60 years. It takes about 15–20 years for the tree to initiate flowering.

Medicinal uses: Strychnine and brucine are the most important alkaloids present in the plant. They are found not only in the seeds but also in roots, wood, bark, fruit pulp and hard fruit shells. The minor alkaloids present in the plant are vomicine, β-colubrine, β-colubrine, pseudostrychnine and N-methyl-sec-pseudobrucine (novacine). in small doses it can also serve as efficacious cure forms of paralysis and other nervous disorders. The seeds are used as a remedy in intermittent fever, dyspepsia, chronic dysentery, paralytic and neuralgic affections, worms, epilepsy, chronic rheumatism, insomnia and colic. It is also useful in impotence, neuralgia of face, heart disease, spermatorrhoea, skin diseases, toxins, wounds, emaciation, cough and cholera. Leaves are applied as poultice in the treatment of chronic wounds and ulcers and the leaf decoction is useful in paralytic complaints. Root and root bark useful in fever and dysentery.

Scientific name : *Tamarindus indica* Linn.

Family : Fabaceae

Common name : Tamarind

Description: It is a tropical evergreen tree, upto 25m tall, attractive, with dark red heart wood and softer yellowish sapwood and compound leaves. Flowers are borne in terminal racemes, pale yellow in colour. Fruits are brown pod, 5 to 15cm long. Fruits contain soft acidic pulp with many hard coated seeds.

Cultivation: It grows under the tropical climate. A sunny location or light shade with a soil pH 4.5-8.7 is ideal for its cultivation. It prefers hot summer and mild winter. Trees are mostly propagated by seeds; and vegetatively propagated through inarching and patch budding. Planting is done at a distance of 10m x10m during June-July and September-October. A dose of 20 to 40 kg FYM, 100-300g N, 200 to 500g P_2O_5 and 100 to 250g K_2O is required per plant per year. Basically, it is a rainfed crop. For initial development, plants need irrigation at fortnightly interval. Generally fully ripe fruits are harvested during summer and spring. A full grown tree can yield 200kg of fruit pods per year.

Medicinal uses: Fruit pulp is rich in tartaric and citric acids, vitamin C and sugar. Fruit pulp is used as spice and an important ingredient in Worcester'shire sauce and H.P sauce. The ripened fruit is sweeter , used in desserts and sweeten drinks or as a snack. Fruit pulp is also used in syrup, juice concentrates, chutneys and pickles. It is used as ingredient in cardiac and blood sugar reducing medicine. Jellulose is obtained from seeds which is used in the cotton and jute industries. The bark is administered against diarrhoea and skin disorders. Fruits are carminative and laxative.

Scientific name : *Terminalia arjuna* (Roxb.ex DC) Wight & Arn.

Family : Combretaceae

Common name : Arjun

Description: It is a large evergreen tree with buttressed trunk and spreading crown with drooping branches. Bark is smooth, grey outside and flesh coloured inside, flaking off in large flat pieces. Leaves are simple, sub-opposite, oblong or elliptic, coriaceous, crenulate, pale dull green above, pale brown beneath, often unequal sided, nerves 10-15 pairs and reticulate. Flowers are white, borne in panicles of spikes with linear bracteoles. Fruits are ovoid or oblong with 5-7 short, hard angles or wings, the lines on the wings oblique and curving upward.

Cultivation: The plant naturally grows in subtropical and tropical moist regions of the country. The tree prefers alluvial loamy or black cotton soils, which are loose, moist, fertile, and have good drainage and water holding capacity. River bank soils, streams, and ravines are its natural habitat. The plant also survives in open sunny and low rainfall areas. The seeds are collected in early summer from trees that are more than six-year-old. About 4–10 kg of seeds are required for raising one hectare of plantation. The seeds are soaked in boiling water and allowed to cool. The seeds should remain soaked for 24 hours. Pits of size 45 cm × 45 cm × 45 cm are dug at a spacing of 6 m × 6 m. About

10 kg FYM (farmyard manure) and NPK (nitrogen, phosphorus, and potassium) @ 75:50:30 g are added per pit and thoroughly mixed with soil as basal dose. About 10-month-old saplings are transplanted in the field in pits in July–August at a spacing of 6 m × 6 m. In all, about 280–300 saplings are required per hectare. Further top dressing is given with 50 g nitrogen per tree after three months. Irrigation is recommended at 15-day intervals in the summer season for young plantations. The bark is removed from well grown trees, preferably 10th year onwards. About 500 kg dried bark can be obtained from one hectare of plantation every year from the 10th year.

Medicinal uses: Fruits yield flavanones - arjunone and 5,7,2', 4' - tetramethoxy flavone and a chalcone - cerasidin. Other constituents are β-sitosterol, friedelin, methyloleanolate, gallic acid, ellagic acid and arjunic acid. Bark contains a triterpene arjungenin, triterpene glucosides I, II and III. The bark is useful in fractures, ulcers, urethrorrhoea, leucorrhoea, diabetes, vitiated conditions of *pitta*, anaemia, cardiopathy, fatigue, asthma, bronchitis, tumours, internal and external haemorrhages, cirrhosis of the lever and hypertension. It is useful in fractures and the powdered bark is taken with milk. The bark powder is diuretic and has a general tonic effect in cases of cirrhosis of liver. The bark has been considered by the ayurvedic physicians as well as by modern practitioners as a cardiac tonic. It is given as a decoction with milk. In Ayurveda, "*Arjunaghrita*" and "*Arjunarishta*" are two important cardiotonic preparations of this drug.

Scientific name	:	*Terminalia chebula* Retz.
Family	:	Combretaceae
Common name	:	Haritaki

Description: It is a medium sized deciduous tree with a cylindrical bole, rounded crown, spreading branches with dark brown bark and brownish gray heartwood. Leaves are simple, alternate or subopposite, ovate or elliptic ovate with short petioles bearing two glands below the blades. Flowers pale yellow or white in 4-10cm long, borne in axillary spikes. Fruit is a drupe, ovoid glossy, glabrous, faintly angled and yellow to orange brown in colour. Seeds are hard and pale yellow.

Cultivation: In the natural habitat, the species grow at temperature ranges between 36°C and 45°C, and rainfall ranges from 1200 mm to 3000 mm per annum. It is suited for growing on different types of soils, but attains best development on loose well-drained soils, such as sandy loam as well as clayey loam. It grows in open areas in the forest, forming top to middle canopy. Seed is the most appropriate material for this plant's propagation. About 5 kg seeds are required for raising stock for planting in one hectare of land. The depulped seeds are be either treated by fermentation process for a period of 15–20 days, or the seeds may be clipped at their broad end and then soaked in water for a period of two days before sowing in the nursery beds.

Alternatively, seeds may be mixed with cow dung slurry and kept in pits for one to two weeks. Pits of size 60 cm × 60 cm × 60 cm are dug at a spacing of 6 m × 6 m. The soil of each pit is mixed with 15 kg FYM (farmyard manure) and a mixture of NPK (nitrogen, phosphorus, potassium) @ 75:30:30 g and refilled before transplanting of seedlings. Transplantation of one-year-old saplings is done in the next monsoon. A spacing of 6 m × 6 m enables a crop stand of 280–300 plants per hectare. Flowering and fruiting generally commence after 8–10 years of planting in the summer season. Approximately 40–50 kg of dry fruits are produced per tree per year after it attains six years of age. This gives an average yield of about 12.6 quintals/hectare.

Medicinal uses: Kernel oil of *Chebulic myrobalan* a number of fatty acids *viz.* Palmitic, stearic, oleic, linoleic, arachidic and behenic acid. The fruits are rich in chebulinic acid, tannic acid, gallic acid, chebulin and tannin. Leaves contain terpenes and saponins and β-sitosterol is present in the bark. The fruit is a common constituent of "*Triphala*" capable of imparting youthful vitality and receptivity of mind and sense. It is a major constituent in the ayurvedic preparations like *Abhayarishta, Abhaya modak, Haritaki khand, Triphaladi churnam and Agastya rasayanam*. In allopathy, it is used in astringent ointments. In unani system, it is used as a blood purifier. The pulp of the fruit is used in piles, chronic diarrhoea, dysentery, costiveness, flatulence, asthma, urinary disorders, vomiting, hiccup, intestinal worms, ascites and enlarged spleen and liver. Powder of the fruit is useful in chronic ulcers and wounds, carious teeth and bleeding ulceration of the gums. The bark is a good cardiac tonic.

Scientific name	:	*Terminalia bellirica* (Craertn.) Roxb.
Family	:	Combretaceae
Common name	:	Bahera

Description: This is a large tree with buttresses. Leaves are large, 10-25cm long, clustered near ends of branches. Flowers are small, pale green, bad smelling, borne in simple spikes. Fruits are 2-3cm long, ovoid, brownish, densely covered with hairs.

Cultivation: Young plants grow well in shade while the matured plants tolerate light frost and drought. It thrives well in hilly areas. This is propagated by seeds. Natural multiplication happens rarely due to the poor seeds germination. Seeds soaked in water for 48 hours before sowing in seedbeds which should be covered with straw after sowing. It is watered immediately. Usually it takes 3-5 months to germinate. It can be transferred to polybags at two-leaf stage. One-year-old seedlings are ready for transplanting. For transplanting, pits are made of 50cm cube at a spacing of 4m. Organic manure, added regularly, promotes growth. Irrigation is required during first year. Weeds are removed regularly. This plant grows slowly. It fruits within 6-7 years. This is continued for many years. It is coppices well. Fruits are collected immediately after falling down or covered with soil to protect it from pests. Fruits dried well in sun and used or stored. The hard seed coat is removed before sowing.

Medicinal uses: Fruits yield β-sitosterol, gallic and ellagic acids, ethyl gallate, galloyl glucose, chebulagic acid and a cardiac glucoside bellaricanin. The fruit is useful in bronchitis, strangury, sore throat, diseases of eye, nose, heart and bladder, hoarseness and piles. It forms an important constituent of the ayurvedic drug '*triphala*'. Bark is used in anaemia and leucoderma.

9

CULTIVATION OF IMPORTANT MEDICINAL SHRUBS

Medicinal shrubs constitute an important natural wealth of a country. They play a significant role in providing primary health care services to rural people. They serve as important therapeutic agents as well as important raw materials for the manufacture of traditional and modern medicines. They are rich sources of bioactive compounds and thus serve as important raw materials for drug production. They constitute a precious natural wealth of a country and contribute a great deal to its health care programs. Judicious and scientific exploration of this wealth can significantly improve the general health of the people. Being a valuable item of commerce, a country can also earn a good amount of foreign exchange by exporting this natural wealth to other countries.

Scientific name	:	*Abroma angusta* Linn.f.
Family	:	Sterculiaceae
Common name	:	Ulatkambal

Description: *Abroma* is a shrub or a small tree, attaining a height of 3–5 m, with horizontal and velvety branches. Leaves are about 10–30 cm long and 6–18 cm broad, ovate or lanceolate, more or less cordate, finely acuminate, membranous,entire, and three to five lobed. Flowers are purple in colour, about 5 cm in diameter, occurring on fewflowered cymes.

Cultivation: Hot and humid climate is favourable for the growth of the crop. The plant thrives well in avariety of soils, like sandy loam to loam type.It grows in open areas in nature. Seeds are the best propagation material. The pre-treated seeds are sown in well-prepared nursery beds during February to mid-March. Seeds may also be directly sown in polybags after pretreatment. Seedlings become ready for transplanting in field during May–Junewhen they attain a height of about 20–25 cm. About 150–200 g seeds are sufficient to produce 12 000–14 000 seedlings for one hectare plantation. Pits of 30 cm × 30 cm × 30 cm size are dug at a spacing of 1 m × 1 m, after light ploughing of the soil. A basal dose of FYM @ 10 tonnes/hectare is mixed with dried soil and filled in the pits for better and faster growth. The basal dose of NPK (nitrogen,phosphorus, potassium) @ 30:40:20 kg/hectare,along with FYM, may also be applied to the pit soil. May–June isthe best time for transplantation of seedlings. Optimum spacing recommended in the field is 1 m × 1 m, that is, an optimum crop stand of 10 000 plants/hectare. After basal dose of manure andNPK, top dressing of N @ 30 kg/hectare is given 120 days after transplanting.The follow-up fertilizer is applied in the second year withdoses of NPK @ 30:40:20 kg/hectare after 12 months at the timeof soil working, followed by applying N @ 30 kg/hectare after18 months. First intercultural manual weeding operation is done 45 days after plantation and second after four months of plantation. The crop is grown under rain-fed conditions in North-East India. About 1.95 quintals of dry root bark is obtained per hectare after two years.

Medicinal uses: Roots are rich in in alkaloids abromine, abromasterol,and digitonide.Root bark of ulatkambal is used as a valuable emmenagogue and uterinetonic, chiefly used in intra-uterine diseases and other gynaecological disorders mostly related to menstrual disorders such as dysmenorrhoea,amenorrhoea, and gonorrhoea. Powdered root is an abortifacient and anti-fertility agent. The leaves and stem are demulcent. Infusion of fresh leaves and stems is useful in treatment of gonorrhoea.

Scientific name : *Adhatoda vasica* Nees.

Family : Acanthaceae

Common name : Vasaka

Description: It is an evergreen shrub grown upto one metre in height with many long opposite branches. Leaves are large and lance-shaped. Stems are herbaceous above and woody below. Leaves are opposite and exstipulate. Flower are spikes or panicles, small irregular zygomorphic, bisexual, and hypogynous. It has capsular four seeded fruits. The flowers are either white or purple in colour.

Cultivation: Alluvial soils are best suited for raising the crop. The plants are tolerant to shade but susceptible to waterlogging. It is propagated by tender stem cuttings. Stem cuttings of 15-20cm long and 3-4 nodes are ideal for planting. Rooted cuttings are planted on mounds or ridges which are prepared at 60cm apart and 30cm plant to plant. In sloppy areas, cuttings are directly planted in pits. Organic manure in the form of FYM, leaf mould or compost is applied at the rate of 10 tonnes per ha as basal dressing. N: P: K is applied each at the rate of 50kg/ha. Entire P is applied basally and N and K are given in two splits. The field is kept weed free and earthing up is done after top dressings with fertilizers. Leaves are harvested from the first year of planting itself whereas roots become ready for harvest after two years of planting. Total yield of root, stems and leaves is 10-11 tonnes per hectare.

Medicinal uses: The plants contain vasicine, vasicinone, deoxyvasicine, maiontone, vasicinolone and vasicinol. The leaves, flowers, fruit and roots are extensively used for treating cold cough, whooping cough, chronic bronchitis and asthma, as sedative, expectorant and antispasmodic.

Scientific name : *Aristolochia indica* Linn.

Family : Aristolochiaceae

Common name : Indian Birthwort

Description: A twining shrub with woody stems and rootstocks. Leaves are simple, alternate, short petioled, the blade ovate or somewhat wedge shaped, very variable in shape and size. The young leaves are light purplish. Flowers are small, greenish white. The fruit is a capsule, roundish or oblong and hexagonal.

Cultivation: It is naturally distributed in lower hilly regions and in plains; more common in southern and eastern India. It grows well in sunny or semi-shade locations in humus rich, loamy, well drained soils. Plants are propagated through seeds or soft wood cuttings.

Medicinal uses: Roots, rhizomes and leaves are used in medicine. The plants contain aristolochic acid, aristolic acid, aristolic acid methyl ester, methyl aristolochate, aristolamide, aristolinic acid and aristolonitrite.The plant is used in skin diseases where there is morbidity of vata, pitta and kapha. It is used as an appetiser, aphrodisiac and anthelmintic. The fresh juice of the leaves is a popular antidote to snake poison. The leaves and bark are used in bowel complaints of children, diarrhoea and in intermittent fevers. In traditional medicine, the underground parts of the plant are rubbed with honey and given to treat leprosy; and, macerated with black pepper, it is prescribed in diarrhoea.

Scientific name	:	*Artemisia maritima* Linn.
Family	:	Asteraceae
Common name	:	Wormseed

Description: This is a stout, much branched, perennial aromatic shrub, 1 m tall. Leaves are 2-5cm long, whitish, divided into numerous fine linear segments; upper leaves are simple, undivided linear. Flower heads are small, borne in short spikes.

Cultivation: It is a short-day, temperate plant, which requires cold winter and moderate summer. The critical photoperiod lies between 12 and 16 hours (average 13:31 hours). It can also be cultivated in subtropical areas as a winter crop. They thrive well in any soil but prefers a poor dry soil with a warm aspect. Easily grown in a well-drained circumneutral or slightly alkaline loamy soil, preferring a sunny position. Plants are longer lived, more- hardy and more aromatic when they are grown in a poor dry soil. Tolerant to a pH in the range 6.0 to 7.6. They do not prefer shade. Established plants are very drought tolerant. Tolerates maritime exposure· Propagated through seeds or divisions. Seeds can be sown in nursery during September, October for the late rainy season crop and during November - December for the summer crop. For seed production, rainy season sown crop is advisable, whereas, November - December planted crop is reported to give maximum oil yield.The seedlings which are 6-8 weeks old, healthy and uniform should be transplanted at a spacing of 30-60 cm between rows and 45-60 cm between the plants. However, planting at a spacing of 45 × 45 cm is the best to obtain maximum biomass, essential oil and artemisinin yield.Well decomposed farm yard manure / compost should be to be applied at 10 t/ha before planting. A fertilizer dose of 60-80 kg N; 40-60 kg P_2O_5 and 60 kg K_2O/ha may be applied depending upon the soil condition for getting good yield. Full dose of P_2O_5 and K_2O and two third dose of N are incorporated into the soil at the time of land preparation. The remaining one third dose of N is applied in two equal split doses at 30 and 60 days after transplanting. *A. annua* is highly sensitive to the boron and iron deficiency. Hence, basal application of borax at the rate of 8 kg/

ha is recommended.The field is to be irrigated frequently from transplanting to establishment of the crop. After establishment only 3-4 irrigations will be required up to harvest period.The crop requires 2-3 weeding and hoeing during the growth period. The first inter-tillage and weeding should be done about 20 days after transplantation. The second weeding is needed before the major branching stage, followed by hilling. The crop is ready for harvesting in 4.5-5 months after transplanting. The crop is to be harvested at full bloom stage, as harvesting at this stage recorded the maximum herb yield, essential oil and artemisisnin content. Artemisia produces fresh herbage yield of about 10-15 t/ha which in turn gives 30-40 kg of essential oil per hectare. The essential oil content generally varies from 0.2-0.4%, which is found to be generally higher (0.34%) in early sown crop (September) and it decreases in late sown crop. Popular varieties are Asha, Suraksha, Jeevanraksha and Cim-Arogya.

Medicinal uses: The main chemical constituents of *A. annua* are sesquiterpenoids, including artemisinin I, artemisinin II, artemisinin III, artemisinin IV, artemisinin V, artemisic acid, artemisilactone, artemisinol and epoxyarteannuinic acid. Immature leaves and flower heads are used in medicine. They are rich in santonin which is used for expulsion of worms from the stomach. The drug is also useful in fevers, dropsy and as stimulant.

Scientific name	:	*Berberis aristata* DC.
Family	:	Berberidaceae
Common name	:	Indian Barberry

Description: This is a large thorny shrub with yellow wood and whitish or pale grey branches. Leaves are fascicled in axils of branched or simple spines, coriaceous, usually sharp toothed. Flowers are borne in short bunches. Fruits are ovoid, bluish purple and few seeded.

Cultivation: The plants are naturally distributed in the Nilgiri hills and the Himalayas between 2000 and 3000 m. It prefers a warm moist loamy soil and light shade but it is by no means fastidious, succeeding in thin, dry and shallow soils. It grows well in heavy clay soils. Plants are propagated by seeds or hard wood cuttings. Stored seeds require cold stratification and should be sown in a cold frame as early in the year as possible. When they are large enough to handle, the seedlings are pricked out into individual pots and grown them on in the greenhouse or cold frame for at least their first winter. Once they attain at least 20cm tall, are planted them out into their permanent positions in late spring or early summer, after the last expected frosts. Cuttings of mature wood of the current season's growth, preferably with a heel, are taken October/November in a frame.

Medicinal uses: Dried stems, root bark and roots are used in medicine. The plants contain alkaloids like Berberine. The drug is useful in opthalmia, ulcers, fevers, stomach disorders and tonic.

Scientific name	:	*Calotropis gigantea* (L.) R. Br. ex Schult.
Family	:	Apocynaceae
Common name	:	Crown Flower

Description: This is a large shrub, 4m tall with rounded stems. Leaves are smaller in branchlets, elliptic ovate to obovate. Flowers are purple or greenish white, borne in clusters. Fruits are saccate, ovoid and many seeded.

Cultivation: It is distributed in tropical and subtropical area of the world and throughout India. The plant grows very well in a variety of soils and different environmental conditions. It thrives well on poor soils particularly where overgrazing has removed competition from native grasses. It is drought tolerant and the pioneer vegetation in desert soil. The seeds freely float in the air and natural regeneration is very common. Vegetative propagation through stem and root cuttings is very useful in large scale multiplication of the superior genotypes. Planted at a spacing of 1.0 to 1.5 m.

Medicinal uses: The plants are rich in Laurane, Saccharose, β-amyrin; α&β calotropeols; holarrhetine, Cyanidin-3-rhamnoglucoside; Taraxsterol isovalerate; Giganteol; Calotroposide;Calactin, Calotoxin; Calotropins DI &DII, Gigantin. The whole plant is used for the treatment of skin diseases, boils and sores and as a tonic and purgative in small doses, and as an emetic in larger doses. The powdered root bark is used in dysentery, elephantiasis, and leprosy. The stem bark is useful as diaphoretic and expectorant, and is used for dysentery, spleen complaints, convulsions, lumbago, scabies, ringworm, pneumonia, and to induce labour. The latex is used on stings, toothache, caries, ringworm, leprosy, syphilis, rheumatism and tumours, and also as an antiseptic, vermifuge, emetic and purgative, as well as for poisoning arrows. The powdered flowers are given for coughs, colds and asthma. The crushed and warmed leaves are applied on burns, headaches and rheumatic pains, and as a tincture for intermittent fever.

Scientific name : *Cassia senna* Linn.

Family : Caesalpiniaceae

Common name : Indian Senna

Description: It is a shrub or undershrub, 60-75cm tall with pale subterete or obtusely angled erect or spreading branches. Leaves are paripinnate. Leaflets are 5-8 in number, ovate-lanceolate and glabrous. Flowers are yellowish, many and arranged in axillary racemes. Fruits are flat legumes, greenish brown to dark brown and nearly smooth.

Cultivation: The plant requires a mild subtropical climate with warm frost free winters for its growth. Semiarid areas with adequate irrigation facilities are good for cultivation. Areas having high rainfall, humidity and poor drainage are not congenial. Light or medium loamy soils with adequate drainage and pH varying from 7.0-8.2 are preferable. In South India both summer and winter crops are possible. The plant is propagated by seeds. The seed rate recommended is 15-20kg/ha. Seeds are sown in October-November (winter rainfed crop) or in February-March (irrigated crop). Higher seed rate is required for unirrigated crop. Seeds are sown in lines 30cm apart. Application of 5-10 tonnes of FYM/ha before planting or raising a green manure crop is beneficial. About 40kg N and 25-50kg P_2O_5/ha applied as basal dressing and 40kg N/ha applied in 2 split doses as top dressing gave better yield. While the rainfed crop is grown without irrigation, the irrigated crop requires 5-8 light irrigations during the entire growing season. The crop succeeds with 2-3 weedings and hoeings in order to keep it free from weeds. *Alternaria alternata* causes leaf spot and dieback but the disease is not serious. In North India, the plant is attacked by the larvae of butterfly *Catopsilia pyranthe* which can be controlled by planting the crop in March-April instead of June-July. Under irrigated conditions, the first crop is obtained after 90 days of planting. The leaves are stripped by hand when they are fully green, thick and bluish-green in colour. The second crop is taken 4 weeks after the first harvest and the third 4-6 weeks after the second one. The last harvest of leaves is done when the entire crop is harvested along with the pods. Yield under irrigated conditions is nearly 1.4 tonnes of leaves and 150kg pods/ha and under unirrigated conditions is 500-600kg leaves and 80-100kg pods/ha. The leaves are dried in thin layers under shade so as to retain the green colour and the pods are hung for 10-12 days to get dried. The leaves and pods are cleaned, graded and marketed.

Medicinal uses: Leaves contain glucose, fructose, sucrose and pinnitol.Pods are rich in sennosides A and B, glycoside of anthraquinones rhein and chrysophanic acid. Seeds contain b-sitosterol. The leaves are useful in constipation, abdominal disorders, leprosy, skin diseases, leucoderma, splenomegaly, hepatopathy, jaundice, helminthiasis, dyspepsia, cough, bronchitis, typhoid fever, anaemia, tumours and vitiated conditions of *pitta* and *vata*.

Scientific name	:	*Clerodendrum indicum* (Linn.) Moon.
Family	:	Verbenaceae
Common name	:	Chingari

Description: This is an erect, less branched shrub, 1.5–3.0 m tall. The roots are light brown in colour and more than 2.5 cm in diameter. Stemis herbaceous, ridged, fluted, and hollow. Leaves are axillary, fascicled orterminal, and hang from the upright branches. Flowers of the species occur in axillary or terminal racemes. Fruit is upto 1.5 cm across,dark bluish green when ripe and seated on the enlarged red fleshy calyx.

Cultivation: Chingari grows well in moist tropical and subtropical climate,which should be free from frost during wintersand dry heat in summers. It is also possible to cultivate thespecies in the dry regions under partial shade. The soil should be deep, fertile,and well drained, having a pH range of 6.0–8.0. Chingari is commercially propagated by stem cuttings as well as root cuttings. The stem cuttings should be taken from semi-matured branches and root cuttings from about 2-cm-thick roots in July–August. Semi-hard woody stem cuttings give higher success rate than soft wooded and hardwooded cuttings. In general, the crop responds well to the basal application of 100 quintals of well-decomposed FYM (farmyard manure) along with 40 kg nitrogenand 75 kg phosphorus per hectare. The transplanting of chingari in the field at a spacing of 45 cm × 30 cm gives good yield. Planting at a spacing of 45 cm × 30 cm can accommodate about 74 000 plants per hectare. Besides, basal application of FYM, and 37.5 kg of nitrogen/hectare, an additional 37.5 kg of nitrogen should be applied in two equal split doses as top dressing: first atone month after transplanting and the second after six months (February–March) of transplanting. In winter,irrigation should be given at an interval of15–20 days. In summer months, irrigation should be given at an interval of 7–10 days,depending on the severity of the summer. The crop becomes ready for harvest in 10–12 months after planting. May–June isthe best time for harvesting the crop. Irrigation of field four to six days before harvesting facilitates digging of roots. An average crop of Chingari produces about 10–12 quintals per hectare of dry roots.

Medicinal uses: The root contains sapogenins,saponins, triterpenes, and D-mannitol. The root of chingari is stomachic, expectorant, anti-inflammatory,anti-bronchitis, febrifuge, hence useful for asthma, cough, and scrofulous affections. In Unani medicine, the root increases appetite and lowers fever. The leaves and roots are used externally to treattumours and certain skin diseases.

Scientific name	:	*Commiphora wightii* (Arn.) Bhandari.
Family	:	Burseraceae
Common name	:	Guggal

Description: Itis a slow growing, highly branched, spiny shrub, 3-5m tall with crooked and knotty branches ending in sharp spines. The stem is covered with silvery white, papery bark that peels off as flakes from the older parts of the stem, whereas, the younger branches are pubescent and glandular. Leaves are trifoliate; leaflets rhomboid, ovate, and entire at the base and serrate at the apex. Flowers are sessile and borne singly or in groups of two to three. Fruits are drupe, ovoid, reddish brown to purple in colour.

Cultivation: The plant thrives well in arid, sandy, and rockytracts of tropical India. Sandy or sandy loamsoils are best for its cultivation. Black coloured seeds are used as planting material. One to four seedlings emerge from a single seed due to polyembryony. The plant can be easily propagated through leafless stem cuttings, prepared in May, or through air layering of 5–8-year-old mother plants in July–August. Approximately, 100 g of blackseeds are required for raising plantation in one hectare of land at a spacing of 2 m × 2 m. Pits of size 45 cm × 45 cm× 45 cm are dug at requisite spacing of 2 m × 2 m and refilled after mixing the sand, soil, and FYM in the ratio of 1:1:1. A crop stand of 2500 plants per hectare is obtained with a spacing of 2 m × 2 m. Guggal are intercropped with crops like pearlmillet and cluster bean. Besides, the basal dose of FYM provided in the pits, the plants may be provided with growth regulators like IBA (10 mg/litre) along with organic manures like hexameal @ 100 g/plant every month or NPK (nitrogen, phosphorus,potassium) @ 75:130:30 g/plant after every three months up to a period of one year, which helps in the better survival of plants. After one year, quarterly dose of compost or hexameal @ 100 g/plant is good for growth and survival of plants. The pits should always be kept weed-free. Manual weeding and hoeing are practised at aninterval of 15–20 days during rainy season. Irrigation with 8 litres of water per plant at an interval of 15 days gives best growth of plant. The plant should be allowed to grow forat least five to six years before commencing incision of thick branches for extracting oleo-gum resin. The oleo-gum resin is tapped during winter, from November–February, by making a 7–10-cm long incisionin the main stem near the base. A yield of 120–130 kg oleo-gum resin isobtained per hectare after about eight years

Medicinal uses: The gum resin is rich in 3.2% gum and 19.5% mineral matter chiefly consisting of silicon dioxide, magnesium, calcium,iron, and aluminium. It also contains about 1.5% essential oil, comprising 6.5% myrcene and 11% dinyrcene. A number of steroids have also been isolated from guggul like Z-guggul sterone and E-guggulosterol I, II, and III. The gum of guggal is anti-inflammatory and useful in the treatment of arthritis, rheumatism, hyperlipidemia, thrombosis,and hypercholesterolemia. It has alterative, carminative, astringent,and antispasmodic properties.

Scientific name	:	*Datura metel*
Family	:	Solanaceae
Common name	:	Datura

Description: This is an erect succulent branched undershrub, divaricate often purplish branches and ovate pubescent leaves. Flowers are large, solitary, short pedicelled, purplish outside and white inside. Fruits are sub-globose capsules covered all over with numerous, fleshy prickles, irregularly breaking when mature. Seeds are numerous, smooth and yellowish brown.

Cultivation: Datura thrives well in a wide range of climate from tropical to temperate conditions. The plant performs best in areas of low rainfall where winter and monsoon rains are followed by long dry periods. Areas with annual rainfall below 1000mm with mean temperature of 10-15°C in winter and 27-28°C in May-June are ideal for its cultivation. The crop is susceptible to frost, high rainfall or high temperature in the plains in May-June. It grows in wide range of soils, however, alkaline or neutral clay loam soil or those tending to saline-alkaline reaction rich in organic matter are ideal for vigorous growth. The clayey, acidic, water-logged or moisture deficient soils do not favour crop.

The plant is propagated by seeds but it is characterised by poor and erratic seed germination which can be improved either by leaching out the inhibitor from the seeds or by alternate freezing and thawing of seeds. The optimum season for raising the crop is Rabi in tropical and subtropical areas while *Kharif* in temperate areas. The seeds are broadcast-sown or seedlings can be raised in nursery and then transplanted. Recommended seed rate is 7-8 kg/ha for broadcasting and 2-3 kg/ha. For transplanting. The field is ploughed and disced adequately to produce fine seed bed. In the case of direct seeding, seeds are drilled in rows taken 45-60 cm apart. The plants are thinned to keep a spacing of 30-45 cm at the time of first weeding. In the case of transplanting 4-6 weeks old seedlings are planted at 45-60 x 30-45 cm spacing.

The field should be irrigated immediately after sowing or planting if soil moisture is inadequate. Thereafter, 3-4 irrigations are given if sufficient rainfall is not received. Application of organic manure at 10-15 t/ha and fertilisers at 60:40:40 kg N, P_2O_5 and K_2O/ha is recommended for the crop for better growth and yield. N is applied in 3-4 equal split doses at planting and after each weeding which is required 2-3 times during the growing season. Application of micronutrients is reported to improve the alkaloid contents. No major insect pest is known to attack this crop. However, leaf spot, wilt and mosaic diseases cause damage to this crop. Leaf spot is caused by *Alternaria tennuissima* (Nees)Wiltshire and characterised by brown round to oval spots, becoming necrotic at later stage which leads to withering and dropping of leaves. Wilt is caused by *Sclerotium rolfsii* Sace which results in dropping of leaves and finally wilting of the entire plant. Root and foot wilt, caused by *Corticium solani,* appears as damping off of seedlings and mature plants. Datura distortion mosaic is characterised by yellowing of the veins followed by inward rolling and distortion of leaves with a reduction in plant size. For reducing the impact of these diseases, field sanitation, use of resistant varieties, crop rotation for 3-4 years and fungicide application should be resorted to. For the purpose of leaf and top, harvesting is done as soon as flowering starts. Entire top containing leaves and twigs is cut, dried in shade and stored in gunny bags. For seed and fruit, fully grown fruits, still green are picked 2-3 times before final harvest when the entire plant is cut from the base and dried in the open. The dried fruits are then thrashed with a stick to separate the seeds. The seed yield is about 1-1.5 t/ha.

Medicinal uses: The alkaloids like hyoscyamine and hyoscine (scopolamine) and meteloidine are found in all parts of the plant. Fruits contain daturaolone and daturadiol while roots contain additionally ditigloyloxy tropane derivatives, tigloidine, apohyoscine, norhyoscine, norhyocyamine, cusiohygrine and atropine. The plant and fruit are spasmolytic, anticancerous and anthelmintic. Leaves and seeds are inhaled in whooping cough, asthma and other respiratory diseases. Root, leaf and seed are febrifuge, antidiarrhoeal, anticatarrhal and are used in insanity, cerebral complications and skin diseases. Leaf is used as antitumour, antirheumatic and vermicide. Flower is antiasthamatic, anaesthetic and is employed in swellings and eruptions on face. Fruit juice is used in earache and seed decoction in ophthalmia. For the rheumatic swellings of joints, lumbago, sciatica and neuralgia, warm leaf smeared with an oil is used as a bandage or sometimes the leaf is made into a poultice and applied. The root boiled with milk is used in insanity. It is also an ingredient in the ayurvedic preparation *Kanakasva* used in bronchial troubles, and the *Unani* formulations "*Roghan dhatura*" used as a massage oil for the paralysed part.

Scientific name	:	*Desmodium gangeticum* (L.) DC.
Family	:	Fabaceae
Common name	:	Salparni

Description: It is an erect diffusely branched undershrub, 90-120cm tall with a short woody stem and numerous prostrate branches provided with soft grey hairs. Leaves are unifoliate, ovatelanceolate, membranceous and mottled with grey patches. Flowers are white, purple or lilac borne in elongate lax, terminal or axillary racemes. Fruits are moniliform, 6-8 jointed, glabrescent pods, joints of pods separately pubescent with hooked hairs, joint separating when ripe into indehiscent one seeded segments. Seeds are compressed and reniform.

Cultivation: Desmodium grows in a variety of climate and soils. However, it prefers tropical and subtropical climatic conditions. Although it can grow on all types of soils, waterlogged and highly alkaline soils are not favourable. Light sandy loam is preferred for commercial cultivation. It is propagated through seeds. Seeds are planted directly in the field or seedlings raised on the nursery beds and transplanted. Transplanting always gives better results in commercial cultivation, as it gives assured crop stand. Planting is done at a spacing of 40 x 20cm on flat beds or ridges. Manure at the rate of 10 tonnes/hectare should be mixed thoroughly with the soil at the time of field preparation. Nitrogen and phosphorus should be applied at the time of planting at the rate of 20 kg/hectare and 40 kg/hectare, respectively. Nitrogen and phosphorus can also be applied in the form of DAP (Di-ammonium phosphate) at the rate of 100 kg/hectare. The inter-row spaces between plants, both in the field and nursery should be kept free from weeds by frequent weeding and hoeing as the plant suffers from weed competition, especially during early stages of growth. Hand weeding is usually done. Irrigation of seedlings just after planting is good for crop establishment. Although it can be cultivated as a rainfed crop under humid tropical conditions, irrigation every month is beneficial during summer. The root is the economic part and harvesting can be commenced after 8-9 months. About 500- 700kg roots can be harvested from a hectare of land per year.

Medicinal uses: The root contains gangetin, gangetinin, desmodin, N,N-dimethyl tryptamine, hypaphorine, hordenine, candicine, N-methyl tyramine and b-phenyl ethyl amine. The plant has febrifuge, aphrodisiac, analgesic, diuretic, anti-inflammatory,and haemorrhagic properties. It is useful inpostnatal complaints, diarrhoea, chronic fever, biliousness, cough,vomiting, and asthma. It is an important ingredient of *dasmoolarishta* and *chyavanprash*.

Scientific name	:	*Ephedra gerardiana* Wallich.
Family	:	Gnetaceae
Common name	:	Somalata

Description: It is a small shrub, 1m tall. Stems are woody, highly branched. Branches are whorled, erect and ascending upwards. Leaves are reduced to minute 2-toothed sheaths. Male flowers are few, borne in ovate spikes. Female flowers are 1-2, borne in short spikes. Fruits are 7-10mm long, ovoid, red, surrounded by succulent bracts.

Cultivation: It can be cultivated at an altitude of 2500-3000m. It prefers a sunny situation and dry soil. Annual rainfall should not exceed 50cm. Plants are propagated by seeds or air layers or division of rootstocks. Seeds are sown early in the spring at a distance of 5cm keeping the distance of one metre between two rows.

Medicinal uses: The drug yields Ephedrine, used in the treatment of asthma, for relieving bronchial spasms. It is useful as a stimulant for heart. It proves to be a very effective cardiac stimulant.

Scientific name	:	*Gaultheria fragrantissima* Wallich.
Family	:	Ericaceae
Common name	:	Wintergreen

Description: This is an evergreen shrub, 3m tall with orange brown highly branched stems. Leaves are 13 cm long, rather broad, leathery, dotted with glands, margins toothed. Flowers are small, greenish white, borne in short axillary branches. Fruits are roundish, enclosed in blue sepals.

Cultivation: It prefers a moist but not boggy humus rich soil in shade or semi-shade. This is a peat and moisture loving species and it requires a lime-free soil. Plant can be propagated by seeds, semi-hardwood cuttings or divisions. Seeds require cold stratification to germinate.

Medicinal uses: The oil contains methyl salicylate, enanthic alcohol and an acid.Used as a counter irritant and treatment of rheumatism.The essential oil obtained from the leaves is antiseptic, aromatic, carminative and stimulant· Used as very effective vermicide against hook worms.The unripe fruits are chewed or made into a juice to treat stomach troubles. Oil is employed as flavouring agent for candies and also in perfumery.

Scientific name	:	*Hibiscus mutabilis* Linn.
Family	:	Malvaceae
Common name	:	Confederate Rose

Description: It is a rounded bushy deciduous shrub. The leaves are dark green, palmately lobed. Flowers are large showy, white, changes to white and then pink.

Cultivation: This is a hardy plant that can tolerate mild frosts. It prefers a warm and wet winter. A moderate, well drained fertile soil and full sun are ideal for its cultivation. Plants are propagated by cuttings. They are pruned to shape in first two years of growth and then trimmed lightly to maintain shape and density. The plants can also cut back all the one year old wood to encourage density when older.Generous dressings of slow release, balanced fertiliser in spring and early summer are required for its successful growth.

Medicinal uses: The leaves are anodyne, antidotal, demulcent, expectorant and refrigerant. With the flowers, they are applied to burns, swellings and other skin problems. The flowers are considered as antiphlogistic, depurative, febrifuge, pulmonary and stimulant. A decoction is used in the treatment of lung ailments.

Scientific name	:	*Hibiscus rosa sinensis* Linn.
Family	:	Malvaceae
Common name	:	China Rose

Description: This is a large shrub, 5m tall, a native of Indo-China region. The leaves are toothed, arranged alternately and vary a lot, but tend to be large, dark green, and shiny. The plants are coarse in texture and may be upright or broad and spreading. It is often many-stemmed. Flowers are glorious and huge at their best — up to 15cm in diameter — and occur in many colours. Most are flared and are bell shaped and may be single or double, smooth or scalloped. They have a long central tube with stamens and pistils at the tip.

Cultivation: The plants are sun loving and they do well in the summer months. They are propagated by seeds, cuttings, layering, budding and grafting depending upon the species and the cultivars. Cuttings of 2- 3 cm diameter and 10 – 15 cm long can be taken from well matured wood and placed in sand after treatment with IBA (5000-6000 ppm) for rooting. Many varieties of *Hibiscus rosa-sinensis* are propagated through grafting by using rootstocks such as 'Single Scarlet', 'Dainty' and 'Apple Blossom' during February to April or June to July. T-budding or air layerings are practiced during March to April, particularly for those cultivars difficult to start by cuttings. Rooted plants are grown in 20 – 25 cm pots containing a mixture of 2 parts of garden soil, 1 part of leaf mould and 2 tablespoons of bonemeal. Application of single superphosphate during monsoons and liquid manure during winter favours better growth and flowering.

Medicinal uses: The plants are rich in alkaloids, terpenoids, flavonoids, glycosides, reducing sugar, Fatty materials, saponins, gums and mucilage. Various parts of this plant like flowers, leaves and roots have been known to possess medicinal properties like oral contraceptive, laxative, aphrodisiac, menorrhagic etc. The flowers of this plant possess antifertility activity. The aqueous ethanolic extract of aerial parts of this plant was reported for its use in constipation and diarrhoea. In traditional medicine, the leaves of the plant are used in fatigue and skin disease. Powdered root of the plant is given for menorrhagia and the fresh root juice for gonorrhoea. Flowers of the plant are used in diabetes, epilepsy, bronchial catarrh and leprosy. An infusion of the petal is widely used in Ayurvedic medicine in India as a demulcent refrigerant drink in fever and decoction is given in bronchial catarrh.

Scientific name	:	*Holarrhena antidysenterica* (Roth) A.DC.
Family	:	Apocynaceae
Common name	:	Kurchi

Description: This is at all shrub, 3m tall. Leaves are10-30cm long, ovate and thin. Flowers are white, fragrant, 1-1.5cm diameter, borne in terminal bunches. Fruits are slender, cylindric, 20-45cm long, dark grey with white specks.

Cultivation: The plant grows well throughout India up to an altitude of 1100m. Conventional methods of propagation like cuttings, air layering, grafting are not sufficiently rapid and sufficient, thus *in vitro* techniques like micropropagation via multiple axillary branching offers a rapid means multiplication of elite germplasm.

Medicinal uses: Root and bark are used in amoebic dysentery. Barks are astringent, anthelmintic, amoebicidal, diuretic. Used in colic, dyspepsia, piles, diseases of the skin and spleen. Seeds are antibilious. Used for promoting conception, also for toning up vaginal tissues after delivery.The bark contains the alkaloids, regholarrhenine-A, -B, -C, -D, -E and-F; pubescine, norholadiene, pubescimine, kurchinin, kurchinine, kurchinidine, holarrifine, holadiene, kurchiidine, kurchamide, kurcholessine, kurchessine, conessine, cones-simile and isoconessimine, and the steroidal compounds kurchinicin and holadyson. Application of this herb is useful in Rh. Arthritis & Oestioarthritis.

Scientific name	:	*Lawsonia inermis* Linn.
Family	:	Lythraceae
Common name	:	Henna

Description: It is a medium sized or large much branched shrub. Leaves are opposite, 2-3cm, acute and sharp pointed. Flowers are small, white or pinkish, fragrant, borne in terminal large bunches. Fruits are small, pea sized, round and many seeded.

Cultivation: The plant is distributed in drier parts of the peninsula. The plants thrive well in arid to tropical and warm temperate conditions. They grow well at a temperature of 30-40°C and annual rainfall of about 420mm. The plants prefer sandy soils but can be grown in poor clayey soils with a pH range of 5.0-8.0. The crop requires intervention of dry sunny periods along with the cool nights for proper ripening and expression of high dye content in leaves. Hennas are propagated by seedlings or cuttings. Levelled nursery beds are applied with 10 tonnes of FYM, 30kg N and 40 kg P per ha for optimum sapling growth. Sowing is done during first fortnight of March. About 6-8kg seeds are required to raise seedlings for one hectare of land. Transplantation is done in furrows during rainy season in the month of July-August at a spacing of 45cm x 30cm. Young plants must be protected from browsing animals and can be watered to increase growth rate. Fields are hoed once or twice in a year and weeded when required. It is grown as rainfed crop and therefore, timely moisture conservation is essential to enhance productivity. Intercropping with legume crops can be practised for better economic return. Termite is the major pest in Henna field and it can be controlled by soil application of chlorpyriphos (10%) or Furadon 10G @ 25kg/ha. Henna takes around three years for economic leaf production. The leaves are harvested when they are fully mature and retain their green colour. Average productivity is 1-2 tonnes dry leaves per ha.

Medicinal uses: The leaves are used in medicine. These are astringent and are applied locally on boils, burns and skin diseases. Roots are regarded as a potent medicine for gonorrhoea and to enhance fertility in women; a decoction of them is considered to be diuretic or for treating blenorrhoea and pectoral for bronchitis. The leaves contain oxynaphthoquinone called lawsone, which has antibiotic properties. Flower oil contains alpha- and betaionone, the latter being the main component. Leaf and flower infusions are applied externally for ulcers and rheumatism or are taken orally for tetanus, epilepsy and stomach pains; leaves are used in treatment of leprosy, jaundice and scurvy. Astringent roots are ground and rubbed on the heads of children to treat boils and eye diseases.

Scientific name : *Leptadenia reticulata* Retz.
Family : Asclepiadaceae
Common name : Jeevanti

Description: This is a much branched laticiferous twining shrub with yellowishbrown, deeply cracked bark. Leaves are ovate, cordate, coriaceous, glabrous above and more or less finely pubescent beneath. Inflorescence is a many flowered cyme with greenish yellow flowers. Fruit is follicle, horned shaped and slender. Seeds are lanceolate and comose.

Cultivation: Jeevanti is distributed in the sub-Himalayan tracts of Punjab, Himachal Pradesh, Uttar Pradesh, and Deccan Peninsula up to an altitude of 900 m. The plant grows well in warm tropical and subtropical regions with moderate rainfall. Sandy-loam and alluvial black soils are found to be best for commercial cultivation. However, the crop can also be cultivated satisfactorily on red laterite soils. Rooted semi-wooded stem cuttings from six-month-old to one-year-old plants having three to four nodes at pre-flowering stage are generally used for planting. Seeds can also be collected in November for raising the plants. Cuttings (12–15 cm long, with three to four nodes) are treated with 200 PPM (parts per million) TRF (triademophon) or 25PPM PBZ (paclobutrazol) or 300 PPM IBA (indole-3-butyric acid) or NAA (naphthalene acetic acid) for obtaining good rooting and for better field establishment of the rooted cuttings.

At a spacing of 2 m × 1 m, about 5000 plantlets would be required for planting in one hectare of land. Pits of 45 cm × 45 cm × 45 cm size are dug at a spacing of 2 m × 1 m and should be filled with topsoil and well decomposed FYM in the ratio 1:1. FYM at the rate of 20 tonnes/hectare and NPK(nitrogen, phosphorus, potassium) at the rate of 80:160:160 g/plant are used as a basal dose in first year. Full dose of FYM should be applied to the soil while preparing the land. Rooted cuttings are planted in the centre of the pits of size 45 cm3 at a spacing of 2 m × 1 m and filled with topsoil and well-decomposed FYM. Staking is done 20 days after planting along with gap filling. Application of 2 tonnes FYM per hectare and NPK in the ratio 100:200:200 per plant in subsequent years is recommended. After two months of planting, full dose of phosphorus and potassium and 50% nitrogen should be applied in a band all around the plant by opening furrows. The remaining 50% nitrogen should be top dressed during June. Top dressing should immediately be followed by earthing up and irrigation. The crop is given furrow irrigation twice a week for about two to three months after planting in the field. Powdery mildew is a serious problem during the winter months. Control measures involve spraying wettable sulphur (sulfex) at the rate of 3%, at weekly interval for three to four times. During the rainy season, wilting is occasionally noticed and can be controlled by taking phytosanitary measures and drenching the affected vines with carbendiazim (0.15%). Aphid and mite infestations are noticed at various stages of crop growth and can be controlled by prophylactic sprays of monocrotophos (0.15%) and dicofal (0.2%), respectively. The crop remains in the field for 10–15years. Best time for harvesting is once in six months at 50% flowering, without taking out whole roots that serve as future planting material/ rootstock in the same field. A yield of 6–7 tonnes dry weight of roots per hectare per year may be obtained.

Medicinal uses:A fructosan of insulin type has been reported from the roots of the plant. Leaves and roots species are useful in treating skin infections and wounds. They are used to prevent prolapse of uterus and vulva in controlling habitual abortion in women. The roots are used in cardiac disease and haemorrhage, as diuretic, and to cure fever and opthalmia.They are also used as a tonic for general debility to strengthen the body. Leaves are also used as galactagogue for nursing mothers.

Scientific name	:	*Nerium oleander* Retz.
Family	:	Apocynaceae
Common name	:	Oleander

Description: The plants are 2 to 2.5 metres tall, glabrous with green and thin shoots. The leaves are arranged in whorls of three, linear lanceolate, thick in texture and with prominent veins. Flowers are single or double, sweet scented, pink or white and borne in a large terminal panicle. Several varieties like 'Album', 'Carnea', 'Carneum flore-pleno', 'Roseum', 'Apple Blossom' 'Scarlet Beauty' and 'Variegated Plenum' are available in this species.

Cultivation: The plants thrive well in tropical and sub-tropical climate and bloom profusely during summer season. They are multiplied by cutting or layering during rainy season. These plants are hardy and cattle proof. Slight pruning is needed to remove old growth after the commencement of new growths to form the clusters of buds. About 20-25cm pot containing two parts sandy loam and one part each of leaf mould and well rotten cowdung manure used for rooted cuttings or layerings.

Medicinal uses: Oleandrin is useful in treating cardiac conditions in patients who cannot tolerate digitalis. In traditional medicine, the leaves are used for a variety of medicinal purposes, including the treatment of heart diseases, as a diuretic, antibacterial, and against snake-bite. The roots are used externally in traditional medicine for treating cancer, ulcers and leprosy.

Scientific name	:	*Peganum harmala* Linn.
Family	:	Rutaceae
Common name	:	Harmal

Description: This is a shrub, 30-90 cm tall. Leaves are 5-8cm long, divided into numerous narrow segments. Flowers are 2-3cm in diameter, white, single borne in axils of leaves. Fruits are capsular, globose, deeply lobed. Seeds are brownish, with reticulated seed coat.

Cultivation: The plant is distributed in northern and north-western India and in drier parts of the Deccan. It is grown easily from seed sown in Spring, taking about two weeks to germinate. The plant self sows freely. Soil should be fairly rich, sandy, and well drained. Full sun is required. The plants die back to the crowns every winter, and should be mulched where temperatures dip below 10°F. An alkaline pH between 7.3 and 8 is preferable.

Medicinal uses: The dried seeds are used in the medicine. The seeds contain several alkaloids like harmaline, yagenine and harmine and are useful in asthma, hysteria, rheumatism, gallstones, colic pains, fever, jaundice and painful menstruation. They are effective as narcotic, anthelmintic and emetic.

Scientific name	:	*Plumbago zeylanica* Linn.
Family	:	Plumbaginaceae
Common name	:	Chitrak

Description: It is a perennial undershrub, 1.5–2.0 m tall, with rambling branches. Its active growth occurs during rainy season and ceases inpost-flowering stage. Leaf is simple, opposite, oval, pointed, smooth, and shiny. Roots are light coloured inside,when fresh and reddish brown, when dry. Flowers are bisexual and whitein colour. Fruits are green coloured, with sticky hairs when young and become dark brown when mature.

Cultivation: Although Chitrak can be grown in a wide variety of soils, ranging from red laterite soil, with very little topsoil, to deep black soil. However, it thrives well drained/ deep sandy loam to clayey loam soil with high organic content.In natural habitats, the plant grows well in moist soil with high organic content and partially shaded locations. Chitrak is easily be propagated through stem cuttings or seeds. Stem cuttings of 10–15 cm length, having at least three nodes, are obtained from mother plants in March–April to raise the stock. About 80,000 rooted stem cuttings or seedlings are required for planting in one hectare of land. FYM at the rate of 10 tonnes/hectare is applied 30 days before planting at the time of ploughing and land preparation. Nitrogen @ 30 kg/hectare and phosphorus and potassium@ 40 kg/hectare and 30 kg/hectare, respectively, are also applied as basal application. Transplanting is done after 60–75 days of growth of cuttings. The optimum spacing recommended in the field is 50 cm × 25 cm, which produces maximum root biomass. Inorganic nitrogen at the rate of 30 kg/hectare is applied one month after transplantation. This is in addition to the basal dose of FYM (10 tonnes/ hectare) and fertilizers applied at the time of land preparations. First weeding should be done one month after planting in August. Second and third manual weedings are done in October and December, respectively. Pruning can be done in May before harvesting the crop. Plant attains maturity in 10–12 months after transplanting. Best time for harvesting is 12 months after sowing. Dry root yield varies from 12 quintals/ hectare to 18 quintals/hectare under optimum conditions.

Medicinal uses: Chitrak roots are rich in a yellow, crystalline, and bitter principle called plumbagin. It is a reputed thermogenic, astringent, anthelmintic, abortifacient,carminative, appetizer, and expectorant. Roots stimulate the central nervous system. Oil prepared from roots is useful in rheumatism, joint pain, and paralysis. Milky juice of leaves is used for external application in scabies.

Scientific name	:	*Premna integrifolia*Linn.
Family	:	Verbenaceae
Common name	:	Agnimanth

Description: Thisis a scandent, erect shrub or small tree, more or less thorny on the trunk and large branches. Leaves are opposite or whorled and entire or serrate. Flowers are small, irregular, arranged in terminal corymbs. Fruits (drupes) and seeds are oblong in shape.

Cultivation: The plant prefers a warm and humid climate. It thrives well in sandy loamtype of soil with good organic content. At least one-year-old stem cuttings are used as planting material, which are procured from mature trees in February–March. About 2500 rooted cuttings are required for one hectare of plantation at a spacing of 2 m × 2 m. Pits of appropriate size, that is, 45 cm × 45 cm × 45 cm, are dug at the recommended distance and filled with a mixture of sand, soil, and FYM in 1:1:1 ratio. They are planted in the prepared pits just before or at theonset of monsoon. NPK (nitrogen,phosphorus, and potassium) fertilizers each@ 50 g per plant should be applied, preferably during rainy season. Intercultural operations including manual weeding and hoeing ate done at regular intervals as and when required. However, three weedings in the first six months of planting are required. Irrigation is required at an interval of 15–30 days during dry season, especially in the first year, that is, from December to May. Harvesting should be done thereafter towards the end of rainy season, that is, September–October. Dry root yield is about 500–850 g per tree and about 1250 kg root/hectare is obtained from a three-year-old plantation.

Medicinal uses: Roots are anti-inflammatory, stomachic, and anti-pyretic, and is prescribed in liver complaints, cold, obstinate fevers, flatulence, urticaria, rheumatism, and neuralgia. The roots of the plant are constituent of Ayurvedic medicine 'dashamula' and used for a various ailments.

Scientific name	:	*Punica granatum*Linn.
Family	:	Punicaceae
Common name	:	Pomegranate

Description: It is a large deciduous shrub, 10m tall with smooth dark grey bark and often spinescent branchlets. Leaves are opposite, glabrous, minutely pellucid-punctuate, shining above and bright green beneath. Flowers are scarlet red or sometime yellow, mostly solitary, sometimes 2-4 held together. Fruits are globose, crowned by the persistent calyx. Rind is coriaceous and woody, interior septate with membraneous walls containing numerous seeds. Seeds are angular with red, pink or whitish, fleshy testa.

Cultivation: Pomegranate is of deciduous nature in hilly areas where winters are cold, but on the plains it is evergreen. A hot dry summer aids in the production of best fruits. Plants are grown from seeds as well as cuttings. Mature wood pieces cut into lengths of about 30cm are planted for rooting. The rooted plants are planted 4.5-6m apart. When planted close, they form a hedge which also yields fruits. Normal cultivation and irrigation practices are satisfactory for the pomegranate. An application of 30-45kg of FYM annually to each tree aids to produce superior quality fruits. The pomegranate may be trained as a tree with a single stem for 30-45cm or as a bush with 3 or 4 main stems. In either case suckers arising from the roots and similar growths from the trunk and main branches are removed once a year. Shortening of long slender branches and occasional thinning of branches should be done.

Medicinal uses: Pomegranate fruit rind yields an ellagitannin named granatin B, punicalagin, punicalin and ellagic acid. Barks are rich in alkaloids such as iso-pelletierine, pseudopelletierine,methyl isopelletierine, methyl pelletierine, pelletierine as well as iso-quercetin, friedelin, Dmannitol and estrone. Flowers contain pelargonidin-3, 5-diglucoside apart from sitosterol,ursolic acid, maslinic acid, asiatic acid, sitosterol-b-D-glucoside and gallic acid. Seeds give malvidin pentose glycoside. Rind contains pentose glycosides of malvidin and pentunidin.Fluoride, calcium, magnesium, vitamin C and phosphate are also reported from fruits. Leaves yield elligatannins-granatins A and B and punicafolin.Pomegranate has long been esteemed a diet inconvalescence after diarrhoea. The rind of the fruit is highly effective in chronic diarrhoea and dysentery, dyspepsia, colitis, piles and uterine disorders. The powdered drug boiled with buttermilk is an efficacious remedy for infantile diarrhoea. The root and stem bark are good for tapeworm and for strengthening the gums. The flowers are useful in vomiting, vitiated conditions of *pitta*, ophthalmodynia, ulcers, pharyngodynia and hydrocele. An extract of the flowers is very specific for epistaxis. The fruits are useful in anaemia, hyperdipsia, pharyngodynia, ophthalmodynia, pectoral diseases, splenopathy, bronchitis and otalgia. The fruit rind is used for treatment of dysentery, diarrhoea and gastralgia. Seeds are used in scabies,hepatopathy and splenopathy.

Scientific name	:	*Rauvolfia serpentina*Benth. ex Kurz.
Family	:	Apocynaceae
Common name	:	Sarpagandha

Description: This is an erect perennial shrub generally 15-90 cm tall. Roots nearly verticle, tapering up to 15 cm thick at the crown and long giving a serpent-like appearance, occasionally branched or tortuous developing small fibrous roots. Roots are greenish-yellow externally and pale yellow inside, extremely bitter in taste. Leaves are borne in whorls of 3-4 elliptic-lanceolate or obovate, pointed. Flowers are numerous borne on terminal or axillary cymose inflorscence.Fruit is a drupe, obliquely ovoid and purplish black in colour at maturity with stone containing 1-2 ovoid wrinkled seeds.

Cultivation: Among the different species of *R. serpentina* is preferred for cultivation because of higher reserpine content in the root. It grows well in tropical humid climate. It is naturally distributed in common habitats receiving an annual rain fall of1500-3500 mm and the annual mean temperature is 10-38 °C. It grows up to an elevation of 1300-1400m from msl. It can be cultivated in open as well as under partial shade conditions. It is grown on a wide range of soils. Medium to deep well drained fertile soils and clay-loam to silt-loam soils rich inorganic matter are ideal for its cultivation. It requires slightly acidic to neutral soils for good growth.The plants are propagated vegetatively by root cuttings, stem cuttings or root stumps andby seeds. Seed propagation is the best method for raising commercial plantation. Seed germination is very scanty and variable from 10-74%. Seeds collected during September to November give good results. It is desirable to use fresh seeds and to sock in 10% sodium chloride solution. Those seeds which sink to the bottom should only be used. Seeds are treated with ceresan or captan before planting in nursery to avoid damping off. Seed rate is 5-6 kg/ha. Nursery beds are prepared in shade,well rotten FYM is applied at 1kg/m^2 and seeds are dibbled 6-7cm apart in May-June and irrigated.Two months old seedlings with 4-6 leaves are transplanted at 45-60 x 30 cm spacing in July –August in the main field. Alternatively, rooted cuttings of 2.5-5cm long roots or 12-20cm long woody stems can also be used for transplanting. Hormone (Seradix) treatment increases rooting . In the main field10-15 t/ha of FYM is applied basally. Fertilisers are applied at 40:30:30kg N: P_2O_5 :K_2O/ha each year. N is applied in 2-3 splits. Monthly irrigation increases the yield. The nursery and the main field should be kept weed free by frequent weeding and hoeing. In certain regions intercroping of soybean,brinjal, cabbage, okra or chilly is followed in Rauvolfia crop. Pests like root grubs (*Anomala polita*), moth (*Deilephila nerii*), caterpillar (*Glyophodes vertumnalis*), black bugs and weevils are reported on the crop, but the crop damage is not severe.The common diseases reported are leaf spot (*Cercospora rauvolfiae*, *Corynespora cassiicola*), leafblotch (*Cercospora serpentina*), leaf blight (*Alternaria tenuis*), anthracnose (*Colletotrichum gloeosporioides*), die back (*Colletotrichum dematrium*), powdery mildew (*Leviellula taurica*), wilt(*Fusarium oxysporum*), root-knot (*Meloidogyne* sp.), mosaic and bunchy top virus diseases. Field sanitation, pruning and burning of diseased parts and repeated spraying of 0.2% Dithane Z-78 orDithane M-45 are recommended for controlling various fungal diseases. Rauvolfia become ready for harvest after 2-3 years of growth. The optimum time of harvest is in November -December when the plants shed leaves, become dormant and the roots contain maximum alkaloid content. Harvesting is done by digging up the roots by deeply penetrating implements.

Medicinal uses: Roots contain 1.4-3% alkaloids. The alkaloids are classsified into 3 groups, *viz*, reserpine,ajmaline and serpentine groups. Reserpine group comprising reserpine, rescinnamine,deserpine *etc* act as hypotensive, sedative and tranquillising agent. Ajmaline, ajmalicine, ajmalinine; iso-ajmaline *etc* of the ajmaline group stimulate central nervous system, respiration and intestinal movement with slight hypotensive activity. Serpentine group comprising serpentine, sepentinine, alstonine *etc* is mostly antihypertensive. The root is a sedative and is used to control high blood pressure and certain forms of insanity. In Ayurveda it is also used in insomnia, epilepsy, asthma, acute stomach ache and painful delivery. It is useful in snake-bite, insect stings, and mental disorders. The juice of the leaves is used as a remedy for the removal of opacities of the cornea.

Scientific name	:	*Ricinus communis*Linn.
Family	:	Euphorbiaceae
Common name	:	Castor

Description: It is a monoecious evergreen shrub growing upto 4m. Leaves are alternate, palmatifid, 6-10 lobed, each 1-nerved with many lateral nerves and peltate.Paniculate racemes are terminal with male flowers below, female ones above.

Cultivation: Castor is cultivated both in the plains and the hills. It is hardy and capable of resisting drought. It does not withstand waterlogging and frost. It requires hard dry climate for proper development of fruits and seeds. It requires a well drained soil, preferably sandy loam or loamy sand. High soil fertility is of less importance as compared to the good physical condition of the soil. It cannot tolerate alkalinity. It is generally grown in red loamy soils, black soils and alluvial soils. The plant is propagated by seeds. The seed rate required is 5-12 kg/ha (pure crop) and 3 kg/ha (mixed crop). Seeds are sown on a hot bed early in March. When the plants come up, individual plant is planted in a separate pot filled with light soil and plunged into a fresh hot bed. The young plants are to be kept in glass houses till early June where they are hardened and kept out. The suitable season of growing is *kharif* season. The crop is usually sown in April and planting is done in early July. The land is ploughed 2-3 times with the onset of rains and is repeated after rain. The spacing recommended is 60 x 90cm in case of pure crop but it is seldom cultivated pure. It is usually grown mixed with other crops such as jowar, arhar, chilli, groundnut, cowpea, cotton, etc. 10-15 tonnes FYM/ha and 50kg N, 50kg P_2O_5 and 20kg K_2O/ha will be sufficient. Addition of neem cake is beneficial as it increases oil content. There should be sufficient moisture in the field at the time of sowing. A month after planting, weeding and earthing up is to be done. The plant is attacked by hairy caterpillar, castor semi-looper, castor seed caterpillar, etc. which can be managed by integrated pest management measures. The leaf blight disease occurring in castor can be controlled by spraying with Bordeaux mixture 2-3 times at 15 days interval. Harvesting of ripe fruits is done from the end of November till the end of February. The fruit branches are picked when they are still green before splitting and scattering of the seeds. The pods are to be heaped up in the sun to dry. Then the seeds are to be beaten with stick and winnowed. Roots, leaves, flowers, seeds and oil constitute the economic parts. The average yield is 500-600kg/ha.

Medicinal uses: The bean coat yields lupeol and 30-norlupan-3b-ol-20-one. Roots, stems and leaves contain several amino acids. Flowers are rich in apigenin, chlorogenin rutin, coumarin and hyperoside. Castor oil is constituted by several fatty acids. Seed coat contains 1. 50-1. 62% lipids and higher amounts of phosphatides and non-saponifiable matter than seed kernel.

It is considered as a reputed remedy for all kinds of rheumatic affections. They are useful in gastropathy such as gulma, *amadosa*, constipation, inflammations, fever, ascitis, strangury, bronchitis, cough, leprosy, skin diseases, vitiated conditions of *vata*, colic, coxalgia and lumbago. The leaves are useful in burns, nyctalopia, strangury and for bathing and fermentation and vitiated conditions of *vata*, especially in rheumatoid arthritis, urodynia and arthralgia. Flowers are used in urodynia and arthralgia and glandular tumours. Seeds are useful in dyspepsia and for preparing a poultice to treat arthralgia. The oil from seeds is useful for all ailments caused by *vata*and *kapha*. It is also recommended for scrotocele, ascites, intermittent fever, gulma, colonitis, lumbago, coxalgia and coxitis.

Scientific name : *Sida corditolia* Linn.

Family : Malvaceae

Common name : Bala

Description: It is a small branched shrub, covered with minute star shaped hair present all over the plant. Leaves are 2-5cm, ovate to round, thick, with toothed margin. Flowers are yellow, small. Fruits are 6-8mm diameter, divided into 7-10 parts , each strongly reticulated and with two awns in tip.

Cultivation: It is widely distributed along with other species are common throughout the tropical and subtropical plains all over India and Srilanka up to an altitude of 1050 m., growing wild along the roadside. It grows as wasteland weed in drier sandy locations. *Sida cordifolia* produces large amounts of seed and can be easily propagated by seed.

Medicinal uses: It contains Ephedrine, hypaphorine, vasicinone, choline, betaine, phytosterol etc. Bala is an ancient Ayurvedic herb, used widely in a variety of Ayurvedic medicines and oils to improve strength of bones, muscles and joints. Root juice is useful in promoting healing of wounds. Seeds are used in gonorrhoea and colic pains.

Scentific name	:	*Solanum viarnum*
Family	:	Solanaceae
Common name	:	Khasi Kateri

Description: It is a large semi-woody and hardy plant, 75-100cm tall. It is covered with straight and curved prickles over the stem and foliage. The leaves are large, ovate and lobed. Flowers are white 1-4 in group in axillary racemes. Fruit is globose, green berry, containing small brown seeds.

Cultivation: It is easily grown in wide range of soils and climatic conditions. Well drained sandy loam solis of medium fertility in plains are most suited for its commercial cultivation. In Central india, submarginal lands with irrigation are used for its commercial cultivation. It is sown in nursery in September-October or February when temperature is between 20-25°C.

Usually, 250g seed is required to sow 4 nursery beds of 10m x 1.2m size which yield enough seedlings to plant one hectare crop. The seeds are treated with 0.1% Bavistin to protect from damping off disease. The nursery beds are treated with 0.25% solution of copper in soil. They are covered with a thin layer of soil. The seeds germinate in 15 days and grow up to 10-15cm height, with 4-6 leaves in next 40-45 days. The seedlings are ready for field planting. The field is prepared to a fine tilth and added 2.5-4.5 tonnes of farmyard manure/ha together with 20kg of Aldrin (10%) at land preparation.About 20, 60 and 30kg of N, P and K are applied basally, whereas 20kg of N/ha is given in rows 60 and 75 days after planting. The plants grow fast under moist, dry, warm atmosphere. Arka Sanjeevani grows well at 30cm x 45cm or 40cm x 40cm spacing, whereas for Glaxo and BARC the recommended spacing is 75cm x 75cm or 60cm x 60cm. Higher N fertilizers (120kg) give good response in growth and berry formation in marginal lands. The crop requires 2-3 weeding-cum-hoeings and 5 light irrigations. The plants produce flowers 3 months after sowing and fruiting occurs in next 2 months. Fruit set is high at temperature of 27° -31°C but it is reduced when temperature rises higher than 35°C, causing premature fall. The berries are harvested when they become yellowish colour. Two to 4 pickings are done at weekly interval. These fresh berries contain 70-75% water content and need to be dried in the sun very fast since they are prone to fungal infection. The drying of berries should be protected from rain water as solasodine is soluble in water and is lost. Drying through hot air blast at 50°C is preferred. The dry berry yield is 1.8-2 tonnes/ha and contains 2.5% solasodine.

Medicinal uses: The berry pulp is rich in solasodine alkaloid which is used in the production of steroids. It is used in preparation of contraceptive pills, corticosteroids and sex hormones.

Scientific name : *Uraria picta* Desv.
Family : Fabaceae
Common name : Dabra

Description: This is an erect, undershrub, 60–75 cm tall, with several branches. Leaves are generally three to five in number, up to nine-foliate. Purple flowers are borne in dense, cylindrical racemes with bracts. Fruits are pods with three to six joints.

Cultivation: The plant grows well in tropical and subtropical areas. Loam to clayloam soil is suitable for its cultivation. It can tolerate a soil pH up to 8.0. The crop is raised successfully by seeds, which can be collected in December–January. About 4–5 kg seeds are required forraising stock for planting in one hectare of land. Overnight soaking ofseeds in water before sowing improves germination. Organic manure, preferably FYM is recommended@ 10 tonnes/hectare at the time of field preparation. DAP (di-ammoniumphosphate) @ 100 kg/hectare is also recommended as basal dose. Mycorrhizal association has been found to be beneficial for the crop.Proper drainage should be ensured in the field to avoid waterlogging, which causes death of plants. Transplanting of 50–60-day-old seedlings is done in the well-prepared field on ridges. Approximately,111 000 saplings are accommodated in 1 hectare of land at an optimum spacing of 30 cm × 30 cm. Manual weeding is practised twice at 25, 45, and 90 days after transplantation. Earthingup of plants is done at the time of second weeding. The plant roots are harvested in December or May after flowering, which occurs twice. However, for good yield, roots may be dug/harvested in May–June after about10 months of growth. Watering the crop is stopped three weeks before the intended harvest time. The yield of dry roots is approximately3–4 quintals/hectare, while the dry weight of herbs is about 4–5 quintals/hectare.

Medicinal uses: This species is useful in quick healing of bone fractures. It is used as acardio and nervine tonic and has anti-inflammatory, expectorant,and diuretic properties. The root of the plant is one of the ingredients of'dasamoola' in Ayurveda.

Scientific name : *Vitex negundo* Linn.
Family : Lamiaceae
Common name : Chinese Chaste Tree/Nirgundi

Description: It is an aromatic shrub with quadrangular, densely whitish, tomentose branchlets. The bark is reddish-brown. Leaves are digitate, with five lanceolate leaflets, sometimes three. The numerous flowers are borne in panicles, 10 to 20 cm in length, whitish or purple in colour. The fruit is a succulent drupe, 4 mm in diameter, rounded to egg-shaped. It is black or purple when ripe.

Cultivation: The plants are found almost everywhere though mostly on wastelands from the seashore to an altitude of about 1,500 m in the outer Himalayas. It thrives well in humid places or along water courses in wastelands and mixed open forests. It often grows gregariously and is abundantly distributed on sandy soils. The mature seeds sown in nursery beds normally germinate within 2–3 weeks. Four- to six-month-old seedlings are used for transplanting in the field. It can be propagated readily from shoot cuttings. *Vitex negundo* roots are strong and deep and suckers profusely. It produces root suckers which can also be utilized as planting material.

Medicinal uses: Leaf juice contain casticin, isoorientin, chrysophenol D, luteolin, p–hydroxybenzoic acid and D-fructose. The main constituents of the oil are sabinene, linalool, terpinen-4-ol, β-caryophyllene, α-guaiene and globulol constituting 61.8% of the oil. The plantis used for treating stored garlic again pests and as a cough remedy in the Philippines. Roots and leaves are useful in eczema, ringworm and other skin diseases, liver disorders, spleen enlargement, rheumatic pain, gout, abscess, backache.Seeds are used as vermicide. It is also used to control population of mosquitoes.

Scientific name	:	*Withania somnifera*Dunal
Family	:	Solanaceae
Common name	:	Indian Ginseng/ Aswagandha

Description: It is an erect, evergreen, tomentose shrub, 30-75cm in height. Roots are stout, fleshy, cylindrical, 1-2cm in diameter and whitish brown in colour.Leaves are simple, ovate, glabrous and opposite. Flowers are bisexual, inconspicuous,greenish or dull yellow in colour borne on axillary umbellate cymes, comprising 5 sepals, petals and stamens each; the two celled ovary has a single style and a bilobed stigma. The petals are united and tubular. The fruit is a small berry, globose, orange red when mature and is enclosed in persistent calyx.The seeds are small, flat, yellow and reniform in shape and very light in weight.

Cultivation: Aswagandha is a tropical crop grown well under dry climate. The areas receiving 600 to 750mm rainfall is best suited for its cultivation. Rainy season crop requires relatively dry season and the roots are fully developed when 1-2 late winter rains are received. Sandy loam or light red soils having a pH of 7.5- 8.0 with good drainage are ideal for its cultivation. It is usually cultivated on poor and marginal soils. *Withania* is propagated through seeds. It is a late *kharif* crop and planting is done in August. Seeds are either broadcast-sown or seedlings are raised in nursery and then transplanted. Seed rate is 10-12 kg/ha for broadcasting and 5kg/ha for transplanting. In direct sown crop plants are thinned and gap filling is done 25-30 days after sowing. Seeds should be treated with Dithane M-45 at 3g/kg of seeds before sowing. Seeds are sown in the nursery just before the onset of rainy season and covered with light soil. Seeds germinate in 6-7 days. When seedlings are six weeks old they are transplanted at 60cm in furrows taken 60cm apart. The crop is mainly grown as a rainfed crop on residual fertility. However, application of organic manure is beneficial for realizing better yields. It is not a fertilizer responsive crop. One hand weeding 25-30 days after sowing helps to control weeds effectively. No serious pest is reported in this crop. Diseases like seedling rot and blight are usually observed. Seedling mortality becomes serious under warm and humid conditions. The disease can be minimized by use of disease free seeds and treatment with thiram or deltan at 3-4g/kg seed before sowing. Further, use of crop rotation, timely sowing and keeping field well drained also protect the crop. Spraying with 0.3% phytolan, dithane Z-78 or dithane M-45 will help controlling the disease incidence. Spraying is repeated at 15 days interval if the disease persists. Aswagandha is a crop of 150-170 days duration. The maturity of the crop is judged by the drying of the leaves and reddening of berries. Harvesting usually starts from January and continues till March. Roots, leaves and seeds are the economical parts. The entire plant is uprooted for roots, which are separated from the aerial parts. The berries are plucked from dried plants and are threshed to obtain the seeds. The yield is 400-500kg of dry roots and 50-75kg seeds per hectare.

Medicinal uses: Aswagandha roots are rich in alkaloids, starch, reducing sugar, hentriacontane,glycosides, dulcital, withaniol acid and a neutral compound. Eight Amorphous bases found are withanine, somniferine, somniferinine, somnine, withananine, withananinine,pseudowithanine and withasomnine. Other alkaloids reported are nicotine, tropine,pseudotropine, 3, α-tigloyloxytropane, choline, cuscudohygrine, anaferine, anahygrine andothers.Roots, leaves and seeds are used in Ayurvedic and Unani medicines, to combat diseases ranging from tuberculosis to arthritis. Roots are used in medicines for hiccup, several female disorders, bronchitis, rheumatism, dropsy, stomach and lung inflammations and skin diseases. Roots and paste of green leaves are used to relieve joint pains and inflammation. It is also an ingredient of medicaments prescribed for curing disability and sexual weakness in male.Leaves are used in eye diseases. Seeds are diuretic. It is a well- known rejuvenating agent capable of imparting long life, youthful vigour and intellectual power. It improves physical strength andis prescribed in all cases of general debility.

Scientific name	:	*Woodfordia fruticosa* (Linn.) Kutz.
Family	:	Lythraceae
Common name	:	Dhataki

Description: It is a deciduous shrub, usually with a much-fluted stem. When in flower, the bush appears twiggy and formless but entirely swathed in red. This is because the small flowers grow singly or in groups all the way along the branches and side twigs, and it is at this time that the leaves fall. Each flower, borne on a tiny stem, is a slender tube, slightly curved, the greenish base of which is the sepal. Swelling slightly, the tube divided into narrow, pointed lobes and from within emerges a bunch of long stamens. The whole length, including the stamens, is not more than 2 cm. The fruit is a small, oblong capsule, covered by the withered sepals.

Cultivation: It is commonly distributed in Sri Lanka, South Konkan and on the Ghats and ascends the Himalayas to 1500 m, but is rarer in South India.North eastern India due to heavy rains forover 8 months the mature seeds fail to germinate or loose its viability for germination and vegetative method is only a sustainable method of propagation. Branch cutting is the mode of propagation for the plant and 4"-6"branches placed in shady and moist soil start rooting and young shoot branches also generate which give rise to young plants suitable for transplantation. Branch cuttings planted during June-July under shade give rise to new plant. After transplantation in field during September-October develops new field for further harvesting. Within 2 years these newly planted plants start flowering and producing flowers for utilization.

Medicinal uses: Stem contains C-glycoside norbergenin, yields gum, Beta-sitosterol. Leaves contain lupeol, betulin, Beta-sitosterol, ursolic acid, betulinic acid, woodfruticosin, lawsone, dyeing principle of henna. Flowers contain woodfordins A-D, oenotherin A. Leaves and flowers contain ellagic acid, cyanidins, Beta- sitosterol. Flowers are astringent, refrigerant, haemostatic, anthelminthic, dessicative, enhance wound healing etc. It is commonly used to cure diarrhoea, menorrhagia, and bleeding piles. The bark is also pungent, acrid, cooling, toxic, alexatric, antihelmintic, antidysentric, and uterine sedative. The bark is used in thirst, dysentery, leprosy, erysipelas and diseases of the blood. Antitumor, DNA inhibitory, immunomodulatory, antioxidant, anti-proliferative, antihyperglycemic, anti-inflammatory, analgesic, cytotoxic, antibacterial, hepatoprotective, antimicrobial and anti-ulcer activities were found in various parts of this plant.

Scientific name	:	*Zanthoxylum armatum* DC
Family	:	Rutaceae
Common name	:	Timru

Description: Timru is an evergreen, thorny shrub, attaining a height up to6 m. Leaves are 4–20 cm long, imparipinnate, pungent, and aromatic with glabrous, narrowly winged petiole having two stipular prickles at the base.Leaflets are lanceolate, glabrous on the underside, and occur in two to sixpairs. Flowers are borne in dense terminal or sparse axillary panicles and are greento yellow in colour. Ripe carpels or follicles are usuallysolitary, pale red, and tubercled. Seeds are globose, shining, and black.

Cultivation: The plant prefers subtropical climate of lower warm valleys of the Himalayas with sufficient rainfall for growth and production. It grows well in open pastures and secondary scrub forests. Loamy or clayey soil rich in organic content is suited for its cultivation. Freshly harvested seeds are best for the large-scale cultivationof *Zanthoxylum* species. Mature seeds are collected in June–July. In the absence of sufficient seeds,terminal stem cuttings may be used as propagules. About 10–12 tonnes per hectare of FYM are mixed with the soil as a basal dose before transplantation. After10–12 months of growth in the nursery, transplanting in the field is done during May–June if irrigation facilities are available. Otherwise, it is done in July–August (monsoon season) underrain-fed conditions. A spacing of 50 cm × 50 cm is recommended for accommodating about 40 000 plants per hectare. Weeding is carried outduring the initial establishment phase, 30–45 days after transplanting, and again at 60–80 days after transplanting. The crop takes five to seven years to mature. The yield of fresh and dry fruits is about 20 quintals per hectare and 6.50 quintals per hectare, respectively.

Medicinal uses: It contains volatile oils with active constituents linalool, limonene and lignin. Fruits, seeds, and bark are used as aromatic tonic in dyspepsia and fever. Fruits and seeds are useful in dental troubles, thus used to prepare dental paste and powder. Tender twigs are used in brush teeth and used as a remedy for toothache. The essential oil from fruits (known as Wartara oil) has deodorant and antiseptic properties.

10

Cultivation of Important Medicinal Climbers

The recognized plants for medicinal uses mostly belonging to trees, shrubs and herbs, but there is an another group of plants known as "Climbers" an important group in angiosperms evolution. Climbing plants are one of the most interesting group but a much neglected group of plants. But, these neglected climbers contribute largely to the attractiveness of our landscapes by the manner in which they climb over trees, hedgerows or rocks. Climbers depend on other plants for mechanical support due to their weak stem and they attach themselves to any neighbouring object by means of some special organs of attachment. They show great diversity in their climbing mechanism depending on which they are classified as root climbers, hook climbers, tendril climbers, leaf or stem climbers or twinners.

They represent a large sector of medicinal, ornamental, and play a key role in horticulture sector. Climbers also form an essential part of diet of many insects at the time of scarcity of flowers and fruits. Climbers are important components of forest ecosystem. Climbers contribute to their environment by purifying air quality, water, conserving water, preserving soil, and supporting wildlife. They also lower down the air temperature and reduce the heat intensity of the greenhouse effect by maintaining the low levels of carbon dioxide. Leguminous climbers have nodulation or nitrogen fixation ability in their roots. Climbers and creepers are important components of plant diversity and are also valuable for their medicinal uses, nutrient recycling, etc.

Scientific name : *Abrus precatorius* Linn.

Family : Fabaceae

Common name : Gunja

Description: It is a climbing, twining, or trailing woody vine with slender herbaceous branches. Leaves are alternate, petioled, 5-13 cm long, even-pinnately compound with 5-15 pairs of leaflets, oval to oblong with entire margins. Flowers are pea shaped, white to pink or reddish, small, in short stalked, dense clusters at leaf axils. Fruit is a short, oblong pod, splitting before falling to reveal 3-8 shiny hard seeds, long, scarlet with black bases.

Cultivation: It grows well in tropical areas in full sun or partial shade. Propagated through seeds. Pre-sowing treatment of the seeds with concentrated sulphuric acid for 120 minutes to break the seed coat dormancy and to achieve 95% germination is a promising approach. A well drained moist soil and frost free area are ideal for its cultivation.

Medicinal uses: Roots and leaves contain glycyrrhizin, isoflavanquinones, abrusquinone A, B & C; Leaves juice are used anti-diabetic drug .Seeds and roots are said to be highly a medicinal for rheumatism and swellings. Small dose taken along with goat milk in empty stomach. Roots are used for treating gonorrhoea, jaundice and haemoglobinuric bile.

Scientific name : *Antigonon leptopus* Linn.

Family : Polygonaceae

Common name : Coral Vine

Description: It is a fast-growing climbing vine that holds by tendrils, and is able to reach 8-10 metres in length. It has cordate (heart shaped), sometimes triangular leaves, 2.5 to 7.5 cm long. The flowers are borne in panicles, clusted along the rachis producing pink or white flowers from spring to autumn. It has underground tubers and large rootstocks.

Cultivation: It grows well in wet tropical and subtropical climate. It prefers to grow in areas with warm temperatures and high precipitation at low elevation upto 1000m. However, it can tolerate drought well by shedding leaves and re-growing strongly after rains. *A. leptopus* has the ability to grow in almost any soil type including sandy soils, seasonally waterlogged soils, and alkaline limestone soils. Plant are propagated through seeds, tubers, root suckers and stem segments.

Medicinal uses: Plants contain alkaloids, quinines, resins, tannins, fixed oils, flavonoids, fats, saponins phenolic compounds, proteins, and carboxylic acids. Bark, fruits, leaves and seeds are used in medicine. The leaves and flowers are used as a tea for coughs,colds and constriction of the throat. It is used as a nectar source for honey production. The infusion of dried leaves are used in jaundice.

Scientific name : *Asparagus racemosus* Willd.

Family : Liliaceae

Common name : Shatavar

Description: This is an armed climbing undershrub with woody terete stems and recurved or rarely straight spines. The tuberous succulent roots are 30cm to 100cm or more in length, fascicled at the stem base, smooth tapering at both ends. Young stems are very delicate, brittle and smooth. Leaves are reduced to minute chaffy scales and spines; cladodes are triquetrous, curved in tufts of 2-6. Flowers are white fragrant in simple or branched recemes on the naked nodes of the main shoots or in the axils of the thorns. Fruits are globular or obscurely 3-lobed, pulpy berries, purplish black when ripe; seeds with hard and brittle testa.

Cultivation: The plant thrives well under a wide range of tropical and subtropical climate. Fertile moist sandy loam soils are ideal for its cultivation. Better root development is observed in soils in increased proportion of sand. However, a decline in the yield of the crop is observed in soils containing previous year's residue of the roots. Asparagus plant is best grown from its tuberous roots even though it can be successfully propagated through seeds. Since root tubers are of commercial value seed propagation provides economic advantage to the farmers. Seeds usually start germinating after 40 days and average germination is 70%. For the cultivation of the crop, the land is ploughed well with pre-monsoon showers and seed nurseries are raised on seed beds of approximately 1m width, 15cm height and suitable length. Seed nursery is irrigated regularly and kept weed free. With the onset of monsoon in June-July the main field is ploughed thoroughly and pits of size 30cm cube are dug at a spacing of 60-100cm. The per hectare yields were highest in the closer spacing of 30cm x 30cm. The pit is filled with a mixture of top soil and well decomposed FYM or compost applied at 10 - 15 t/ha and the seedlings are transplanted. Application of N, P_2O_5 and K_2O at 60:30:30 kg/ha increases the root yield. Regular irrigation and weeding are needed to realize higher yields. Standards are to be provided for training the plant. Few pests and diseases are observed on this crop. Harvesting the crop after two years provided higher root yield than annual harvests in pots as well as in field experiments. Irrigating the field prior to harvest enables easy harvesting of the root tubers. The average yield is 10 - 15 t/ha of fresh root tubers though yields over 60t/ha have been reported.

Medicinal uses: Asparagus roots contain protein 22%, fat 6.2%, Carbohydrate 3.2%, Vitamin B 0.36%, Vitamin C 0.04% and traces of Vitamin A. Alcoholic extract yields *asparagin*- an anticancer agent. It also contains a number of antioxytocic saponins, viz. Shatavarisn - I to IV. Leaves are rich in rutin, diosgenin and a flavonoid glycoside identified as quercetin - 3 - glucuronide. Flowers contain quercetin hyperoside and rutin.

Asparagus has its widespread applications as diuretic, cooling agent and an excellent safe herbal medicine for ante-natal care. It is useful in nervous disorders, dyspepsia, diarrhoea, tumours, inflammations, vitiated conditions of *vata* and *pitta*, burning sensation, hyperdipsia, ophthalmopathy, nephropathy, hepatopathy, strangury, scalding of urine, throat infections, tuberculosis, cough, bronchitis, gleet, gonorrhoea, leucorrhoea, leprosy, epilepsy, fatigue, hyperacidity, colic haemorrhoids, hypertension, abortion, agalactia, cardiac and general debility. It contains an anticancer agent asparagin which is effective against leukaemia. It also contains active antioxytocic saponins which have got antispasmodic effect and specific action on uterine musculature. In Ayurvedic classics it is used as a cooling agent and uterine tonic. It is the main ingredient in ayurvedic medicines like *shatavari gulam* and *shatavari ghrtam*.

Scientific name	:	*Cissus quadrangularis* Linn.
Family	:	Vitaceae
Common name	:	Hadjod

Description: Plants are1.5m tall with quadrangular-sectioned branches with internodes 8 to 10 cm long and 1.2 to 1.5 cm wide. Along each angle, there is a leathery edge. Toothed trilobe leaves 2 to 5 cm wide appear at the nodes. Each has a tendril emerging from the opposite side of the node. Flowers are racemes of small white, yellowish, or greenish colour. Globular berries are red when ripe.

Cultivation: The plant has a wider adaptability and can be grown thought the year. *Cissus quadrangularis* are mainly propagated by cuttings. The plants are cut between the two nodes 30 cm and dipped in cow dung slurry and dried in the sun for one day. The next day the plants are planted in the prepared pits. A pit of 15×15×15 cm spacing at 5 ft is prepared. The pit is filled with one part of vermi-compost, one part of dry cow dung, one part of sand and one hand full of bone meal. The plants which are dried planted in the pit. The plants are watered. Usually, the plants are planted in straight line in both direction of the field. For first 15 days, they should be watered every alternate days after that only once in every four days (hot climate). Water stagnation near base should not be allowed The plants require rope and bamboo supports for staking. The plants should be cut between the two nodes in the center and dried in the shade, which will give more medicinal value.

Medicinal value: *Cissus quadrangularis* is known for its ability to help broken bones heal faster. When it is combined with a healthy diet and exercise plan, it can help promote weight loss. It can also reduce the risk of type 2 diabetes. It helps treat and prevent obesity. The plant is very high in Vitamin C. Vitamin C is an antioxidant nutrient that has been shown to reduce inflammation.

Scientific name	:	*Clitorea ternatea*Linn.
Family	:	Fabaceae
Common name	:	Blue Bell Vine

Description: It is a perennial herbaceous plant, arranged with elliptic, obtuse leaves. The most striking feature of plant is the color of its flowers, a vivid deep blue; solitary, with light yellow markings. They are about 4 cm long by 3 cm wide. Some varieties yield white flowers. The fruits are 5–7 cm long, flat pods with six to ten seeds in each pod.

Cultivation: The plants show best performance in the area of 1500 mm rainfall. They are fairly drought-tolerant, adapted to a wide range of soil conditions from sandy to deep alluvial loams and heavy black cracking clays. These grow on high pH soils. Seed is sown in spring to midsummer at 1 to 3 kg./ha on a well prepared seed bed at a depth of 1.5 to 4 cm and lightly covered. They can be planted on stakes for seed production. Dormancy can be broken by treatment with concentrated sulphuric acid for 20 minutes or by soaking for 12 hours in water, followed by 12 hours in a freezer at-15°C, then defrosted for planting.

Medicinal uses: This is one of four herbs traditionally used as *Shanka Pushpi*, an Ayurvedic medicine used to promote neurological health. It shows promise in animal models for its memory enhancing effects, and has a wide spectrum of neurological benefits (anti-depression, anxiolytic, anti-pyretic).

Scientific name	:	*Coccnia indica*Linn.
Family	:	Cucurbitaceae
Common name	:	Ivy Gourd

Description: Itis a perennial twin with tuberous roots; stem smooth or scabrous, grooved or angled; Leaves are palmately 5-nerved from a cordate base, 3-5 lobed, shining, margins glandular tipped; Flowers are white. Male flowers having peduncles one-flowered; Female flowers solitary; Fruits are fusiform-ellipsoid, slightly beaked, young marked with white steaks, bright scarlet when ripe, seeds embedded in red pulp, yellowish grey.

Cultivation: The plants thrive well in coastal climate. Since the plant is a long vine, it has to be trained over a trellis. The site elected must be free from all weeds & big stones. Pits around 1 meter wide and up to 2 or 3 feet deep are dug. 2 to 3 baskets of organic manure are added to each pit and the soil is mixed well so that the manure added is evenly mixed in the soil. Healthy stumps of the coccinia plant are chosen, and a pointed cut is given to the base of the stump for polarity. The selected stumps must be dipped in a solution containing 2 gram per litre carbofuran granules. This is done to prevent the soil pathogens, nematodes from attacking the stumps. In each pit 4 stumps must be planted preferably at the corners of the pit and soil added. Top part of the stumps must be supported by rope or nylon thread and tied to the trellis.!it is highly recommended to apply 2 baskets of fresh organic manure to every pit, every month along with 100 to 200 gramNPK. The most important in cultivation of coccinia is the timing of pruning the vines. Repeated prunings are recommended as the newly developing vines produce more flowers and yield more. Pruning the vines must be done every 3 to 4 months to maximise yield. It reequires good quantity of water, but cannot tolerate stagnation. During fertilizer application every month, earthing up of the soil is done to break the clods and crust formation.

Medicinal uses: The leaves are anti-diabetic, cooling and effective against eye, gonorrhoea, hypolipidemic, skin diseases, urinary tract infection. Fruits are hypoglycemic, analgesic, antipyretic, hepatoprotective, tuberculosis, eczema and anti-inflammatory. Stems are Expectorant, antispasmodic, asthma, bronchitis, urinary tract infection, skin diseases. Roots are hypoglycemic, antidiabetic, skin diseases, removes pain in joint, urinary tract infection.

Scientific name	:	*Dioscorea alata*Linn.
Family	:	Dioscoreaceae
Common name	:	Greater Yam

Description: It is a vigorously twining herbaceous vine, from massive underground tuber. Stems are upto 10 m or more in length, freely branching above; internodes square in cross section, with corners compressed into "wings," these often red-purple tinged. Aerial tubers (bulbils) are formed in leaf axils, elongate, with rough, bumpy surfaces. Leaves are long petioled, opposite; blades to 20 cm or more long, narrowly heart shaped, with basal lobes often angular. Flowers are small, occasional, male and female arising from leaf axils on separate plants, male flowers occur in panicles to 30 cm long, female flowers in smaller spikes. Fruit are 3-parted capsule; and wing seeded.

Cultivation: It prefers a tropical climate without extremity in temperature. It is adapted to moderate to heavy rainfall area. The plants can be grown in a variety of soils, but light soil is good. The ideal soil pH is 5.5-6.5 but tolerates fairly wide variation in soil pH. *Dioscorea* are propagated by tuber pieces, single node stem cuttings or seed. Commercial planting is normally established by tuber pieces only. Propagation through seed progeny is variable and it may take longer time to obtain tuber yields. IIHR, Bangalore has released two improved varieties, FB(c)-1, a vigorously growing strain relatively free from diseases and Arka Upkar, a high yielding clone. Three types of tuber pieces can be distinguished for propagation purpose, viz. (1) crown (2) median and (3) tip, of which crowns produce new shoots within 30 days and are therefore preferred. Dipping of tuber pieces for 5 minutes in 0.3% solution of Benlate followed by dusting the cut ends with 0.3% Benlate in talcum powder in moist sand beds effectively checks the tuber rot. The treatment is very essential for obtaining uniform stand of the crop. The best time of planting is the end of April so that new sprouts will grow vigorously during the rainy season commencing in June in India. Land is to be prepared thoroughly until a fine tilth is obtained. Deep furrows are made at 60cm distance with the help of a plough. The stored tuber pieces which are ready for planting is to be planted in furrows with 30cm between the plants for one year crop and 45cm between the plants for 2 year crop at about 0.5 cm below soil level. The new sprouts are to be staked immediately. After sprouting is complete, the plants are to be earthed up. Soil from the ridges may be used for earthing up so that the original furrows will become ridges and vice versa. *Dioscorea* requires high organic matter for good tuber formation. Besides a basal doze of 18-20 tonnes of FYM/ha, a complete fertilizer dose of 300kg N, 150kg P_2O_5 and K_2O each are to be applied per hectare. P and K are to be applied in two equal doses one after the establishment of the crop during May-June and the other during vigorous growth period of the crop (August- September). Irrigation may be given at weekly intervals in the initial stage and afterwards at about 10 days interval. *Dioscorea* vines need support for their optimum growth and hence the vines are to be trailed over *pandal* system or trellis. Periodic hand weeding is essential for the first few months. Intercropping with legumes has been found to smother weeds and provide extra income. The major pests of *Dioscorea* are the aphids and red spider mites. Aphids occur more commonly on young seedlings and vines. Young leaves and vine tips eventually die if aphids are not properly controlled. Red spider mites attack the underside of the leaves at the base near the petiole. Severe infestations result in necrotic areas, which are often attacked by fungi. Both aphids and spider mites can be very easily controlled by Kelthane sprays. The tubers grow to about 25-30 cm depth and hence harvesting is to be done by manual labour. The best season for harvesting is Feb-March, coinciding with the dry period. On an average 50-60t/ha of fresh tubers can be obtained in 2 years duration. Diosgenin content tends to increase with age, 2.5% in first year and 3-3.5% in the second year. Hence, 2 year crop is economical.

Medicinal uses: Presence of high level of Vitamin B6 in *Dioscorea alata* reduces depression among women. It proved beneficial for human urinary, respiratory and nervous systems. It increases the energy level in human body. *Dioscorea alata* has been used as a vermifuge and laxative in various healing practices. It is widely known for curing burned haemorrhoids, fever, gonorrhoea, tumours, leprosy and many more. It solves a number of digestive problems, controls high blood pressure impacts and provides relief in muscle tension, nerve tension, neuralgia and cramps.

Scientific name	:	*Dioscorea bulbifera*Linn.
Family	:	Dioscoreaceae
Common name	:	Ratalu

Description: It isa clockwise twining herb with glabrous stem. Leaves are alternate, simple, three to five veined from base, glabrous, ovate-triangular with deep cordate base.

Axillary bulbils are spherical, tubercled, and dark brown-green in colour. Flowers are arranged in hanging, axillary, solitary or fascicled spikes. Capsules are oblong and winged, while seeds are winged at the base. Perianth is tubular and fruit is three-valved.

Cultivation: A subtropical or sub-temperate and humid climate with distinct two to three months of cold winters is favourable for the growth of the plant. Sandy loam soil is most suitable for its cultivation and better yield of corm. The species can be easily propagated by underground corms or aerial bulbils. Vegetatively propagated plants have relatively faster growth than the seedlings. Larger bulbils that are more than 4 cm in diameter sprout about two days earlier than the smaller ones. Direct planting of the corms or bulbils in the field is more effective than transplanting nursery-raised seedlings. The suitable time for planting corms or bulbils is April–May in mid-hills. For raising seedlings, seeds are separated from mature fruits collected in November and are sown in spring or early summer. About 80–100 kg well-developed bulbils or corms are required for plantation on 1 hectare of land at a spacing of 50 cm × 50 cm. A basal dose of 15–20 tonnes/hectare of FYM should be applied to the soil at the time of pre-planting tillage. Plant-to-plant spacing of 50 cm × 50 cm is considered best for good growth and yield. This spacing gives an optimum crop stand of 40 000 plants per hectare. Only light irrigation to maintain humidity during dry season is recommended through sprinklers. Bulbils become ready for harvest within 180–200 days from sprouting. The mother corm may be harvested after two to three years. Hence, October–November are the ideal months to harvest bulbils/ underground corms. The crop yields an average of 12 tonnes of fresh corms per hectare in the second year.

Medicinal uses: The main active ingredient of corm is diosgenin along with furanoid norditerpenes, sinodiosgenin, diosgenin ß, smilagenone, and epismilagenins. Corms are aphrodisiac, tonic, and used for treating sore throat, boils and swellings, dysentery, piles, and syphilis. They are also used against tumour. Various steroidal drugs derived from diosgenin by artificial synthesis include corticosteroids, sex hormones, anabolic steroids and oral contraceptives.

Scientific name	:	*Gloriosa superba*Linn.
Family	:	Liliaceae
Common name	:	Glory Lily

Description: It is a glabrous climbing herb with tuberous root stock grows over hedges and small trees. Stems are 6m long which grows to a height of 1.2-1.5m before the stem branches. Leaves are simple, alternate or whorled, sessile, ovate-lanceolate, tip elongating into a spirally coiled tendril, base cordate and margin entire. Flowers are large, borne in terminal racemes; perianth segments 6, linear, flexuosus and deflexed, basal half bright yellow, upper half red. Fruits are capsules, linear-oblong, 3 equal lobes, one or two lobes shorter in malformed fruits; green dried to pale and then black colour, dehisced into three sections. Seeds are oval in shape, containing 30-150 seeds per capsule, pale orange attached to the sutures. Tubers are cylindric, large, simple, 'V' shaped with the two limps equal or unequal in length pointed towards end brownish externally and yellowish internally.

Cultivation : It is a rainy season plant and sprouts well in warm, humid and tropical conditions. It grows well in sun. *G. superba* is a shallow rooted plant and grows well in a variety of soils either clay or sand through out India. It prefers a light porous soil with good drainage. For vigorous growth, greater blooms and strong tuber, a mixture of soil, sand and compost manure is recommended. The propagation is mainly by tubers, by division of rhizomes. Seeds remain dormant for 6-9 months and due to hard seed coat, about 20-30 days are required for germination and seeds may take 3-4 years before it matures to flower. Treatment of seeds by gibberellin (1-3 ppm) resuls in higher yield of colchicine in the plant and higher production of tubers. In tissue culture, young sprouts are cultured on Murashige and Skoog's medium supplemented with kinetin (1-4 mg/l). Direct regeneration of the explants are obtained. The seeds and rhizomes are sown usually in the last week of June to mid July. The rhizomes are planted by splitting carefully into two from their 'V' shaped joints (two buds being at the extreme end of each rhizome) in lines 20cm apart at a distance of 20cm (while seeds are sown in lines at a distance of 4-6cm apart). They are watered regularly when the plants are growing. After green shoots appear 2-3 showers are given weekly. The irradiation of the plant at 42% natural sunlight intensity increased the production of tuber and colchicine. They usually takes 6-10 weeks to flower after sprouting and then set on fruits. The fruits ripen at the end of October and after that aerial shoot eventually dies, leaving the fleshy tubers underground. The tubers are dug out with great care. An individual plant produces 50g tubers on an average. The average yield is approximately 4000-5000kg of rhizomes and 1000 kg of seed per hectare. The content of colchicine is usually 0.358% and 1.013% in tubers and seeds, respectively.

Medicinal uses: The flowers, leaves and tubers are rich in colchicine, superbin, N-formyl deacetyl colchicine, demethyl colchicine and lumicolchicine. Tubers also contain gloriosine.

The roots and rhizomes are used in traditional system of medicine. Its abortifacient and antipyretic properties have been mentioned in ancient classics "*Charaka*". The name *Garbhaghatini* is due to this abortifacient activity. They are useful in the treatment of inflammations, ulcers, scrofula, haemorrhoids, pruritus, dyspepsia, helminthiasis, flatulence, intermittent fevers and debility. The root is given internally as an effective antidote against cobra poison. A paste of the root is also used as an anodyne; applications in bites of poisonous insects, snake bites, scorpion sting, parasitic skin diseases and leprosy.

Scientific name : *Gymnema sylvestre* (Retz.)R. Br.

Family : Asclepiadaceae

Common name : Small Indian Ipecacuanha

Description: This is a large, woody much branched climber with pubescent young parts. Leaves are simple, opposite, elliptic or ovate, more or less pubescent on both sides, base rounded or cordate. Flowers are small, yellow and arranged in umbellate cymes. Fruits are slender and follicles are upto 7.5cm long.

Cultivation: The plant grows well in tropical and subtropical humid climate. It is common in hills of evergreen forests. Sandy loam soil is best for its cultivation. It may be cultivated on a variety of soils including gravelled soil. The plants are propagated both by seeds and stem cuttings. Seedlings are to be raised in polybags. Pits of size 50cm cube are to be taken, filled with 10kg dried cowdung or FYM and covered with topsoil. On these pits about 3-4 months old seedlings are to be transplanted from polybags at a distance of 1.0 x 1.5m.. Trailing can be facilitated by erecting poles and tying the plants to the poles. The plant will attain good spread within one year. Regular weeding, irrigation and organic manure application are beneficial. The plant is not attacked by any serious pests or diseases. Leaves can be collected from the first year onwards at an internal of one week. This can be continued for 10-12 years. Fresh or dried leaves can be marketed. About 1250 kg of dry-weight leaves can be obtained per hectare every three months.

Medicinal uses: The produce contains 7%–9.6% of gymnemic acid as active principle. Besides, alanine, aminobutyric acid, isoleucine, valine, adenine, choline, gymnamine (alkaloid), and many other ingredients are isolated from leaves. It is administered to cure cough, dyspnoea, ulcers, *pitta, kapha* and pain in the eyes. The plant is useful in inflammations, hepatosplenomegaly, dyspepsia, constipation, jaundice, haemorrhoids, strangury, renal and vesical calculi, helminthiasis, cardiopathy, cough, asthma, bronchitis, intermittent fever, amenorrhoea, conjunctivitis and leucoderma. The drug is said to be a destroyer of *madhumeha* (glycosuria) and other urinary disorders. Root is reputed as a remedy for snakebite. Leaves triturated and mixed with castor oil are applied to swollen glands and enlargement of internal viscera as the liver and spleen. The drug is used to strengthen the function of heart, cure jaundice, piles, urinary calculi, difficult micturition and intermittent fevers.

Scientific name	:	*Hemidesmus indicus* Linn.
Family	:	Asclepiadaceae
Common name	:	Indian Sarasaparilla/Anantmool

Description: It is a perennial, slender, laticiferous, twining or prostrate, wiry shrub with woody rootstock and numerous slender, terete stems having thickened nodes. Leaves are simple, opposite, from elliptic-oblong to linear-lanceolate, variegated with white above and silvery white and pubescent beneath. Flowers are greenish purple crowded in sub-sessile cymes in the opposite leaf-axils. Fruits are slender follicles, cylindrical, 10cm long, tapering to a point at the apex. Seeds are flattened, black, ovate-oblong and coma silvery white. The tuberous root is dark-brown, coma silvery white, tortuous with transversely cracked and longitudinally fissured bark.

Cultivation: The plant is grown throughout the tropical and subtropical parts of India. Loam to silt-clay loam soils with appropriate humus are suitable for its cultivation. The soil should be slightly alkaline with a pH of 7.5–8.5. The plant can best be propagated from stem and rootstock cuttings obtained from more than one-year-old plants. The cuttings are planted in polythene bags or styrofoam trays between July and September. About 28 000 rooted cuttings are required for planting in one hectare of land. Rooted plantlets are transplanted in the field at three- to five-leaved stage in August and September under rain-fed conditions. The propagules are planted in 30 cm^3 pits with a spacing of 60 cm × 60 cm for better harvest. Only organic manure @ 1–2 kg/plant is recommended to be mixed with the soil at the time of field preparation. A minimum period of two-and-a-half years is required for root maturity. Harvesting is be done in December and January. A total dry root biomass of approximately 1.2 tonnes per hectare is obtained.

Medicinal uses: The roots contain coumarin and two sterols – hemidosterol and hemidesmol – besides resins and tannins. Aerial parts of the plant contain carotenoid (22.4 mg/100 g), provitamin A, vitamin C (27.2 mg/100 g), tannins (1.7%), phenolics, anthocyanins, reducing and non-reducing sugars (17.6%), and anti-nutritional factors.

The roots are useful in vitiated conditions of *pitta*, burning sensation, leucoderma, leprosy, skin diseases, pruritus, asthma, bronchitis, hyperdipsia, opthalmopathy, hemicrania, epileptic fits, dyspepsia, helminthiasis, diarrhoea, dysentery, haemorrhoids, strangury, leucorrhoea, syphilis, abcess, arthralgia, fever and general debility. The leaves are useful in vomiting, wounds and leucoderma. The stems are bitter, diaphoretic and laxative and are useful in inflammations, cerebropathy, hepatopathy, nephropathy, syphilis, metropathy, leucoderma, odontalgia, cough and asthma. The latex is good for conjunctivitis.

Scientific name	:	*Holostemma ada-kodien* Schult.
Family	:	Asclepiadaceae
Common name	:	Chirvel

Description: It is a large woody climber and glabrous. Its latex is milky white. The leaves are opposite and simple, with petioles decussate, ovate-oblong to broadly ovate, cordate, caudate or aristate at apex. Basal lobes are rounded, more or less pubescent beneath lateral nerves that are prominent and arched. The flowers of the plant are arranged in umbellate axillary cymes; peduncles are terete and glabrous; bracts are linear acute, grooved on ventral surface, glabrous. Seeds are many, comose, small ovate, thick acute, brown, coma shaped, silky-white.

Cultivation: The plant grows well in a tropical humid climate and partially sunny locations. Sandy-loam soil is ideal for its cultivation. The plants are propagated through seeds. The crop can also be propagated by root and stem cuttings. The crop is raised in a nursery in February. The seeds are sown on sand beds. About one-month-old seedlings are transplanted in polybags of size 14 cm × 10 cm, which are filled with soil, sand, and well-decomposed FYM in the ratio 1:1:1. Polybags should be kept in shade and irrigated regularly. Seedlings become are ready for planting in the field in May–June. About 1.5 kg of seeds are required to raise a plantation in one hectare of land. About 28 000–30,000 seedlings per hectare would be required for planting at a spacing of 60 cm × 60 cm. The sowing is done at the start of rainy season in May–June in South India. In North India, it should be sown just after pre-monsoon showers. Pits of size 30 cm × 30 cm × 30 cm are dug at a spacing of 60 cm × 60 cm, and are filled with thoroughly mixed surface soil and sand in 1:1 ratio along with FYM @ 2 kg for each plant, and mounds are formed. The roots are used as drug and it takes one to two years for its good development. The crop yield is maximum (400–450 kg/hectare of fresh roots) after about one year of planting. The yield of dry roots varies in the range 10–15 tonnes per hectare.

Medicinal uses: The roots are used in treating ophthalmopathy, or chitis, cough, burning sensation, stomach pain, constipation, fever, and *tridoshas*. The root can also be employed as a rejuvenative drug, imparting resistance to diseases.

Scientific name : *Humulus lupulus* Linn.

Family : Cannabaceae

Common name : Common Hop

Description: It is a dioecious, perennial, herbaceous climbing plant which sends up new shoots in early spring and dies back to a cold-hardy rhizome in autumn.

Cultivation: A proper support system is required γ for successful hop growing since the vine can grow 5 m or more each season. Planting is done in a well-drained, full-sun bed along a tall trellis or fence, or with a wire support system on T-posts. Hops grow best in a loose sandy loam soil with a pH between 6 and 7.5. Hops remove from the addition of 2 tablespoons of all-purpose fertilizer, such as a 10:10: 10 blend, per plant before you plant them. Trenches of 30cm are dug for each root and mixed the fertilizer into the soil at the bottom. After laying the root in the trench, it is covered with to a 2.5cm depth. The hops require 1m-1.2m of space between them. They are planted in a single row along the base of the support. Once planted, a covering of mulch laid to a 5cm depth keeps down weeds and retains soil moisture. The plants are watered once or twice weekly with enough water to moisten the top 15- to 20 cm of soil provides sufficient moisture. The plants can get benefit from an application of ½ tablespoon of 21-0-0 fertilizer per plant in early summer. The fertilizer is sprinkled onto the soil 15cm from the plant and scratched it into the soil. The vines are needed to be cutting down after harvesting of all the cones. They are trimmed back 60cm height and the old vines are removed from the trellis.

Medicinal uses: It has sedative, anti-stress, and sleep-inducing properties, estrogenic activity, bacteriostatic and anti-inflammatory activities, stimulation of digestive tract, diuretic and against bladder complaints like kidney stones and an aphrodisiac activities.

Scientific name : *Ipomoea aquatica*Forssk.

Family : Convolvulaceae

Common name : Water Spinach

Description: It is a semi-aquatic climber. Stems are 2–3 metres or more long, rooting at the nodes, and they are hollow and can float. The leaves vary from typically sagittate (arrow head-shaped) to lanceolate, 5–15 cm long and 2–8 cm broad. The flowers are trumpet-shaped, 3–5 cm in diameter, and usually white in colour with a mauve centre.

Cultivation: It grows well in tropical and subtropical climate in water or on nutrient rich moist soils. The plants require full sun or partial shade. Propagation is done either by planting cuttings of the stem shoots that will root along nodes or planting the seeds from flowers that produce seed pods. Deep and regular watering is essential to keep the soil nearly wet. First harvest date of nutritious water spinach is usually 4 to 6 weeks after planting – depending more on the temperature and growing conditions. Only a few leaves or entire plant with leaves and stems leaving only 7.5 cm of growth are cut and the plant will regrow again.

Medicinal uses: It has been used extensively as a medicinal plant: as a mild laxative; in the treatment of ringworm and as a poultice in febrile delirium. Leaves are used to treat diabetes in pregnant women. Kangkong is also used as a sedative to promote relaxation and sleep.

Scientific name	:	*Ipomoea hederacea* Jacq.
Family	:	Convolvulaceae
Common name	:	Ivy Leaved Morning Glory

Description: This annual vine is up to 2m long, branching occasionally. The terete stems are light green to dull red, and more or less covered with white hairs. The stems twine about surrounding vegetation, or sprawl about haphazardly. The alternate leaves are up to 10cm long and 7.5cm across; they are medium to dark green, deeply 3-lobed, and indented at their bases. Each lobe is widest in the middle and tapers to a blunt tip. The margins of the leaves are smooth and somewhat undulating, while the upper surface is sparsely hairy. The flowering stalks develop from the axils of the leaves and are quite short, producing 1-3 flowers. The corollas of the flowers are funnelform and about 5cm across; they are initially blue-violet, fading to pale purplish pink with age.

Cultivation: They prefer full or partial sun, mesic conditions, and fertile loamy soil, although they will adapt to other kinds of soil. The seeds don't germinate until the soil becomes warm during early summer. These plants can spread aggressively in open areas with exposed ground.

Medicinal uses: This medicinal herb is rich in ecdysteriods, steroidal glycosides, aromatic acids, triterpenes, amino acids, organic acids, mineral elements and vitamins. A number of pharmacological properties such as diuretic, anthelmintic, blood purifier, deobstruent, laxative, carminative and anti-inflammatory actions have been ascribed to this plant. Besides its useful to treat abdominal diseases, fevers, headache and bronchitis.

Scientific name	:	*Marsdenia tenacissima* (Roxb.) Moon
Family	:	Asclepiadaceae
Common name	:	Murva

Description: It is a perennial climber with grey bark. Leaves are 7–15 cm long and 7–10 cm wide, broadly ovate, acuminate, deeply cordate at the base with rounded lobes; both surfaces are densely velvety tomentose when young, but become almost glabrous above when old. Petioles are pubescent and about 5–10 cm long. Flowers are greenish yellow, arranged in much branched, corymbose cymes. Fruit is a follicle, about 10–15 cm long, paired, ovoid, lanceolate, and longitudinally wrinkled. Seeds are about 1–2 cm long, flattened, and ovate–oblong in shape.

Cultivation: Plants are widely distributed in tropical and subtropical dry and moist deciduous forests having annual rainfall between 1000 mm and 1500 mm. It grows in moist places in nature and is a shade-loving plant. Sandy-loam soil is ideal for its cultivation and it can grow in nutritionally poor soils. Both seed and vegetative cuttings are suitable for raising planting material at commercial scale. The best time for raising the crop is May–June. For propagation through vegetative means, fresh and healthy leafy cuttings, having two to three nodes with axillary buds, are treated with 1000 ppm solution of IBA, by quick dip method. Treated cuttings are planted at 10 cm × 10 cm spacing in nursery or mist chambers to hasten rooting. About 100 g of seeds are required for a nursery bed of size 10 m × 1 m, which is sufficient to raise seedlings one hectare. Freshly collected seeds are recommended always to be used for sowing. About 30 000 plantlets would be required for one hectare if rooted cuttings are used. FYM @

5 tonnes/hectare should be added at the time of field preparation and thoroughly mixed in the soil. Seedling/rooted cuttings are transplanted at a spacing of 60 cm × 60 cm. The plant needs humid soil for better growth. After the rainy season is over, the crop could be irrigated twice a week in October and November. Thereafter, in the ensuing dry season, irrigation may be required three to five times a week. The crop matures in six to eight months after plantation. March–April is the ideal period for harvesting the crop. The estimated yield of dried roots is about 528 kg/hectare.

Medicinal uses: Roots and seeds are reported to be rich in pregnane glycosides of 2-deoxysugars. Roots are purgative, alterative, and given in colic and gonorrhoea.

Scientific name	:	*Merremia hederacea* Burm. f.
Family	:	Convolvulaceae
Common name	:	Ivy Woodrose

Description: It is a climbing or prostrate herb, with stems slender, hairless or velvety, often minutely spotted, occasionally rooting at the nodes. Leaves are ovate, 1.5–5 cm long, 1.25–4 cm wide, blunt at the tip, heart-shaped at the base, entire, toothed, or shallowly to deeply 3-lobed. Leaf stalks are 0.5–6 cm long. Flowers are borne singly or several in lax, branched clusters. Flower stalks are 2–4 mm long. Sepals are obovate to spatula-shaped, outer 3.5–4 mm long, inner up to 5 mm long, notched at the tip. Flowers are yellow or white, bell-shaped, 6–12 mm long, smooth outside, hairy inside at the base. Capsule is round or conic.

Cultivation: It is a tropical and tender perennial plant, suitable for growing in containers in full sun or partial shade. Propagated by dividing rhizomes, tubers, corms or bulbs (including offsets) and from herbaceous stem cuttings.

Medicinal uses: The plants can be used to treat colds, febrile disease, sunstroke, oliguria, tonsil inflammation, laryngitis as well as leucorrhoea. The seeds are used to treat fevers, colds, sore throats, haematuria, conjunctivitis and boils. Leaves are used in the treatments of chapped hands and feet.

Scientific name	:	*Mucuna pruriens* Linn.
Family	:	Fabaceae
Common name	:	Velvet Bean

Description: The plant is an annual climber with long vines that can reach over 15 m in length. When the plant is young, it is almost completely covered with fuzzy hairs, but when older, it is almost completely free of hairs. The leaves are tri-pinnate, ovate, reverse ovate,rhombus-shaped or widely ovate. The flower heads take the form of axially arrayed panicles. They are 15–32 cm long and have two or three, or many flowers. In the fruit-ripening stage, a 4–13 cm long, 1–2 cm wide, unwinged, leguminous fruit develops. There is a ridge along the length of the fruit. The husk is very hairy and carries up to seven seeds. It bears white, lavender, or purple flowers. Its seed pods are about 10 cm long and are covered in loose, orange hairs that cause a severe itch if they come in contact with skin. The itch is caused by a protein known as mucunain.

Cultivation: They are presently cultivated in many tropical and subtropical areas. A warm equable temperature of 20-30°C throughout the growing period is required;

night temperatures of 21°C are reported to stimulate flowering. Velvet beans are susceptible to frost and require a frost-free period of 180 to 240 days. A well drained soil is ideal for its cultivation. Velvet beans are frequently grown mixed with other vigorous growing crops such as sugarcane or maize. They are also grown as a green manure or cover crop and are particularly valuable as an anti-erosion crop. In India, the wild species is often used as a cover crop in coconut plantations. When grown as a mixed crop with cereals, such as maize or millet, it is usual to plant every alternate or third row with velvet beans. Mucuna is grown as kharif crop in India. Seeds are sown at rate of 50 kg/ha with plant spacing of 60 × 60 cm. Plant needs support for growth. Increase in yield upto 25% and reduction in pest infestation is obtained by providing support. Normally, flowering begins in 45 to 50 days after sowing. When grown for seed, velvet beans are frequently planted in rows 90-180 cm apart, with 15-90 cm between the plants. When grown for seed in the tropics many cultivars are normally ready for harvesting approximately 180 to 270 days after planting.

Medicinal uses: The seeds contain high concentrations of LDOPA, an unusual non protein amino acid and a direct precursor to the neuro transmitter dopamine, an important brain chemical involved in mood, sexuality and movement. Besides, it also contains some other amino acids, glutathione, lecithin, gallic acid and beta sitosterol. *This is* an established herbal drug used for the management of male infertility, nervous disorders, and also as an aphrodisiac. It is also used for the treatment of menstruation disorders, constipation, oedema, fever, tuberculosis, etc.

Scientific name	:	*Operculina turpethum* (Linn.) Silva-Manso.
Family	:	Convolvulaceae
Common name	:	Nishothra

Description: It is a perennial climber with latex. Root is long, slender, fleshy, and much branched. Stems are very long, fast growing, twining and much twisted, angled, and winged. Leaves are ovate or oblong, more or less pubescent on both sides, especially when young, minutely reticulate veined. Inflorescence is a cyme with few flowers. Peduncles are stout, 2.5–5 cm long. Capsules are about 1.6 cm in diameter, globose, enclosed in the enlarged, brittle sepals.

Cultivation: The plant grows well in tropical dry and moist deciduous regions in central and peninsular India. Sandy loam to clay loam soil in moist tropical regions is ideal for the cultivation of this species. It prefers shady and moist places. The plants can be raised by vegetative propagation as well as from seeds. Stem cuttings having two nodes may either be planted directly in the field in monsoon (July) or may be rooted first in a mist chamber during March–June. Best method for raising the plants is to use vegetative cuttings of length 10 cm, which should have two nodes. These cuttings may be planted in polybags or directly in mist chamber for rooting. The potting mixture should contain sand, soil, and FYM in equal quantities. Best time for raising poly pot plants is March–June in mist chambers. Vegetative cuttings are directly be planted in the field in July. About 2 kg seeds are required for raising planting stock for one hectare plantation. About 2 tonnes of FYM per hectare is mixed in the soil during field preparation in June. Rooted cuttings or seedlings are transplanted in the well-prepared field at a spacing of 30 cm × 30 cm. Therefore, the optimum crop stand per hectare is 110 000. In hot summer, flood irrigation at an interval of five to seven days is given till December–January. Crop duration is 10–12 months. Best time for harvesting the roots, stem, and leaves is March–April when the plants begin to dry. Root yield is about 1500 kg/hectare after 10 months.

Medicinal uses: Turpeth contains 9%–13% resin, which is a mixture of α and β turpethin, glycosides, and turpethin, besides coumarin, scopoletin, and sugars. is a source of a drug known as turpeth or Indian jalap. Root and root bark are cathartic and laxative. The tuberous roots are also effiective in dropsy, melancholia, gout, leprosy, rheumatism, and paralysis.

Scientific name	:	*Paederia foetida*Linn.
Family	:	Rubiaceae
Common name	:	Stink vine/Gandha Prasarini

Description: This is a perennial, semi-woody, herbaceous vine that twines upward on supports and creeps across the ground. Stems are light green, slender, up to 10 m long, with fine, short hairs in lines or glabrous; flexible, climbing by twining on other species or onto its old stems from the previous season; stems are creeping along the ground will root at the nodes. Leaves are opposite, simple, broadly lanceolate to elongated-ovate, tip pointed, bases rounded, truncate or heart-shaped. Petioles 2-5 cm long, slender, with fine, short hairs in lines. Stipule pairs broadly wedge-shaped, extending between the bases of opposing petioles (interpetiolar). Flowers are arranged on lax scorpioid or double scorpioid inflorescences (cymes) with long stalks; terminal and axillary; small, greyish pink or lilac. Fruit is a capsule, light brown to yellow or reddish-brown, glossy, nearly globose, 4-6 mm diameter crowned by persistent calyx lobes; containing one or two seeds. Seeds are ellipsoidal, plano-convex or convexo-concave, without conspicuous wings, black, often covered with raphides.

Cultivation: Plants have a wide ranging adaptability to different levels of light, types of soil, and salt conditions. The plants grow well in humid tropical and subtropical climate. Suited to nutrient poor and acid soils. A sunny sheltered position is ideal for its cultivation. Propagated by seeds and cuttings.

Medicinal uses: The major chemical constituent present in this plant are iridoid glycosides, sitosterol, stigmasterol, alkaloids, carbohydrates, protein, amino acid, dimethyl disulphide, and volatile oil. It is considered as alterative, anti-arthritic, anti-spasmodic, diaphoretic, hepato-protective, anti-ulcerous, vermifuge, anti-tumorous, expectorant and stomachic. It is also used in asthma, bowel complaints, diarrhoea, diabetes, seminal weakness. It is effective in gout, vesical calculi, diarrhoea, dysentery, piles, inflammation of the river and emetic.

Scientific name	:	*Passiflora edulis*Linn.
Family	:	Passifloraceae
Common name	:	Passion fruit

Description: It is a vigorous woody perennial vine clinged with tendrils having alternate, deeply 3-lobbed and toothed glossy and dark green leaves and young stems and tendrils tinged with red or purple. Flowers are single, fragrant borne at each node on the new growth. Blooms are clasped by three large green bracts, consists of 5-greenish white sepals, 5-white petals white tipped rays, rich purple at the base. Purple forms are self-fruitful while the flowers of the yellow form perfect but self sterile. Fruits are round or oval and deep purple when ripe. Rind is tough and cavity is filled up with an aromatic mass of double walled, membranous sac containing orange coloured pulpy juice and numerous small dark brown or black seeds.

Cultivation: Passion fruits vines, thrives well in full sun or partial shade upto an elevation of 1600m from sea level. A well drained sandy loam soil with pH of 6-5 is ideal for its growth. Plants need regular watering especially during maturity. Plants are sensitive to frost. Favourable temperature for good yield of fruits is 20 to 30°C. For vegetative growth promotion 30 to 32°C is ideal. Passion fruits vines are generally raised through seeds extracted from ripe fruits. Plants can also be propagated by layers or cuttings of matured wood of 30-35cm long and with 3-4 nodes. Cleft or side wedge grafting is advised to perpetuate hybrids by using the resistant yellow passion fruit as rootstock. Rooted cuttings are transplanted to the main field at a spacing of 2m row to row and 4.5 m distance plant to plant. Two arm kniffin system is ideal for training.

Medicinal uses: The passion fruit is rich source of antioxidants, minerals, vitamins, and fiber. 100 g fruit contains about 97 calories. Being a good bulk laxative, it also helps protect the colon mucous membrane by decreasing exposure time to toxic substances in the colon and wiping off cancer-causing toxic substances from the colon. Consumption of fruits like passion fruit rich in vitamin C helps the body develop resistance against flu-like infectious agents and scavenge harmful, pro-inflammatory free radicals. Current research studies suggest that compounds like vitamins A and flavonoid anti-oxidants (ß-carotene and cryptoxanthin-ß)available in fruits have antioxidant properties, and along with vitamin A are essential for good eye-sight. helps regulate heart rate and blood pressure.

Scientific name	:	*Piper longum* Linn.
Family	:	Piperaceae
Common name	:	Long Pepper

Description: The plant is a glabrous perennial vine with erect or sub-scandent nodose stem and slender branches, the latter are often creeping or trailing and rooting below or rarely scandent reaching a few metres height. Leaves are simple, alternate, stipulate, and petiolate or nearly sessile; lower ones broadly ovate, cordate; upper ones oblong, oval, all entire, smooth, thin with reticulate venation; veins raised beneath. Inflorescence is spike with unisexual small achlamydeous densely packed flowers and form very close clusters of small greyish green or darker grey berries. Female spikes with short thick stalk varying from 1.5 to2.5 cm in length and 0.5 to 0.7 cm in thickness.

Cultivation: Long pepper is a tropical plant, suited to high rainfall areas with high humidity. An elevation of 100-1000 m is best. It needs partial shade to the tune of 20-30% for good growth. The natural habitat of the plant is on the borders of streams. It is successfully grown in well drained forest soils rich in organic matter. Laterite soils with high organic matter content and moisture holding capacity are also ideal for cultivation. Long pepper is propagated by suckers or rooted vine cuttings.15-20 cm long 3-5 nodded rooted vine cuttings establishes very well in polybags. The ideal time for raising nursery is March-April. Normal irrigation is applied on alternate days. The rooted cuttings will be become ready for transplanting in 2 months time. With the onset of monsoon in June the field is ploughed well and brought to good tilth. 15-20 cm raised beds of convenient length and breadth are taken. On these beds, pits are dug at 60 x 60 cm spacing and well decomposed organic manure at 1kg/pit is applied and mixed with the soil. Rooted vine cuttings from polybags are transplanted to these pits. Gap filling can be done after one month of planting. The crop requires heavy manuring at the rate of 20 tonnes FYM/ha every year. Application of heavy dose organic matter and mulching increase water retention in the soil and control weeds. Small doses of chemical fertilisers can also be applied. The crop needs irrigation once a week. Sprinkler irrigation is best. With irrigation the crop continues to produce spikes and off-season produce will be available. However, it is reported that unirrigated crop after the onset of monsoon grows vigorously and shows much hardiness than the irrigated crop. Crop losses can be heavy due to pests and diseases. Mealy bugs and root grubs, attack the plant particularly during summer. Infested plants show yellowing and stunted growth. Application of systemic insecticides like nuvacron or dimecron will control the pests. Adults and nymphs of *Helopeltis theivora* severely feeds on the foliage which can be controlled by 0.25% neem kernel suspension. Rotting of leaves and vines during monsoon season is caused by *Colletotrichum gloresporiodes* and necrotic lesions and blights on the leaves during summer is caused by *Colletotrichum* and *Cercospora spp*. These diseases can be controlled by spraying of 1% Bordeaux mixture repeatedly. A virus like disease characterised by yellowing and crinkling of leaves, stunted growth and production of spikes of smaller size and inferior quality was also recently reported.

The vines start flowering six months after planting and flowers are produced almost round the year. The spikes mature in 2 months time. The optimum stage of harvest is when the spikes are blackish green. The pungency is highest at this stage. Spikes are hand picked when they become mature and then dried. The yield of dry spike is 400 kg /ha during first year, increases to 1000kg during third year and thereafter it decreases. Therefore, after 3 years the whole plant is harvested. The stem is cut close to the ground and roots are dug up. Average yield is 500 kg dry roots/ha. *Piper longum* can also be cultivated as an intercrop in plantations of coconut, subabul and eucalyptus.

Medicinal uses: The spike of long pepper contains 4-5% piperine, piplartin, piperolactam, N-isobutyl deca trans-2-trans-4-dienamide and piporadione alkaloides, besides 0.7 % essential oil. Roots are rich in alkaloids piperine, piperlongumine (piplartine) and piperlonguminine; sesamine, methyl 3, 4, 5-trimethoxy cinnamate. Stem contains triacoutane 22, 23 - dihydrostigmasterol. spike is widely used in ayurvedic and unani systems of medicine particularly for diseases of respiratory tract. The root is used in bronchitis, stomachache, diseases of spleen and tumours . Fruit is effective in *vata* and *kapha,* asthma, bronchitis, abdominal complaints, fever, leucoderma, urinary discharges, tumours, piles, insomnia and tuberculosis. Root and fruit are useful in gout and lumbago. The infusion of root is prescribed after parturition to induce the expulsion of placenta. The root and fruit decoction are used in acute and chronic bronchitis and cough.

Scientific name : *Rubia cordifolia* Linn.

Family : Rubiaceae

Common name : Manjit

Description: This isa perennial, herbaceous climber with very long, cylindrical roots having a thin red bark. Stem is long, rough, slightly woody at the base, quadrangular, and glabrous. Branches climb by means of numerous prickles. Leaves are heart shaped, five-nerved from the leaf base, and occur in whorls of four. Flowers are small, yellow, and scaly, and borne in terminal cymes. Fruits are globose, smooth, shining, violet or purple black in colour with grey black seeds.

Cultivation: The species is distributed throughout the hilly subtropical to sub-temperate regions of India, between 300 m and 2000 m altitudes.It occurs mostly in loamy soil rich in humus in high rainfall regions. The plant is a climber and requires support for growth. The plant is propagated by seeds and two-node root cuttings. About 350 g of seeds are required to raise a nursery for planting in one hectare of land. It is recommended to treat the seeds with 0.02% Bavistin 50 WP before sowing. FYM @ 10 tonnes/hectare and NPK (nitrogen, phosphorus, potassium) @ 30:40:20 kg, respectively, are applied to the soil as basal dose during land preparation. Seedlings/rooted cuttings are transplanted to the main field in April–May. An optimum spacing of 60 cm × 75 cm is recommended in the field, which gives an optimum crop stand of 22,000 plants per hectare. FYM @ 5 tonnes/hectare or nitrogen @ 20 kg/hectare is applied 120–130 days after transplanting by top dressing. In the second and third years, NPK may each be applied @ 20 kg/hectare by top dressing with the onset of rainfall after the intercultural operation. In the first year, first manual weeding cum hoeing is done 45 days after planting, and the second one is carried out 120 days after planting. Bamboo/shrub staking is done 30–45 days after transplanting.The plants start flowering after one year in August and seeds mature in October–November. Roots are harvested after two years at preflowering stage in October or late fruiting stage by the end of November or even in early December when seed is required for next crop. About 3 tonnes of dry root is obtained per hectare of cultivated crop.

Medicinal uses: Roots are rich in resinous and extractive matter, gum, sugar, colouring matter, and salts of lime. The colouring matter consists of purpirin, manjistin, garancin, xanthine, besides several anthraquinone derivates. It is a bitter, antiseptic, styptic, anodyne, depurative, and hypotensive drug. The plant is also used against blood dysentery, inflammations, and urino-genital disorders. It is also an alterative, pigment stimulator, and tonic.

Scientific name : *Tinospora cordifolia*(Willd.) Miers ex HK. f. & Th.

Family : Menispermaceae

Common name : Giloe

Description: This is a gregarious glabrous, twiner. Older stems are up to 2 cm in diameter having corky bark. Aerial roots arise from nodal scars of branches. Stem and branches are specked with white vertical lenticels. Bark is grey-brown or creamy white, warty, papery thin, and peels off easily. Leaves are ovate, and acute. They are membranous when young but become more or less leathery with age. Flowers are yellow, unisexual and minute in size. Male flowers are grouped in axillary racemes, while female flowers are solitary. Fruit is an ovoid and succulent drupe, lustrous, red in colour, and pea sized having a single seed. Seed is fleshy and curved.

Cultivation: The plant thrives well in subtropical and tropical climate. Light medium sandy loam soil rich in organic matter, and with adequate drainage, is ideal for its cultivation. It does not tolerate high rainfall or waterlogged conditions. Stem cuttings are the best planting material for raising commercial crop. The cuttings are obtained from mother plants in June–July. The plant can also be raised through seeds. About 2500 cuttings are required for plantation in one hectare of land. A basal dose of FYM @ 10 tonnes per hectare and half dose of nitrogen (75 kg) are applied at the time of land preparation. The stem cuttings with nodes are sown directly in the field. An optimum spacing of 3 m × 3 m is recommended for better yield. The plant needs support to grow, which can be provided by raising wooden stakes or trellis. Already growing shrubs or trees can also support the plant. About two to three weedings and hoeings are required for good growth of twiner. The crop is usually grown under rain-fed condition. The stem is harvested during autumn when it develops to a diameter of more than 2.5 cm. Basal part is kept for regrowth. The plant yields about 1500 kg of fresh woody stem, reduced to 300 kg of dry weight per hectare in about two years.

Medicinal uses: The stems are rich in bitter substances, of which tinosporine, a bitter principle, is a marker compound. Other compounds include gilonin, gilosterol, gilenin, and furanoditerpenes.It is a tonic and has alterative, diuretic, and aphrodisiac properties. It is a febrifuge useful in malarial and chronic fever. It is also used a liver tonic. The plant is used in general debility, loss of appetite, fevers, urinary disorders, diabetes, rheumatism, and dyspepsia. Fresh plant is more efficacious than dried plant.

Scientific name : *Tylophora indica* (Burm. f.) Merr.

Family : Asclepiadaceae

Common name : Antamool

Description: This is a profusely branching climber, which attains a length of 15 m or more. It has short, knotty, thick rootstock. The roots are numerous, fine, very brittle and consist of fibres or fibrils. Leaves are ovate to orbicular, elliptic–lanceolate, shinning green, leathery and cordate at base. The climbers are be recognized by its ovate–oblong, shiny leaves and divaricated fruits. Flowers are small, greenish yellow outside and purplish within. They are arranged in umbellate racemes. Fruit is a striated, divaricated follicle up to 10 cm long and 1–2 cm broad. Seeds are ovate and elongate into 2–2.5-cm-long coma.

Cultivation: It is grown wild mostly in forests in plains and hills up to a height of 1000 m throughout the eastern and southern parts and subtropical regions. Annual rainfall of about 1000–1500 mm is best for *Tylophora* plant. It thrives partial shade conditions of the forests and soil rich in humus. It needs the support of host vegetation for climbing to a sunny location. It performs well when cultivated in plains. For its cultivation, loamy soil rich in organic matter is preferable. The seeds have a high germination percentage (more than 90%) but fruit set is rare. Vegetative propagules as cuttings from stem coppices or ground layering are prepared in spring season. About 250 g of seeds are required to raise a nursery for planting in one hectare of land. These seeds are sufficient to be sown in six beds of size 10 m × 1 m to raise about 20 000 seedlings. The entire quantity of FYM (20 tonnes per hectare), phosphorus (50 kg per hectare), potash (40 kg per hectare), and half of nitrogen (45 kg per hectare) are applied as basal dose at the time of field preparation. Three-month-old plantlets are ready for transplanting in the main field. Transplanting is done during the rainy season. The crop gives maximum yield if row-to-row and plant-to-plant distance is maintained as 75 cm × 75 cm. About 17,500 propagules per hectare are required as a pure crop, while about 10,000–14,000 plants may be required in an intercropping system. it can be intercropped with non-climbing summer and winter vegetables. Besides the recommended basal dose, rest of the nitrogen (45 kg per hectare) should be applied in two equal split doses, one in March and the other after six months, that is, in September. Broadcasting is followed for manure application. Fertilizer application should always be followed by irrigation. Staking should be done to the crop to allow support for the growth of the plant. Weeding is required once a month. Plants are earthed up and stakes should be checked after every three months. It requires 15–18 irrigations per year at an interval of 20 days, depending upon the moisture availability and water holding capacity of soil. The flowering appears in the second year of growth. The leaves should be harvested once a year when the plant attains a height of 1.5–2.5 m. The crop is cut 10–15 cm above-ground level with the help of sickle. October is ideal time for harvesting. Gross maximum dry matter yield of aerial parts averages about 3.25–3.75 tonnes/hectare.

Medicinal uses: *Tylophora indica* contains 0.2%–0.3% alkaloids but the alkaloid content (w/w) in commercial samples is standardized to about 0.1% of tylophorine in the sun-dried material. It is used in treating bronchial asthma and allergic rhinitis. It has diaphoretic and expectorant properties. The dried leaves and roots are often used as substitute for Ipecacuanha.

Scientific name	:	*Vigna trilobata.*
Family	:	Fabaceae
Common name	:	Banmoong

Description: It is annual, spreading, and herbaceous runner. Numerous deep purple stems emerge out from a woody rootstock, which attain the length of 50–60 cm. The stem is prostrate, wiry, slender, not twining, glabrous or somewhat hairy. The leaves are trifoliate, with grooved petiole. Flowers arise in few flowered racemes at the apex. Pods are slightly recurved, green when immature and turn black on maturity, dehiscent, with small, black, blue or sometimes whitish seeds.

Cultivation: The species is found in tropical and subtropical climate up to 2000 m altitude in the Himalayas, extending southwards up to sea level. The plants require warm climate for its growth and can tolerate drought to some extent, but waterlogging conditions are detrimental to the crop. It is a kharif crop grown under hot and humid climate and high rainfall condition in open as well as partial shade. The crop can be cultivated in a variety of soils including red lateritic soil, black cotton soil, and sandy loam soils. However, well-drained, rich in humus, sandy loam soils are most suitable for its growth. Seed is the best material for propagation. Seeds are collected in November after the colour of the pods changes to black or brownish black in September–October. The plant is raised as a rainy season crop and is sown directly in the prepared field. Seeds are sown in the furrows and covered with a thin layer of soil. The field is divided into beds of convenient size with the help of bund maker to facilitate irrigation, drainage, and intercultural operations. Seed requirement rate is 15–20 kg per hectare. The seeds are sown on ridges/rows at a spacing of 20 cm × 15 cm. About 10 tonnes of FYM /compost per hectare is mixed uniformly in the field and harrowed twice to mix it well in the soil. After exposing the field for 15–30 days, the field is harrowed twice or thrice and levelled to get a good tilth and facilitate proper drainage. Manures and fertilizers as basal dose, at the rate of 30 kg of nitrogen and 60 kg phosphorus per hectare, are applied at the time of sowing, if sufficient FYM is not available. Alternatively, 100 kg per hectare of DAP (di-ammonium phosphate) may be applied. Experimental results suggest highest biomass yield with 30:60:40 kg per hectare of NPK. A critical period of crop–weed competition is up to 30 days of sowing. After this period, only perennial weeds need to be removed; other weeds are smothered by the growing canopy. Flowering starts after 40–45 days of sowing. Within four to five days of flowering, fruiting also starts. Flowering and fruiting continue till October–November. For 'panchang' production, 65–70-day-old crop is harvested by uprooting the plants. Pods are collected in October for collecting seeds. A well-managed crop yields 30–40 quintals per hectare of dry herbage. A seed yield of 2.5–5 quintals per hectare has been recorded.

Medicinal uses: The chemical constituents of seeds include friedelin, epifriedelin, sigmasterol, tannins, epifriedelinos, and γ glutamylphenylalanine. anti-oedema, anti-inflammatory, sedative, and used in urinogenital disorders. The roots and whole plant (panchang) are used as *tridosh shamak*, especially against *vat* and *pitta*, urogenital disorders, and intermittent fever.

❑❑❑

11

Cultivation of Important Medicinal Herbs

The People's Republic of China is the leading country for incorporating traditional herbal medicine into a modern health care system and as a result, a blend of herbal medicine, acupuncture, and Western medicine is introduced.

In India, traditional systems have remained quite separate from Western medicine. Spain, with a high medicinal herb biodiversity has been a good supplier of high quality material commonly from wild harvesting. The development of the medicinal plant production requires a trade regulation framework that has been delayed by the lack of understanding between pharmacist and herbalist. On the other hand, 150 species are threatened in Europe and especially in Albania, Hungary, Spain and Turkey by unselective collection and overuse. The most threatened species are Pheasant's eye, *Adonis vernalis*, Bearberry *Arctostaphylos uva-ursi*, Arnica, *Arnica montana*, Island moss *Cetraria islandica*, Sundew *Drosera rotundifolia*, Yellow gentian, *Gentiana lutea*, Liquorice *Glycyrrhiza glabra;* some species belong to *Gypsophila* spp (*Ankyropetalum gypsophiloides*), Bogbean *Menyanthes trifoliata*, orchid family , peony family, *Paeonia* spp., prímula family *Primula* spp., Butcher's broom, *Ruscus aculeatus*, some species of genus *Sideritis* spp., and species of genus of thyme and oregano (*Thymus* spp., *Origanum* spp., *Thymbra* spp.).

Medicinal Herbshave curative powers and are used in making medicines because of their healing properties as a result of containing active ingredients.

Scientific name : *Abelmochus moschatus* Medic.

Family : Malvaceae

Common name : Musk Mallow

Description: It is an erect, hirsute or hispid annual or biennial herb, upto 2 metres tall. stems are woody at bases, hispid, pubescent throughout. Leaves are palmately 3-7 lobed, with long petioles. Flowers are solitary in upper leaf axils. Bracts are involucrate, persistent into fruits. Sepals are five toothed at the apex. Petals are yellow in colour with deep purple spots at the base. Fruit capsule is oval to fusiform. Seeds are musk scented and black.

Cultivation: Musk mallow is a hardy plant, grown in varied climate under tropical and subtropical conditions. It can be grown both as a rainfed crop as well as an irrigated crop. It grows well in drained loamy and sandy loam soils. Loamy soils with neutral pH and plenty of organic matter are ideal for its cultivation.

Musk is propagated through seeds. The optimum time of sowing is June-July with pre-monsoon showers. The land is prepared well by ploughing, harrowing and levelling. Well decomposed FYM or compost is incorporated into the soil at 10 - 15 t/ha. Ridges and furrows are formed giving a spacing of 60 - 100 cm. Seed rate is 2-3 kg/ha. Seeds are soaked in water before sowing for 24 hours. Two to three seeds are sown per hole at 60 cm spacing on one side of the ridge at a depth of 1 cm and covered with a pinch of sand or loose soil. It takes 5-7 days for proper germination. After germination, extra seedlings are thinned out leaving one healthy growing plant per hole within 20 days. Fertilisers are applied at 120:40:40 kg N, P_2O_5, K_2O/ha generally. However, a dose 160:80:80 kg/ha is recommended for best yields of seed and oil. Phosphorus is applied fully as basal. N and K are applied in 3 equal doses at planting, 2 and 4 months after planting. Fertilizers are applied 10 cm away from the plants. For irrigated crop, field is irrigated soon after sowing. Irrigation is given twice a week during the initial period and once a week thereafter. The field is kept weed free by regular weeding during the growing period. Musk plants suffer from pests like spider mites, fruit bores and leaf eating caterpillars. Diseases like powdery mildew and wilt are also observed on the plant. Spider mites and powdery mildew are controlled by spraying 30g wettable sulphur in 10 litres of water. Pod borers can be controlled by spraying 20ml oxydemeton methyl in 10 litres of water.

The crop starts flowering about 75 days after sowing. The flowers set into fruits in 3- 4 days and the pods take nearly a month to mature. Flowering and fruit setting extends from October to April. Fruits have to be plucked as soon as they attain black colour; otherwise, they split and seeds scatter. Therefore, weekly collection of pods is recommended and in all 20-25 pluckings are required as it is a 170-180 days duration crop. The fruits are further dried and threshed to separate seeds. The seed yield is about 1-1.5 /ha.

Medicinal uses: The fatty oil of seeds contain phospholipids as 2 - cephalin, phosphatidylserine and its plasmalogen and phosphatidyl choline plasmalogen. Seeds are effective aphrodisiac and antispasmodic, and used in tonics. They check vomiting and cure diseases due to *kapha* and *vata* and are useful in treating intestinal disorders, urinary discharge, nervous disorders, hysteria, skin diseases, snake bites, pruritus, leucoderma and general debility. Flower infusion is contraceptive. The leaves and roots are used for gonorrhoea and to treat boils and swellings.

Scientific name : *Acalypha indica* Linn.

Family : Euphorbiaceae

Common name : Indian Acalypha

Description: It is an annual herb with ovate thin leaves and axillary flowers borne on erect spikes. Fruits are small, hairy, hidden in the bracts.

Cultivation : It is naturally distributed throughout plains of India.

Medicinal uses: Dried flowers are useful in asthma, bronchitis, pneumonia and rheumatism. Leaves are laxative, emetic and anti-ulcerous. Roots are effective in bowel activities.

Scientific name	:	*Aconitum falconari* Stapf.
Family	:	Ranunculaceae
Common name	:	Aconite

Description: It is an erect biennial herb and grows upto 1 meter in height. The flowers are hooded and foliage delphinium like.

Cultivation: This species is found in Garhwal hills between 3000-4000 metre altitude. These plants are grown in ordinary soil in partially shaded location and are propagated by division.

Medicinal uses: This plant is useful as anti-diabetic, anti-pyretic and diuretic. Roots are poisonous. Flowers and roots are used for the treatment of skin diseases, rheumatism, neuralgia fever and diarrhoea.

Scientific name	:	*Aconitum nepellus* Linn.
Family	:	Ranunculaceae
Common name	:	Aconite

Description: A herbaceous perennial with a short fleshy rootstock. Tubers tapering below leaves are alternate, long stalked, spreading, very deeply cut into 5 to 3 segments, dark green and shiny. Flowers are large, stalked, with blusih purple, petaloid sepals.

Cultivation: It is widely distributed in Europe, North and South West America and in the Himalayan range.

Medicinal uses: The roots contain alkaloids like 'aconitine' and 'pseudoaconitine'. Leaves contain 'aconitia' and 'aconitic acid'. Aconite is a strong sedative, anodyne, diuretic and anti-pyretic. It can be given in acute and chronic rheumatism, gout and neuralgia; in heart diseases, to relieve pain in various inflammatory diseases, hypertrophy and carcinomatous infections.

Scientific name	:	*Aconitum heterophyllus* Wall Ex. Hoyal 'Atees'
Family	:	Ranunculaceae
Common name	:	Aconite

Description: It is a perennial herb with oblong ovoid or fusiform rootstock. Leaves are numerous, reniform or roundish, 5 lobbed. Flowers are large blue, closely appressed to the stem.

Cultivation: Atees generally prefers sub-alpine and alpine climate, cultivation up to 2000 m altitude in sandy (10 cm deep) soils with rich organic matter. In Garhwal Himalayas, altitudes above 2000 m above mean sea level is found to be suitable for the cultivation of atees. Sandy loam and slightly acidic soil, with pH about 6, is found to be the best for seed germination, survival, better growth, and yield. Addition of humus or leaf litter to the soil increases survival rate and growth of seedlings at all altitudes. Forest leaf litter also helps in retaining moisture content in the soil. The plant prefers open, sunny sites, and abundant air and soil moisture during summer months. Seeds, tuber segments or young leafy stems are used as propagules. Seeds are sown in sand and FYM (farmyard manure)/compost mixture (1:2) at a depth of 0.5 cm in styrofoam trays in a mist house. Nursery beds of 2 m × 2 m or even smaller size are ideal for raising seedlings inside the polyhouses. Seeds are sown during October–November or March–April in polyhouses at middle altitudes (1800–2200 m), during February– March in glass houses at lower altitudes (600–1000 m), and during May–June at alpine sites in open beds or in hot houses. Plants raised from seeds have very slow growth and the cotyledonary phase (pseudomonocotyle) persists for at least one growth season (three to four months). Under polyhouse conditions and sandy textured soil, germination takes about two months to complete. However, diurnal temperature variations (25–25°C) and mulching of soil promote germination within 15–25 days. Approximately 1.5 kg seeds are required for obtaining seedlings for 1 hectare. However, about 111 000 tuber segments are required for plantation on 1 hectare. Pretreatment of seeds with GA_3 (200 PPM) results in early germination (9–20 days) and up to 55% germination. A high leaf litter dose (150 quintals/hectare) is added to the soil before transplanting for good seedling growth and survival (up to 56%). In subsequent years, a dose of 60–70 quintals/hectare/ year can be added for maximum yield. Propagules/ seedlings are transplanted after three months of the first true leaf initiation during March–April at middle altitudes and during May–August at alpine sites. A spacing of 30 cm × 30 cm in field is considered optimum for better vegetative growth and resultant high tuber yield. *A. heterophyllum* tubers harvested in May–June contain lower quantity of atisine (0.35%) as compared to those harvested in November and December (0.43%), which also contain traces of aconitine. The tubers harvested in May, however, show higher quantity of aconitine and hypoaconitine as compared to those harvested in other seasons. Actual production of 518 kg/hectare and 579 kg/hectare, respectively, from seedlings and tuber cuttings has been recorded after third year of cultivation under experimental conditions.

Medicinal uses: Tubers are rich in atisine (0.19-0.27%), heteratisine (0.3%), histisine, heterophyllisine, heterophylline, heterophyllidine, atidine, and hitidine. Roots are non-poisonous and used as appetizer, astringent, carminative and anthelmintic. It is used in diarrhoea, dysentery and gastric pain. A small dose of root powder is given for stomachache and cold.

Scientific name	:	*Acorus calamus* Linn.
Family	:	Araceae
Common name	:	Sweet Flag

Description: It is a semi-aquatic rhizomatous perennial herb. Rhizome is creeping, much branched, cylindrical or slightly compressed, light brown or pinkish brown externally, white and spongy within. Leaves are bright green, distichous, ensiform, base equitant, thickened in the middle and with wavy margins. Flowers are light brown and densely packed in sessile cylindric spadix. Fruits are oblong, turbinate berries with a pyramidal top. Seeds are few and pendant from the apex of the cells.

Cultivation: *Acorus* is a hardy plant, grown in tropical to subtropical climates. It needs a good and well distributed rainfall throughout the year. It needs ample sunlight during the growth period as well as after harvest for drying the rhizomes. It may be cultivated in any good but fairly moist soil. It comes up well in clayey soils and light alluvial soils of river bank. The field is laid out and prepared exactly as for rice, irrigated sufficiently and after ploughing twice, watered heavily and again ploughed in the puddle. Sprouted rhizome pieces are used for planting and pressed into the mud at a depth of about 5cm at a spacing of 30 x 30cm. The rhizomes are planted in such a way that the plants in the second row comes in between the plants of the first row and not opposite to them. FYM is to be applied at 25t/ha. Fertilisers are applied at 25:50:60 kg N:P_2O_5:K_2O/ha/yr. Whole of FYM and 1/3 of N, P_2O_5 and K_2O are to be added in the field during March - April as a basal dose. The remaining 2/3 of nutrients is to be given in two equal split doses at 4 months and 8 months after planting. The field is to be regularly irrigated. About 5 cm of standing water is to be maintained in the field in the beginning. Later, it is to be increased to 10 cm as the plant grows. The field is to be regularly weeded. About 8 weedings are to be carried out in all. At each weeding the plants are pressed into the soil. The plant is attacked by mealy bugs. Both shoot and root mealy bugs can be controlled by spraying the shoot and drenching the roots of grown up plants with 10 ml Methyl parathion or 15ml Oxydemeton methyl or 20ml Quinalphos in 10 litres of water. The crop is ready for harvest at the end of first year. The field is to be dried partially so that sufficient moisture is left in the soil to facilitate deep digging. The leaves start turning yellow and dry, indicating maturity. The rhizome will be at a depth of 60cm and having about 30-60cm spread. Therefore, harvesting is to be done carefully. The rhizomes are to be cut into 5-7.5cm long pieces and all the fibrous roots are to be removed. Yield of rhizome is about 10t/ha.

Medicinal uses: Rhizomes, roots and leaves yield essential oil. The important constituents of the Indian oil are asarone and its b-isomer. The sweet flag is an important *medhya* drug, capable of improving memory power and intellect. It is used in vitiated conditions of *vata* and *kapha*, stomatopathy, hoarseness, colic, flatulence, dyspepsia, helminthiasis, amenorrhoea, dismenorrhoea, nephropathy, calculi, strangury, cough, bronchitis, odontalgia, pectoralgia, hepatodynia, otalgia, inflammations, gout, epilepsy, delirium, amentia, convulsions, depression and other mental disorders, tumours, dysentery, hyperdipsia, haemorrhoids, intermittent fevers, skin diseases, dumbness and general debility. It is reportedly useful in improving digestion, clearing speech and curing diarrhoea, dysentery, abdominal obstruction and colic. It is also useful in infantile fever, cough bronchitis and asthma. The drug is reported to cure hysteria, insanity and chronic rheumatic complaints.

Scientific name	:	*Allium sativum* Linn.
Family	:	Liliaceae
Common name	:	Garlic

Description: It is a scapigerous foetid perennial medicinal herb with underground compound bulbs covered over by outer white thin scales and with simple smooth, round stem, surrounded at the bottom by tubular leaf sheath. The leaves are simple, long, flat and linear. The flowers are small and white arranged in rounded umbels mixed with small bulbils. The entire umbel is enclosed in a tear-drop-shaped membranous spathe. Flowers are usually sterile.

Cultivation: Garlic is grown under a wide range of climatic conditions. It prefers moderate temperature in summer as well as in winter. Short days are very congenial for the formation of bulbs. Garlic requires well drained loamy soils rich in humus, with fairly good content of potash. Garlic is propagated by cloves or bulblets. In the hills, sowing is done in April and May. Types with bold and compact cloves and thick white covering sheath are selected for planting. Garlic may be broadcast, planted in furrows or dibbled at the rate of 150-200kg cloves/ha. In furrow planting, cloves are dropped 7.5-10cm apart in furrows 15cm deep and covered lightly with loose soil. Cloves may be dibbled 5 to 7.5cm deep and 7.5cm apart in rows which are 15cm apart with their growing end upwards and then covered with loose soil. A basal dose of 60kg N and 50kg each of P_2O_5 and K_2O are applied along with 25t/ha of FYM. 60kg N is given as topdressing 45 days after planting. First irrigation is given immediately after sowing and subsequent irrigations are given at 10-15 days interval depending upon the soil moisture availability. The last irrigation should be given 2-3 days before harvesting to facilitate easy harvest and minimum damage to bulbs. First weeding and hoeing is to be done at one month after sowing followed by a second weeding one month after first interculture. Hoeing at about two and a half months from sowing loosens the soil and helps in setting of bigger and well-filled bulbs. Garlic is attacked by *Thrips tabacii* which causes withering of leaves. Application of methyl demeton 25EC or dimethoate 30EC at 1ml/l will check the pest incidence. Leaf spot caused by *Alternaria solanii* can be controlled by spraying Dithane M.45 at fortnightly intervals at 2.5g/l of water. Garlic is harvested when the tops turn yellowish or brownish and show signs of drying up. The plants are uprooted, tied into small bundles and kept in shade for 2-3 days for curing. Average yield of garlic is 6-8t/ha.

Medicinal uses: Garlic bulb is reported to contain volatile oil, alliin (S-allyl-L-cysteine sulfoxide), S-methyl-L-cysteine sulfoxide and allinase. It is rich in vitamins like thiamine, riboflavin and niacin. Volatile oil contains allicin (diallyl thiosulphinate), an active odour principle of garlic. Other major compounds present are diallyl disulphide, diallyl trisulphide, allyl methyl trisulphide and allyl methyl disulphide. It possesses high nutritive value. Its preparations are useful in vitiated conditions of *kapha* and *vata*, cough, whooping cough, bronchitis, asthma, fever, facial paralysis, flatulence, colic, constipation, atonic dyspepsia, helminthiasis, duodenal ulcers, pulmonary and laryngeal tuberculosis, opthalmopathy, cardiopathy, fatigue, leucoderma, leprosy, hysteria, haemorrhoids, sciatica, otalgia, lumbago, swellings, splenopathy, hepatopathy, pneumonopathy, anthralgia, sore eyes, ear ache and dental caries.

Scientific name : *Aloe vera* Linn.

Family : Liliaceae

Common name : Barbados Aloe

Description: It is a stemless or very short-stemmed succulent plant grown upto 60–100 cm tall, spreading by offsets. The leaves are thick and fleshy, green to grey-green, with some varieties showing white flecks on their upper and lower stem surfaces. The margin of the leaf is serrated and has small white teeth. The flowers are produced in summer on a spike up to 90 cm tall, each flower being pendulous, with a yellow tubular corolla 2–3 cm long. Like other *Aloe* species, *Aloe vera* forms arbuscular mycorrhiza, asymbiosis that allows the plant better access to mineral nutrients in soil.

Cultivation: This succulent enables the species to survive in areas of low natural rainfall, making it ideal for rockeries and other low water-use gardens. It is intolerant of very heavy frost or snow. In pots, the species requires well-drained, sandy potting soil and bright, sunny conditions; however, *Aloe* plants can burn under too much sun or dry when the pot does not drain water. The use of a good-quality commercial propagation mix or packaged "cacti and succulent mix" is ideal, as they allow good drainage.Terra cotta pots are preferable as they are porous.Potted plants should be allowed to completely dry prior to re-watering. When potted, aloes become crowded with "pups" growing from the sides of the "mother plant", they should be divided and repotted to allow room for further growth and help prevent pest infestations. During winter, *Aloe vera* may become dormant, during which little moisture is required. In areas that receive frost or snow, the species is best kept indoors or in heated glasshouses.

Medicinal uses:*Aloe vera* leaves are rich in acetylated mannans, polymannans, anthraquinone C-glycosides, anthrones, other anthraquinones, such as emodin, and various lectins. *Aloe vera* as the miracle herb that can be used to treat wounds, minor cuts, dry skin and severe burns. It improves digestion and cures any kind of stomach trouble. *Aloe vera* can improve the effectiveness of diet and maximise the weight-loss potential. It promotes hair growth, prevents itching on the scalp, reduces dandruff and conditions your hair.

Scientific name	:	*Alpinia galanga* Linn.
Family	:	Zingiberaceae
Common name	:	Greater Galangal

Description: It is a perennial herb, about 2m high with lower portion covered with smooth leaf sheaths. The leaves are broadly lanceolate, 30-60cm long and 10-15cm broad. The flowers are arranged in erect, terminal panicles. composed of numerous spreading dichotomous branches each with two to six, pale greenish-white faintly fragrant flowers. Fruits are oblong, constricted in the middle or even pear shaped, three sided and deep orange red in colour. Seeds are ash coloured, three angled, finely striated towards the hilum. Both the seeds and rhizomes have pungent aroma.

Cultivation: It comes up well in tropical climate. It grows on a wide range of climates and soils. Well drained hilly areas and places of 1400m high altitude are favourable for its cultivation. This is commercially propagated vegetatively by rhizomes. The field should be ploughed to a good tilth. All the stones and pebbles should be removed. Organic manures at 10t/ha are applied during land preparation. Seedbeds are prepared with 1m breadth, 2m length and 15cm height. Small pits are made at 25cm spacing above the seedbeds and 5cm long rhizomes are planted. Seedbeds are covered with dried leaves. It is irrigated immediately after planting. Regular weeding is needed during the initial stages of crop growth. This is cultivated also as an intercrop in coconut or rubber plantations. Rhizomes are dug out after cutting the top portions when the crop reaches 1.5-2 years of maturity. The average yield is 10-15 tonnes of fresh rhizomes/ ha and the driage is 25-30%. The collected rhizomes are washed and cut into pieces of 5cm long and dried in sun for 4 days before sale.

Medicinal uses: The rhizomes are rich in tannins and flavonoids, such as kaempferide, galangin and alpinin. The rhizomes are bitter, acrid, thermogenic, aromatic, nervine tonic, stimulant, revulsive, carminative, stomachic, disinfectant, aphrodisiac, expectorant, broncho-dilator, antifungal, febrifuge, anti-inflammatory and tonic. Rhizome is CVS and CNS active, diuretic, hypothermic. Seed is anti-ulcerative.

Scientific name : *Amaranthus spinosus* Linn.

Family : Amaranthaceae

Common name : Prickly amaranth

Description: A prickly annual herb, erect in nature, greenish flowers

Cultivation: Amaranth is highly tolerant of an arid environment. Amaranth seeds need soil temperatures of between 18 °C and 25 °C to germinate and an air temperature above 25 °C for optimum growth. The growth ceases at temperatures below 18 °C. The number of growing degree days during the growing season is a major determinant of amaranth plant growth. Lower temperatures and shorter days will induce flowering with a subsequent reduction in leaf yield. Loose and friable soils with high organic matter content are ideal for an early and heavy yield. A soil with a pH of 6.4 could produce high yields. Amaranth grows naturally but can also be propagated from seed. The seedlings are fragile, so it is important to have a fine, firm seedbed. Prepare the soil well and mix with cattle, chicken or compost manure at a rate of one bucket per 1 m^2 or one to two teaspoonfuls of mineral compound fertilisers per metre row. Seeds should be planted not more than 1.25 cm deep, depending on soil texture and surface moisture at planting time. An optimum plant population has not been established, but approximately 2-3 kg of seed per hectare is considered suitable. Seeds can be planted in seed trays. Transplant them after approximately four weeks when the plants should be about 15 cm tall. Transplant into rows 1, 5 m apart and with a spacing of 30 cm in the row. Nitrogen requirements may vary from 50 to 100 kg N/ha and the requirement also differs, depending on the species. Plants can be fertilised by using cow manure at 10-15 t/ha as well as commercial fertilisers with a high nitrogen content. In sandy soils, an irrigation frequency of four to five days is maintained in the summer season, while in the rainy season the irrigation frequency is based on soil moisture level. Early weeds are controlled by tillage or a contact herbicide prior to planting the amaranth.

Medicinal uses: Leaves and young plants are used for home consumption as a cooked, steamed or fried vegetable, especially during periods of drought. Plants are used as a tenderizer in cooking tough vegetables such as cowpea leaves and pigeon peas. Leaves are occasionally found for sale on markets. *Amaranthus spinosus* is also used as forage and said to increase the yield of milk in cattle. *Amaranthus spinosus* has many medicinal properties like astringent, diaphoretic, diuretic, emollient, febrifuge, galactogogue etc. It is used in the treatment of internal bleeding, diarrhoea, excessive menstruation, snake bites, boils, stomach disorders, ulcerated mouths, vaginal discharges, nosebleeds and wounds. A paste of the root is used in the treatment of menorrhagia, gonorrhoea, eczema and colic. The juice of the root is used to treat fevers, urinary troubles, diarrhoea and dysentery. Plant sap is used as an eye wash to treat ophthalmic and convulsions in children.

Scientific name : *Ammi majus* Linn.

Family : Apiaceae

Common name : Bishop's Weed

Description: It is an annual or biennial herb growing to a height of 80 to 120 cm. It has a long tap root, solid erect stem, decompound leaves, light green alternate, variously pinnately divided, having lanceolate to oval segments. Inflorescence is axillary and terminal compound umbels with white flowers. The fruits are ribbed, ellipsoid, green to greenish brown when immature, turning reddish brown at maturity and having a characteristic terebinthinate odour becoming strong on crushing with extremely pungent and slightly bitter taste.

Cultivation: Ammi is relatively cold loving and it comes up well under subtropical and temperate conditions. It does not prefer heavy rainfall. A mild cool climate in the early stages of crop growth and a warm dry weather at maturity is ideal. It is cultivated as a winter annual crop in *rabi* season. A wide range of soil types from sandy loam to clay loam are suitable. However, a well drained loamy soil is the best. Waterlogged soils are not suitable. Being a hardy crop, it thrives on poor and degraded soils. The plant is propagated by seeds. Seeds germinate within 10-12 days of sowing. The best time of sowing is October and the crop duration is 160-170 days in north India. Crop sown later gives lower yield. The crop can be raised either by direct sowing of seed or by raising a nursery and then transplanting the crop. Seed rate is 2 kg/ha. The land is brought to a fine tilth by repeated ploughing and harrowing. Ridges and furrows are then formed at 45-60 cm spacing. Well decomposed FYM at 10-15 t/ha and basal fertilisers are incorporated in the furrows. Seeds being very small are mixed with fine sand or soil, sown in furrows and covered lightly with a thin layer of soil. A fertilizer dose of 80:30:30 kg N, P_2O_5 and K_2O/ha is generally recommended for the crop while 150:40:40 kg/ha is suggested in poor soils for higher yields. The furocoumarin content of *Ammi majus* is increased by N fertiliser and the N use efficiency increases with split application of N at sowing, branching and at flowering.

For obtaining high yields it is essential to give one or two hoeings during November to February which keeps down the weeds. In winter, one irrigation is essential during November to January. Due to long harvesting season, two irrigations during March and April meets the requirements of the crop. White ants and cut worms are reported to attack the crop which can be controlled by spraying the crop with 40g carbaryl in 10 litres of water. Damping off and powdery mildew are the common diseases of the crop. Seed treatment with organo-mercuric compounds is recommended for damping off. To control powdery mildew the crop is to be sprayed with 30g wettable sulphur in 10 litres of water whenever noticed.

The primary umbels and the early maturing secondary umbels are the major contributors to yield. A little delay in harvesting results in the shattering of the seed which is the main constraint in the commercial cultivation of the crop and the main reason for poor yields in India. It has been reported increased yield by 50 - 60% by the application of planofix at 5 ppm at flower initiation and fruit formation stages. The optimum time of harvest is the mature green stage of the fruit in view of the reduced losses due to shattering and maximum contents of furocoumarins. The primary umbels mature first within 35-45 days. These are harvested at an interval of 2-4 days. Later, the early appearing secondary umbels are harvested. Afterwards, the entire crop is harvested, stored for a couple of days and then threshed to separate the seeds. The seed yield is 900-1200 kg/ha.

Medicinal uses: Fruits are rich in amorphous glucoside 1%, tannin 0.45%, oleoresin 4.76%, acrid oil 3.2%, fixed oil 12.92%, proteins 13.83% and cellulose 22.4%. This is one of the richest sources of linear furocoumarins. Furocoumarins have bactericidal, fungicidal, insecticidal, larvicidal, moluscicidal, nematicidal, ovicidal, viricidal and herbicidal activities.

Scientific name	:	*Andrographis paniculata* Wall ex Nees.
Family	:	Acanthaceae
Common name	:	Kalmegh

Description: It is an erect branched annual herb, 0.3-0.9m tall with quadrangular branches. Leaves are simple, lanceolate, acute at both ends, glabrous, with 4-6 pairs of main nerves. Flowers are small, pale but blotched and spotted with brown and purple distant in lax spreading axillary and terminal racemes or panicles. Calyx-lobes are glandular pubescent with anthers bearded at the base. Fruits are linear capsules and acute at both ends. Seeds are numerous, yellowish brown and sub-quadrate.

Cultivation: The best season of planting Kalmegh is May-June. The field is to be ploughed well, mixed with compost or dried cowdung and seedbeds of length 3m, breadth 1/2m and 15cm height are to be taken at a distance of 3m. The plant is propagated through seeds. Seeds are to be soaked in water for 6 hours before sowing. Sowing is to be done at a spacing of 20cm. Seeds germinate within 15-20 days. Two weedings, first at one month after planting and the second at 2 month after planting are to be carried out. Irrigation during summer months is beneficial. The plant is almost free from any serious pests or diseases. Flowering starts from third month onwards. At this stage, plant are to be collected, tied into small bundles and sun-dried for 4-5 days. Whole plant is the economic part and the yield is about 1.25t dried plants/ha.

Medicinal uses: Leaves are rich in "andrographolid" and "kalmeghin". The ash contains sodium chloride and potassium salts. Plant is very rich in chlorophyte. Kalmeghin is the active principle that contains 0.6% alkaloid of the crude plant. It is useful in hyperdipsia, burning sensation, wounds, ulcers, chronic fever, malarial and intermittent fevers, inflammations, cough, bronchitis, skin diseases, leprosy, pruritis, intestinal worms, dyspepsia, flatulence, colic, diarrhoea, dysentery, haemorrhoids and vitiated conditions of *pitta*. It is used to overcome *sannipata* type of fever, difficulty in breathing, hemopathy due to the morbidity of *kapha* and *pitta*, burning sensation, cough, oedema, thirst, skin diseases, fever, ulcer and worms. It is also useful in acidity and liver complaints.

Scientific name	:	*Anethum sowa* Kurz.
Family	:	Umbelliferae
Common name	:	Indian Dill

Description: It is a herbaceous annual reaching a height of 1.2m. Leaves are palmately divided. Flowers are pale yellow borne in compound umbels. The fruit is subelliptical. Ripe seeds are light brown in colour with aromatic odour.

Cultivation: Successful cultivation requires warm to hot summers with high sunshine levels; even partial shade will reduce the yield substantially. It also prefers rich, well drained light sandy to loamy soils. The seeds are viable for three to ten years. The seeds are sown in mid October directly in well prepared fields in rows when temperature is about 20°C. Seeds are sown in 1.0-2.0 cm deep in lines in 45cm spacing followed by light planking. Seed rate for line sown and broad cast crops are 5-6 kg/ha and 10 kg/

ha respectively. Usually, 30 kg of N and 40 kg of P_2O_5 are applied at land preparation and 30kgN/ha is drilled in rows at flower bud initiation stage. It is grown on border of wheat fields or as an intercrop in vegetables to get 2-3 pickings of foliage crop before seed crop is harvested. It requires 2-3 weedings and 6 light irrigations. The seed is harvested by cutting the flower heads off the stalks when the seed is beginning to ripen during early April. The seed heads are placed upside down in a paper bag and left in a warm, dry place for a week. The seeds then separate from the stems easily for storage in an airtight container. The seed yield is 1-1.5 tonnes /ha.

Medicinal uses: The seeds yield 2 to 4% essential oil. The herb yield 0.06% oil. This is use for treatment of digestive disorders, diarrhoea, dysentery and respiratory disorders.

Scientific name	:	*Anthemis nobilis* Kurz.
Family	:	Asteraceae
Common name	:	Common Chamomile

Description: It is a hardy perennial herb. Foliage is fully cut, strongly scented. flower heads are terminal, pedunculated, white in colour.

Cultivation: Plenty of sunlight is required for the proper growth of these plants. The soil should be well drained with a pH between 5.6 and 7.5. Regular watering is essential to keep the soil moist. However, care should be taken to prevent accumulation of water. Using plenty of compost and applying fertilizers every five to six weeks help in the growth of these plants. Harvesting of the flowers should be done immediately after blooming as they start to lose flavour once their colour starts to fade. One can collect the fresh leaves when required.

Medicinal uses: Dried flowers are medicinal. it constitutes azulene, dihydrocinnamic acid, apigenin and angelic acid. an infusion of this plant is used in cases of fever, hysteria, nervousness and dyspepsia. Lotion is used to treat neuralgia, earache and toothache. It is used for making chamomile tea.

Scientific name	:	*Anthriscus cerefolium* Hoffm.
Family	:	Umbelliferae
Common name	:	Chervil

Description: This is an annual herb, attains a height of 60cm. Leaves are pinnate and leaf segments are ovate and deeply cut.

Cultivation: Transplanting of chervil is very difficult, due to the long taproot.It prefers a cool and moist location; otherwise, it rapidly goes to bolting. It is usually grown as a cool-season crop, like lettuce, and should be planted in early spring and late fall or in a winter greenhouse. Regular harvesting of leaves also helps to prevent bolting.If plants bolt despite precautions, the plant can be periodically re-sown throughout the growing season, thus producing fresh plants as older plants bolt and go out of production. The leaves are ready for harvesting in about 8 weeks from sowing, the plant responds well to cut and come again harvesting.

Medicinal uses: Chervil is sometimes employed as a 'spring tonic' for cleansing the liver and kidneys, is a good remedy for settling the digestion and is said to be of value in treating poor memory and mental depression. The fresh plant, harvested just before flowering, is digestive, diuretic, expectorant, poultice and stimulant. The juice is used in the treatment of dropsy, arthritis and chronic skin ailments. The bruised leaves are used as a poultice for slow-healing wounds and a warm poultice is applied to painful joints. An infusion of the fresh leaves is also used as an eyewash to treat sore or inflamed eyes

Scientific name	:	*Aristolochia bracteolata* Lam.
Family	:	Aristolochiaceae
Common name	:	Worm Killer

Description: It is a perennial prostrate herb with weak, glabrous stems. Leaves are simple, alternate, reniform or broadly ovate, cordate at the base with a wide sinus upto 7.5cm in diameter, reticulately veined. Flowers are solitary with a large sessile orbicular bract at the base. Perianth tube is cylindric with dark purple tip having revolute margins. Fruits are oblong-ellipsoid, 12-ribbed glabrous capsules. Seeds are deltoid with slightly cordate base.

Cultivation: It prefers shady areas and well-drained soils. The plants are seed propagated. 3-month-old seedlings raised in polybags are required for transplanting. Pits of size 50cm cube are to be taken at a distance of 3m and filled with sand, topsoil and dried cowdung. To these pits, the seedlings are to be transplanted. Regular irrigation and organic manure application is beneficial. The plant is to be trailed on iron wires tied to poles. The plant is not attacked by any serious pests or diseases. Plant attains good spread within one year. Leaves can be collected for the next 10 years. Roots and leaves constitute the economic parts.

Medicinal uses: Leaves and fruits are rich in ceryl alcohol, β-sitosterol and aristolochic acid. Root contains aristolochic acid. Seeds give an alkaloid magnoflorine, aristolochic acid, fatty oil comprising palmitic, stearic, lignoseric and oleic acids and β-sitosterol. It is a killer of intestinal worms especially roundworms. It is also used in vitiated conditions of *kapha* and *vata*, constipation, inflammations, amenorrhoea, dysmenorrhoea, foul ulcers, boils, syphilis, gonorrhoea, dyspepsia, colic, skin diseases, eczema, artheralgia and intermittent fevers. The plant is an insect repellent due to the presence of aristolochic acid, which is poisonous to man and livestock. Plant is also used against scorpion sting. Seeds ground in water to form a lotion and used for softening hair. Powdered root is used in fertility control.

Scientific name	:	*Atropa belladonna* Linn.
Family	:	Solanaceae
Common name	:	Belladonna

Description: It is a spreading perennial herb, reaches a height of 1.5metres with a long, cylindrical, branched, fleshy creeping root. The leaves are alternate, usually on the upper branches each leaf is accompanied by smaller stipule like leaf. Leaves are ovate-oblong-ovate, 8-20cm long and 5-10cm wide. Flowers are large bell shaped, borne singly on the leaf axils, dull purple within and yellowish on the outsides. Fruits are globular. Seeds are kidney shaped, many in number and embedded in the pulp.

Cultivation: Under temperate climate, these plants grow as a perennial plant in an open situations. In subtropical climate, they may be grown for a short duration in winter with poor yield. Rich fertile soils of medium texture is preferred. Belladonna can be propagated by stem cutting, rootstock division, but commercially from seeds. Seeds are very small and about 50g seeds can produce more than 10,000 plants. Seeds are sown in a well -drained raised bed. Seed germination is generally poor because of hard seed coat and seeds with underdeveloped embryos. Pre-treatment seeds with Captan is recommended to prevent damping off disease. Seeds may be sown in Kashmir conditions during May-June or September-October. Seeds raised in October planted in May, and seeds raised in June are planted in September and placed in a distance of 7.5cm x 7.5cm. Well prepared land with adequate fertilizer is required for cultivation of this crop. Fertilizer dose comprising 80kg N, 40kg P_2O_5 and 40kg K_2O/ha/year is recommended. Nitrogen is applied in four equal split doses during different phases of growth, while phosphorus and potash is applied at the time of planting. Proper irrigation and weeding is needed wherever required.

Three months after planting, the first crop of leaves is available. Harvesting is done when the plants are in full bloom. Terminal branches which are tender are cut along with the leaves. During the first year, three to four harvests are made and about 6 quintals yield can be obtained. In the second and third years, the yield increases upto15 quintals per hectare. The young seedlings suffer from damping off disease if proper care is not taken. Mature plants sometimes die because of wilting of plants. Such plants are to be pulled up and destroyed. After three years, roots of the plants are also harvested. This crop should not be grown continuously in the same plot. Crop rotation after three years may be followed. The cut tops are immediately dried after harvest in shade or with artificial heat. Prolonged drying reduces alkaloid content through enzyme action. Leaves are dried upto one-fifth of their fresh weight. Roots of uprooted plant are also washed, processed, plantation is around 300kg/ha.

Medicinal uses: All plant parts contain alkaloids chiefly hyoscyamine, atropine and hyoscine. People who require the homeopathy remedy belladonna usually suffer from symptoms like acute fever, excruciating pain, especially in the head owing to quick blood circulation, dilated pupils and a dry, flushed skin. Normally, belladonna is prescribed for people enduring severe pain, inflammation or any infection, especially on the upper part of the respiratory tract. In addition, this homeopathic medicine is also given to cure menstrual pain wherein there is excessive and painful menstrual flow. In addition, belladonna is given to patients to treat febrile convulsions (convulsions in children and infants that are accompanied by high fever), sunstroke as well as mastitis (breast inflammation) during breastfeeding. Belladonna is most suitable to treat disorders of the heart, blood vessels, lungs as well as the nervous system. The other symptoms endured by people requiring belladonna include excruciating headaches and hard throbbing pulse. Homeopathic physicians prescribe belladonna to cure pulsation headaches wherein even the least movement of the eyes exaggerates the pain, distended, red breasts owing to breastfeeding, labour pain, boils, convulsions, nephritis or inflammation of the kidneys, cystitis (inflammation of the urinary bladder), and sleep disorders or insomnia.

Scientific name : *Bacopa monnieri* Linn.

Family : Scrophulariaceae

Common name : Brahmi

Description: This is a glabrous, creeping herb, leaves are oblong, spathulate and obtuse. The flowers are upto 1 cm in diameter, pale bluish, pimple veined, fading to white, arises from the axil of leaves and are borne on short pedicles. The roots appear from the nodes of the stem.

Cultivation: The plant grows throughout the warm humid tropics upto 1200m elevation. Brahmi gets established well in water logged fields. The plant is propagated vegetatively by stem cuttings. Land is prepared by ploughing 2 or 3 times. Two to three tonnes/ha of cowdung or compost is applied and the field is again ploughed and levelled. Stem cuttings, 10cm long are spread at a spacing of 20cm. Waterlogging to height of 30cm is always required. Rooting may start within 15-20 days. It will spread over the field within 6 months. Regular application of organic manure will take care of the manurial requirement. Weeding once in a month is required. Care should be taken to maintain water level at a height of 30cm during the growth period. No serious pests or diseases are noted in this crop. Harvesting commences from sixth months onwards. Brahmi leaves can be collected once a month. After 3 years, the whole crop is harvested and removed. Fresh cultivation can be carried out in the same field.

Medicinal uses: The entire plant is used as drug. The plant is rich in brahmine, hespestine, saponins. This is an important drug in Ayurveda for the improvement of intelligence and memory and revitalisation of sense organs. It clears voice and improves digestion. It is recommended against dermatosis, anaemia, diabetes, cough, dropsy, fever, arthritis, anorexia, dyspepsia, emaciation, and insanity. It dispels poisonous affections, splenic disorders and impurity of blood. It is useful in vitiated conditions of *kapha* and *vata*, biliousness, neuralgia, ascites, flatulence, leprosy, leucoderma, syphilis, sterility and general debility. The plant is anti-cancerous and improves learning ability.

Scientific name	:	*Boerhaavia diffusa* Linn.
Family	:	Nyctaginaceae
Common name	:	Punarnava

Description: It is a deep rooted perennial creeping herb. Leaves appear two in one node, one smaller than the other, upper surface green and lower surface whitish. Base of the leaves is cordate. Flowers arise in short clusters on long axillary stalks, very small in size and reddish in colour. Fruits are glandular with fine ridges.

Cultivation: It grows best in sandy-loam to loam soils. Seeds are sown in well-ploughed raised beds in the month of July-August. Later, these plants are transplanted to the ridges at the spacing of 15x15 cm or 30x30 cm. Irrigation the field is essential immediately after transplanting or as and when required. Plants can be harvested after 2-3 years of plantation. Roots are dug out cleaned and cut longitudinally for drying. Seeds can be collected in the month of October-November.

Medicinal uses: The *Boerhaavia diffusa* plant contains a large number of such compounds as flavonoids, alkaloids, steroids, triterpenoids, lipids, lignins, carbohydrates, proteins, and glycoproteins. Punarnavine, boeravinone A.F, hypoxanthine 9-L-arabinofuranoside, ursolic acid, punarnavoside, liirodendrin are also found in the plants. The roots are widely used for the treatment of dyspepsia, jaundice, enlargement of spleen, abdominal pain, abdominal tumors, and cancers. The root powder, when mixed with mamira (*Thalictrum foliolosum*), is used to treat eye diseases. It cures corneal ulcers and night blindness, and helps restore virility in men. People in tribal areas use it to hasten childbirth. The juice of Boerhaavia leaves serves as a lotion in ophthalmia. It is also administered orally as a blood purifier and to relieve muscular pain.

Scientific name : *Calendula officinalis* Linn.

Family : Asteraceae

Common name : Pot Marigold

Description: It is a hardy annual, 30cm tall, with bright orange-yellow flowers. Leaves are oblong-lanceolate. Propagated by seeds.

Cultivation: *Calendula officinalis* is widely cultivated and can be grown easily in sunny or semi-shady locations in most kinds of soils. Seeds are sown during August to September and covered with the soil and one month old seedlings are transplanted during November at a distance of 30-40cm. Dead head spent flowers are pinched back to make the plants bushy. Flower production can be increased by increasing doses of phosphorus levels.

Medicinal uses: Plants are rich in triterpene glycosides and aglycones, carotenoids, essential oils, resin, sterols, flavanoids,calendic acid. The plant is reported to be stimulant, sudorific and febrifuge. It is valued for healing wounds, ulcers, burns and other breaches of skin surface.

Scientific name : *Catharanthus roseus* G. Don.

Family : Apocynaceae

Common name : Sadabahar/Periwinkle

Description: It is a perennial branched herb, 45 to 120cm tall with flowers of white, pink and intermediate colour.

Cultivation: Periwinkle grows well under tropical and subtropical climate. A well distributed rainfall of 1000 mm or more is best. In north India, the low winter temperatures adversely affect the crop growth. It can be grown on any type of soil ,except those which are highly saline, alkaline or waterlogged. Light soils, rich in humus are preferable for large scale cultivation since harvesting of the roots become easy . Catharanthus is propagated by seeds . Fresh viable seeds should be used for cultivation. Seeds can be either sown directly in the field or in a nursery and then transplanted. Seed rate is 2.5 kg/ha for direct sowing and the seeds are drilled in rows 45 cm apart or broadcasted. For transplanted crop the seed rate is 500g/ha. Seeds are sown in nursery and transplanted at 45x 30cm spacing after 60 days when the seedlings attain a height of 15-20cm Nursery is prepared two months in advance so that transplanting coincides with the onset of monsoons. Application of FYM at the rate of 10-15 t/ha is recommended. An alternate approach is to grow leguminous green manure crops and incorporate the same into the soil at flowering stage. Fertilisers are recommended at 100: 120: 100kg $N:P_2O_5:K_2O$/ha for irrigated crop and 80:40:40 kg/ha for rainfed crop. N is applied in three equal splits at planting and at 45 and 90 days after planting. 4 or 5 irrigations will be needed to optimise yield when rainfall is restricted. Fortnightly irrigations support good crop growth when the crop is grown exclusively as an irrigated crop. Weeding is carried out before each top dressing. Alternatively, use of fluchloraline at 0.75 kg a.i. /ha pre-plant or alachlor at 1.0 kg a.i. per ha as pre-emergence to weeds provides effective control of a wide range of weeds in periwinkle crop. De-topping of plants by 2cm at 50% flowering stage improves root yield and alkaloid contents. Fungal diseases like twig blight caused by *Phytophthora nicotianae*., *Pythium debaryanum*, *P. butleri* and *P. aphanidermatum*; leaf spot due to *Alternaria tenuissima, A. alternata, Rhizoctonia solani* and *Ophiobolus catharanthicola* and foot-rot and wilt by *Sclerotium rolfsii* and *Fusarium solani* have been reported. Three virus diseases causing different types of mosaic symptoms and a phyllody or little leaf disease due to mycoplasma-like organisms have also been reported; the spread of which could be checked by uprooting and destroying the affected plants. The crop allows 3-4 clippings of foliage beginning from 6 months. The flowering stage is ideal for collection of roots with high alkaloid content. The crop is cut about 7 cm above the ground and dried for stem, leaf and seed. The field is irrigated, ploughed and roots are collected. The average yields of leaf, stem and root are 3.6, 1.5and 1.5 t/ha, respectively under irrigated conditions and 2.0, 1.0 and 0.75t/ha, respectively under rainfed conditions on air dry basis. The harvested stem and roots loose 80% and 70% of their weight, respectively. The crop comes up well as an undercrop in eucalyptus plantation in north India. In north western India a two year crop sequence of *periwinkle-senna-mustard* or *periwinkle-senna-coriander* are recommended for higher net returns and productivity.

Medicinal uses: Dry leaves contain vinblastine (vincaleucoblastine or VLB) 0.00013-0.00063%, and vincristine (leurocristine or LC) 0.0000003-0.0000153% which have anticancerous activity. Other alkaloids reported are vincoside, isovincoside (strictosidine), catharanthine, vindolinine, lochrovicine, vincolidine, ajmalicine (raubasine), reserpine, serpentine, leurosine, lochnerine, tetrahydroalstonine, vindoline, pericalline, perivine, periformyline, perividine, carosine, leurosivine, leurosidine and rovidine. It is a chief source of patented cancer and hypotensive drugs. It is one of the very few medicinal plants which has a long history of uses as diuretic, antidysenteric, haemorrhagic and antiseptic. It is known for use in the treatment of diabetes in Jamaica and India. Vincristine is used in acute leukemia, lymphosarcoma and *Wilm's* tumour. Its roots are a major source of the alkaloids, raubasine (ajmalicine), reserpine and serpentine used in the preparation of antifibrillic and hypertension-relieving drugs. It is useful in the treatment of choriocarcinoma and *Hodgkin's* disease-a cancer affecting lymph glands, spleen and liver. Leaves are used for curing diabetes, menorrhagia and wasp stings. Root is tonic, stomachic, hypotensive, sedative and tranquilliser.

Scientific name	:	*Centratherum anthelminticum* Cass.
Family	:	Compositae
Common name	:	Somraj

Description: These are tall erect herbs. Leaves are upto 10cm long, covered with minute hair, and margins toothed. Flower heads are upto 2.5cm across, purplish, appear in small clusters of 30 to 40 minute flowers. Fruits are achenes, cylindrical hairy with 10 narrow ridges.

Cultivation: Thi plant is distributed throughout India mostly in waste places near habitations upto 1500 metres altitude.

Medicinal uses: Seeds are medicinal. It is stimulant, antiseptic and anthelmintic. It destroys thread worms and promotes urination. It is also used to treat skin diseases.

Scientific name : *Cephaelis ipecacuanha* Sw.
Family : Rubiaceae
Common name : Ipecac

Description: These are small perennial herbaceous plants, upto 75 cm tall. Stem is slender, cylindrical, trailing, with few to no branches. Leaves are in opposite pairs near the top of the plant, alternate below up to 10 cm long and 5cm wide, pointed tips, margins entire, dark green above, pale green underneath, roughly hairy, short downy petioles. Flowers are white, funneliform, sessile, one cm wide, arises in dense cymose. The fruit is oval, purple when immature and black when ripe, with two stones containing a single white seed. Roots are slender, spreading horizontally, mature roots moniliform, in bunches, root bark brick red to dark brown when mature.

Cultivation: Ipecac grows well with an average rainfall ranging between 2000-3000mm. Maximum temperature should not exceed 38°C and the minimum not below 10°C. It thrives well in tropical mild humid climates similar to Malaysian rain forests. Virgin forest soils rich in humus are ideal for Ipecac. It prefers deep medium fertile soils which are acidic and rich in humus, potash and magnesium. Site should be protected from wind and storm. As Ipecac grows only in shade, it can be cultivated as an intercrop, or planted in artificially shaded beds. The plant is propagated both by seeds and vegetatively by root, stem and leaf cuttings. Vegetative propagation is preferred to maintain genetic uniformity of the plant. Most of the commercial plantations are raised by seeds. Raised seed beds of 2x6m size are made and are mixed with well rotten leaf compost and sand. These are provided with shade on the top as well as on the sides. Seeds are drilled or broadcasted in the beds and watered regularly. Seeds take 3-5 months to germinate. Seed treatment with limewater for 48 hours or H_2O_2improves germination. It has been observed that providing mulch or black polythene in nursery beds improves germination as well as results in control of weeds. The suitable season of planting is January-March in West Bengal. Seedlings are planted in production beds at a spacing of 10x10cm after they are 8-12 weeks old. In West Bengal, it is a practice to transfer seedlings to other nursery beds before being transferred to final production seedbeds. FYM and leaf compost application is required during second and third year. Super phosphate applications is found to improve root growth. Frequent irrigation is required. Waterlogging should be avoided. Both the seedbeds and production beds should be kept free from weeds7. Seedlings are often attacked by damping off fungi like *Rhizoctonia* sp. in nursery. It is better to treat the seeds with a suitable seed dressing fungicide before planting. *Fusarium* wilt caused by *F. moniliforme* has been reported from India. The plants are ready for harvesting after 4 years. The roots should be dug out, washed and dried in the sun.

Medicinal uses: Rhizomes and roots are medicinal. The main alkaloids are cephaeline and emetine. In addition, it also contains psychotrine and psychotrine ethyl ether. The drug also contains a crystalline glucosidal tannin, starch and calcium oxalate (60-70% of the alkaloids is emetine). Root contains minor amounts of O-methyl psychotrine, emetamine, protoemetine and others. It is used in powdered form or as liquid total extract, syrup and tincture. Ipecac syrup in small doses is used as an expectorant, as it is well tolerated by children. It is used in treatment of whooping cough. Ipecac with opium as in Dover's powder is used as a diaphoretic, tincture and syrup. Emetine hydrochloride in the form of injection is used for treatment of amoebic dysentery. Emetine bismuth iodide is also given orally for amoebic dysentery. Ipecac is also used as gastric stimulant and as an anti-inflammatory agent in rheumatism.

Scientific name	:	*Chenopodium album* Linn.
Family	:	Chenopodiaceae
Common name	:	Wild Spinach

Description: These are hardy annual, 120cm tall with whitish, fuzzy surface. The lower leaf surface is whiter than the upper surface. The flowers are arising in spikes from upper leaf axils and greenish in colour.

Cultivation: An easily grown plant that prefers a moist sandy loam soils of pH in the range of 4.5 to 8.3. A good companion crop for potatoes, corn and cucurbits. Seeds are sown *in situ*. It can not grow in the shade. In plains, seeds are sown in Oct-Nov whereas April-May in hills. It can be grown upto an elevation of 4700m.

Medicinal uses: Leaves contain protein (3.9%), 0.76% fat, 8.9% carbohydrate and 3% ash. Seeds are rich in carbohydrate (49%), protein(16%) & ash (7%). Besides, these are rich in protein, vitamin A, calcium, phosphorus and potassium. Leaves are eaten raw or cooked and a good substitute for spinach. Young inflorescences are also taken after cooking. The leaves are anthelmintic, antiphologistic, anti-rheumatic, laxative and odontalgic. The juice of stem is applied on freckles and sunburn. The juice of the root is used in the treatment of blood dysentery.

Scientific name	:	*Chlorophytum arundinaceum* Baker
Family	:	Liliaceae
Common name	:	Safed Musli

Description: It is a rhizomatous herb. The leaves are sub-erect, lanceolate and many nerved. The inflorescence is an erect, dense-flowered raceme. The flowers are star like, white up to 2cm across, sepals are acute, anthers are longer than filaments and green or yellow in colour, bracts are long. Seeds are black in colour with angular edges.

Cultivation: It grows well in tropical and subtropical climate. Organic rich, well drained sandy loam soil and warm humid climate is suitable for cultivation of its rhizome. Seeds remain dormant for nearly two months. Germination of seeds takes two weeks time, and about 20 per cent germination is reported. Vegetative propagation is easy and the plants obtained through fleshy root with attached stem disc are more vigorous. Plants are planted in June-July at a distance of 30cm x 30cm and for one hectare of land around 300 kg of fleshy roots are required. While land preparation, 10 tonnes of FYM/ha is to be incorporated with the soil. It is a three month crop if cultivated and crop matures in September-October when the leaves have dried. Fleshy roots are harvested through careful digging. The estimated yield is one tonne of fresh roots/ha and 200 kg of dry root/ha. The dry roots are hard and bony in appearance and white in colour.

Medicinal uses: The dry roots contain carbohydrates, proteins, root fibres, saponins and minerals. It is a well known tonic and aphrodisiac and used to treat general debility. Leaves are eaten as vegetable. The tubers of about 20g is boiled with milk and taken twice a day for a month in cases of impotency and weakness.

Scientific name : *Colchicum luteum* Baker

Family : Liliaceae

Common name : Colchicum/Suranjan

Description: Suranjan is an annual alpine herb. It ts recognized by an almost conical corm with a longitudinally grooved flat side, rapier-like leaves, short scape, and golden yellow flowers. Corms are almost conical, brownish in colour, with one side flat and other rounded, and are either translucent or opaque. Leaves are 15–30 cm long and 0.8–1.5 cm broad and few in number. They appear with flowers and are narrow but broader towards the tip. They increase in size as the plant approaches fruiting stage. The scape is very short and sessile. Flowers occur in large, erect terminal cymes with one to three flowers.

Cultivation: Natural habitat of Colchicum is characterized by physiologically temperate conditions like severe winter, snow, and low humidity. The plant grows well in well drained sandy-loam soil. The plant is successfully propagated through corms, which are extracted from natural habitats when snow melts in April–May. Corms do not require any pre-planting treatment. About 0.5 million corms are required for planting in 1 hectare of land. FYM (farmyard manure) @ 6 tonnes/hectare is broadcast and properly mixed with soil while ploughing the fields at pre-planting stage. The corms are directly planted in rows in small beds laid out in the field. The crop is sown in rabi season, generally in October, in temperate zones. The corms should be placed at an optimum spacing of 10 cm × 20 cm. The crop takes about 16–18 months to mature. The seeds should be harvested immediately after ripening of fruits in April. By following the standard package and practices, corm yield of 500–750 kg dry weight is obtained from plantation on 1 hectare land.

Medicinal uses: Seeds contain 0.41%–0.43% poisonous alkaloids colchicine and cornigerine. The dried corms contain alkaloid colchicine in a range of 0.21%–0.25%. It is used as a carminative, laxative, and an aphrodisiac. Colchicines are effective in the treatment of gout, rheumatism, and diseases of liver and spleen. Externally, the corms are applied as paste to lessen inflammation and pain.

Scientific name : *Coleus barbatus* Briq.

Family : Lamiaceae

Common name : Patharchur

Description: Patharchur is a large, succulent, densely hairy, herb with aromatic roots. Plants are 70 cm tall. The roots are perennial, while the shoots are annual. Stems are highly branched, scandant, hairy, fragile, and knotted at nodes. Leaves are thick, fleshy, ovate, hairy, with serrate margins. Flowers are pink-white appear in winter season, from October to February.

Cultivation: The plant is sun loving and grows well in tropical regions. The ideal soil for commercial cultivation is sandy loam or sand-rich loam. The species is mainly propagated through stem cuttings of length 15 to 20 cm, containing four to five nodes. The cuttings remain viable up to four days, thereafter, the survival and rooting percentage declines. The crop can be raised round the year under ambient conditions. About 28 000 rooted cuttings are required for planting crops on 1 hectare of land, at a spacing of 50 cm × 60 cm or 60 cm × 60 cm. FYM (farmyard manure) or vermicompost at the rate of 15 tonnes per hectare should be incorporated as a basal dose 15 days before planting. The crop responds well to fertilizer application @ 30 kg per hectare each of nitrogen, phosphorus, and potash, after about 10 days of transplanting. This may be followed by at least two irrigations, if sufficient soil moisture is not available. Ridges are made in the field at a distance of 60 cm and the rooted propagules are planted in the soil up to a depth of 15 cm, generally by using the crowbar method. Phosphate and potash fertilizers are given as basal application. Nitrogenous fertilizers can be used in two equal split doses, the first being the basal dose. Frequent earthing-up once in 15–20 days is essential for good root growth. The soil from the furrow is heaped on to the base, resulting in proliferation and good growth of the roots. Weeding of crop is done after 30 days of transplanting followed by weeding after 60 days of transplantation. The crop can be maintained for three to four years under good phytosanitary conditions. The roots are, however, harvested between fifth and sixth months after planting. Yield of roots is about 250 kg per hectare.

Medicinal uses: The root of *Coleus* species is considered analgesic, ophthalmic, and febrifuge. It is very effective in epilepsy. Fresh aromatic root is offered in temples for its characteristic odour and is also used as a flavouring agent.

Scientific name	:	*Coptis teeta* Wall.
Family	:	Ranunculaceae
Common name	:	Tita

Description: It is hardy evergreen bog plant and a small stemless herb. Leaves are compound and glabrous. Flowers are small, regular, white in colour. Fruits are many seeded follicles and seeds are black in colour.

Cultivation: It requires temperate climate, moist soils and shady locations. It can be cultivated by seedlings raised from seed, or from division of wildings. The crop mature in 6-10 years depending upon source of planting materials. Harvesting is done during winter season. On an average 124 kg dry rhizome per hectare is harvested with a spacing of 30 × 30 cm.

Medicinal uses: The dried rhizomes constitute the drug. It contains alkaloids of berberine group. It is tonic and stomachic. The rhizome is administered in intermittent fever and debility.

Scientific name	:	*Costus speciosus* Koenig.
Family	:	Zingiberaceae
Common name	:	Kebu

Description: *Costus speciosus* consists of two varieties viz., var. *nepalensis* Rose., found only in Nepal and Arunachal Pradesh and var. *argycophyllus* Wall., having a wide distribution in India. The plant is a succulent herb with long leafy spirally twisted stems, 2-3m tall and horizontal rhizomes. Leaves are simple, spirally arranged, oblanceolate or oblong, glabrous above, silky pubescent beneath with broad leaf sheaths. Flowers are white, large, fragrant, arranged in dense terminal spikes. Bracts are bright red. The single stamen present is perfect, lip large with incurved margins. Fruits are globose or ovoid capsules with obovoid or sub-globose seeds.

Cultivation: Costus can be grown under a wide range of agroclimatic conditions. It prefers sandy loam soil for good growth. Propagation is by rhizomes. The best season for planting is April- May. The seed rate recommended is 2-2.4t/ha. The spacing adopted is 50x50cm. After an initial ploughing FYM or poultry manure should be applied at the rate of 30t/ha and the field is to be ploughed again irrigated and prepared to obtain a fine seed bed. Furrows are opened and the rhizome pieces are placed horizontally at a depth of 8-10cm and covered with soil. Care is taken to place the eye buds facing upwards. After 70-75 days about 90-95% sprouting is obtained. Desiccation of the young sprouts have been observed in the hot summer months. As September-November is the period of maximum tuberization at least two irrigations should be given at that time. One during the sprouting period of the crop followed by two more keeps the crop fairly free of weeds. Application of 37t/ha of poultry manure and fertilizers, 60kg P_2O_5 and 40kg K_2O /ha as a basal doze, along with 80kg N/ha applied in 3 equal split dozes will take care. Crop is harvested at the end of seven months. Harvesting includes 2 operations, cutting the aerial shoots and digging out the rhizomes.

Medicinal uses: Tubers and roots contain diosgenin, 5a-stigmast-9(11)-en-3b ol, sitosterol-β-D-glucoside, dioscin, prosapogenins A and B of dioscin, gracillin and quinones. It is widely used as starting material in the commercial production of steroidal hormones. The rhizomes are useful in vitiated conditions of *kapha* and *pitta*, burning sensation, flatulence, constipation, helminthiases, leprosy, skin diseases, fever, hiccough, asthma, bronchitis, inflammation and anaemia. It is used to make sexual hormones and contraceptives.

Scientific name	:	*Curculigo orchioides* Gaertn.
Family	:	Hypoxidaceae
Common name	:	Kalimusali

Description: Roots of kalimusali are straight, cylindrical, tuberous and 5–20 cm long. The external surface is brownish, marked with loosely spaced, prominent, transverse wrinkles. Lateral roots are stout, fibrous, dull white in colour, and spongy externally. The freshly cut surface of tuberous rootstock has a starch-white colour within and is mucilaginous. Leaves are simple, sessile, crowded on the short stem with sheathing leaf bases, tapering into a short petiole, almost radical. They are linear or linear–lanceolate membranous, glabrous or sparsely soft haired. The leaf tip, when in contact with the soil, develops roots and produces adventitious buds. Flowers are epigynous, bright yellow, bisexual or unisexual, with a lanceolate and membranous bract.

Cultivation: Basically, it is a tropical plant and is found in almost all districts of India, from near-sea level up to 400 m altitude, especially in rock crevices and laterite soil. It is a shade-loving plant and grows well in areas that receive high rainfall. Tuber segments of 1.5–2 cm size, containing the apical bud, are collected during February–March and are planted directly in the main field at the onset of south-west monsoon, which breaks over South India in May–June. The tuber segments are planted at an optimum spacing of 10 cm × 10 cm. About 70%–80% sprouting is obtained after two months of planting. The propagule rate is 600–750 kg of root segments per hectare. FYM (farmyard manure) at the rate of 20 tonnes/hectare is applied at the time of land preparation. Alternatively, FYM at the rate of 15 tonnes/hectare may be applied at the time of land preparation and NPK at the rate of 25:15:10 kg/hectare can be applied as top dressing during October–November. An optimum crop stand of 0.6–0.65 million is desirable for a pure crop with an optimum spacing of 10 m × 10 cm or 10 cm × 15 cm, while intercropping with a coconut gives a crop stand of approximately 0.2 million with a spacing of 20 cm × 25 cm. If it is to be raised as a pure crop, artificial shade has to be provided using shade nets of 25% density. Seedling rot is observed during the rainy season and can be controlled by spraying and drenching the soil with 1% bordeaux mixture. Black rot disease is also observed and can be controlled by spraying 0.05% tridemorph. The plant starts flowering one month after planting and maximum number of flowers are noted during second and third months of planting. Flowering takes place throughout the year. Roots mature in the field in seven to eight months and may be harvested by digging. A dried tuber yield of 1000–1700 kg/hectare is obtained.

Medicinal uses: *Curculigo* roots contain acetone extractives (1.5%–1.8%), ash (3.3%–3.9%), and curculigosides (0.2%). It is used as a rejuvenating tonic, aphrodisiac drug, and diuretic. It is useful in general debility, cough, jaundice, asthma, and piles. It is also used as a contraceptive, laxative for stimulation of piles and treatment of tuberculosis.

Scientific name	:	*Curcuma caesia* Roxb.
Family	:	Zingiberaceae
Common name	:	Black Turmeric

Description: Black turmeric is an erect, rhizomatous herb, about 1.0–1.5 m high. The rhizomes are ovoid in shape and acute at tip.Leaves are about 30–60 cm long and up to 15 cm broad, broadly lanceolate or oblong, glabrous, with a deep ferruginous purple cloud down the middle, which penetrates to the lower surface. Leaves arise from the underground rhizome. Inflorescence is a spike, about 15 cm long or altogether about 30 cm high on basal peduncle. Flowers are pale yellow, reddish at the outer border and shorter than their bracts. Petiole and sheath are about as long as the blade. Spikes appear before the leaves.

Cultivation: *Curcuma caesia* grows well in sandy loam, acidic soils of pH 4.5–6.5. It is a partial shade-loving species; although it grows well in open sun under cultivated conditions. Rhizome is the propagation material. The land is ploughed, harrowed, and planked, mixed with FYM (farmyard manure) @ 5 tonnes per hectare together with NPK (nitrogen, phosphorus, potassium) @ 33:80:60 kg per hectare as basal dose during land preparation. Middle of April is the best time for raising the crop in North-East India, while in other regions, it is the pre-monsoon period. The crop is propagated vegetatively through rhizomes. The whole or parts of rhizome, weighing approximately 20 g, should be planted in rows directly in the field during April. Planting at a spacing of 30 cm × 30 cm is found optimum, for which 0.11 million propagules (rhizome segments) are required per hectare. The rhizomes sprout in about 15–20 days. A complete dose of NPK is to be applied @ 100:80:60 kg per hectare as urea, single super phosphate, and potash, respectively. One-third of urea, and full recommended quantity of single super phosphate and potash are applied at the time of final land preparation. The rest two-thirds of nitrogen as urea is top dressed in two split doses, first at the time of first earthing up and the rest at the time of second earthing-up. Earthing-up is carried out at 45 days and 60 days after planting. To reduce the crop–weed competition during the early stages of growth, manual weeding at 60, 90, and 120 days after planting is recommended. The crop is usually grown under rain-fed conditions in high rainfall tracts of Assam and Kerala states. Constant humidity is to be maintained in other areas through regular irrigation. Sprinkler irrigation is the most suitable method. The crop takes about nine months to mature. Harvesting is done during mid-January. Estimated yield of fresh rhizomes is 48 tonnes per hectare while dry rhizome yield is about 10 tonnes per hectare.

Medicinal uses: Dried rhizomes of *Curcuma caesia* are reported to contain 1.6% essential oil containing 76.6% *d*-camphor; 8.2% camphene and bornylene; and 10.5% sesquiterpenes, curcumine,ionone, and turmerone. Rhizome of kali haldi has a bitter, sharp, hot taste, and a pleasant odour. It has anti-bacterial and anti-fungal properties, and is laxative. It is used as a tonic for the brain and the heart. Rhizomes are useful in treating leucoderma, piles, bronchitis, asthma, tumors, tuberculous glands of the neck, enlargement of the spleen, epileptic seizures, inflammations, and allergic eruptions.

Scientific name	:	*Curcuma zedoari* (Christ). Rosc.
Family	:	Zingiberaceae
Common name	:	Karchur

Description: Zedoary is an annual or biennial, aromatic, rhizomatous, tall herb. It attains a height of 1.8 m. Rhizomes are large, pale yellow or whitish inside. Roots are aromatic and end in ellipsoid tubers. The edible root of zedoary has a white interior and a fragrance reminiscent of mango; however, its flavour is more similar to ginger, except for the fact that it leaves a very bitter after taste. There is no distinct aerial stem, but the shoot has a pseudo stem formed of long and closely overlapping four to six sheathing leaf bases. Leaves are large, reaching upto 1 m in length, oblong and deeply veined, often coloured purplish in the centre. Inflorescence is a spathe arising from the rhizome. Flowers are whitish or pale-yellow with bright reddish-green bracts.

Cultivation: Karchur is a semi-domesticated crop of moist tropical, subtropical, and temperate regions. Loam and sandy loam soil and rainfall of about 1100 mm per annum are ideal for its cultivation. It can be grown in open as well as shady conditions. About 150 quintals/hectare of FYM (farmyard manure) and 100:80:60 kg/hectare of NPK (nitrogen, phosphorus, potassium) are mixed thoroughly in the soil before planting. The crop can be planted anytime from April to June, depending upon the availability of soil moisture. Rhizome pieces are planted directly in the field in rows at a spacing of 20 cm, row-to-row distance being 40 cm. Rhizomes sprout in 10–12 days in moist soil conditions; otherwise, they remain dormant in soil and sprout after the first shower. For planting, about 125 000 propagules per hectare at a spacing of 40 cm × 20 cm are needed. The crop requires two to three weedings at intervals of 30, 60, and 90 days after planting. Overall crop maturity occurs in six to eight months. Harvesting of rhizomes is done during November–December. Fresh rhizome yield varies from 285 quintals/hectare to 315 quintals/hectare, while the yield of leaves is in the range 70–80 quintals/hectare on fresh weight basis. The essential oil yield from leaves is 7–8 kg/hectare, while the essential oil yield from rhizomes is 80–90 kg/ ha.

Medicinal uses: The rhizomes contain an average of 0.18% of essential oil when collected during February, March, and November. The major ingredients of essential oil are cineole, pinene, camphor, and so on. Rhizomes of Karchur are aromatic, cardio-tonic, carminative, diuretic, and anti-spasmodic. They are used to cure jaundice, cough, hiccups, and respiratory disorders. Essential oil extracted from leaves is used in perfumery and aromatherapy.

Scientific name	:	*Datura straumonium* Linn.
Family	:	Solanaceae
Common name	:	Jimson weed

Description: It is a bushy herb, 1m tall. Leaves are large, ovate, toothed. Flowers are very large, white. Fruits are ovoid, deeply divided into four, covered with long or short prickles.

Cultivation: Datura can easily be cultivated, in virtually any type of soil, but generally it prefers slightly acidic and calcareous conditions. A dry, sunny location is ideal. Seeds are sown at 30-45 cm apart, and kept free of weeds in the early stages. Adding nitrogen fertilizer to the soil will increase the concentration of alkaloids present in the plant. The plant is harvested when the fruits are ripe, but still green. During harvest, the entire plant is cut down, the leaves are stripped from the plant, and everything is left to dry. When the fruits begin to burst open, the seeds are harvested. For intensive plantations, leaf yields of 1,100 to 1,700 kilograms per hectare and seed yields of 780 kg/ha.

Medicinal uses: Dried leaves, flowering tops and seeds are medicinal. They are rich in Hygrine, Hyoscyamine, Scopolamine, Tropinone, Scopine, Tropine, 7-Hydroxyhyoscyamine and Littorine. Some medicinal uses of the plant are its anti-inflammatory property, stimulation of the central nervous system, respiratory decongestion, treatment of dental and skin infections and also in the treatment of toothache and alopecia.

Scientific name	:	*Dendrobium nobile* Linn.
Family	:	Orchidaceae
Common name	:	Singapore Orchid

Description: This is a perennial herb. The pseudobulb is erect or arching, 60-90cm tall caduceus, leathery, glossy green leaves. Inflorescence is 1 to 3 flowered, short and arises from the upper nodes of the old leafless pseudobulbs. The flowers are fragrant, long lasting, 10cm in diameter, waxy, lip with white margin. The flowers are produced during April- May.

Cultivation: The cool growing Dendrobium orchid group thrives well temperatures ranging between 10 and 24°C. Low temperature and short days could change the concentration of endogenous growth regulators leading to the induction of flowering in sympodial orchids. All types of Dendrobium orchids require warm bright light (2500-3000 foot candles). They should get at least 12-14 hours of light each day year round. Conventionally, Dendrobiums are easy to propagate through keikis that produces along old canes or by division of pseudobulbs. 10-12 cm long cuttings also can be taken from a healthy, older and leafless cane keeping three nodes on each cuttings and placed in moist sphagnum moss for rootings. Commercially, Dendrobium hybrids are usually either seed-propagated or clonally propagated through tissue culture of apical and lateral buds that proliferate as protocorm like bodies. Fresh air and good circulation are essential for orchid production. Full of continual light breezes make a good source of carbon dioxide for photosynthesis. In orchids, foliar feeding is found to be ideal. Frequent application of fertilizers in low concentrations is the best way of feeding orchids. A concentration of 0.2 to 0.3 % of 30:10:10 (N:P: K) at vegetative stage and 10:20:20 (N:P:K) at blooming stage are applied for quality flower production. Sometimes, fresh coconut water, diluted cow urine are also useful as foliar sprays. The potting medium of Dendrobium orchid should be loose, friable as well as well drained. Fresh air and good circulation are essential for orchid production. Full of continual light breezes make a good source carbon dioxide for photosynthesis. Orchids are light feeders and they require nitrogen from beginning to two-third of their life cycle. During rest period, they do not need any fertilizers. During flower initiation and inflorescence development plant are fed with less nitrogen, more phosphorus and potassium. During the blooming time, a small level of nitrogen and phosphorus and high levels of potassium are maintained. In orchids, foliar feeding is found to be ideal. Frequent application of fertilizers in low concentrations is the best way of feeding orchids. A concentration of 0.2 to 0.3 % of 30:10:10 (N:P: K) at vegetative stage and 10:20:20 (N:P:K) at blooming stage are applied for quality flower production. Sometimes, fresh coconut water, diluted cow urine are also useful as foliar sprays. A potting medium consisting of charcoal, brick pieces and coconut fibre in equal proportions is ideal for vegetative growth and flowering of epiphytic orchids like Aerides, Dendrobium etc. In Dendrbium, a potting mixture of cocochips/ cocopeat + brick pieces + tree bark (1:1:1) is ideal. Most orchids are damaged by overwater rather than under watering. Over watering leads to root rot and many other diseases. Most orchids prefer water of pH 5.0-6.5. Watering with lower or higher pH or with high levels of dissolved minerals can hamper nutrient uptake. Frequent watering is essential under high sunlight and high temperature conditions. The single dominant factor which affects the cultivation of orchid is humidity which should be as high as 50-75%.

Medicinal uses: Fresh or dried stems are ideal. These are rich in dendrobine, nobilonine, 6-hydroxydendrobine, dendramine, dendroxine, 6-hydroxydendroxine, 4-hydroxydendroxine, dendrine, 3-hydroxy-2-oxydendrobine. They enhance the physical fitness. It is a commonly used drug for stomachache and epigastric pain. It is a major medicine for the treatment of hepatobiliary diseases, such as hepatitis, cholecystitis, gallstones and others; it aids to strengthen tendons and bones, relieve joint pain, and reduce rheumatism. Modern pharmacological research also showed that it can improve stress ability, prevent or stop fatigue, increase hypoxia tolerance, promote circulation, dilate blood vessels, and lower blood cholesterol and triglycerides; Clinical studies have indicated that it not only enhances activity of insulin, but also significantly reduce and change blood sugar levels back to normal; The drug kills some certain cells of malignancies, such as lung cancer, ovarian cancer, and promyelocytic leukemia. It was also experienced as one of the best eye tonics in the eyes of ancient physicians because of its yin-nourishing and eyesight-improving properties. It nourishes the skin. Modern pharmacological studies confirmed that it contains many trace elements, which are closely related to human health and longevity. Hence, The drug has more comprehensive anti-aging effect compared to other herbs.

Scientific name	:	*Digitalis purpurea* Linn.
Family	:	Scrophulariaceae
Common name	:	Foxglove

Description: A biennial or perennial herb, 1.2m tall. Lower leaves are long stalked, hairy, ovate, 15-30cm long, the upper leaves almost stalkless, becoming smaller in size upwards. Flowers are 5-8cm long, white or purple, with ciliate lobes, borne in 30-60cm long, one sided bunches. Fruits are ovoid.

Cultivation: It is mainly grown in hilly regions. The plants grow in almost any soil under partial shade, but avoid very wet or very dry situations. It prefers a humus-rich soil in partial shade but will grow in full sun. Deadhead after flowering to avoid problems with excess numbers of seedlings. Propagated through seeds. The plants are cut back after flowering.

Medicinal uses: Foxgloves are a source of digitoxin, a glycoside used in the drug digitalis, which has been used as a heart stimulant since 1785. Foxglove tea (an infusion of the leaves) is taken for colds, fevers and catarrh, and compresses are used for ulcers, swellings and bruises. Its most common use is as a diuretic against dropsy.

Scientific name	:	*Eclipta prostrata*Linn.
Family	:	Asteraceae
Common name	:	Kesut

Description: This is an annual, erect or postrate herb, often rooting at nodes. Leaves are sessile, 2.5-7.5cm long with white appressed hairs. Floral heads are 6-8 mm in diameter, solitary and white. Fruit is an achene, compressed and narrowly winged.

Cultivation: It germinates over a wide range of pH, salt, and temperature conditions and preferentially germinates when the seed is on or near the soil or media surface. Eclipta germination is greater in warm, moist soils and as a result it typically occurs in poorly drained fields, irrigated areas, or during periods of heavy rainfall. Eclipta can also reproduce vegetatively through adventitious rooting of nodes or from stem fragments left on the soil or media surface following hand-weeding or cultivation. Eclipta grows aggressively in containers. Eclipta is not drought tolerant and germinates poorly in dry soils. Grouping plants by water requirement, improving irrigation efficiency, installing rain sensors, and other practices that improve water-use efficiency may reduce incidence. The whole plant is plucked after 9 to 10 month. The fresh plants are chopped and dried in shade. The seeds are collected when it changes black in colour. The average matured plants give 2500 kg per acre of dry materials.

Medicinal uses: The leaves are rich in stigmasterol, α-terthienylmethanol, wedelolactone, dismethylwedelolactone and dismethylwedelolactone-7-glucoside. The roots contain hentriacontanol and heptacosanol. It is used in hepatitis, spleen enlargements, chronic skin diseases, tetanus and elephantiasis. The leaf promotes hair growth and use as an antidote in scorpion sting. The root is used as an emetic, in scalding of urine, conjunctivitis and as an antiseptic to ulcers and wound in cattle. It is effective to prevent abortion and miscarriage and also in cases of uterine pains after the delivery. The juice of the plant with honey is given to infants for expulsion of worms. For the relief in piles, fumigation with Eclipta is considered beneficial. A decoction of the leaves is used in uterine haemorrhage. The paste prepared by mincing fresh plants has got an anti-inflammatory effect and may be applied on insect bites, stings, swellings and other skin diseases. In Ayurveda medicine, it is mainly used in hair oil.

Scientific name : *Enhydra fluctuans*Lour.

Family : Asteraceae

Common name : Helancha

Description: It is a glabrous marshy herb with sessile, linear-oblong, acute leaves and axillary, terminal and sessile heads.

Cultivation: It is a tropical herb, more sensitive to cold especially when very young. The species grows in and along ditches, water courses, margins of fish ponds and rice fields in the open, from sea-level up to 1,800 m. It is able to reproduce by fragmentation and may be so abundant that it clogs water course.

Medicinal uses: Plant is rich in protein and is a good source of β-carotene. It also contains saponins, myricyl alcohol, kaurol, cholesterol, sitosterol, glucoside, sesquiterpene lactones including germacranolide, enhydrin, fluctuanin and fluctuandin. Plant is used in ascites, dropsy and anasarca. It is cooked with fish curry and taken to revive appetite after long weakness due to fever or typhoid. Leaves are laxative and antibilious and are used to cure inflammation, leucoderma, bronchitis and biliousness; effective in skin and nervous affections; also useful in tropidity of the liver. Leaf paste is applied over head as a cooling agent and around the inflamed breast to reduce inflammations.

Scientific name : *Erigeron canadensis*Linn.

Family : Asteraceae

Common name : Canadian Fleabane

Description: It is a herbaceous perennial plant, 1.8m tall with bristle leaves. Flowers are of different colours.

Cultivation: Ordinary soil and sunny locations are ideal for their growth. Plants are propagated by seeds and division of roots. Stems are cut down after flowering.

Medicinal uses: Mature leaves and flowering tops are useful. The oil extracted from the plant is used to treat pulmonary problems, haemorrhoids and diarrhoea. A decoction of the tops of plants is used to cure sore throat and dysentery.

Scientific name : *Eulophia campestris*Wall.

Family : Orchidaceae

Common name : Salep

Description: It is a terrestrial orchid with fleshy tubers or rhizomes. The leaves are plicate. The flowers are racemose and purple in colour.

Cultivation: The plants are grown in well drained and rich compost in pots or beds. The growing medium should consist of one part of each loamy soil, sand and well rotted leaf mould and one eighth each of shredded osmunda and chopped tree fibre. Charcoal and broken crocks are also used in the growing media.

Medicinal uses: Tubers are medicinal and used as aphrodisiac. Tubers are used to treat pulmonary and cardiac problems.

Scientific name	:	*Euphorbia hirta*Linn.
Family	:	Euphorbiaceae
Common name	:	Dudhi

Description: This is a herb of erect up to 50cm height with greenish yellow or white flowers. Stems covered with yellowish hair. The leaves are small in opposite pairs.

Cultivation: Naturally distributed in cultivated fields, gardens, roadsides and waste places, from sea-level up to 2,000 metres in in tropical and subtropical climate. It prefers a light well-drained moderately rich loam in an open sunny position· Seed germination usually takes place within 2 - 3 weeks at 20°C

Medicinal uses: Traditionally been used in Asia to treat bronchitic asthma and laryngeal spasm, though in modern herbalism it is more used in the treatment of intestinal amoebic dysentery The plant is anodyne, antipruritic, carminative, depurative, diuretic, febrifuge, galactogogue, purgative and vermifuge. The stem, taken internally, is famed as a treatment for asthma, bronchitis and various other lung complaints.

Scientific name	:	*Ferula narthex*Boiss..
Family	:	Umbelliferae
Common name	:	Asafoetida

Description: It is a perennial herb. Lower leaves simple up to 60cm long, upper leaves divided into segments and young leaves are densely hairy. Flower appear in large terminal clusters, small and yellow in colour. Fruit is one cm long, roots are robust, carrot shaped.

Cultivation: The perennials prefer a sunny situation on moderately moist soil. The substrate should be sandy loam. They tolerate temperatures down to -10°C. Propagated through seeds or division.

Medicinal uses: Oleo-gum resin from rhizomes and roots constitute the drug. The gum resin obtained from the root is anthelmintic, antispasmodic, expectorant and nervine. It is used in the treatment of asthma, whooping cough, flatulent colic, and in pneumonia and bronchitis in children. The leaves are used as carminative and diaphoretic. The plant has recently been investigated as a potential contraceptive.

Scientific name	:	*Gentiana kurroo*Royle.
Family	:	Gentianaceae
Common name	:	Indian Gentian

Description: It is a small perennial herb up to 30 cm height with tufted decumbent stems. Flowers are blue with white spots, 4.5cm long and 2.5 cm in diameter. Fruits are 2 cm long.

Cultivation: Naturally distributed to north-western Himalayas between 1,500-3500 metres altitude in Garhwal, Kashmir and adjoining hills. It can be propagated through seeds, stem cuttings or by root divisions. It requires a cool temperate climate and well-drained moist soil. It requires sunshine and plenty of water when growing.

Medicinal uses: Rhizomes and roots are used for medicinal purposes. They contain bitter glycosides like gentiopicrin, gentianin and gentiomarin. The rootstocks are used as bitter tonic, antiperiodic, expectorant, antibilious, astringent, stomachic, anthelmintic, blood purifier and carminative. It is used for curing skin diseases, bronchial asthma and urinary infections.It is also used for improving appetite, stimulating gastric secretion and in the treatment of digestive disorders.

Scientific name	:	*Geum rivale*Linn.
Family	:	Rosaceae
Common name	:	Water Avens

Description: It is a hardy rhizomatous perennial herb, one metre tall. The leaves are three lobed. The flowers are saucer shaped, borne in loose clusters and purple. It has thick rhizomes.

Cultivation: The plants require rich moist organic soil and sunny locations. Generally propagated by seeds or divisions. The old stems are pruned after flowering.

Medicinal uses: Above ground parts of these plants before flowering and roots are useful. Roots and tops of the plants are chopped and mixed with milk and sugar and taken internally for indigestion, run down conditions and dysentery. For treatment of haemorrhages, stomach ulcers, intestinal worms and menstrual disorders, a root decoction is used.

Scientific name	:	*Glycyrrhiza glabra*Linn.
Family	:	Papilionaceae
Common name	:	Liquorice

Description: This is a tall perennial, self -pollinated herb about 1m tall with long cylindrical burrowing rootstock and horizontal creeping stolons which reach 1.5-1.8m in length. Leaves are alternate, pinnate with 9-17 leaflets. Leaflets are yellowish-green in colour , 2.5-5cm long, ovate and obtuse. Flowers are pale blue arranged in a raceme and 1.25cm long. Calyx is glandular and pubescent. The pods are glabrous, red to brown having 3-4 seeds. Rhizome is soft, flexible and fibrous with light yellow colour and a characteristic sweet taste.

Cultivation: This plant thrives well in subtropical areas with very warm summers and cool winter. Semi-arid and arid areas in subtropical zones are not favourable for the cultivation of this crop. It does not tolerate high humidity and waterlogged conditions. Well drained light loam soils which are rich in calcium and magnesium with slightly alkaline pH and free from stones are suitable for this crop. There are a number of varieties of this crop among which *Spanish, Russian* and *Persian* liquorice are quite common. Commercial varieties are *Typica, Regel* and *Herd*. The plants are propagated by seed, but usually multiplied vegetatively either through crown cuttings or stolon pieces. In the case of crown cuttings, 10-15cm long crown pieces with 2-3 buds are planted vertically at a distance of 0.6-0.7m in rows 1-1.5m apart. However, most of the liquorice is propagated through stolon pieces of the above size planted horizontally, preferably on ridges during spring at the same distance as mentioned above. Rapid clonal propagation is also possible by tissue culture technique. *Murashige and Skoog's* medium supplemented with 6-benzylaminopurine and indole-3-acetic acid favours multiple shoot production without any intervening callus phase. These regenerate plantlets which can be transferred to earthen pots in the glass house and after a brief hardening phase, these are transplanted in the field with a high rate (90-95%) of survival. This plant normally does not require much fertilizers but in deficient soils, it is advised to apply 10-15 tonnes FYM per hectare before planting. The field should be immediately irrigated after planting in spring and after the crop has sprouted, it requires very little irrigation. Space between the rows should be kept free from weeds. Short term vegetables like carrot or cabbage can be planted between the rows for additional income. In order to produce good rhizome, flowering shoots are clipped. No serious disease except leaf spot caused by *Cercospora cavarae* has been reported in this crop. Roots are ready for harvesting after 3-4 years. The root is dug when the top has dried during autumn (November- December). A trench 60cm deep is dug along the ridges and the entire root is lifted. Broken parts of the root left in the soil, sprout again and give another crop after 2-3 years. Thus liquorice once planted properly can be kept for harvesting for 10-15 years.

Medicinal uses: Roots rich in glucoside like glycyrrhizin. Root also contains flavans, flavones, iso-flavanoes and coumarins including a 4-methyl coumarin, liqcoumarin, glabridin, glabrene, 4'-0-methyl and 3'-methoxyglabridin, formononetin, salicylic acid, 0-acetyl salicylic acid. Its dried peeled or unpeeled underground stems and roots constitute the drug which is an important constituent of all cough and catarrh syrups, throat lozenges and pastilles. Hippocrates (400 BC) mentioned its use as a remedy for ulcers and quenching of thirst. Dioscorides, the father of Greek medicine described this drug in detail and considered it useful for maintaining shape of arteries and in burning stomach, trouble of liver and kidney, scabies, healing of wounds and as a remedy for eye diseases.

Scientific name	:	*Habenaria intermedia* D. Don.
Family	:	Orchidaceae
Common name	:	Riddhi

Description: It is a semi-terete orchid of 25cm length. The leaves are rounded at the base, long acuminate to 10cm long and 5 to 7 nerved. The inflorescence is 4 to 6 flowered. The flowers are 5cm across, white or greenish.

Cultivation: The mean annual rainfall is 100 to 150 cm and mean annual temperature is between 10-15^{0}C. Tubers are recommended for the propagation of this species. The compost consisting of rich loamy soil, well rotten manure and one sixth each of shredded osmunda and chopped tree fern fibre is ideal for the growth and flowering of these plants. Direct raising of plant from tubers either by half or full tuber with apical portion intact gives the best results. Planting is done on raised beds in rows at an optimum spacing of 20 X 20 cm.

Medicinal uses: The drug belongs to the group of the "Eight Tonic Herbs", known as Ashtavarga, which is rejuvenating and age sustaining. It is an ingredient of 'Chywanaprasha'.

Scientific name	:	*Hedychium spicatum* Ham. ex Smith.
Family	:	Zingiberaceae
Common name	:	Kapoor Kachri

Description: It is a rhizomatous, fragrant leafy herb with robust stem, and is up to 1.5 m tall. Rhizomes are strongly aromatic and have bitter camphoraceous taste. They are white, starchy, and fragrant within. Bark is rough, reddish-brown with few deep-seated fibrous rootlets. Leaves are up to 30 cm long, lanceolate, with green, obtuse bracts and leaf sheath clasping the stem. Flowers are fragrant, white with orange-red base, present in a dense, terminal, 15–25 cm long spike. Fruit is a spheroid, three-valved capsule with orange-red lining. Seeds are black with a red aril.

Cultivation: Kapoor kachri prefers shady slopes. Waterlogging in the soil is fatal due to rotting of rhizomes. Moderate temperature and well-spread rainfall are suitable for better growth. Sub-temperate to temperate climate with annual precipitation of 1000–1500 mm, and well-drained, deep sandy loam, and humus-rich soil with good moisture retaining capacity are most ideal for its cultivation. Both seeds and rhizomes may be used as propagation material. Rhizome pieces with apical buds are buried in 10 cm × 20 cm polybags containing soil, sand, and FYM (farmyard manure) in equal amounts and irrigated intermittently. About 25 quintals of healthy rhizomes, segregated into pieces with one bud in each and weighing about 40–50 g, are required for raising plantlets in 1 hectare of land. With second ploughing, well-decomposed FYM at the rate of 20 tonnes per hectare should be spread well and thoroughly mixed. The rhizomes are planted in furrows at a depth of 10–12 cm, at an optimum spacing of 45 cm × 30 cm. At this spacing, about 64 000 propagules per hectare will be required. If required, inorganic fertilizer, such as NPK (nitrogen, phosphorus, potassium), can be applied at the rate of 100:120:60 kg/hectare in three split doses. The entire amount of phosphorus and potassium along with one-third of nitrogen should be applied in furrows at the time of land preparation. The remaining nitrogen should be applied in two equal split doses: first after two months of planting and the second in the next rainy season after new sprouting. In rhizome rot infection, leaves of the affected plant become pale and the affected rhizomes become soft and pulpy, and ultimately rotten. Rhizome rot can be controlled by dipping the rhizomes in 0.01% bavistin solution for 25–30 minutes followed by shade-drying before planting. Second year harvesting yields about 115–120 quintals/ hectare dry weight of rhizomes.

Medicinal uses: The plants contain pinene, limonene, camphor, linalyl acetate, terpineol, caryophyllene, benzyl cinnamate, benzyl acetate, terpinene, phellandrene, methyl paracumarin acetate, cinnamic ethyl acetate, ethyl cinnamate, sesquiterpene alcohols, and hydrocarbons. Rhizomes are aromatic, acidic, bitter, pungent, carminative, stomachic, stimulant, expectorant, anti-asthmatic, antiseptic, and anti-inflammatory. It is useful in asthma, bronchitis, vomiting, dyspepsia, and inflammations. It has insect repellent properties, and is also used as a dye and perfume for making 'abir' powder used in 'Holi'.

Scientific name	:	*Hygrophila auriculata* Schum.
Family	:	Acanthaceae
Common name	:	Talmakhana

Description: It is a stout erect herb, 60-150cm tall with unbranched straight, erect four angled, hairy stems. Leaves are whorled, 6 at a node, each having a sharp, yellow spine in its axil. Flowers are 8 in a whorl, purple blue, two lipped where lower lips are folded.

Cultivation: The plants are widely distributed in marshes, fallow lands, dried ditches and roadside water channels. It grows in areas where the mean annual rainfall is in the range 700 - 1,600mm. It thrives well in yellowish sediments and black-cotton soil.

Medicinal uses: Plant are rich in flavonoids, alkaloids like asteracanthine and asteracanthicine and triterpenes.The seeds are used as ingredients in various aphrodisiacs and tonic confections, and in the treatment of blood disorders, biliousness, gonorrhoea, spermatorrhea and fever. The seeds are ground into a paste and given in buttermilk to cure diarrhoea. AKSIR-ULIMRAZ, a preparation having Talamkhana (seeds) as one of the ingredients, is used to prevent leukorrhea. The ashes of the plant are also used against dropsy and gravel. A tincture of the whole plant is beneficial in urinary affections, dysuria, and painful micturition. A root decoction drunk to combat rheumatism, gonorrhoea, and hepatic obstruction. The leaves are diuretic, sweet, tonic, aphrodisiac, hypnotic and useful in the treatment of cough, diarrhoea, thirst, urinary calculi, urinary discharges, inflammations, joint pain, eye diseases, pains, ascites, anaemia, and abdominal disorders.

Scientific name	:	*Hyoscyamus niger* Linn.
Family	:	Solanaceae
Common name	:	Black Henbane

Description: This is an erect, biennial herb, densely covered with glandular hairs and attains a height of 1.5 metres. Leaves are 15-25cm long, coarsely dentate and pinnately lobed. Upper leaves are smaller and divided into many segments. Flowers are 2-3 cm in diameter, yellowish pale, some borne solitary at the place of branching of stems, others borne on terminal cymes. Fruits are globose containing numerous small seeds.

Cultivation: They can be cultivated throughout the year with equitable climate. A well drained, sandy loam soil rich in organic matter is suitable for this crop. The crop is grown from seeds and can be raised directly sown seeds or transplanted seedlings. For directly sown, around 5 kg seeds/ha are required. Seeds germinate within a week after sowing in nursery beds and can be transplanted when the seedlings attain 4 to 5 leaves. Generally, the seeds are sown during October. After thorough ploughings, about 20 tonnes of FYM is added to the soil during land preparation. A fertilizer dose consisting of 40kgN, and 80 kg P_2O_5 is recommended for this crop. Seedlings are planted in final site at a distance of 30 cm x 45cm in the month of November. In directly sown seeds, the crop should be thinned after germination of seeds. This crop is grown as a short duration crop for 90 to 100 days winter season crop in subtropical plains while in temperate and hilly regions the crop is raised during summer. During the growing period, the crop requires 3-4 irrigations and two weeding operations.

When the plants reach flowering stage, they are then cut from the ground level. The entire herbage is spread in shade for complete drying. The dried material should be stored in a dry place. From one hectare, 15 quintal dried leaves can be obtained and the total herbage yield including stalks, pedicels and leaves is around 30 q/ha.

Medicinal uses: The leaves and flowering tops contain 0.1 to 0.15% alkaloids and important alkaloids extracted are hyoscyamine and hyoscine. The drug is useful for relieving certain painful spasmodic conditions of muscles and in nervous irritations as in hysteria and cough. It dilates the pupil of the eye. This has a sedative effect on brain. The drug is also administere against asthma and whooping cough. The seeds are pasted and applied locally on pains. Externally, It may be used as a poultice. It is maturative and narcotic.

Scientific name : *Inula racemosa* Hook.f.

Family : Asteraceae

Common name : Puskarmool

Description: This is a perennial herb up to 1.5 m tall with fragrant prominent root and rootstock. Stems are many in number, ascending from the base of the rootstock. Leaves are leathery, rough above and densely hairy below, 25–50 cm long and 10–12 cm broad, and elliptic–lanceolate in shape. Flower heads are yellow in colour, have bisexual florets, and occur in terminal racemes. Fruits are achenes, slender and about 0.5 cm long.

Cultivation: Temperate and sub-alpine regions are most suited for the cultivation of *Inula* species. Well-drained, clay-loam soils are suitable for the crop. It prefers open and sunny locations. Propagation can be done both by seeds and rootstocks. The crop can be raised through seeds in nursery in November or early March by broadcasting the seeds in well-prepared nursery beds. Germination is completed in about 50 days. Seedlings are transplanted in the field with a ball of earth after about two months of growth. About 1 kg seeds are required to raise about 40 000 seedlings required for planting 1 hectare of land. Usually, FYM @ 15 tonnes/hectare is applied at the time of land preparation. Optimum spacing recommended in the main field is 50 cm × 50 cm. An optimum crop stand of approximately 40 000 plants per hectare is accommodated with the above-mentioned plant density. NPK (nitrogen, phosphorus, potassium) @ 200 kg/ hectare, 100 kg/hectare, and 50 kg/hectare, respectively, may be used per year to achieve high biomass yield. Nitrogen is usually applied in two split doses, one as a basal dose and the other (100 kg/hectare) as top dressing in six to eight weeks after transplantation. Only manual weeding is recommended twice during the establishment phase of the crop. Light irrigation at an interval of three to four weeks is sufficient to maintain the crop. The crop is harvested at one-and-a-half years of growth. About 80 quintals of dried roots are harvested from 1 hectare cropped area.

Medicinal uses: The rootstock and roots are rich in inulin, alantolactone, sitosterol, isoalantolactone, dihydroalantolactone, and its glucosides. Puskarmool is an aromatic tonic, febrifuge, and expectorant with anti-inflammatory, carminative, diuretic, and antiseptic properties. The plant is used in chronic bronchitis and rheumatism. Dried rhizomes and roots are effective to cure loss of appetite and stomach troubles.

Scientific name : *Kaempferia rotunda* Linn.

Family : Zingiberaceae

Common name : Indian Crocus

Description: This is an aromatic herb with tuberous root-stalk and very short stem. Leaves are simple, few, erect, oblong or ovate-lanceolate, acuminate, 30cm long, 10cm wide, variegated green above and tinged with purple below. Flowers are fragrant, white, tip purple or lilac arranged in crowded spikes opening successively. The plant produces a subglobose tuberous rhizome from which many roots bearing small oblong or rounded tubers arise.

Cultivation: The plant grows well in tropical climate. Well drained rich loamy soil is ideal for the plant. Laterite soil with heavy organic manure application is also well suitable. Planting is followed in May-June with the receipt of 4 or 5 pre-monsoon showers. The recommended seed rate is 1500-2000kg rhizomes/ha. Whole or split rhizome with one healthy sprout is considered as the planting material. Well developed healthy and disease free rhizomes having attached root tubers are selected for planting. Rhizomes can be stored in cool dry place or pits dug under shade plastered with mud or cowdung. The field is ploughed to a fine tilth, mixed with organic manure at 10-15t/ha. Seed beds are prepared at a size of 1m breadth and convenient length. Pits are made at 20cm spacing in which 5cm long pieces of rhizomes are planted. Pits are covered with organic manure. They are then covered with rotten straw or leaves. Apply FYM or compost as basal dose at 20 t/ha either by broadcasting and ploughing or by covering the seed in pits after planting. Fertilisers are applied at the rate of 50:50:50 kg N, P_2O_5 and K_2O/ha at the time of first and second weeding. After planting, mulch the beds with dry or green leaves at 15 t/ha. During heavy rainy months, leaf rot disease occurs which can be controlled by drenching 1% Bordeaux mixture. The crop can be harvested after 7 months maturity. Drying up of the leaves is the indication of maturity. The crop is carefully harvested without cutting the rhizome and after removal of dried leaves and roots. The rhizomes are washed in water. They are stored in moisture-proof sheds. Prolonged storage may cause insect and fungus attack.

Medicinal uses: The tubers are rich in crotepoxide and β-sitosterol. The tubers of Indian crocus are used as a local application for tumours, swellings and wounds. They are useful in gastric complaints. They help to remove blood clots and other purulent matter in the body. The juice of the tubers is given in dropsical affections of hands and feet, and of effusions in joints. The juice causes salivation and vomiting. It also improves complexion and aids to cure burning sensation, mental disorders and insomnia.

Scientific name	:	*Lobelia nicotianaefolia* Hyne ex Roth.
Family	:	Campanulaceae
Common name	:	Wild Tobacco

Description: It is a herbaceous plant with stout and hollow stem, which grows to a height up to 5 metres. The leaves are very large, 45 cm length in lower part of stem, main nerve of the leaf is whitish. The flowers are borne in terminal branches, white in colour. The fruits are up to 1 cm in diameter containing small, yellowish brown small seeds.

Cultivation: It grows well on rich, moist loam soil in full sun or partial shade. It requires 7 to 19 degrees centigrade and a soil pH of 4.8 to 6.8 for its growth. Plants are propagated through seeds.

Medicinal uses: The plant contains alkaloids such as lobeline, lobelidine, lobelanine, nor-lobelaine, lobelanidine, nor-lobelanidine and isolobenine. It is used in India to treat bronchitis, asthma, and insect and scorpion bites and to induce nausea and vomiting.

Scientific name : *Lycopodium clavatum* Linn.

Family : Lycopodiaceae

Common name : Staghorn Clubmoss

Description: These are low growing evergreen perennial plants. The plant is clothed with scale like, dark green leaves. The leaves are often with bristles at the apex. The foliage is ornamental. The plant produces yellow spores and no flowers. Spores appear in short cylindrical spikes. Capsules are kidney shaped enclosing a powder known as lycopodium or commonly called vegetable sulphur or vegetable brimstone. The spores are very light, fine, odourless and highly inflammable.

Cultivation: This species in naturally distributed in cold and temperate regions of Old and New World. It requires deep moist and shaded location for cultivation. Well drained organic rich soil with charcoal and lime added is preferable. They require plenty of water in dry weather.

Medicinal uses: The plants are rich in fatty oil and a minor amount of palmitic acid and linoleic acid. Spores of this plant are known as 'vegetable sulphur' which are used in medicine for many purposes. It is used as a local desiccative to prevent chapping children's skin and to dust the bed of patients who develop sores from prolonged rubbing of sheets. It is administered as a decoction as anti-rheumatic, diuretic, anti-epileptic and anti-diarrhoeic. It is orally taken for urinary and digestive complaints. The decoction of spores is used to treat severe diarrhoea, to increase the appetite and has been considered as aphrodisiac.

Scientific name : *Malaxis muscifera* Ridl.

Family : Orchidaceae

Common name : Jeevak

Description: It is a terrestrial orchid, pseudobulbs are stem like, rather small, clustered. The leaves are paired, sessile or short stalked upto 10cm long. The inflorescence is up to 45cm long, many flowered raceme. The flowers are very small, yellowish green in colour.

Cultivation: This species is found in sub-alpine areas and is frequently distributed in meadows, forests, shrubberies and grassy slopes in the temperate Himalayas. It grows in loose sandy loam soil, rich in humus, chiefly on upper stratum of organic layer, in the wet localities. The requirement of mean annual rainfall ranges between 1000 mm and 1500 mm, and the optimal mean annual temperature range is 10–15 °C. Mature bulbs may be collected from the wild sources in the rainy season. About 250 000 nodal segments or 125 000 bulbs are required for planting as a sole crop in 1 hectare of land at a spacing of 20 cm × 20 cm. Mycorrhizal association is necessary for increasing nutrient uptake efficiency in this orchid. Application of FYM and leaf mould @ 40–50 tonnes per hectare is optimum for the crop growth. The crop matures in five months and the tubers are ready to be harvested when dormancy sets in during the last week of October or first week of November.

Medicinal uses: It is used as tonic. The root is traded in the name of "Rsabhakah" or "Jeevak". In Ayurvedic texts, Chyawanprash is classified under the group of Rasayana, used to maintain the body's integrity for delaying the ageing process, enhancing longevity and improving digestion. It is a polyherbal formulation comprising of more than 50 medicinal plants ingredients. Rsabhaka *(Malaxis muscifera)* is one of the important ingredient. of 'Chywanaprasha'.

Scientific name	:	*Malva parviflora* Linn.
Family	:	Malvaceae
Common name	:	Little Mellow

Description: It is an annual plant, 1.2m tall. Pink flowers of small size appear in clusters in leaf axils.

Cultivation: An organic rich well drained soil and sunny locations are ideal for these plants. Propagated through seeds. Seed germination takes place 10-15 days.

Medicinal uses: The mature leaves and flowering tops are useful. It is used to treat swollen glands of throat, fever, pneumonia, headache and also to induce perspiration and menstrual flow.

Scientific name	:	*Melilotus officinalis* Mill.
Family	:	Papilionaceae
Common name	:	Cock's Head

Description: The plants are annual or biennial herbs, 1.2–1.8 m tall. Leaves are alternate on the stem and possess three leaflets. Yellow flowers bloom in spring and summer and produce fruit in pods typically containing one seed. Seeds remain viable for up to 30 years. Plants have large taproots and tend to grow in groups.

Cultivation: It commonly grows in calcareous loamy and clay soils with a pH above 6.5 and can tolerate cold temperatures and drought and it does not tolerate standing water.

Medicinal uses: This is emollient, resolutive, anodyne and carminative. In a fomentation on the abdomen, it is good for colic, for gout and rheumatism; in a rectal, it is vermifuge and carminative. In the form of an herbal tea, sweet clover is employed to alleviate headaches, muscles pains as well as gastrointestinal disorders.

Scientific name	:	*Nardostachys grandiflora* DC.
Family	:	Valerianaceae
Common name	:	Jatamansi

Description: This is an erect, hairy, perennial rooted herb, 10–60 cm high, with stout woody main root. Rootstock is thick, long, covered with remnants of petioles of withered leaves. Stems are generally pubescent upward, and glabrate below. Leaves are radical as well as cauline. Rhizome is dark brown, tapering, and densely covered with the remains of old leaf bases, giving the appearance of a heavy beard, and hence the name given as jatamansi. The fibrous rhizomes are 6–9 cm long with about 20–30- cm-long yellowish taproot. A single root bears 30–50 rhizomes. Flowering shoot is 30–40 cm high and bears 40–50 flowers at summit in three to seven condensed cymes. Flowers are campanulate, pinkish red to bluish white.

Cultivation: The plant generally grows in steep hills with 40°–70° inclined slopes and is very frequent on open, stony, and grassy slopes. It is more distributed in the western cooler slopes. Thick humus layer promotes rapid growth and profuse branching. The crop is grown by raising a nursery in May, from seeds or vegetative rhizomes separated from the mother plants. Seeds are sown manually in small plots of 1 m × 1 m size or in styrofoam trays in the polyhouse. Approximately, 600 g of seeds are required for raising seedlings for transplanting in 1 hectare of land. Both seeds and rhizomes are treated with GA_3 (gibberellic acid; 100 ppm and 200 ppm for 48 hours for rapid germination/ sprouting. A minimum of about 40 quintals FYM/forest leaf litter as basal dose is required for the better growth of plants and good biomass production. The seedlings are transplanted to the main field about 50–60 days after germination at a spacing of 20 cm × 20 cm or 20 cm × 30 cm. Plants raised through seeds may take three to four years to mature, while the plants raised through splitting of rhizomes mature in two to three years and are ready for harvesting earlier. At the experimental site at an elevation of 3600 m, the recorded yield was 835 kg/hectare dry roots; plantation was raised through seedlings. At lower altitudes (2200 m), the recorded yield was 670 kg/hectare dry weight after third year of cultivation; plantation was raised through seedlings.

Medicinal uses: The dried rhizomes contain 1.5%–1.9% of a pale yellow essential oil, commercially known as spikenard oil, emitting a pleasant odour. It is a nervine tonic, hypotensive, antiseptic, stomachic, carminative, tranquilizer, sedative, anti-bacterial, antispasmodic, and stimulant. It is useful in the treatment of insomnia, hysteria, convulsions, vertigo, and is also used as a cardiac tonic.

Scientific name	:	*Orchis latifolia*Linn.
Family	:	Orchidaceae
Common name	:	Cuckoo Flower

Description: These are herbaceous perennial, leafy orchids. The tubers are entirely oblong or palmately lobed. The stems are fistular up to 90cm long. The tubers are paired, lobed and palmate. The leaves are 15cm long. Spike is cylindrical, densely flowered, up to 15cm long. The flowers are 2cm long, variable in colour from pink to purple to almost pure white.

Cultivation: This species is distributed in the western Himalayas and Kashmir between 3000-4000 m altitude. In temperate region, they are grown in the open field. Otherwise, they grow well in full sun or partial shade. A rich and porous compost is suitable. Division of the tubers is done as the flowers fade out. Division can also be done when the plant has a fully developed rosette of leaves but before it comes into flower.

Medicinal uses: Tubers are rich in starch, mucilage, sugar, phosphate, and loroglossin. Tuberous roots are cooling, emollient, aphrodisiac, rejuvenating and nervine tonic. These root are used to cure dysentery, diarrhoea, chronic fever, cough, wounds, cuts, burns, fractures and general weakness.

Scientific name	:	*Panax gingseng*C.a. Meyers.
Family	:	Araliaceae
Common name	:	Ginseng

Description: This is an erect perennial herb and grows to a height of 60cm. The leaves are compound, palmate. Petiole is longer. Leaflets are usually, five, sometimes three, upper three layers up to 15cm long and lower up to 4cm long, elliptic or ovate. The inflorescence is simple, umbelliform. The flowers are green-yellow. Fruits are berry like. Roots are thick, fleshy, short 1.0 to 2.5 cm diameter, cylindrical or fusiform.

Cultivation: The plants are propagated by seeds for commercial purposes. It can also be propagated by root cuttings or grafting. Plants grow better in light loamy soil rich in organic and humus, in a temperate climate at a temperature of 20-25°C in summer and 4°C in winter. It is a greenhouse herb and requires shade for cultivation. Generally, stratified seeds are sown at 2.5cm depth in a well drained raised bed which takes 8 to 9 months to germinate. One year old seedlings are transplanted under shaded locations at a distance of 20 x 30cm. Harvesting of root is generally done when the plantation is 4 to 5 years old. The roots are dug, thoroughly washed and dried at a temperature of 45°C. Plants cultivated under artificial shade can give a yield of 1000kg/ha after drying, roots are packed in cardboard boxes and stored in dry and cool atmospheres.

Medicinal uses: The roots contain panaxosides, steroids, vitamins and trace elements. The drug is said to be tonic, stimulant and aphrodisiac. It is claimed that ginseng increases the natural resistance and recuperative power of the body. It is used for the treatment of amnesia, headache, convulsion, dysentery and cancer. It is also prescribed for neurasthenia, dyspepsia, palpitation and asthma.

Scientific name	:	*Papaver somniferum*Linn.
Family	:	Papaveraceae
Common name	:	Opium Poppy

Description: It is an erect herbaceous plant yielding the commercially viable drug, opium. Plants are 120cm tall with large serrated leaves and attractive coloured flowers. Flowers are red, pink, white, purple and multi-coloured. Common varieties available under the species are Mandsaur 'MOP3' and 'MOP 16', Dhola Chota Gothia', 'Telia' and 'Ranjhatak'.

Cultivation: They grow well in well drained fertile soil, neutral or slightly acidic in reaction. Deep clay rich in organic matter, adequate sunshine, temperature below 20°C during active vegetative phase and 30-35°C during reproductive phase are conducive for high yield. Warm and dry weather conditions during February and March are favourable for good flow of latex. Hailstorms and high gusty winds, frost and cloudy weather during lancing period are harmful. The crop should be grown in well pulverised soil with four to five ploughings mixed with 10-15 tonnes of FYM. As basal dose, 50 kg each of phosphorus and potash should be added. If the seedlings are transplanted, about 5 kg seeds per hectare is sufficient. For broadcasting, however, 8 kg seeds per hectare is required. Before sowing, seeds may treated with Thiram or Brassicol for protecting from soil borne diseases. Seeds are generally sown between late October to mid-November. Nitrogen at the rate of 75kg/ha to be applied in three equal split doses, for example, at sowing, 30-40 days and 60-65 days after sowing. Zinc sulphate at the rate of 15 kg/ha may be added in zinc deficient soil. Depending upon the types of soil, the crop requires 8-12 irrigations. Lancing is a very important operation for opium poppy cultivation. Depending on the variety, it takes 90 to 95 days for flowering from sowing and the flower remains open for nearly a week. The capsule comes to lancing stage when it is green and immature after 20 to 25 days of falling of petals. At this the capsule is grown, however, to its full size studded inside with seeds. Lancing operation is done on a sunny day, between 12 noon and 4 pm at this stage by giving a longitudinal incision on the capsule. For giving incision, a special knife known as 'naka' which produces equal number of incisions to a depth of 0.4 cm. At an interval of three to four days, each capsule is incised four times to obtain maximum latex.

When incised, the milky juice rapidly exudes and coagulates on exposure, turning into a brownish mass over the capsule surface. The exude is then scraped, as raw opium, next day during morning hours and collected in containers. To make a homogenous mixture, the latex is stirred and churned every day and its water content is decanted off. The get the mature seeds, the lanced crop is left in the field for further three weeks. The dried capsules are then picked and threshed and the seeds dried to be utilized for the next three year's crop. The seed are rich sources of fatty oils and protein, employed as a major source of cooking media in some European countries. The average yield of opium is 30 kg/ha and 8 quintals seeds/ha.

The crop is severely infested by powdery mildew and requires critical attention. Spray of 0.3% Dithane M-45 at fortnightly intervals three to four times is recommended. To keep the infection under control crop rotation is essential. Some of the insects like aphids, thrips, cut worms and capsule borers also cause damage to opium poppy. For control of aphids and thrips, spary of 0.03% Dimecron or Nuvan is effective. Spraying of 0.25% carbaryl at fortnightly intervals after formation of capsule controls the capsule borer. Soil application of Aldrin during land preparation is useful for eradication of cutworms.

Medicinal uses: The latex of immature fruits are rich in several alkaloids like morphine, codeine, thebaine, noacapine, narceine, and paparine. Other minor alkaloids are aporeine, codamine, cryptopine, gnoscopine, hydrocotarnine, lanthopine, laudanidine, laidanine, meconidine, narcotoline, neopine, papaveramine, porphyroxine, protopine, reticuline, rhoeadine and xanthaline. Presence of meconic acid and methyltransferase enzyme has also been reported from the plant. Opium alone or along with quinine salt, is used in cases of delirium, spasmodic coughing or spasms. It reduces all secretions except for perspiration. It is prescribed for diarrhoea and diabetes. Alkaloids extracted from the plants are used as pain killers, cough depressant, sedative, anti-tussive and smooth muscle reluctant. Poppy seeds are rich in oleic acid and linoleic acids which reduces bad cholesterol levels and increases good cholesterol levels. Seeds are excellent sources of vitamin B complex and minerals and beneficial to health. Seeds are used in bakery products and in confectionery.

Scientific name : *Pergularia daemia* (Forsk.) Chiov.

Family : Asclepiadaceae

Common name : Vtaran

Description: It is a twining herb with milky and hairy stem and ovate or round leaves. Flowers are pale or white, small and clustered. Fruits are up to 8cm long and 1.5cm broad, covered with spinous outgrowths.

Cultivation: *Pergularia daemia* is distributed in dry bushland, savanna or forest margins, as well as near seasonal watercourses on shrubs in more arid localities, from sea-level up to 1500 m altitude. It can easily be grown from seed. *In-vitro* culture techniques have also been developed.

Medicinal uses: The whole plant is used in medicine. The juice of the leaves is useful in catarrhal affection and diarrhoea. The plant is purgative and used in rheumatism and amenorrhoea. It is useful as tonic, expectorant and emetic. It is effective in curing gynaecological problems.

Scientific name	:	*Pholidota articulata* Lindl.
Family	:	Orchidaceae
Common name	:	Rattle Snake Orchid

Description: These are epiphytic herbaceous orchids with jointed, two leafed, cylindrical, 10 cm long pseudobulbs. Leaves are thickly membranous, many nerved, elliptic, acute, short stalked up to 10cm long and 4cm broad. The inflorescence is up to 15cm long, drooping, many flowered. The flowers are upto 1.25cm across, musk scented and yellowish white.

Cultivation: It is grown in a compost mixture consisting of equal parts of osmunda fibre, sphagnum moss, coarse sand and broken crocks. They require moist atmosphere and semi-shaded locations.

Medicinal uses: The whole plant is used a tonic and as bone jointer. Bulbs are used in the treatment of bronchitis, stomachic and toothache.

Scientific name	:	*Picrorhiza kurroa* Royle ex. Benth.
Family	:	Scrophulariaceae
Common name	:	Katki

Description: It is a prostrate, perennial herb with bitter rhizome and leaves. The leaves turn black on drying. Flowers are light blue in colour.

Cultivation: The plant thrives well in cool and moist climate. Sandy clay textured soil is ideal for its growth. It requires porous soil layers, which facilitate horizontal spreading of the rhizomes underneath, that produce aerial sprouts from the nodes. The plants prefer sufficiently moist and shaded locations. The planting stocks are initially be raised through seedlings, but can also be propagated through rhizomes/stolen/offsets. The nursery beds are planted using rhizomes/stolen/offsets in October–November. The planting stock can be raised in polybags, styrofoam trays or mother beds.

However, sunken beds should be prepared in the nursery to conserve moisture during the period of water scarcity. Each offset or fragment of rhizome should have two to three intact nodes for better establishment in the field. Stolen and offsets are also recommended for multiplication under controlled cultivation conditions with or without hormonal treatment. Seeds can also be planted in mother beds or polybags for germination. Seeds do not show any dormancy and germinate without any pre-treatment. However, seeds treated with 100-200 ppm GA3 for 24 hours have 90- 95% survival rate, respectively. About 165 000 seedlings per hectare are needed, which means 1–1.5 kg seeds are required for raising saplings for 1 hectare of land. Forest leaf litter or well-decomposed FYM is mixed with the soil at the rate of 6 tonnes/hectare at least 15 days before transplanting. Approximately 110 000 plants are required at the time of transplantation that is done at a spacing of 20 cm × 30 or 30 cm × 30 cm. If the agronomic practices are carried out properly, plant density may increase to 300 000 after the first growth season. In the absence of FYM, urea should be sprayed at a concentration of 1.5% in July and August to fulfil the fertilizer needs of the plants. At lower altitudes (1800–2500 m), plants are infected by powdery mildew during early growth period (March–May), which can be controlled by spraying Topsin-M (thiophinate methyle 0.1%) about 15–20 days after initiation, followed by another spray after 15 days. Powdery mildew also appears due to excess manuring. At the time of flowering and seed formation, spraying of insecticide (Ekalux, 0.5%) twice at 10-day interval prevents seed loss due to insects and aphids. The plant requires one year for complete maturity of the seed. The roots and rhizomes are manually harvested in September when the shoots or the aerial parts begin to wither and dry. Plants raised through stem cuttings mature almost an year earlier than those raised from the seedlings. The plant gives about 11 quintals/hectare of dry roots and rhizomes in the third year when the crop is raised through rhizomes.

Medicinal uses: Roots and stems are used in medicine. Rhizomes are rich in a brown resinous glucoside, picrorhizin, and a glycone, picrorhizetin. Roots contain bitter principle kutkin, a non-bitter product kurrin, vanillic acid, and kutkoli in varied quantities. This is an anti-pyretic, anthelmintic, carminative, stomachic, hepato-protector, anti-malarial, and laxative in small doses but cathartic in large doses. It improves appetite and stimulates gastric secretions. It is useful for asthma, cold, and cough, and also administered in liver complaints, anaemia, and jaundice.

Scientific name	:	*Plantago ovata* Forsk
Family	:	Plantaginaceae
Common name	:	Isabgol

Description: It is a small stemless herb, 30-50cm tall, covered with dense or soft hairy growth. Leaves are narrow, 10-20 cm long. Flowers are small, in oval or cylindrical spikes. Fruits are 8mm long, its upper half separates like a lid. Seeds are boat shaped, pinkish grey brown in colour. Each seed is crested in a thin, white, translucent membrane called called husk. The husk is colourless and tasteless.

Cultivation: The plants grow well in sandy loam to loamy soils at pH 7.2 to 8.5 but avoid soils with poor drainage. Being a rabi crop, Isabgol needs cool and dry weather during growth period. By ploughing and harrowing the soil is brought to fine tilth. For good germination fresh seeds should be used @ 4 kg per hectare after treatment with Mercuric compounds @ 3 mg/kg. Since the seeds are small and light in weight, they should be mixed with sufficient quantity of fine sand for uniform germination. Popular varieties available are G1, G2, Nikarika. Generally, November to December, However, sowing in the first week of December is ideal. Early sowing with high seed rate makes the crop susceptible to downy mildew. Line sowing (20-22 cm between rows) is preferred rather than broadcasting. Immediately after sowing light irrigation is required. Seed germination takes place within 6-7 days. If the germination is poor second irrigation is given. Last irrigation is given when the maximum spikes milk stage. Crop requires 6-7 irrigations. 1-2 hand weedings keep the crop weed free during early vegetative growth. 50 kg N + 25 kg P_2O_5 per hectare is recommended. N is given in 2 splits. Second dose of N should be given 30 days after sowing. Downy mildew at the lime of spike initiation, intensity is high where there is humid weather condition. Downy mildew can be controlled using Bordeaux mixture 6:3: 100 or Dithane M - 45/ Z -78 @ 2.0 to 2.5 g/ litre. Crops become ready for harvest in March or April (100- 130 days) duration). When the crop is mature and husk turns yellowish and spikes turn brownish seeds are shed. Crop should be harvested when there is dry atmosphere by cutting the plants 15 cm above the ground late in the morning. The produce is. threshed and seeds are separated. Dried-harvested material is bundled in cloth and threshed preferable in the early morning for easy separation of spikes. Yield is about 800 to 1000 kg seeds per ha.

Medicinal uses: The husk of the seed yield a colloidal mucilage which contains xylose, arabinose and galacturonic acid. Besides, presence of polysaccharides, rhamnose and galactose has also been reported. The husked seeds contain semi-drying oil, glycoside aucubin, tannin and an active principle resembling acetylcholine. Plantago seeds serve as a bulk laxative which promotes regular bowel movement. For chronic dysentery of amoebic and basillary origin, this is administered. The seeds and husk are used to cure inflammation of the mucous membrane of gastro-intestinal and genito-urinal tracts, duodenal ulcer, gonorrhoea and piles. It can also be used as cervical dilator for termination of pregnancy. The embryo oil of seeds has 50 per cent linoleic acid which prevents artetiosclerosis.

Scientific name	:	*Podophyllum emodi* Wall.
Family	:	Berberidaceae
Common name	:	May Apple

Description: This is a succulent erect herb, upto 30 cm tall, with a creeping long knotty rhizome. Stem is one or two, simply leafy without top. The leaves are alternate, palmate, up to 25 cm in diameter, deeply divided in three to five lobes, toothed, purple spotted. The flowers are white to pinkish, up to 4cm across, appear in the fork of the stem; petaloid; stamens usually six. The fruits are ovoid, pulpy, upto 5cm long, scarlet when ripe, many seeded.

Cultivation: It is grown upto an elevation of 3000m. Plants are propagated by seeds or by division of rhizomes. Seed germination is poor. The species are grown in temperate climate. It prefers shady, moist forest areas. Seeds are sown in spring season. Seedlings are transplanted in a well pulverised and fertilized soil at a distance of 30 c, x 20 cm when the seedlings attain a height of 10-12cm. In autumn, when the tops. of the plants dry, the roots and rhizomes are harvested. Uprooted roots and rhizomes are cut into pieces of 15-20cm length, washed and dried in the shade.

Medicinal uses: The rhizomes and roots contain a resinous mixture called podiphyllum resin or podophyllin. The primary constituents are lignin glycosides, podosides, podophyllotoxin, podophyllic acid and picropodophyllin and peltalin. The rhizomes also contain gum, resin , starch, albumin, gallic acid, calcium oxalate, lignin and flavones. The drug of rhizome is employed as a vermifuge and in treating constipation, typhoid fever, jaundice, dysentery, chronic hepatitis, scrofula, rheumatism; kidney and bladder and prostrate problems; gonorrhoea and syphilis. The podophyllum is used as a purgative and also for treatment of vaginal warts. Two derivatives, podophyllotoxin, called eloposide and teniposide are employed for treatment of cancer. The root paste is applied on ulcers, cuts and wounds. It is also used for curing skin diseases and tumour growth.

Scientific name : *Polygonum recumbens* Royle ex Bab.

Family : Polygonaceae

Common name : Vishalya Karni

Description: It isa much branched decumbent herb with small pale white flowers in axillary whorls. The leaves are lanceolate and elliptic.

Cultivation: Plants are distributed in different parts of Himachal Pradesh between 1500-2500 m elevation. Sunny situation and ordinary rich soils are preferred by these plants. These are propagated by seeds or by cutting.

Medicinal uses: A paste of leaves and apical parts is used in weeping eczema and wounds between fingers and toes. It makes emollient poultice.

Scientific name : *Potentilla fulgens* Wall.

Family : Rosaceae

Common name : Vajradanti

Description: These are herbaceous perennials up to 30 cm tall with crimson flowers.

Cultivation: This is distributed in Uttar Kashi between 1500-3000 m elevations. Deep, rich and sandy soil and sunny rockeries are preferred. Annual mulching with decayed manure is beneficial. The plants are propagated by seeds or by cuttings.

Medicinal uses: The paste of root is filled in the hollow cavity of teeth to check infection.

Scientific name : *Rheum emodi* Wall ex Meissn

Family : Polygonaceae

Common name : Rewand chini

Description: Thisis a perennial stout herb, 1.5–3.0 m tall. Roots are very stout and thick. Leaves are radical, orbicular, or broadly ovate, very large, and 30–45 cm in diameter with long petioles. Flowers are small, dark purple or pale red in axillary panicles.

Cultivation: The growing of this crop is restricted to the temperate, sub-alpine, and alpine zones of the Himalayas. Well-drained, porous, humus-rich soil is favourable for its cultivation. It prefers exposed or partially shaded habitat and can be cultivated on altitudes above 1800 m in height. Both seeds and top rhizome segments from mother plants may be used for propagation. Seeds are sown in lines 15 cm apart. In temperate conditions, seeds are sown in the spring season (March–early April). About 600 g of seeds are required to raise a nursery for planting at a spacing of 50 cm × 50 cm for one hectare of land. Forest litter/FYM @ 10 tonnes /hectare may be used as a basal dose for soil containing 1%–1.5% of organic carbon at 2200 m, applied at the time of land preparation. At lower altitudes (1800 m), higher dose of manure (about 15–20 tonnes/ hectare) is desirable for maximizing the yield. The optimum spacing for the crop is 50 cm × 50 cm, which accommodates 40 000 plants per hectare. While the entire quantity of FYM is applied as basal dose (10 tonnes/hectare), half of nitrogen and full dose of phosphorus and potash are also applied as a basal dose before transplanting the crop. Rest of nitrogen is applied as top dressing six to eight weeks after transplantation. Light sprinkler irrigation during the summer season in found to be useful for the growth of the crop. Regular weeding/hoeing operations, at an interval of 15–30 days, during the establishment and initial growth phase of the crop as well as during monsoons have shown better results in terms of plant growth. The crop is often infected with *Fusarium* sp. Soil drenching with Carbendazim 50 at a rate of 2 g/litre of water is recommended as a control measure. Insects and aphids, which attack the plant at early stage, may be checked by spraying Ekalux 0.5% twice at an interval of 10–15 days. Crop raised through seedlings needs four to five years for maturity; The expected yield ranges from 37 quintals/hectare to 66 quintals/hectare at an altitude of 1800 m, and between 65 quintals/ hectare and 108 quintals/hectare at an altitude of 2200 m, after three years.

Medicinal uses: Roots are rich in emodin, emodin-3-monomethyl ether, chropophenol, als-emodin, rhein, sennoside, eugenol, rutin, and glucoglatin.It is a mild purgative, astringent, tonic, laxative, stomachic, and aperient. Powdered rhizomes are sprinkled over ulcers for quick healing. It is also reported to be a potent anti-inflammatory drug.

Scientific name	:	*Rhyncostylis retusa* Bl.
Family	:	Orchidaceae
Common name	:	Cat's Tail Orchid

Description: It is an epiphytic herb. Stems are stout and woody, clothed with sheaths of fallen leaves, up to 60 cm long. The leaves are leathery, strap shaped, linear, deeply channelled, gracefully arching, up to 30 cm long and 2.5 cm broad. The inflorescence is pendulous, compact, many flowered, cylindrical, up to 60cm long. The flowers are one cm across, fragrant, long lasting, white, spotted with bluish purple.

Cultivation: These plants require high humidity, moderate shade and plenty of water during active phase of growth. They need bright light with the intensity of 3000-4000 foot candles and strong air movements all the times. These orchids are ideal for hanging baskets. They grow well in large chunks of tree fern fibre or in a media of chunks of hardwood charcoal. A day temperature of 30-32°C and night temperature of 24-25°C and relative humidity of 80% in summer and 60-70 % in spring and winter are ideal for their active vegetative growth and flowering. Plants should be watered heavily while actively growing, but aeration around the roots must be excellent, allowing the roots to dry rapidly after watering. For plants grown in pots or baskets, the medium must never become water logged or soggy. These plants are heavy feeders and respond very well with the dilute solution of 0.3% NPK (20: 20: 20) sprayed twice a month.

Medicinal uses: Roots of *Rhyncostylis retusa* are effective against rheumatisms. Plants are used against asthma, tuberculosis, cramps, epilepsy, vertigo, palpitation, kidney stone and menstrual disorders.

Scientific name	:	*Ruta chalepensis* Linn.
Family	:	Rutaceae
Common name	:	Garden rue

Description: It is an aromatic perennial herb growing upto 75cm height. Leaves are compound, shortly petiolate with ultimate segments, oblong or obovate-oblong. Flowers are yellow. Fruits are capsules and shortly pedicelled.

Cultivation: The plant is grown at about 1000m above mean sea level and with moderate rainfall and sunlight. The plants are propagated either by seeds or stem cuttings. Seeds are to be sown in seedbeds. Stem cuttings of length 20-25cm are to be planted in polybags for rooting. About 3-4 months old seedlings can be transplanted to pots and harvested when plants attain 6-8 months age. In highlands land is to be ploughed to a fine tilth, mixed with organic manure and seedlings are to be transplanted at a spacing of 45cm between plants. Irrigation is essential during summer months. Regular weeding is to be done. The plant is free from attack by any serious pests and diseases. Harvesting commences from sixth month onwards.

Medicinal uses: Roots contain coumarins-xanthyletin and (-)-byakangelicin. The alkaloids are rutacridone-epoxide, gravacridonol and its monomethyl ether, gravacridonchlorine, furacridone, 1-hydroxy-3-methoxy-N-methylacridone, iso-gravacridonechlorine, dictamine, r-fragarine and skimmianine. Skimmianine is also present in leaves and stem. Leaves and stem also contain graveolinine. Aerial parts give coumarins bergapten, xanthotoxin and psoralen. The plant is useful in strangury, fever, flatulence, colic, amenorrhoea, epilepsy and hysteria. The oil acts as a stimulant for uterine and nervous systems. The fresh leaves are used for rheumatalgia. The juice obtained from the leaves is given to children for helminthic infections and is good for odontalgia and otalgia The dried leaves, powdered and combined with aromatics, are given as a remedy for dyspepsia and with the fresh leaves a tincture is made which is used as an external remedy in the first stages of paralysis.

Scientific name	:	*Saussurea lappa* C.B. Clarke
Family	:	Asteraceae
Common name	:	Kurh, Kur

Description: It is a robust perennial herb, 2m tall with bitter roots. The leaves are very large, the lower basal ones up to one metre long, borne in winged stalks. The flowers are bluish purple, borne on round flower heads.

Cultivation: Rich soil and sunny rockeries are ideal. Plants are propagated by seeds. Seedlings are transplanted when three to four leaves are formed.

Medicinal uses: The dried root of this plant is antiseptic and disinfectant. It is administered in bronchitis, asthma, flatulence and certain cardiac complaints.

Scientific name	:	*Solanum indicum*
Family	:	Solanaceae
Common name	:	Bari kateri

Description: It is a stiff, much-branched, prickly herb growing upto a height of 0.3–1.5 m. Prickles are sharp, often slightly recurved, short hooked, and have a broad compressed base. Stem is stout and the branches are covered with minute stellate brown hairs. Leaves are ovate in outline and sparsely prickly on both sides.They are clothed above with simple hairs and have bulbous base intermixed with small stellate ones, while small stellate hairs cover them on the lower surface. Flowers occur in racemose, extra-axillary cymes. Fruit is a globose berry, green with white lining when young and becomes yellow when ripe.

Cultivation: It grows well in tropical regions where annual rainfall ranges from 1000 mm to 1500 mm. Sandy loam soil is found to be ideal for the cultivation of this plant. The species is xeric in nature and can grow well in shady places and in areas that receive low rainfall.

Planting materials are raised in nursery in May–June. One- to one-and-half-month-old seedlings are planted in the field during July–August. Direct seed sowing in a plantation may also be done. Seed is sown in well-prepared nursery beds (size 10 m × 1 m) in June in shady places or in temporary mist chambers of size 10 m × 15 m. FYM @ 50 kg and poultry manure @ 2 kg are mixed in soil at the time of bed preparation before sowing seed. Irrigation is done twice a day after sowing to maintain proper humidity. About 4 kg of seeds are required for sowing in the nursery for planting one hectare of land. Five tonnes of FYM per hectare is mixed thoroughly at the time of field preparation. A spacing of 30 cm × 30 cm in the field is optimum for good growth and productivity. The optimum crop stand is about 111 000 plants/hectare. First weeding is done 15 days after transplantation or about a month after direct sowing in the field. Best time for harvesting is April after the species is 9–10 months old. About 8 quintals of fruits and 3 quintals of seeds are obtained as fresh yield per hectare. Approximately 20 quintals of dried root is obtained from a two-year-old crop.

Medicinal uses: Fruits and roots contain wax and fatty acids. Alkaloids like solanine and solanidine are present in the roots and leaves. Fruits contain 1.8% of alkaloids and can find use in cortisone and sex hormone preparations. The plant is reported to be bitter, acrid, astringent, carminative, stomachic, resolvent, demulcent, diuretic, emmenagogue, febrifuge, and cardiotonic. It is useful in the treatment of asthma, catarrh, dropsy, chest pain, chronic fever, colic, dry and spasmodic cough, oedema, scorpion stings, difficult urination, and worm infestation.

Scientific name : *Solanum khasianum* Dunal

Family : Solanaceae

Common name : Night shade

Description: It is a stout, much-branched spiny herb up to 1.5 metres tall. The plant produces berry like green fruits with white streaks which turn yellow after ripening. Numerous small brown coloured seeds are produced which remain embedded in a sticky mucilage of fruits.

Cultivation: It can be grown in all types of soils but well drained lateritic soil is most suitable for its cultivation. These plants are propagated from seeds. Both broadcasting of seeds or transplanting of seedlings are followed. Freshly harvested seeds show little germination. A storage period of atleast one month is necessary to ensure better and quicker germination. The seeds washed with water give a higher germination than unwashed seeds. By manipulating the conditions of seed treatment and regulating temperature, very high rate of germination can be achieved.

Seeds are sown during June-July. For one hectare of transplanted crop 1.5 kg seeds are required. Seeds 1.5 kg seeds are required. Seeds germinate within a week and after a month when the seedlings reach a height of 10-15cm, they are transplanted in the main site. Well prepared land after several ploughings and soil rich in organic matter is preferred. Green manuring before planting increases berry yield. It is always better to plant the seedlings in rows at a distance of 75cm x75cm. A fertilizer dose comprising 60kg nitrogen and 40 kg each of phosphorus and potash is recommended. Entire quantity of phosphorus and potash may be added at the time of soil preparation and nitrogen may be applied in three split doses.

Harvesting of berries is possible for June planted crop from October onwards. Berries are handpicked at regular intervals when they begin to change colour from green to yellow. The berries are dried in sun by cutting the fruits into halves and spreading in thin layers, on the floor and turned frequently.

Medicinal uses: Berries are important source of steroid hormone solasodine. The steroidal hormones are used as anti-inflammatory, anti-arthritic and hormonal.

Scientific name	:	*Swertia chirayita* (Roxb.) Karst.
Family	:	Gentianaceae
Common name	:	Chirayata

Description: It is an erect annual herb and can attain 1.5 metres height, much branched and robust. The leaves are opposite in pairs, without stalks, pointed at tips up to 10cm long. Flowers are greenish yellow, tinged with purple. Fruits are ovoid, 8 mm long.

Cultivation: The plant is distributed in temperate regions in the Himalayas. Loamy to sandy loam, friable, and well-drained soils are suitable for its cultivation. The soil should be enriched with FYM and if soils are clayey, addition of sand is recommended. The crop can be grown in areas having mild rainfall (100 cm) in rainy season and in areas with long cold winter, receiving snowfall frequently.

The crop is raised through nursery seedlings. Sowing is done in October–November, as seeds sown in March–April *Chirata* show very poor germination and low survival rate. Seeds of chirayata germinate well under nursery conditions in the media having FYM, sand, and soil in 2:2:1 ratio. Seeds are sown in rows 10–15 cm apart and covered with 0.5-cm thick layer of sand or fine soil. It takes about 25–28 days for complete germination to take place under nursery conditions. About 200 g of seeds are required for raising planting stock for one hectare of land. About 50 000 plants would be required for planting one hectare of land as a pure crop. Chilling treatment of seeds at 3°C or below for 15 days is required for good germination in the nursery. Application of vermin-compost in the soil @ 3.75 tonnes/hectare and forest leaf litter @ 2 tonnes/hectare at the time of field preparation is recommended. Transplanting of seedlings is done in March–April, and the seedlings are planted with the ball of earth at a distance of 45 cm × 45 cm in the field. Plants are collected when the capsules are fully formed during summer or in October–November. Plants flower within six to eight months, and thus provide yield and seeds for propagation every year. The whole plant is collected and dried. About 3.75 tonnes/hectare of dried herbage yield is estimated in cultivation in two years.

Medicinal uses: The plant contains a bitter glycoside chiratin, which on hydrolysis yields two bitter principles: ophelic acid, an amorphous bitter hygroscopic principle, and chiratogenin, a yellow bitter glycoside, insoluble in water. It is a bitter tonic, carminative, laxative, anti-pyretic, febrifuge, anti-periodic, anti-inflammatory, stomachic, and antihelmintic. It is used in treating piles, skin diseases, ulcers, and diabetes.

Scientific name	:	*Trianthema monoegyna* Linn.
Family	:	Aizoaceae
Common name	:	Lalsabuni

Description: It is a spreading, much branched, succulent herb. Leaves are broader towards tip. Flowers and fruits are minute and concealed in the base of the leaf stalk. Seed are black, covered with minute outgrowth, kidney shaped.

Cultivation: It is a weed, distributed throughout the tropical and subtropical countries. It occurs in wastelands, roadsides, lawns, gardens, cultivated crops, and in paddy fields if the water supply is low.

Medicinal uses: The leaves have medicinal properties which contain punarnavine alkaloid. It is used to promote urination and useful in dropsy and kidney diseases.

Scientific name	:	*Trigonella foenum-graecum* Linn.
Family	:	Papilionaceae
Common name	:	Common Fenugreek

Description: It is an annual herb, 30-60cm in height. Leaves are light green, pinnately trifoliate, leaflets toothed. Flowers are white or yellowish white, papilionaceous and axillary. Fruits are legumes, 5-7.5cm long, narrow, curved, tapering with a slender point and containing 10-20 deeply furrowed seeds per pod.

Cultivation: Fenugreek is successfully cultivated both in the tropics as well as temperate regions. It is tolerant to frost and freezing weather. It thrives well in places receiving moderate or low rainfall areas but not in heavy rainfall area. It can be grown on a wide variety of soils but clayey loam is relatively better. The optimum soil pH should be 6-7 for its better growth and development. Some of the improved cultivars available for cultivation are *CO1* (TNAU), *Rajendra Kanti* (RAU), RMt-1(RAU) and *Lam Selection-1*(APAU). Land is prepared by ploughing thrice and beds of uniform size are prepared. Broadcasting the seed on the bed and raking the surface to cover the seeds is usually followed. But to facilitate intercultural operations, line sowing is also advocated in rows at 20-25cm apart. Sowing in the plains is generally in September-November while in the hills it is from March. The seed rate is 20-25kg/ha and the seeds germinate within 6-8 days. Besides, 15 tonnes of FYM, a fertiliser dose of 25:25:50kg NPK/ha is recommended. Entire P, K and half N are to be applied basally and the remaining half N 30 days after sowing. First irrigation is to be applied immediately after sowing and subsequent irrigations at 7-10 days interval. Hoeing and weeding are to be done during the early stages of plant growth and thinning at 25-30 days to have a spacing of 10-15cm between plants and to retain 1-2 plants per hill. Root is a serious disease and can be controlled by drenching carbendazim 0.05% first at the onset of the disease and another after one month of first application. In about 25-30 days, young shoots are pinched off 5cm above ground level and subsequent cuttings of leaves may be taken after 15 days. It is advisable to take 1-2 cuttings before the crop is allowed for flowering and fruiting when pods are dried, the plants are pulled out, dried in the sun and seeds are threshed by beating with stick or by rubbing with hands. Seeds are winnowed, cleaned and dried in the sun. They may be stored in gunny bags lined with paper. An yield of 1200-1500kg of seeds and about 800-1000kg of leaves may be obtained per hectare in crops grown for both the purposes.

Medicinal uses: Seeds contain sapogenins-diosgenin, its 25-epimer(yamogenin), tigogenin, gitogenin, yuccagenin, 25-2-spirosta-3-5-diene and its b-epimer. Other chemical constituents are sterols-β-sitosterol and cholesterol, flavone Cglycosides- vitexin, iso-vitexin, vitexin-2"-O-P-coumarate and vicenin-2. Flavonoidsquercetin and luteolin, flavonoid glycoside-vicenin-I. Seeds are bitter, mucilaginous, aromatic, carminative, tonic, diuretic, thermogenic, galactagogue, astringent, emollient, aphrodisiac, anti-rheumatic, CNS depressant and antiimplantation.

Scientific name	:	*Trillium erectum* Linn.
Family	:	Liliaceae
Common name	:	Birth Root

Description: It is tuberous rooted perennial herb, 60cm tall. Three broad leaves are arranged in a whorl at the top of the stem. From the centre of the whorl, flower arises which is purplish-maroon in colour.

Cultivation: Sandy soil and rich organic matter, moist and shady locations are preferable. It is propagated by division of bulbs during dormant period.

Medicinal uses: Roots are used in medicine and said to be aphrodisiac. Roots are used to treat heart palpitation, haemorrhages, skin infections and also to ease pains of childbirth. The root is used as an alternative medicine and is antiseptic, antispasmodic, diuretic, emmenagogue (to promote menstruation), and ophthalmic.

Scientific name	:	*Typha latifolia* Linn.
Family	:	Typhaceae
Common name	:	Cat-Tail

Description: This is a hardy aquatic perennial monocotyledonous plant up to 2 metres tall. The lower part of the stem is a thick rhizome, the upper part projects high out of water. The leaves arise from the base of the stem and grow to a height of 1.2m. The flowers are clustered in a dense spike, brown in colour.

Cultivation: Ordinary rich soil, full sun and margins of shallow rivers or ponds are ideal. Commonly grown in ponds and still water. It may be grown in 5 to 15 cm water depth with a pH range of 4.5 to 7.5. Plants are propagated by seeds or by division.

Medicinal uses: Roots and seeds are useful. Roasted seeds and seeds are eaten. Leaves are diuretic. Crushed roots are applied on burns and wounds. The roots are used to treat worms, diarrhoea, diuretic and gonorrhoea. The flowers are used in the treatment abdominal of pain, amenorrhoea, cystitis, dysuria and vaginitis.

Scientific name : *Urginea indica* HK.F.

Family : Liliaceae

Common name : Indian Squill

Description: It is a bulbous herbaceous plant producing racemed flowers. The bulb is the size of an apple, bitter and nauseous. The leaves are 15 to 45 cm long and 1 to 2 cm broad, linear, flat, active. Raceme is erect, 15cm to 30 cm long. Flowers are very distinct, long pedicelled and drooping. The capsule is oblong.

Cultivation: Plants of this species are found globally from Tropical Africa to Myanmar. In India, it is found throughout the plains and in the dry hills of the lower Himalayas. It has potential to regenerate itself by using its adventitious new bulbs developed near the main bulb of the plant underground.

Medicinal uses: Bulbs are rich in hentriacontanol, octacosanoic acid, glycosides and scillaren A and B. Roots, bulb and leaf contain sitosterol, stigmasterol and campasterol. Tubers contain scillarenin. The syrup made from bulbs is employed in chronic bronchitis, bronchial catarrh. It is also considered as diuretic and cardiac stimulant. The bulbs are said to be poisonous and is a source of raticide.

Scientific name : *Urtica dioica* Linn.

Family : Urticaceae

Common name : Stinging Nettle

Description: The plants are 2 m tall. The leaves are 15cm long, ovate, acuminate, deeply serrate. The flowers are borne in loose clusters.

Cultivation: Three cultivation techniques are used for the stinging nettle: 1) direct sowing, 2) growing seedlings in nurseries with subsequent transplantation and 3) vegetative propagation via stolen or head cuttings.

Direct sowing: The seedbed should have a loose and friable structure, but should be reconsolidated by ploughings prior to sowing. Sowing can be done either in autumn or in spring. Seed rate should be 6 kg/ha with row spacing of 30 cm and 42–50 cm in autumn and spring, respectively.

Growing seedlings: For this technique, pre-germinated seeds are sown between mid-/end-February and beginning of April and grown in nurseries. Seedlings are grown in tuffs with 3-5 plants / tuff and a seed density of 1.2-1.6 kg / 1000 tuffs. A fastened germination is achieved by alternating high temperature during daytime (30 °C for 8 h) and lower temperature during night time (20 °C for 16 h). Before transplanting, the seedlings should be fertilized and acclimated to cold temperatures. Transplantation should start around Mid-April with row spacing of 42–50 cm and plant spacing within rows of 25–30 cm.

Vegetative propagation: Stolen (with several buds) of 10 cm should be planted from Mid-April in a depth of 5–7 cm. Head cuttings are grown in nurseries starting between mid-May and mid-June. Growing tips with two leaf-pairs are cut from the mother plant and treated with root-growth inducing hormones. Transplantation can be delayed in comparison to the growing seedling technique.

Medicinal uses: The decoction of the plant is administered in jaundice, nephritis and menotthagia and haematurica and it is considered astringent, anthelmintic, emmenagogue and diuretic. The root is also diuretic.

Scientific name	:	*Valeriana officinalis*Linn.
Family	:	Valerianaceae
Common name	:	Common Vallerin

Description: It is a perennial glabrous herb, attains a height up to one metre. Stem is solitary, erect, furrowed and hollow. Leaves are opposite, pinnate, lower leaves with long petioles and upper are small. Flowers appear in cymose panicle, corymb, white or dull white in colour, fragrant, small, calyx many toothed and corolla 5 lobbed. Fruits are oblong ovate, small, smooth, without hair, single seeded. Rootstock in thicker., short, with many fasciculate rootlets.

Cultivation: It grows well in temperate climates of Kashmir, Himachal Pradesh and Uttar Pradesh. Medium fertile soil, rich in humus is suitable for this crop. The plant is propagated through seeds or through rootstocks. Seeds are sown in a well drained raised bed in the nursery during April-May or July-August. In a well prepared and levelled land, seedlings are transplanted at a spacing of 40 cm each way. Roots are harvested after two years during theit dormancy in November. Roots are dug thoroughly, washed properly and dried.

Medicinal uses: The root contains epoxy-iridoid esters called valepotriates which include valrate, didrovaltrate, acevaltrate, isovaltrate and isovaleroxy-hydroxy didrovaltrate. It also contains chatinine and valerine alkaloids and volatile oils. The drug is said to be sedative, hypotensive, anti-spasmodic and stomachic. The juice of the fresh rhizomes is used in the treatment of hysterical fits and other nervous disorders and flatulence. The root preparation is also known as tonic and stimulant. The fresh juice is administered in caes of insomnia and in certain cardiac preparations.

Scientific name	:	*Vanda tessellata* HK.f.
Family	:	Orchidaceae
Common name	:	Rasna

Description: It is an epiphytic herb with scandent stem, 60 cm long. Leavse are keeled, linear-oblong, leathery, recurved, channelled to 22.5 long and 2.5 cm broad. Inflorescence is sub-erect to 45 cm long, 6 to 8 flowered. Flowers are 5 cm across, sepals and petals tessellated, yellow with brown lines and white margin, lip bluish dotted with purple.

Cultivation: All Vanda orchids require high light levels. On an average they require a light intensity of 4000 foot candles. Flat leaved vandas need a winter night temperature of 10-12°C and a summer day time temperature of 22-25°C They are capable of tolerating higher and lower temperatures for short periods only. Adult plants are watered once a week in the winter time and every day during summer. Vandaceous orchids require humidity upto 70-75%. Easily propagated by cuttings of the shoots along with roots or air layerings. Rapid large scale multiplication is possible through tissue culture using shoot tips, axillary buds and roots as explants. Potting mixture should consist of Cocochips + brick pieces + leaf fern (1:1:1).

Medicinal uses: The paste of leaves is effective against fever. It is ingredient of Rasna Panchaka Quatha. In Ayurvedic medicine, it is used in the treatment of arthritis and rheumatism. The leaf juice extract is used in the treatment of otitis media. The root is an effective against scorpion sting and bronchitis.

Scientific name	:	*Verbascum thapsus* Linn.
Family	:	Scrophulariaceae
Common name	:	Common Mullein

Description: This is a hardy perennial plant, 2 m tall. Leaves are hairy, lance shaped and oblong. The flowers are arranged densely along a club like spike, yellow in colour.

Cultivation: The plants can be grown up to an elevation of 2500m in western Himalayan regions. These plants require ordinary rich soil and sunny situations. Plants are propagated by seeds. Seeds are sown in light rich soil in sunny locations. The seedlings are transplanted at four leafed stage.

Medicinal uses: Mature leaves, flowers and roots are used as medicine. Flowers contain caffeic acid, coumaric acid and ferulic acid. Leaves contain aucubin, luteolin, flavone and verbacoside. The leaves are smoked for asthma and sore throat. Leaves and flowers are considered soothing for mucus membrane. Small pills of crushed leaves are given to cure constipation. Infusion of plant is orally given as a cure to snake bite. Decoction of its leaves with onion forms a remedy for stomach complaints. Tea prepared from leaves is used to treat colds and dysentery. The oil extracted from flower is used to treat earache. The leaves are also useful for local applications in Inflammation, haemorrhoids and sunburn.

Scientific name	:	*Veronica officinalis* Linn.
Family	:	Scrophulariaceae
Common name	:	Speedwell

Description: It is a herbaceous perennial with hairy green stems, 10–50 cm long that cover the ground in mats and send up short vertical shoots which bear soft violet flowers. The leaves are 1.5–5 cm long and 1–3 cm broad, and softly hairy.

Cultivation: The plants grow well in ordinary rich soil and sunny locations. These are propagated by seeds or by division of roots. It requires adequate watering during dry weather. after every third year, plants are to be lifted, divided and replanted.

Medicinal uses: The plant contains leptandrin, cosmosiin, cynaroside and glucoside. This plant is used as a light tonic in the convalescence of certain fevers, asthma and bronchitis. It acts as a stimulant, provoking urination and facilitating expectoration. It is also used for the treatment of chest diseases such as chronic pulmonary catarrh, phthiasis and bronchitis.

Scientific name	:	*Viola serpens* Linn.
Family	:	Violaceae
Common name	:	Banafsha

Description: This isa small perennial herb with a short semi-subterranean fleshy stem. The older parts bear fibrous roots. The upper younger portion is marked with close circular scars of the leaves of the previous years, and gives off elongated, rather fleshy, leafless runners. Leaves are ovate and crenate–serrate. Flowers are borne on long, slender, axillary stalks (scapes) exceeding the leaves. Flowers are lilac, blue, white or purple in colour and borne on the main stem, Fruit is a few seeded, globose capsule.

Cultivation: The plant prefers a cool and moist climate but heavy and frequent rain is fatal during blooming. Sandy loam soil is suitable for its cultivation. The plant is generally multiplied through separation of the new plantlets that develop from the runners. As much as 50–60 new plantlets may be separated from one mother plant. This crop is also propagated by seeds that show about 80% germination. About 110 000 plantlets are required for planting in one hectare of land. When seeds are used as propagation material, about 1–1.5 kg seeds per hectare are required. Well-decomposed FYM @ 12 tonnes/hectare should be applied to the soil before transplantation of the seedlings to obtain maximum yield. The crop is generally grown in kharif season during May–June, with the onset of pre-monsoon rains. This is the ideal period for establishment of plantlets in the field. When the crop is propagated through division of the mother plants, the separated plantlets can be planted directly in the main field in May–June. About one-month-old seedlings are transplanted in the main field, at an optimum spacing of 30 cm × 30 cm. Irrigation is given as and when required during hot weather. For promoting the growth of the plants, 50 ppm solution of GA_3 is sprayed at an interval of one month from September to May. Manual hoeing and weeding once in a month are practised to control the weeds. Leaves are harvested after two months of establishment of the crop and subsequent harvesting may be done at an interval of one month till December. The plant starts flowering after 9–10 months of growth during February– March, corresponding with the increase in temperature. Fruits are harvested in April and May. Flowers and seeds are harvested either in February– March or April–May, depending on the climatic conditions of the area.

Medicinal uses: Voiline, ionone, and ionine are the main active ingredients of the plant. Roots, leaves, and blossoms also contain methyl-salicylate in the form of a glucoside. Leaves of *Viola* contain rutin content in the range 0.15%–0.45%. The species is antipyretic, diaphoretic, diuretic, aperient, antipyretic, and febrifuge in action. It is used in respiratory track congestion, asthma, sore throat, cold, coryza, and cancer of the throat. It is also very useful in bleeding piles, constipation, fever, headache, and skin diseases. In the Unani system, this plant is the main ingredient of Joshanda, consisting of a mixture of drugs, used mainly for cough and cold in the form of a decoction.

12

CULTIVATION OF IMPORTANT AROMATIC PLANTS

Aromatic plants possess odorous volatile substances which occur as essential oil, gum exudate, balsam and oleoresin in one or more parts, namely, root, wood, bark, stem, foliage, flower and fruit. The characteristic aroma is due to a variety of complex chemical compounds. The term *essential oil* is considered to fragrance or perfumes because these fragrances are *oily* in nature and they represent the *essence* or the active constituents of the plants. They are called *volatile or ethereal oils* which evaporate when exposed to air at ordinary temperatures. Essential oils are highly concentrated, low volume, high value products. Most essential oils are primarily composed of terpenes and there oxygenated derivatives mostly monoterpenes and sesquiterpenes: The most characteristic group present in essential oils contain hydrocarbon as a rule of the formula $C_{10}H_{16}$ and group of oxygen containing compounds with empirical formula $C_{10}H_{16}O$ and $C_{10}H_{18}O$ (hydrocarbons, alcohols, aldehydes, ketones,acids, esters, oxides and ethers). Out of 1500 species of aromatic plants which serve as a source of raw materials for the perfumery, about 50 species are used for commercial source of essential oils and aroma chemicals. Essential oils and aroma chemicals constitute a major group of industrial products. These oils are adjuncts of cosmetics, soaps, pharmaceuticals, perfumery, confectionery, ice-creams, aerated waters disinfectants, tobacco, agarbathis and a host of related products.

Presently the total world production of essential oil is about 110,000-120,000 m. tones out of which India produces 15%, holding, third place after China&Brazil in the world production, while in terms of value its share is much better with 21-22%. But considering the total share in world export of essential oils and perfumery material it is only 0.4% and therefore there is a bright future for India in the fast charging global economy as per production and trade of natural raw material are concerned.

Essential oils and aroma chemicals constitute a major group of industrial products. These oils form indispensable ingredients of the necessities in many spheres of human activity. They are adjuncts of cosmetics, soaps, pharmaceuticals, perfumery, confectionery, ice-creams, aerated waters disinfectants, tobacco, agarbattis and a host of related products.

Use of essential oils in therapeutics is becoming popular in Japan and European countries. Aromatherapy involves the use of essential oils and aromatics derived from plants to cure diseases. Some of the essential oils are reported to be in many ways better than antibiotics due to their safety and wide spectrum of activity. Synergistic activity of essential oils needs further probe. Application of essential oils in agriculture as anti-feedants, repellents, botanical insecticides, natural herbicides and growth boosters are still open to fascinating realms of research. Production of secondary metabolites in bioreactors under controlled conditions using cell and tissue culture offers exciting frontiers of future research.

Scientific name	:	*Cymbopogon flexuosus*
Family	:	Poaceae
Common name	:	Lemongrass

Description: This is a tufted robust perennial grass, 2 m tall. It is floriferous. Leaves are linear- lanceolate. Panicles are very large, drooping, lax, greyish-green, rarely with a purplish tinge, with the raceme pairs in dense masses, spreading, 100-135 cm long, slightly hairy, lower glumes of the sessile spikelets 3-4, 4-5 mm long, 1 mm wide with 1-3 definite or obscure intracranial nerves, shallowly concave with one to two depressions.

Cultivation: Lemongrass prefers a warm humid climate with plenty of sunshine and rainfall ranging from 2500-3000 mm, uniformly distributed over the year. The grass requires an average temperature of 23-30° C. The plant is hardy and drought tolerant. This crop is well suited for rainfed agriculture. It thrives well at altitudes between 100 and 1200 m above MSL. It is usually grown on poor soils along hill slopes, though it flourishes on a wide variety of soils ranging from rich loam to poor laterite. The grass performs best on well drained sandy loam soil. It is propagated by seeds and root divisions or slips. Clonal propagation is better as seed propagation leads to considerable genetic heterogeneity resulting in gradual deterioration of yield and oil quality. Improved varieties such as 'Sugandhi' or 'OD-19' 'LS-48' and 'Pragati' are available for cultivation.

Fresh seeds are to be sown as they are short viable and the viability is drastically reduced after six months of storage. Recommended seed rate is 3-4 kg/ha for transplanting and 25-30 kg/ha for broadcast sowing. For transplanting, seeds are sown in a well prepared nursery during April-May. Seeds germinate in 5-7 days. Seedlings are transplanted when 45-60 days, old at a distance of 20-50 cm in rows, 20-75 cm apart depending on the soil fertility and tillering nature of the variety. In high rainfall areas, planting on ridges or beds is recommended. Ridges or beds are prepared across the slope on hilly areas. Lemongrass is a soil exhausting crop. Spent-lemongrass in the form of compost at 10 tonnes/ha and wood ash at 2 tonnes/ha, which are obtained as by -products of grass distillation, supply sufficient amount of nitrogen and potash to the crop. Inorganic fertilizers are recommended at 30 kg/ha each of N, P_2O_5 and K_2O as basal dose at the time of planting. 60 kg N/ha is applied as top dressing in 3-4 splits after every harvest during a year. 4-6 irrigations are essential for getting optimum yield in areas where rainfall is erratic. Crop-weed competition period is the first 25-30 days after planting or harvest.

Generally, 2-3 weedings are necessary during a year. Among herbicides diuron at 1.5 kg/ ha and oxyfluorfen at 1.5 kg ai/ha are effective for weed control. Spraying Paraquat at 2-2.5 l/ha in 500 litres of water immediately after cutting the grass for distillation is an excellent method of weed control. Burning of stubbles in summer is practised in some areas to get rid of pests, diseases and weeds. Earthing up of plants after weeding and fertilizer application is beneficial as the roots have a tendency to grow above the soil. A stem borer, *Chilotrea* species bores into the stem and remains there feeding on the shoot. It is recognized by the drying up of the central shoot followed by the death of the whole shoot. The stem borer attack can be controlled by timely application of any systemic insecticide. *Helminthosporium cymbopogi* and *H. sacchari* cause leaf spot diseases in lemongrass which can be controlled by fungicide application. *Tolyposporium christensenii* causes a long smut in Kerala and completely destroys the inflorescence, for which effective control measure is not available at present. The first cutting of grass is taken about 3 months after planting, thereafter every 6-8 weeks. Harvesting is done with the help of sickles, the plants being cut close to the base about 10 cm above ground level. Generally, 3 harvests are possible during the first year and 4-5 during subsequent years. The plantation remains economical for 4-6 years. The fresh grass yield ranges from 3 to 10 tonnes/cut/ha depending on soil fertility, agro-climatic conditions and management practices. The grass yield is the lowest in the first year and highest in the third or fourth year of planting.

Importance and uses: *C. flexuosus* oil contains 75-85% of aldehydes consisting largely of citral. Other constituents in the oil are linalool, geraniol, citronellol, nerol, 1,8-cineole, citronellal, linalyl acetate, geranyl acetate, limonene, caryophyllene, ß-pinene, ß-thujene, myrcene, ß-ocimene, terpinolene, methyl heptanone and terpineol.

Lemongrass oil is one of the most important essential oils being widely used for the isolation of citral which can be converted into ionones having the odour of violets. They are used in flavours, cosmetics and perfumes. ß-ionone is used for the commercial synthesis of vitamin A. In some Far Eastern countries like Java, Japan, China and India the leaves are used for flavouring foods, drinks and tea and for scenting bathwater. The oil is used as a repellent against flies and mosquitoes. The spent grass is an excellent fuel, manure and mulch and forms raw material for the production of silage, mushroom and paper boards.

Scientific name : *Cymbopogon martinii*

Family : Poaceae

Common name : Palmarosa

Description: It is a perennial grass growing to a height of 2-3 m. In this species, the following two varieties are recognized which are morphologically very similar but differ considerably in their habitat and also in the essential oils. *Cymbopogon martinii var. motia,* called as Palmarosa and *C. martinii var. sofia,* called as Gingergrass. *Jamrosa* is an interspecific hybrid between *Cymbopogon nardus var. confertiflorus* and *C. jwarancusa.* Its oil contains 75-80% geraniol and 15-20% geranyl acetate. The oil has more terpenic, limy or grassy by-odours in contrast to the sweet, rosy odour of palmarosa oil.

Cultivation: Palmarosa grass is a tropical plant and it grows well in warm humid areas. It is susceptible to frost and hence is not recommended for frost-prone areas for its cultivation. Although it grows best on neutral soils, it can survive and give economic yields on alkaline soils of pH upto 9.0. Motia grass prefers well drained soils and it performs well in separate clumps on open dry hill sides with a rainfall of 800-900 mm. It does not respond stagnant water. Sofia grass grows well densely and abundantly at lower altitudes in moist soils. It is propagated best through seeds and also through slips. 'IW 31245', 'IW 3629', 'IW 3244' and 'Trishna' are popular varieties available for cultivation. Seeds are sown on nursery beds prepared in May. About 5 kg seeds are required to give seedlings for planting one hectare. As the seeds are very small and light, they are usually mixed with fine sand or soil in the ratio of 1:10 for even distribution and ease of sowing. The beds are watered lightly and regularly. Germination starts in 10-15 days. In about 4-6 weeks, seedlings are become ready for transplanting. The field is prepared well before the onset of monsoon and the seedlings are transplanted during June-July. Healthy and established seedlings, about 15 cm tall are carefully removed from the nursery and transplanted in rows, 20-60 cm apart with the plants spaced at 20-60 cm. Spacing is increased on fertile soils. Farm yard manure is given at 10 tonnes/ha before planting. Fertilizers at 20 kg N, 50 kg P_2O_5 and 40 kg K_2O/ha are given at planting as a basal dose. About 40 kg N/ha is applied in two splits during the growing season. The NPK application should be repeated each year at the time of appearance of fresh leaves. Application of micronutrients like iron (as $FeSO_4$) and manganese (as $MnSO_4$) improves the growth, herbage and oil yield. Palmarosa plantations are to be irrigated at 10-14 days interval during the summer period. The plantation should be kept weed free by regular weeding and hoeing. Diuron at 1.5 kg ai/ha and oxyfluorfen at 0.5 kg ai/ha are recommended to control weeds in palmarosa. No serious insect pests have been reported in this crop. *Curvularia trifolii* (Kauff.) Boed causes leaf blight. Yellowing of leaves and necrosis are the symptoms. *Dreschlera cymmartinii* sp. nov. also infects palmarosa. Application of a suitable fungicide can be resorted to the cases of severe attacks. The optimum stage of harvest is the initial seed setting stage. This stage will be reached about 10-15 days after flowering. The grass is cut at a height of about 10 cm from the ground level and the whole plant is used for distillation. During the first year one or two cuttings are obtained depending upon the climatic conditions. After the first harvest, subsequent harvests are made at 70-80 days interval and 3-4 cuttings are taken a year. The plantation remains economic for 4-6 years. The yield of grass and oil starts decreasing from the third or fourth year onwards. The grass yield is 6-10 tonnes/cut/ha.

Importance and uses: Motia oil contains upto 95% of geraniol. The chemical composition of motia oil is reported as limonene 0.1%, p-cymene 0.1%, methyl heptanone 0.1%, 2-nonanol 0.1%, linalool 2.4%, citronellol 6.4%, farnesene 0.6%, ß-terpineol 1.0%, ß-humulene 0.6%, -terpineol 0.4%, geraniol 81.7%, geranyl acetate 5.7% and farnesol 0.4%.

The flowering tops and foliage of Palmarosa contain a sweet-smelling oil of *rose-like odour*. Two types of palmarosa grasses have been recognized, *viz*, *Rosha or Russo grass and Gingergrass*. Rosha grass yields a superior oil which is used in perfumery, particularly for flavouring tobacco and blending soaps. In soap perfumes, it has a special importance because of geraniol, being stable to alkali. It is a source of very high grade geraniol. Geraniol is highly valued as a perfume and used as a starting material for a large number of synthetic aroma chemicals like geranyl esters which have a permanent rose-like odour. It is also used as an adulterant of Turkish attar of roses.

Scientific name : *Cymbopogon nardus*

Family : Poaceae

Common name : Citronella

Description: *Cymbopogon nardus var. lenabatu* known as *Ceylon citronella*. It is the older form and less important than Java type. Ceylon type is mainly cultivated in Sri Lanka. It is hardier and can be grown on poor soils. The leaves of these types are smaller than that of Java type yield an inferior oil with a lower citronellal content. The plant is a robust, stoloniferous perennial grass, 0.5-1.0 m tall, leaves broad, panicles large and mostly compound, spikelets awnless, lanceolate and flattened on back.

Cymbopogon nardus var. mahapengiri is sometimes recognized as a distinct species, *C. winterianus* Jowitt. It is known as *Java citronella*. Java citronella is a tufted aromatic perennial grass with fibrous roots. Plants are more than 2 m in height. Clumps are stout, erect, terete, leafy and glabrous at the nodes. Leaf blades are linear, with long acuminate tip and serrate margins. Ligules are scarious. Inflorescence is a very large compound panicle. Racemes are 20 mm long, one sessile and the other pedicelled with the two lower spikelets, homogeneous, male or neuter, the remaining pairs in both racemes are heterogeneous. Ovary is superior with one carpel, unilocular. Stigma is bifid, lateral and leathery. Ovule is single and erect.

Cultivation: Java citronella is more extensively and commercially cultivated than Ceylon type, though the former is less hardy compared to the latter. Citronella performs well under tropical and subtropical conditions. It requires plenty of sunshine and moisture for good growth. Even though it grows upto 1000 m above MSL its growth is restricted, resulting in low yields, when grown above an altitude of 400 m. Annual rainfall of 2000-2500 mm and high atmospheric humidity are ideal for better growth of the plant and quality of the oil. Citronella thrives well under varying soil conditions but optimum growth and yield is obtained in sandy loam soil with abundant organic matter. Heavy clay and sandy soils do not support good growth. It grows best on soils with a pH around 6.0 though a pH range of 5.8-8.0 is preferable. The crop is very sensitive to water-logging. Citronella grass is propagated only vegetatively by slips which are obtained by dividing well grown clumps. Clumps are separated in a manner that each slip contains 1-3 tillers. Roots and leaves are trimmed off before planting. About 50 slips are obtained from one year old healthy clump. 'Jorlab C2', 'RRL JOR-3-1970', 'CIMAP/Bio-13' and 'CIMAP/73-1' are the improved high oil yielding varieties recommended for cultivation. With the onset of monsoon, the land is brought to fine tilth by ploughing and harrowing. Ridges and furrows or beds are made. The slips obtained from healthy vigorously growing plants are planted during June-July at 60-90 cm spacing and 10 cm deep. Delay in planting results in drying of slips and poor establishment and plant population. Farm yard manure is applied at 10 tonnes/ha before planting. A fertilizer dose of 200 kg N, 80 kg P_2O_5 and 40-80 kg K_2O is recommended per hectare per annum for optimum growth and yield. Better results can be obtained by applying N in 4 equal split doses at an interval of about 3 months. P and K are applied fully as basal. Irrigation is given within 24 hours of planting if there is no rain. Supplemental irrigation is not required in areas with 2000-2500 mm rainfall, well distributed though out the year. However, in drier months irrigation may be provided for better yield. Depending on weather and soil conditions, 8-10 irrigations are required in dry areas during the rain free period. The field is to be kept weed free till a complete cover of the crop is obtained. Weeding in the inter-row spaces can economically be achieved using cultivators between the rows. Earthing up is done after about 4 months of planting and again after every harvest as the citronella rootstock has a tendency to work out of soil by itself. Termite attack on planted slips as well as on the live clumps can be controlled by soil application of aldrin or heptachlor at 25 kg/ha. Leaf blight disease is caused by *Curvularia andropogonis* (Zimm.) Boedijn. in the beginning of monsoon. It appears as small brownish spots which enlarge into long patches along the tips and margins of the leaves. Another fungus *Colletotrichum graminicola* (Ces.) Wilson has been found to affect the crop. These diseases can be controlled by prophylactic spraying of Dithane M-45 or Dithane Z-78 at 2g/l at an interval of 10-15 days during the disease prone period.

The crop becomes ready for the first harvest after about 9 months of planting and subsequently at an interval of 3 months. Harvesting is usually done by sickle above the first node at 20-45 cm above the ground. Generally, leaf blades are cut and sheath are left out. Flowering is discouraged as it causes aging and reduces the life span of the plantation. Harvesting can be done 4 times a year. Generally the Java crop once planted yields profitable returns for 4-5 years and needs replanting only afterwards. Ceylon citronella remains productive for 10-15 years. Uprooting of the plantation after its life span and rotation with any legume species is recommended. Horse gram, cowpea and sunhemp are good rotational crops.

Importance and uses: Java citronella oil contained citronellal 32-45%, geraniol 12-18%, citronellol 11-15%, geranyl acetate 3.8%, citronellyl acetate 2-4%, limonene 2-5%, elemol and other sesquiterpene alcohols 2-5%, ß-elemene and -cadinene 2-5%, and traces of cubebene, calamenene, bourbonene, bisabolene, eugenol, methyl eugenol, isopulegol, nerol, linalool, geraniol, methyl heptanone, myrcene and -pinene. Good quality citronella oils contain more than 38% of citronellal, 16% of geraniol and 12% of citronellol. Citronella is a tufted perennial grass, the leaves of which on distillation give a yellowish-brown essential oil with *citrus odour*. Two types of citronella are identified, viz, *Ceylon and Java citronella*, the former gives inferior quality oil while the latter yields a superior oil of commercial importance. Java oil is used as a starting material for the extraction of geraniol and citronellal which can be converted into aroma chemicals such as citronellol, hydroxy citronellol, synthetic menthol and esters of geraniol. These find extensive application in soap, perfumery, cosmetic and flavouring industries throughout the world. Soaps, soap flakes, detergents, household cleansers, insecticides and other technical products are often perfumed exclusively with this oil. Citronellol is used in many perfumery blends of the soap and cosmetic industries when rosaceous notes are required. Hydroxy citronellol is a key ingredient in compounding and in floralizing perfume materials. Citronellol esters like formate and acetate are used in a wide range of fragrances. The oil is also used in the manufacture of deodorants, mosquito repellent creams and allied products. The oil of Ceylon citronella is employed chiefly for perfuming low priced technical preparations such as detergents, sprays, polishes and insect repellents. The spent grass can be used as a source of raw material for cellulose pulp and paper production by using sulphate, sulphite and cold caustic soda.

Scientific name	:	*Vetiveria zizanioides*
Family	:	Poaceae
Common name	:	Vetiver

Description: Vetiver belongs to the family Graminae (Poaceae). It was earlier known as *Andropogon muricatus* Retz. Later it was recognized as *Vetiveria zizanioides* (L.) Nash with chromosome number 2n=20. Vetiver grows in large clumps from a much branched spongy rootstock with erect culms, 0.5-1.5 m high. Leaf blades are stiff, long, narrow, upto 75 cm long, 8 mm or less in width, glabrous, but rough on the edges. Panicles are 15-30 cm long, very narrow; branches 2.5-5.0 cm long, whorled; spikelets in pairs, narrow, acute, appressed, awnless; one sessile and hermaphrodite, somewhat flattened laterally, with short sharp spines, 3 stamens and 2 plumose stigmas; the other spikelet pedicelled and staminate.

Cultivation: Khus thrives well in tropical and subtropical climate for its proper growth, development and essential oil yield. It grows luxuriantly in places upto an altitude of 600 m, with an annual rainfall of 1000-2000 mm, temperature ranging from 21 to 44° C and with moderately humid climate. In places of scanty rainfall, it can be grown as an irrigated crop. The plant is hardy and grows on almost all types of soils. Light soils, however, should be avoided as the roots obtained produces very low percentage of oil. Red lateritic soils with abundant organic matter are considered best as the roots produced in such soils are thick and contain more essential oil. Heavy soils make harvesting of the roots difficult, with a loss of the finer roots which contain most of the oil. It can also be grown on saline and alkaline soils with a pH range of 8.5-10.0, being suitable otherwise. The grass is capable of both vegetative and sexual reproduction in nature. However, under cultivation, it is generally multiplied vegetatively through slips. While preparing the slips for planting, fibrous roots and leaves should be trimmed off. 'Pusa hybrid-7', 'hybrid-8', 'CIMAP/KS-2', 'Sugantha' and 'ODV-3' are improved varieties recommended for commercial cultivation. With the onset of monsoon, land is prepared. Farm yard manure or compost is applied at 10-15 tonnes/ha and slips collected from good healthy and disease free clumps are planted during June-July vertically about 10 cm deep at a spacing of 30-60 cm. Late planted crop yields coarse roots which yield inferior quality oil. Normally, vetiver crop is not fertilized on rich soils. On poor soils, N, P_2O_5 and K_2O may be applied each at 25-50 kg/ha. It is beneficial to apply N in 2-3 split doses. In areas where rainfall is good and well distributed over the year and humidity is high, supplemental irrigation is not required. In other areas, 8-10 irrigations are required to get optimum yield. 2-3 weedings at an interval of about 25-30 days are needed during the initial period of plant growth. Once the plantation is established very well, weed problem is not very severe because of the thick and robust shoot cover. Earthing up after weeding is also beneficial to the crop. Leaf blight caused by *Curvularia trifolii* and *Fusarium* disease, affect the growth and yield of the crop, which can be controlled by repeated spraying of copper fungicides and drenching with copper oxychloride or 1% Bordeaux mixture. Roots are harvested after 15-24 months of planting. Harvesting can be done even earlier. Although the young roots give a higher yield of oil, this will be of low specific gravity and lacking in the valuable high boiling constituents. If the roots stay in the ground for over two years, the yield of oil diminishes considerably and the oil becomes very viscous with a dark colour but of a high quality. In general, the crop is harvested after 18 months during the dry season from December to February by digging out the bush along with its roots manually. The length of the roots varies from 10-35 cm according to the condition of growth, soil, climate, etc of the locality. The roots are separated from the plants, washed to remove the adhering soil and dried under shade for 1-2 days which improves the olfactory quality of the essential oil. Prolonged drying in the sun may also reduce the oil yield. The root yield is 3-5 tonnes/ha.

Importance and uses: Vetiver oil is light brown to deep brown in colour with a characteristic aroma and persistent odour of sweet woody note. The oil contains more than 150 complex compounds including elemol 0.4-2.3%, 10-epi-eudesmol 1.1-2.2%, ß-eudesmol 5.5-8.5%, vetiverol + cyclocopacamphenol 6.1-7.5%, vetiselinenol 11-20%, khusimol 13-28%, ß-vetivone 2-5%, among others. Vetiver oil enjoys world wide reputation as one of the finest aromatic oils. It is used in perfumery and as a fixative for more volatile constituents for scenting soaps and in the preparation of cosmetics. It is also used for the extraction of vetiverol, vetiverone and vetiveryl acetate which are widely used aroma chemicals. The aromatic roots of vetiver have been used since ancient times. They are cleaned, dried and used for making mats, fans, screens, pillows and sachet bags. Roots are often put with clothes because of their scent and in the belief that they keep out insects. Young leaves are used as fodder and they make good bedding material for horses and cattle. Leaves are also used for thatching purposes. Stem and inflorescence peduncles are used for making brooms and fancy baskets. Leaves made into pulp are suitable for making straw boards. The plant has gained much recognition as one of the best soil binders and is being used to check soil erosion.

Scientific name : *Mentha* spp.

Family : Labiatae

Common name : Mints

Description: Mints are a group of plants belonging to the family Labiatae (Lamiaceae) and the genus *Mentha*. Four major mints are recognized in this genus.

1. *Mentha arvensis* L. Japanese Mint
2. *M. piperita* L. Pepper Mint
3. *M. spicata* L. Spear Mint
4. *M. citrata* Ehrh. Bergamot Mint or Lemon Mint

In some cases, Japanese Mint is specifically recognized as *Mentha arvensis* L. subsp. *Haplocalyx* Briquet var. *piperascens* Holmes.

In the strict sense, mints are temperate crops. However, Japanese Mint can be grown in subtropical and even tropical areas with cooler climate. It is a downy perennial herb with rootstock creeping along or just under the ground surface. Red, purple and green varieties are known. Branches are rigid, pubescent, 60-100 cm tall; leaves are lanceolate to oblong, 3.7-10 cm long, sharply toothed, sessile or shortly petiolated and hairy; flowers are arranged in cyme which are usually sessile or rarely pedunculate, purplish, minute; calyx 2.5-3.0 mm long, narrowly deltoid, acuminate; corolla white to purple, 4-5 mm long.

Cultivation: Mints grow well in a wide range of climatic conditions. Japanese Mint grows well under subtropical conditions while others prefer temperate climate. Adequate and regular rainfall during the growing period and good sunshine during harvesting are ideal for its cultivation. Medium deep soil rich in humus is best suited for the cultivation of Japanese Mint. The soil should have a pH range of 6-7.5 with good water holding capacity but waterlogging is detrimental to the crop. Japanese Mint is propagated through stolons. 'CIMAP/MAS-1' and 'Hybrid-77' are improved varieties available for its cultivation. Seed rate is 400 kg/ha. One hectare of well established mint provides enough planting materials for 10 hectares of land. Stolons are planted either on flat land or ridges. In plains, they are planted in shallow furrows of 7-10 cm deep at a spacing of 45-60 cm after incorporating compost or farm yard manure at 10-12 tonnes/ha. Inorganic fertilizers upto 160 kg N and 50 kg each of P_2O_5 and K_2O/ha are applied; nitrogen being applied in 2-3 split doses. Irrigation enhances growth and improves the yield. The field usually should be kept weed free, particularly during the initial stages of growth till proper establishment and coverage of the ground. Termite attack observed during the dry months can be controlled by the soil application of 3% heptachlor at 50 kg/ha. Hairy caterpillars cause rapid defoliation. Cut worms, semi-loopers and red pumpkin beetle also attack the crop. These insect pests can be controlled by 5% DDVP, 2% methyl parathion dust or any other suitable insecticide. Nematode attack has also been reported for which application of Fenamiphos at 10-12 kg/ha is effective. Mentha rust is caused by *Puccinia menthae* which results in severe leaf shedding. Powdery mildew caused by *Erysiphe cichoracearum* can be controlled by wettable sulphur application. *Macrophomina phaseoli* and *Thielavia basicola* cause stolon rot which is effectively controlled by the application of Maneb or PCNB. Japanese Mint is first harvested after about 4 months of planting when the lower leaves start turning yellow. Subsequently, two more harvests can be taken generally at an interval of 80 days. The fresh herb yield ranges from 25 to 50 tonnes/ha obtained in 3 cuttings annually.

Importance and uses: The chemical constituents in Japanese mint oil are reported to be l-menthone, d and l- iso-menthone, methyl acetate, camphene, caryophyllene, esters of formic, iso-valeric and caproic acids, ethyl carbinol, hexanol l-limonene, ß-pinene, cineole, 3-octanol, linalool, menthofuran, neo-menthol, pulegone, piperitone and piperitone oxide. *Japanese Mint* is as a primary source of menthol, which is widely used for flavouring toothpastes, candies, beverages, confectionery, chewing gums, pan parag, and mouth washes and for scenting shaving creams, tobacco, cigarettes, aerosols, polishes, hair lotions and lipsticks. It is employed as a soothing ingredient in cosmetic preparations, colognes, deodorants, aftershave lotions and perfume bases. It is also employed in a number of medicinal preparations like ointments, pain balms, cough syrups, cough lozenges and tablets.

Scientific name	:	*Ocimum* spp.
Family	:	Labiatae
Common name	:	Tulsi

Description: Ocimums belong to the family Labiatae (Lamiaceae) and the genus *Ocimum*. Owing to a high degree of polymorphism exhibited by the species as also abundant cross pollination, a large number of species, subspecies, varieties and strains have come into existence which make the botanical nomenclature extremely difficult. In view of the great diversity, the various species have been classified into two broad groups, viz, *basilicum* and *sanctum* groups, based on the geographical sources, morphological and cytological features and chemical constituents. The additional distinguishing features between the two groups are that in basilicum group, bracts are petiolate, flowers more conspicuous and seeds are black and ellipsoid which become mucilaginous when wetted whereas in sanctum group, the bracts are sessile, flowers less conspicuous and seeds brownish, globose to ellipsoid but do not become mucilaginous when wetted. Sweet or French basil is an erect, almost glabrous herb, reaching a height of 30-90 cm. Leaves are ovate to lanceolate, flowers are 0.72- 1.25 cm long, borne on long terminal racemose inflorescence; calyx 5-toothed; corolla long, white to purplish, 2-lipped tube; stamens 4, protruding; ovary bicarpellary, syncarpous, bilocular; stigma bifid; fruits are nutlets. *Clocimum* is a hybrid strain of *Ocimum gratissimum var. clocimum* obtained by crossing *O.gratissimum* race 1 and *O. gratissimum* race 2, developed at the Regional Research Laboratory, Jammu, India. The name *clocimum* means *clove scented ocimum* and its odour is similar to that of clove oil. Clocimum oil contains 70-75% eugenol and 10-15% myrcene.

Cultivation: The crop grows well under tropical climate upto an altitude of 1800 m. The growth is poor in heavy rainfall areasl. Frost is harmful to the plant and hence, frost prone areas are to be avoided. Basil can be cultivated on a wide variety of soils, though moderately fertile well drained loamy or sandy loam soils are considered best for its cultivation. Basil is tolerant to higher concentration of copper and zinc but is susceptible to cobalt and nickel. The plant is propagated by seeds. Seedlings are first raised in the nursery and then transplanted in the field. The seed rate is about 125 g/ha for transplanting. Seeds start germinating 3 days after sowing and germination is over in 7-10 days. When 6-10 cm tall, the seedlings are transplanted in the field at 40-60 cm spacing in rows. At the time of planting, 10-15 tonnes of compost or farm yard manure is to be applied. A medium fertilizer dose of 40:40:40 kg/ha of N, P_2O_5 and K_2O is recommended for economic yield though good response has been received upto 120:100:100 kg/ha. Irrigation is required once a week when it is raised as a summer crop. The field should be kept weed free for the first 20-25 days, till the crop canopy completely covers the ground. Weeding is usually carried out once or twice. A number of diseases are reported in basil crop. *Corynespora cassicola* (Berk. and Curt.) Wie. causes leafspot disease which appears as small water soaked spots turning brown. *Elsinoe arxii* sp. nov. causes scab disease. The symptoms are little defoliation with pluckering, clipping of the leaves and distortion of the tender twigs. Blight caused by *Alternaria species* and *Colletotrichum capsici* (Sy.) Butler & Bisby can be controlled by spraying 0.2% zineb or maneb. Wilt is caused by *Fusarium oxysporum* at all stages of growth. But the attack is more pronounced in rainy season. This is controlled by dipping the seedlings in a solution of organo mercurial fungicide. Basil is harvested when the plant is in full bloom and lower leaves start turning yellowish. The crop comes to full bloom 9-12 weeks after planting. For high quality oil, only the flowering tops are harvested. 4-5 crops are obtained per year. In some areas it is possible to get four floral harvests. The first harvest is done when the plants are in full bloom and the subsequent ones after every 15-20 days. The last harvest comprises the whole plant. Floral harvests yield 3-4 tonnes of flowers and the final harvest of the whole plant is 13-15 tonnes of herb per hectare. While harvesting the whole herb, plants are cut not less than 15 cm from the ground for enabling regeneration of the crop.

Importance and uses: Ocimums are an important group of aromatic and medicinal plants which yield many essential oils and aroma chemicals and find diverse uses in the perfumery and cosmetic industries as well as in indigenous systems of medicine. Ocimum species with oil rich in camphor, citral, geraniol, linalool, linalyl acetate methyl chavicol, eugenol and thymol are important and can be harnessed for successful utilization by the industry. Among the various *Ocimum* species *Ocimum basilicum* L. is commercially and extensively cultivated for essential oil production. Its oil is employed for flavouring of food stuffs, confectionery, condiments and in toiletry products such as mouth washes and dental creams. It is also used in the flavouring of foods such as spiced meats, sausages, tomato pastes, various kinds of sauces, fancy vinegars, pickles, ketchups and beverages. In the perfumery industry, the oil is used for compounding certain popular perfumes notably jasmine blends. It is recognized as a febrifuge and antimalarial plant. The juice obtained from the leaves gives relief to irritation of throat ear ache and ringworm infections. Seeds are used internally for the treatment of constipation and piles.

Scientific name	:	*Pogostemon patchouli*
Family	:	Labiatae
Common name	:	Patchouli

Description: Patchouli *Pogostemon patchouli* Pellet. var. *suavis* Hook. f. which is synonymous to *Pogostemon cablin* Benth comes under the family Labiatae (Lamiaceae). The plant is herbaceous, erect, branched, pubescent, 0.5-1.0 m tall and aromatic when crushed. Leaves ovate to oblong, serrate with dotted glands beneath; stems are swollen on the nodes; spikes terminal or axillary, dense, sometimes interrupted, 2.5-14 cm long; flowers are small, irregular, bisexual, hypogynous; corolla lobes are obtuse, 6-9 mm, white to purple, blotched on all segments; filaments violet; bracts as long as calyx.

Cultivation: Patchouli performs well in warm humid climate with a fairly heavy and evenly distributed rainfall of 2500-3000 mm per annum, a temperature of 24-28° C and an average atmospheric humidity of 75%. It grows successfully upto an altitude of 1000 m above MSL. The crop thrives well under irrigation in less rainfall areas, and in partially shaded conditions. It is relatively a hardy plant and adopts itself to a wide range of soil conditions. A well drained deep fertile loamy soil, with a loose friable structure and with no impervious hard layer at the bottom is ideal. A pH range of 5.5-6.2 is suitable. Patchouli is a shade loving plant. It can be grown as an intercrop in orchard crops like fruit trees, coconut or arecanut. The plant is propagated vegetatively by stem cuttings having 4-5 nodes and 15-20 cm length. Improved varieties commonly cultivated are 'Johore', 'Singapore' and 'Indonesia'. Cuttings are prepared from the apical region of healthy stocks. The basal 2-3 pairs of leaves are carefully removed and the cut ends are treated with IBA, IAA or NAA at 500, 1000 or 1500 ppm respectively for better rooting. Cuttings are planted 3.5 cm apart in nursery beds, seed pans or polythene bags. It is advised to provide aeration, partial shade and regular watering in order to get early and good rooting. Rooting occurs in 4-5 weeks and they are ready for transplanting in 8-10 weeks. Before transplanting, the field is prepared well and laid into beds of convenient size. These beds are incorporated with organic manure at 12-15 tonnes/ha and N, P_2O_5 and K_2O at 25:50:50 kg/ha and levelled. Rooted cuttings are transplanted at 40-60 cm spacing and irrigated immediately after planting. After 2 months, 25 kg N is applied.

Subsequently, 50 kg N/ha is applied in two split doses; the first dose just after the harvest and the other about two months later. Totally, 150 kg N/ha/year is applied to the crop. Constant watering, regular weeding and light cultivation after every harvest are required for proper growth and yield of the crop. The crop is highly susceptible to root-knot nematode, *Meloidogyne incognita* (Kofoid & White) Chitwood. An integrated approach consisting of crop rotation, application of neem oil cake, carbofuran and systemic nematicide proved effective. Phytosanitory measures are adopted from the nursery stage itself. This is done by passing steam through the seed pans or poly bags for about an hour. Nematicides like carbofuran and carbofenthion are also effective in controlling the nematode problem. Leaf blight caused by *Cercospora* sp. is controlled by spraying 0.5% solution of zineb. Yellow mosaic disease is transmitted by white fly, *Bemisia tabacci* (Gen.). Caterpillar and leaf webber attacks can be controlled by spraying with Malathion. The crop is harvested when the foliage becomes pale green to light brown and the stand emits a characteristic patchouli odour. The first harvest of the leaves is taken after about 5 months of planting. Subsequent harvests are done after every 3-4 months depending on the local conditions and management practices. Harvesting is done in the cool hours of the morning to avoid loss of essential oil. Young shoots of 25-50 cm length which contain at least 3 pairs of mature leaves are cut. In practice, a few shoots are always left unplucked to ensure better growth for next harvest. The crop usually stands for 3-4 years.

Importance and uses: Patchouli oil contains ß-elemene 1%, caryophyllene 20%, -guaiene 15%, -bulnesene 25%, -guaiene oxide 1%, -bulnesene oxide 4%, caryophyllene oxide 2%, nor-patchoulinol 0.5%, patchouli alcohol 30% and pogostol 1%. In fact, it is a perfumed by itself and is highly valued in perfumes, soaps, cosmetics and flavour industries. The oil is extensively used as a flavour ingredient in major food products, including alcoholic and non-alcoholic beverages, frozen dairy desserts, candy, packed foods, gelatin, meat and meat products. It blends well with the oils of sandal wood, geranium, vetiver, cedarwood, clove, lavender, bergamot and many others. The oil emits one of the finest attars when blended with sandal wood oil. Tenacity of odour is one of the great virtues of patchouli oil and is one of the reasons for its versatile use. The oil possesses antibacterial and insect repellent activity. In Chinese medicine, patchouli leaves are used as decoction with other drugs to treat nausea, diarrhoea, cold and headaches. The dried leaves are used for scenting wardrobes. The leaves and tops are added in bath water for their anti-rheumatic action. It is also used as a masking agent for alcoholic breath.

Scientific name	:	*Rosmarinus officinalis*
Family	:	Labiatae
Common name	:	Rosemary

Description: It is an evergreen dense highly branched herb or undershrub grows to a height of 1 m. The long slender branches are arranged with many sessile opposite leaves, smooth and green, woolly whitish and glandular beneath, 2-4 cm long, almost cylindrical and folded inwards; flowers are situated in small clusters towards the ends of the branches; calyx is 2-lipped, with an upper single broad oval lobe and a lower two segmented triangular lobe; corolla is two lipped with two violet stamens and a long style projecting from it; fruit is an oval in shape and 4-sectioned cremocarp.

Cultivation: The rosemary comes up well in Mediterranean climate. It is vulnerable to frost injury. In cooler areas, it can be cultivated in summer season. It requires light dry soil, with a neutral to alkaline pH. The plant is propagated through seeds and vegetatively by cuttings, the latter being generally adopted. The cuttings should be 15 cm long and leaves removed from the basal half portion. The cuttings are put in nursery beds of sandy soil at a depth of about 10 cm. The main field is prepared well incorporating 10-15 tonnes/ha of organic manures. The rooted cuttings are transplanted in rows, 120 cm apart with a plant to plant spacing of 30-40 cm. Fertilizers are applied at 100:40:40 N, P_2O_5 and K_2O/ha, N being applied in 4-5 split doses during each year. Irrigation is needed when the soil is depleted of water during non-rainy period. The field of rosemary is kept weed free by regular weeding and hoeing. Inter-culture keeps the soil loose and clean from weeds and promotes proper plant growth and development. *Phytocoris rosmarini* sp. nov. and *Ortholylus ribesi* sp. nov. are reported to infest rosemary crop. The shoots are cut for distillation when they have reached their maximum size but before they have become woody. The hard wood imparts an undesirable turpentine odour to the essential oil. Harvesting is usually done during May-June. Frequent cutting of the bushes after 2-3 years keeps them free from becoming leggy and encourages the formation of numerous young shoots.

Importance and uses: The essential oil contains -pinene 7-24%, camphene 3-9%, ß-pinene 0.6-1.5%, sabinene trace, myrcene 2-19%, -phellandrene 0.1-1.8%, -terpinene 0.5-1.5%, limonene 3.3-5.4%, 1, 8-cineole 15-20%, -terpinene 0.2-1.8%, p-cymene 2-3.7%, terpinolene 0.1-0.6%, camphor 0.1-0.6%, copaene 0.1-0.3%, linalool 14-17%, terpinen-4-ol 0.9-1.8%, caryophyllene 0.9-2.9%, -terpineol 1.8-3%, thymol 0.1-0.7% and carvacrol 0.1-2%. Its essential oil is used almost wholly in the perfumery industry in the production of soaps, detergents, household sprays and other such products. It is an excellent fixative material. The oil contributes a strong, fresh odour, which blends well with various other oil odours and also serves to mask the unpleasant smells of certain other ingredients. It is used in shampoo, toilet soaps and medicine. Rosemary oil is known to have antimicrobial activity. It can kill 90-100% of mosquito and larvae of *Culex quinquefasciatus*. It is carminative and mildly irritant. It is used in formulations of compounded oils for flavouring meat, sauces, condiment and other food products, It is used as a culinary herb. A distilled water is obtained from flowers which is used as an eyewash.

Scientific name	:	*Salvia sclarea*
Family	:	Labiatae
Common name	:	Clarysage

Description: Sage oils are obtained from different species of *Salvia*. Inferior Spanish and Dalmatian materials are obtained from *Salvia officinalis* L., whereas a superior and more expensive oil is obtained from *Salvia sclarea* L. The latter is commonly cultivated. *Salvia sclarea* is 60-90 cm tall herb with a widely branched deep root system. Stem is branched, 4-ribbed with numerous epidermal glands on younger parts; leaves are petiolate, opposite, broadly oval toothed and thinly hairy above; flowers are many, blue to purple, in 15-80 cm long peduncle, bisexual and zygomorphic; calyses are amosepalous, inferior; corolla bilipped, upper 3-lobbed, lower 2-lobbed; stamens two, inserted in corolla tube; ovary bicarpellary, syncarpous; superior; stigma bilobbed; placentation axile. Honey discs are present at the base of the ovary and 138 kg/ha of honey is reported to be recovered from these honey discs. Anisophylly is observed in clarysage. *Xylocopa violacea* is reported as the only effective pollinator of clarysage in Yugoslavia. The cells of the pericarp and skin of the fruit contain oil drops.

Cultivation: Clarysage is tolerant to cold and drought and adaptable to a wide variety of situations. Higher altitude with ample sunshine and few good showers in spring produces a good yield of oil having superior quality. The plant is generally grown on poor soils. Slightly acidic soils of pH 4.0-5.5 are better. It is propagated through seeds. Few high yielding hybrid varieties have been developed in Bulgaria.

'Zarya' is a medium early variety with 0.24% oil and 78% linanyl acetate in the oil. 'Lazur" is cold resistant with 0.23% oil and 73% linalyl acetate. A new biennial variety with 0.31% oil and 81.5% linalyl acetate has also been developed. The seeds are directly sown in the field or transplanted either in November or March-April depending upon the weather conditions. Seed rate is 3-4 kg/ha for transplanting. Seedlings appear in 10-15 days and are transplanted when 30-35 days old, at 100 cm row spacing after incorporating 10-12 tonnes of organic manure in the field for optimum growth of the plants. 100-120 kg N and 30 kg each of P_2O_5 and K_2O are recommended per hectare. N is applied in 4 equal splits. One or two irrigations may be given in case of a drought situation. One or two weeding should be done during March-April. Pre-emergence application of fluometuron or diuron at 2 kg/ha and post-emergence application of preforan or introchlor at 3 kg/ha is recommended for weed control. 2-3 hoeings should be done before the flowering season. An aphid *Acyrtosiphon salviae* is found in colonies on clarysage which can be controlled by a mild insecticide. Rootknot nematode *Meloidogyne incognita* infests the plant heavily. The fungus *Rhizoctonia solani* causes rot disease. All the above ground portions got found infected. Under humid and wet condition, the whole plant collapses within 2-3 days. Drenching with copper oxychloride or Bordeaux mixture is recommended for control of rot disease. The flowering tops and leaves are harvested twice a year during July and September. Excessive stalk growth is removed as it contains less amount of oil. After harvest, a hoeing is done. Plants remain productive for 5-6 years and thereafter yields decline and new plantation is started in a different location.

Importance and uses: The chemical constituents of the oil are ß-ocimene, p-cymene, terpinolene, cis-3-hexen-1-ol, linalool (the major constituent) and its acetate, terpinen-4-ol, caryophyllene, -terpineol, citronellol, nerol, geraniol, and their acetates, trans ß-terpineol ß-gurjunene, caryophyllene oxide, tricyclene, camphene, 1, 8-cineole, methyl heptanone, camphor, ß-thujone, ß-humulene, -thujone, delta-cadinene and citral a and b.

The plant was used in middle ages for *clearing the vision* and for this reason it received its popular name *clarysage or clary-eye*. The essential oil is used in perfumery because of its coriander-like notes. It is used as a flavour in liquors and as a modifier in spice compounds. The oil is also used in preparations of non-alcoholic and alcoholic beverages, ice-creams, candy and baked goods.

Scientific name	:	*Thymus vulgaris*
Family	:	Labiatae
Common name	:	Thyme

Description: Thyme belongs to the family Labiatae (Lamiaceae) and genus *Thymus*. There are many species of thyme. *Thymus serpyllum* L. and *T. satureioides* Coss. & Bal. are some wild species. Commercial supplies are derived from *T. zygis* L. (*white thyme*) and *T. vulgaris* L. (*garden thyme*). The demand for garden thyme is more and hence commonly cultivated. The plant *T. vulgaris* is an evergreen perennial aromatic herb, 20-30 cm tall. Roots are fairly robust and stem very much branched. The sessile leaves vary in shape from elliptic to linear or diamond-shaped towards the apex; young leaves are slightly woolly; flowers, united in spikes at the top of the branches, have a bilabiate, tube-like calyx and corolla, lower lip of corolla 3-lobed; fruit consists of a smooth, dark coloured, 4-sectioned nutlet found in the remains of the calyx tube.

Cultivation: The plant grows well in a warm humid climate at an elevation of 1500-4000 m from MSL. Light loamy fertile and calcareous soils are prefered. On heavy wet soils, the leaves become less aromatic. Thyme is propagated by divisions of the old plant cuttings, layering or by seeds. Cuttings and layers are prepared during summer months. Seeds are sown on well prepared nursery beds. Seedlings are very small and remain inconspicuous for several weeks after germination. Planting of rooted cuttings, layers or transplanting of seedlings is done during late summer at a spacing of 30-45 cm between plants and 60 cm between rows. In autumn, a light dressing of farm yard manure is required. Fertilizers are applied at 100:40:40 kg N, P_2O_5 and K_2O/ha. Top dressing of N in the spring promotes the formation of numerous leafy shoots. Irrigation is given as and when required. The field is to be kept weed free. Not much insect-pests and diseases are reported in this crop. About 15 cm long shoots, in the early flowering stage, are harvested during May-June. The lower portions of the stem, together with any yellow or brown leaves are rejected.

Importance and uses: The essential oil of thyme has a powerful fresh odour masking other unpleasant smells. This plant is extensively used as a pot herb in cooking, perfumery and in liquor distillery. Thyme oil is mainly used in the perfumery industry in soap and detergent work. Thymol has a powerful medicinal odour and applied more in flavours than in perfumes. Owing to the presence of thymol the oil shows germicidal properties and is effective against a variety of pathogenic bacteria. It is employed in dental preparations, oral hygiene products, vermifuges and antigastro-intestinal products. In aromatherapy, garden thyme is regarded as one of the most important elements because of its antiseptic properties. The essence is effective in treating whooping cough as well as parasitic infestations. The dried leaves and floral tops constitutes the thyme of commerce known as *Thymi Herba* in pharmacy. Dried flowers and leaves are used to preserve linen from insects and to impart characteristic smell.

Scientific name : *Artemisia pallens*

Family : Asteraceae

Common name : Davana

Description: The plant davana *Artemisia pallens* Wall.ex DC., belongs to the family Compositae (Asteraceae). It is an erect branched annual herb, 45-60 cm tall and covered with greyish white tomentum. Leaves are alternate, exstipulate, petiolate, lobed; inflorescences are capitulum, axillary, peduncled to sessile, heterogeneous with yellow glabrous florets; involucre two or more, seriate, ovate to elliptic-linear; inner florets are 5-lobed, bisexual; stamens 5 with free epipetalous filaments; style bifid.

Cultivation: Davana is a delicate plant and hence cannot tolerate heavy rains. It prefers light drizzles, bright sunshine, and a mild frostless winter and heavy morning dew during the growing season. Cloudy weather and rains during flowering and seed ripening stages adversely affects the yield. The crop grown during November gives the maximum herb and oil yield. However, the crop can be grown round the year for use in garlands and bouquets. The plant can be grown on various types of soils ranging from sandy loam to medium black soils, but humus rich red loam soils are ideal. The plant is propagated by seeds. Seeds are short-viable and hence cannot be stored for long time. Transplanting is generally practised in the crop. A nursery area of 500 m^2 sown with about 1.5 kg seeds is sufficient for planting one hectare of land. The seeds are mixed with fine sand, broadcast over the nursery bed, covered with a thin layer of sand and watered regularly. Seeds germinate in about 3-4 days. When the seedlings are 10-12 cm tall they are transplanted to the main field at 15 x 7.5 cm spacing. Before transplanting, 12-15 tonnes of well decomposed FYM and 40 kg/ha each of phosphate and potash are incorporated into the soil. N is applied at 120 kg/ha in 4 equal splits, 3 for the main crop and one for the ratoon crop at 15 days interval. The crop is irrigated at weekly intervals. Two weedings are carried out in the main crop and one in the ratoon crop. The crop is often infected by damping off at the tender early stage, particularly in the nursery. This disease, caused by *Rhizoctonia* species is common during cloudy and rainy period. Hence adjusting of sowing time is important for avoiding the disease. The crop is harvested during February-March when a large number of flower buds start opening. Flower to plant ratio at the time of harvest is reported to be important in davana. Harvesting is done by cutting the whole plant with sickle at a height of 10 cm from the ground. The herb yield is 8-10 tonnes/ha.

Importance and uses: Davana oils are rich in davanone, fenchyl alcohol, cinnamyl cinnamate, caryophyllene, cadinene, linalool, dehydro—linalool, davanafuran, isodavanone, dihydrorosefuran, n-alkanes, hydroxy davanone, geraniol and nerol. Davanafurans are responsible for the characteristic odour of davana oil though they constitute only 0.8% of the oil. The flowering top of Davana yields an essential oil which is extensively used in high grade fine perfumes. The oil is used for flavouring cakes, pastries, tobacco, beverages, sausages and preserved products. The leaves form an important component of garlands and bouquets.

Scientific name : *Matricaria chamomilla*

Family : Asteraceae

Common name : Chamomile

Description: The Chamomile plant belongs to the family Compositae (Asteraceae) and three species of chamomile are recognized .

(1) *Matricaria chamomilla* L. known as *German or Hungarian Chamomile* and is the most common. It is cultivated in Europe, Germany, France, Hungary and Russia.

(2) *Anthemis nobilis* L. known as *Roman Chamomile*

(3) *Ormenis multicaulis* Braun Blanquet & Maire termed as *Moroccan Chamomile*.

Matricaria chamomilla is the most widely and commonly cultivated species. It is an annual herb, much branched, erect spreading, 60-90 cm tall and glabrous. Leaves are pinnate; leaflets narrow and linear; inflorescence is capitulum, 1.3-2.5 cm in diameter; flowers are borne on hemispherical or conical hollow receptacles, surrounded by involucre of 2-3 rows of small imbricate bracts; ray florets are 10-20, whitish or yellowish, later becoming reflexed, disc florets, numerous, yellow, tubular, dark brown or greyish yellow achenes with 3-5 faint ribs.

Cultivation: The plant is usually grown as a winter crop in plains and as a summer crop on hills. For good seed germination, the optimum temperature is 18-20° C. Temperature and light conditions play a greater role in essential oil production. It grows on any type of soil, but comes up well on moist, moderately heavy humus rich soils. The optimum soil pH is 7, though saline and alkaline soils having pH as high as 9.0 also support good growth. The crop is raised through seeds. Seedlings are raised in nursery during September-October. Seed rate is about 1 kg/ha. The seedlings are transplanted when 6 weeks old at a spacing of 30-40 cm Adequate manuring is required for good growth and yield of the crop. Application of 15-20 tonnes/ha. of well-rotten farm yard manure and 80:40:20 kg N, P_2O_5 and K_2O/ha are recommended. On normal soils, 3-4 irrigations are sufficient during the entire growing season. Saline soils require frequent light irrigations. Generally, one or two weedings and hoeings are required for raising a good crop. Black bean aphids (*Aphis fabae*) is a serious pest. Flowers are attacked by an insect, *Nysius minor*, which causes shedding. Another insect, *Antographis chryson* also attacks the plant and causes defoliation. These insect pests can be effectively controlled by spraying suitable insecticides. The plants start flowering from February to April. The flowers are harvested at full bloom stage. Generally, 4-5 harvests are taken at an interval of 10-15 days. The yield is 4000-7000 kg fresh flowers which gives 1000-1500 kg/ha on drying. Drying is done under shade between 22-24°C as the flowers are delicate. Fully dried flowers are packed and stored in moisture free environment.

Importance and uses: The essential oil consists of 1-15% of chamazulene, which is responsible for the blue colour of the oil; azulene, farnesene, -bisabolol oxide a and b and a dicyclo ether. German or Hungarian chamomile is a much branched erect spreading annual herb from which essential oil, infusions, tinctures and fluid extracts are prepared for diverse uses. The essential oil is used in alcoholic and non-alcoholic beverages, ice-creams, ice candy, baked goods and chewing gums as flavouring agent. It is used in high class perfumes in low concentrations. Medicinally, it acts as antispasmodic, expectorant, carminative, anthelmintic, sedative and diuretic. It also possesses antimicrobial activity. It is used in infant ailments such as teething troubles and stomach disorders. The infusion is a common beverage which is a mild sedative and digestive. The dried herb and the tinctures are used in bitter tonic beverages and elixirs.

Scientific name	:	*Pelargonium graveolens*
Family	:	Geraniaceae
Common name	:	Geranium

Description: It is a suffrutescent bushy aromatic perennial shrub with cylindrical stem. Leaves are simple, alternate, stipulate, broadly cordate with palmatisect primary lobes and pinnatisect secondary lobes, pubescent and aromatic; inflorescence is umbellate, bracteate; flowers are pentamerous, basically bisexual, hypogynous; calyx free quinquincial; corolla pink, zygomorphic, polypetalous; stamens 10; filaments subequal, united at the base; ovary is superior, pentacarpellary, syncarpous, hairy; stigma 5 lobed.

Cultivation: A Mediterranean type of mild climate with a low humidity, warm winter, mild summer temperature and an annual rainfall of 1000-1500 mm is best for the crop. It grows successfully at an altitude of 1000- 2100 m. Well drained porous soils are suitable for its cultivation. Saline, alkaline and damp soils are not suitable. There is no seed setting in geranium. The plant is propagated by stem and root cuttings. Terminal cuttings root earlier than middle and basal cuttings. IAA is better than IBA for inducing rooting. Rooted cuttings are raised in the nursery during November-January and transplanted after 2 months at 60 x 40 cm spacing in the main field after applying well decomposed FYM or Compost at 10-12 tonnes/ha. Inorganic fertilizers like phosphate and potash are applied at 40-60 kg/ha as basal while N is applied upto 200 kg/ha/year in six equal splits to cover the 3 harvests. Application of micronutrients such as Cu at 20 kg/ha/year and Mo at 3 kg/ha/year in 3-4 split doses is found to be beneficial. Irrigation is provided daily for the first 3-4 days, on alternate days till two weeks and weekly thereafter. The crop requires weeding 20 and 40 days after planting and hoeing after harvest. Attack of root-knot nematodes *Meloidogyne incognita* and *M. hapla* are common in this crop. These are controlled by applying Aldicarb at 20 kg/ha. *Fusarium oxysporum* and *Botryodiplodia theobromae* cause wilt disease for which the systemic fungicide benomyl is effective. The crop is ready for harvest after 4 months from transplanting when the leaves begin to turn light green and exhibit a change from lemon like odour to that of rose. The green leafy shoots are harvested with a sharp sickle. Three harvests are taken per annum and the crop remains in the field for 4-6 years. The yield of fresh herbage/ha/year from the 3 harvests is about 15 tonnes.

Importance and uses: The leaves and branches are steam distilled to get "oil of geranium" (Aporrox. yield 0.8%-0. 1%), used in high-grade perfumery product and soaps. Geranium oil blends well with all kinds of scents, floral and oriental bouquets and is extensively used in perfumery and cosmetic industries. It is widely used for flavouring soaps due to its stability in the slightly alkaline medium. The oil is also used for the production of rhodinol used in the manufacture of perfume compounds.

Scientific name	:	*Kaempferia galanga*
Family	:	Zingiberaceae
Common name	:	Kacholam

Description: The plant is a stemless perennial herb with tuberous, aromatic rootstock having fleshy cylindrical nonaromatic root fibres. Leaves are round to ovate, deltoid, acuminate, 10-12 ribbed, 6-12 cm long, 4.5-9.0 cm wide, horizontally spreading; inflorescence is scape with 6-12 flowers, figatious, fragrant, open successively and seldom set fruits.

Cultivation:*Kaempferia galanga* prefers a warm humid climate. It thrives well upto an elevation of 1500 m. A well distributed annual rainfall of 1500-2500 mm during the growing period and dry spells during land preparation and harvesting are ideal. Rich loamy soil with good drainage is suitable for the cultivation of the crop. Laterite soil with heavy application of organic matter is also suited. It cannot withstand waterlogging. The plant is propagated by division of rhizomes. For planting, mother rhizomes are better than finger rhizomes. Seed rhizomes are stored in cool dry place or in pits dug under shade. Smoking of rhizomes prior to planting is beneficial for better germination and establishment of sprouts. With the receipt of pre-monsoon showers in May, land is ploughed and beds of 1-2 m width, 25 cm height and convenient length are taken and sprouted seeds are planted at 15-30 cm spacing. Recommended seed rate is 500-750 kg/ ha. Kacholam responds well to organic manuring. Application of 30 tonnes/ha of FYM or compost and mulching with leaves or straw at 15-20 tonnes/ha are recommended. Application of 50-75 kg each of N, P_2O_5 and K_2O in 2-3 splits is beneficial. Weeding is done 45 and 90 days after planting which is to be followed by fertilizer application and earthing up. It is a shade loving plant. Growth is better in partial shade offering great potential for its cultivation as intercrop in coconut, arecanut, banana and other widely spaced perennial crops. Insect pests are not commonly reported in this crop. Leaf spot and rhizome rot diseases occur particularly during the rainy months which could be controlled by drenching and spraying with 1% Bordeaux mixture. The crop is harvested 6-7 months after planting when the leaves start drying up. The rhizomes are dug out, cleaned and washed to remove the adhering soil particles.

Importance and uses: The essential oil comprises of over 54 components of which the major constituents are ethyl-trans-p-methoxy, cinnamate 16.5%, pentadecane 9%, 1,8-cineole 5.7%, -carene 3.3%, and borneol 2.7%. Terpenoid constituents amounted to 16.4%. The oil is used in the manufacture of perfumes and in curry flavouring. It is also utilized in cosmetics, mouth washes, hair tonics and toiletries. The pungent, hot, sharp, bitter and aromatic rhizomes play a vital role in indigenous medicine as stimulant, expectorant, diuretic and carminative. It promotes digestion and cures skin diseases, piles, phantom tumors, coughs, oedema, fever, epilepsy, spleenic disorders, wounds, asthma and rheumatism. The rhizomes are used for protecting clothes against insects and are eaten along with betel and arecanuts as a masticatory. The rhizomes and leaves are attached to neck laces and added to bath water for perfume.

Scientific name	:	*Vanilla planifolia*
Family	:	Orchidaceae
Common name	:	Vanilla

Description: Vanilla is an orchid, belonging to the family Orchidaceae. The genus *Vanilla* comprises of more than 100 species of stout, scandant, terrestrial or epiphytic herbs. But only three are important source of vanillin.

They are:

1. *Vanilla frangrans* (Salisb.) Ames syn. *V. Planifolia* Andrews. This produces short thick pods.
2. *Vanilla pompona* Schiede (F.). It is known as West Indian vanilla. It is cultivated in Trinidad and Central America. It has the largest pods.
3. *V. tahitensis* J.W. Moore (F & B). It is called Tahitian Vanilla. It is found in Hawaii. It is fleshy, herbaceous, perennial vine.

Vanilla fragrans is the most popular and commercially cultivated species. It is a climber grown as wild in the forests, aerial roots are seen in nods by means of which the plant clings and climbs. Leaves are opposite, sessile, oblong and 10-23 cm long.

Racemose inflorescences emerge from leaf axils about 8 cm long, generally unbranched bearing 20-25 flowers. It takes 50-60 days from bud initiation to the appearance of fully developed flowers. Synchronous flowering is lacking. Each flower remains open for a single day. Blooming period continues for 14-30 days in an inflorescence. In a vine 7-15 inflorescences are produced. Flowers are large pale green, bisexual and zygomorphic. Sepals and petals look alike and are referred to as perianth. The lower petal is short, broad and is modified into a labellum which envelopes a central structure called the column (gynostemium). The column bears a single stamen with two pollen masses (pollinia) covered by a hood. Below it is the stigma covered by a flap like rostellum, which prevents natural pollination. Ovary is slender and 4-5 cm long. Flowers bloom from the base upward. Pollen fertility is 72- 87%. Flowers are pollinated through melapone bees and humming birds. Over 70% of flowers develop into fruits. Fruits are capsules or pods and commercially known as beans. They are pendulous, cylindrical, three angled and up to 21cm long. Fruits mature in 10-12 months. Seeds are black in colour.

Cultivation: Vanilla prefers humid moist tropical climate having temperature range of 25-32 °C. It grows well at an elevation of 700 – 1500 m with an annual rain fall of 2500 mm well distributed for a period of 9 months and a dry period for 3 months. It is adaptable to a wide range of soil types provided there is plenty of organic matter and proper drainage. It prefers a pH range of 6-6.5. Vanilla is generally propagated by stem cuttings. Vines of 60-120 cm long are selected as planting material. The vines are coiled and buried inside the soil. Plants raised from lengthy cuttings commence early flowering whereas the plants raised from short cuttings take three to four years for bearing. Therefore, cutting with less than 5-6 internodes and 60 cm length may not be used for planting. As the availability of planting materials is a limiting factor, recently tissue cultured planting materials are made available by some private companies and spices board. Vanilla needs support for growing. It flourishes well in partial shade. Vines may be trained on trellises or trees having low branching with a rough and small leaves. Trees like *Jatropha, Plumeria alba, Casuarina equisetifolia, Erythrina, Glyricida, Bauhinia* or Silver oak are now used for this purpose. The supports are planted at a spacing 2.5 to 3.0 m between rows and 2m within the row making a population of 1600 to 2000 per hectare. If limb cuttings are used for planting, they should have roughly 4-6 cm diameter and about 1.5 to 2 m length. The supporting saplings may be established 6 months before planting vanilla cutting. Vanilla is generally planted in a slight wetting weather. While planting the cuttings, 3-4 basal leaves in the cutting should be removed and this defoliated portion is laid on the loose soil and covered with a thin layer of about 2-3 cm soil. The growing end is gently tied to the support for climbing by aerial roots. Artificial shade with any suitable material may be provided to the cuttings. It makes 4-8 weeks for the cutting to strike roots and to show initial signs of growth. Any operation done in the plantation should not disturb the roots, which are mainly confined to the mulch and surface layer of the soil. When the support is grown up they are pruned early to introduce branching so as to give more shade and protection to the growing vines. If the trees are evergreen types they are to be pruned before the commencement of heavy rain to allow in more sunlight. The pruned vegetation is chopped and applied as mulch in the plantation. The decomposed mulch is the main source nutrients to vanilla. Animal sources of manure are not generally applied. Annually, vine may be fed with 40 to 60 g

N, 20 to 30 g P_2O_5 and 60-100g K_2O. The above quantity may be given in two or three splits for efficient uptake. Part of the above fertilizers may also give through foliar spray since they respond well to it. If the vine is permitted to grow up on tree, it will rarely blossom, so long as it is growing upward. Hence the vines are allowed to grow up to 1.50 m and then tied horizontally on the branch of support and later coiled round them. This induces more flower production in this portion of the vine. Anthracnose (*Calospora vanillae*) is the most serious disease. It attacks almost all parts. Root rot, *Fusarium betatis* var. *vanillae* is a limiting factor in certain areas. They can be controlled by spraying suitable fungicides. The bug, *Trioza litseae* is the major pest attacking the buds and flowers of vanilla, which is controlled by any systemic insecticide. The vines commence flowering in the second or third year depending on the length of cutting used. Due to the peculiar structure of the flower described earlier artificial pollination by hand is the rule of fruit setting. The procedure involved is simple. Using a pointed bamboo splinter or pin anther is pressed against stigma with the help of thumb and thus smearing the pollen over it. Generally, 85-100% success is obtained by hand pollination. The stigma is highly receptive during morning hours. The ideal time for pollination is 6 am to 1 pm. Unfertilized flowers fall off within two or three days. Normally 5-6 flowers per inflorescence and a total of not more than 10-12 inflorescence per vine are pollinated. The excess flower buds are nipped off to permit the development of good pods. Pods take six weeks to attain full size from fertilization and 4 to 10 months to reach full maturity depending upon the locations. When immature, the bean is dark green in colour, but when ripe yellowing commences from its distal end. This is the optimum time for harvesting the bean. If left on the vine the bean turns yellow on the remaining portion and start splitting, giving out a small quantity of oil, reddish brown in colour, called *the balsam of vanilla*. Eventually they become dry, brittle and finally become scentless. The yield of vanilla varies depending upon the age of vines and the method of cultivation. Normally it starts yielding from the third year and the yield goes on increasing till the seventh or eighth years. Thereafter, it slowly declines till the vines are replanted after 14 to 15 years. Under reasonable level of management a middle aged plantation may yield 300-400 kg cured beans per hectare.

Importance and uses: The cured vanilla beans contain vanillin, organic acids, fixed oil, wax, gum, resins, tannins, pigments, sugar, cellulose and minerals. Vanillin content of cured seed is about 2.41%. The cured beans contain very little steam volatile (essential) oil. The fat content is 4.5 – 15%. The major components of fatty acids have been identified as oleic and palmitic acids. Sugar content is 7 – 20%, glucose and fructose being the major components. The essential oil comprises of aromatic carbonyls, alcohols, acids and esters. The vanillin content of a conventional, single fold vanillin extract, containing the soluble extractives of 1 part of beans by weight to 10 parts of solvent by volume is usually around 0.2%. Vanilla beans and derivatives are important in food flavouring, especially, confectionery, ice-creams, liquors and baked goods. Vanilla flavoured ice-creams, custards, milk shakes, cakes, puddings, chocolates, beverages and different confectioneries are very popular in the market. Extracts and tinctures are prepared from beans. The usual concentration of vanilla extract in various items are: non-alcoholic beverages 200 ppm, ice-cream 3000 ppm, candy 4000 ppm, baked goods 1900 ppm, icings 2000-4800 ppm, syrups 10-50 ppm and toppings 2700 ppm. Vanilla essence made from extracts of the pod or vanilla sugar (castor sugar placed in a closed jar together with a vanilla pod thereby absorbing its aroma) is used for flavouring. Vanilla is also used in perfumery and to a smaller extent in medicine. Nowadays, synthetic vanillin, which is cheaper and more convenient, is often used instead of vanilla. The aroma of genuine vanilla, however, cannot be matched because it is the result of the natural balance of vanillin and small quantities of other aromatic components contained in the pods.

The various types of vanilla products employed in flavouring applications are:

a) **Vanilla extract:** This is a hydro alcoholic solution containing the extracted aroma and flavour. Principles of vanilla beans and may also contains added sweetening/thickening agents like sugar and glycerin. Vanilla extract may be prepared by direct extraction (maceration) or by dilution of concentrated vanilla extract or vanilla flavouring or vanilla oleoresin.

b) **Vanilla flavouring:** This is similar to vanilla extract but contains less than 35% ethyl alcohol by volume.

c) **Vanilla tincture:** This is prepared by maceration from one part vanilla beans by weight to 10 (ten) parts of aqueous alcohol by volume. It contains added sugar. It differs from other two products mentioned above, in respect of its alcohol percentage, which is around 35%. It is mainly used for pharmaceutical purposes.

d) **Vanilla Oleoresin:** This is a semi solid concentrate obtained by complete removal of its solvent from a vanilla extract. Aqueous iso-propanol is frequently used instead of ethyl alcohol. Owing to unavoidable evaporation loses during solvent stripping step, this product is inferior in aroma.

e) **Vanilla powder:** This is a pure powdered vanilla but is more usually a mixture of vanilla powder or vanilla Ole with sugar, food starch or gum acacia. The vanilla beans are grounded to give powder. Besides these, for perfumery applications a number of advent extracted products are employed, among which vanilla perfumery tincture may contain 90% alcohol. However due to high cost and known availability of natural vanillin most of the consuming countries including India now depends on low priced synthetic vanillin of very poor quality for making vanilla based products.

Scientific name	:	*Santalum album*
Family	:	Santalaceae
Common name	:	Sandal Wood

Description: Sandalwood is a small evergreen tree growing to 18 m in height and 2.4 m in girth, with slender drooping branches. The sapwood is white and odourless while the heartwood is yellowish brown and strongly scented. Leaves are elliptic, lanceolate glabrous and petiolate; inflorescence terminal or axillary, paniculate cyme; flowers are bisexual, many, brownish purple; perianth campanulate; stamens 4, exerted, alternating with 4 rounded obtuse scales; fruit is drupe, globose, 1.25 cm in diameter, purple black, endocarp hard and ribbed. Sandal tree is a plant parasite and its roots thrive on many types of host plants such as *Cassia siamea, Pongamia pinnata, Lantana acuminata, Cajanus cajan,* etc.

Cultivation: Warm tropical climate is best suited to sandalwood tree. It grows well between altitudes of 600 m and 1350 m above MSL though it may grow between 360 m and 1850 m altitude. Annual rainfall of 600-1600 mm is ideal for its growth. More than 1800 mm of rain is not very conducive to its growth. It performs well on laterite soils on the slopes of hills exposed to the sun. The plant is propagated through seeds. Barring a few tissue culture attempts, vegetative propagation has not been very successful. Seeds are obtained from plants over 20 years old. Fresh seeds are obtained from October fruiting are depulped, dried and sown on seed beds. Gibberellic acid is used to bring down the dormancy period and to induce quick and uniform germination. After germination, seeds are put in polybags of size 15 cm x 25 cm. A host plant is sown in the polybag when the seedling reaches 15 cm in height. The optimum stage for planting is when the seedlings are 25-50 cm high and the basal portion becomes darker. Pits of 30-50 cm cube are dug and the sandal seedlings along with the host seedlings are planted from May to October at 2.5-4.0 m spacing. Weeds are removed as and when necessary. In case the sandalwood seedlings are overtopped by the host plant the host is lopped to provide sufficient light to the seedlings. The heartwood formation is at its peak when the trees are 30-60 years and the trees attain a girth of 40-60 cm. Sandal spike disease is reported to be caused by mycoplasma-like organisms which causes severe reduction in leaf size and shortens internodes. As the roots are richest in oil, sandalwood tree is harvested by uprooting and not by cutting to avoid the loss of root system.

Importance and uses: The major constituents of sandalwood oil are reported to be α-santalol 60%, ß-santalol 30%, and ß santalene, and ß curcumene, ß-farnesene, santene, santenol, santenone, teresantalol, teresantalic acid, santalic acid, nor-tricyclockasantalol, borneol and isovaleraldehyde. The term sandalwood, in the world market, is frequently used for a variety of woods that yield oils similar in smell to that of the *East Indian Sandalwood* which is the *true sandalwood*. The world famous *East Indian Sandalwood Oil* is extracted from the strongly scented heartwood of this tree. Sandalwood oil is used primarily in perfumery because of its outstanding fixative properties. It is used in preparing all types of perfume compositions especially Indian attars like Henna, Gulab, Kewda and Jasmine in which the natural essential oils from distillate of floral distillation is absorbed in sandalwood oil. With neem oil, it is used as contraceptive. It is used for healing wounds and blisters caused by the smallpox vaccination. Sandalwood is also one of the finest woods for carving. The wood is smooth with uniform fibres. Saw dust from heartwood is mostly used in incense for scenting cloths and cupboards.

Scientific name	:	*Eucalyptus spp.*
Family	:	Myrtaceae
Common name	:	Eucalyptus

Description: The genus *Eucalyptus*, belonging to the family Myrtaceae, comprises more than 700 species. Some of them possesses medicinal value or volatile oils having fragrances varying from camphor, thymol, peppermint to rose and lemongrass. *Eucalyptus citriodora* Hook. and *E. globulus* Labill. are the most common species. *E. citriodora* commonly known as *citron (lemon) scented gum or spotted gum* is a tall, graceful tree, 25-40 m tall, with a crown of leaves and branches at the top. Bark smooth, white to pinkish. distinct cotyledonary, seedling, juvenile, intermediate and adult leaves are observed at various stages of the plant. Adult leaves are alternate, lanceolate; inflorescence is usually axillary, corymbose panicle, umbels 3-5 flowered on terete 5-7 mm long peduncles; flowers are pedicellate; calyx tube hemispherical or cylindrical; anthers ate adnate; fruits are ovoid, truncate, contracted at the edge, rim thin and valves opening downwards. *E. globulus* grows to a height of 55 m. Bark blue-grey; inflorescence axillary, usually solitary, but occasionally in 3-flowered umbels on a very short or rudimentary peduncle; flowers sessile, operculum, flattened, hemispherical; fruit sessile, globular to broadly conical, 4-ribbed and warty.

Cultivation: Eucalyptus plants prefer tropical and subtropical climate but grow satisfactorily in temperate regions as well. They are susceptible to frost in early stages. *E. citriodora* grows in plains upto 600 m while *E. globulus* prefers an elevation of 1300-2700 m. The former grows well on poor gravelly soils in rainfed areas but can be grown on any soil whereas the latter is chiefly grown on moisture rich loamy soils. The plant is propagated by seeds only. Seeds are sown directly in the field or seedlings can be raised in the nursery during February-September. Germination takes upto a fortnight depending upon atmospheric humidity and temperature. Cold stored seeds germinate in 2-4 days. Seedlings are maintained in the nursery for 2-3 months until they attain a height of 20-30 cm when they are transplanted to the mainfield at 70-100 cm spacing. 10-12 tonnes/ha of organic manure and 120:60:60 kg/ha of N, P_2O_5 and K_2O are applied. Weeding is required during the initial period until the plants are established well. Seedlings, leaves and stems of *E. citriodora* are severely damaged by blight caused by *Cylindrocladium scoparium* Morgan. Drying and defoliation are caused by *Physalospora latitans* Sace. *Fusarium orthoceros* attacks the seedlings at the hypocotyl level. These fungal diseases can be controlled by spraying and drenching with 1% Bordeaux mixture or copper oxichloride. Harvesting is done when the weather is clear and oil content in leaves is maximum. Accordingly, harvesting leaves during February, April, July and October has been recommended. In Kerala, harvesting twice a year, in May and November, is practised. Plants can be pollarded to promote vigorous sprouting of side shoots. Fresh shoots sprout in about four weeks after pollarding, which are ready for harvesting after 4-5 months. First pruning is done at 30-45 cm above ground and the subsequent ones at 75-90 cm above ground.

Importance and uses: The major constituents of the oil are cineole (60-70%), caryophyllene, camphene, sabinene, myrcene, p-menthane, ß and -terpinene, fenchone, and ß-thujone, citral, verbenone, iso-amyl alcohol, trans-pinocarveol, borneol, myrtenol, eudesmol, thymol, bornyl acetate, caproic acid, piperitone and globulol. Eucalyptus is an essential oil yielding tree which has perfumery, industrial and medicinal uses. The essential oil is utilized in soap and cosmetic industries. It is an effective substitute for Java citronella oil and a source of citronellal for the manufacture of citronellol, hydroxy citronellal and menthol. The oil is used in germicides and disinfectants to improve the odour. It is used as an antiseptic especially in the treatment of infections of the upper respiratory tract and in certain types of skin diseases. It is used as a stimulating expectorant in chronic bronchitis and as a vermifuge against hookworm. It is internally administered or inhaled with steam for asthma and respiratory disorders. Eucalyptus wood is used as mine props, railway sleepers, for paper manufacture and in house construction.

Scientific name : *Rosa damascena*
Family : Rosaceae
Common name : Damask Rose

Description: Roses belong to the family Rosaceae and the genus *Rosa*. Out of about 120 species of roses, three species are commercially used for the production of rose oil, namely, *Rosa damascena* Mill., *R. gallica* L. and *R. centifolia* L. Out of these, *R. damascena* is the most important species with a delicate fragrance and considerably larger percentage of oil. Most of the high grade rose oil is obtained from this species. *Rosa damascena* is a perennial erect or climbing shrub with a life span ranging upto 50 years. It grows upto 3 m, stems are spiny, leaves are pinnate, leaflets are serrate, stipules are adnate to the petiole. Flowers are terminal, solitary or corymbose, pine red; bracts rarely persistent; calyx tube persistent, globose, ovoid or pitcher shaped, mouth contracted, lobes leafy, imbricate in bud, petals many, large, stamens many, inserted on the disc; carpels many, rarely few, in the bottom of the calyx tube; styles subterminal, free or connate above, stigma thickened, ovule 1, pendulous. Achenes are cariaceous or bony, enclosed in the fleshy calyx tube.

Cultivation: Roses grow in almost all climates. The plant requires plenty of sunshine and protection from strong winds for proper growth. Humidity above 60% and a temperature of 15-20°C is congenial for good harvest. A temperature of 0-5° C for a fortnight prior to blooming enhances the quality and quantity of flowers. At the time of flowering the temperature should be 25-30°C and the relative humidity 60%. The plant grows on a wide range of soils, but light and well drained soils are considered ideal. Acidic soils inhibit growth and reduce flower yield. Alkaline soils with pH range of 7.0-9.0 are quite suitable. Rose is propagated vegetatively by cutting and budding. Cuttings taken from mature plants in January are planted at a spacing of 10 x 30 cm in a nursery or in poly bags. Treatment of cut ends with 200 ppm IBA induces profuse rooting. The rooted cuttings are ready for transplanting when they are about 9-12 months old. The rooted cuttings are planted in trenches, about 1 m deep and 0.5 m wide, spaced 100 cm apart. The trenches are filled with well rotten FYM at 8-10 tonnes/ha. About 10,000 plants are required for planting one hectare of land. Rose is a soil exhausting crop. A good crop requires 200 kg N as calcium ammonium nitrate, 50 kg P_2O_5 as superphosphate and 30 kg K_2O as muriate of potash per hectare in 2-3 split doses annually. Rose plants need frequent irrigation during the period of vegetative growth, flowering and just after pruning. In a year, 10-12 irrigations may be required. Plants are regularly pruned, once or twice a year, for getting higher yield of flowers. October-February is the best time for pruning. Plants should be pruned upto a height of 50 cm from the ground level. It takes 75-90 days for flowering after pruning. Weeding and hoeing should be done after pruning. Generally, 3 weedings and hoeings are required annually. Annual weeds can be controlled by herbicides like simazine or atrazine applied at 3 kg/ha in light soils and 5 kg/ha in medium or heavy soils. Insect pests of roses are *Macrosiphum rosae* (Aphidae), *Eulecanium caryli* (Coccidae) *Agrilus chrysoderes* (Buprestidae), *Podophylla fulho* (Scarabaedae) and *Rhynchites hungaricus* (Curculionidae) which can be controlled by suitable insecticides. Leaf spot is the main fungal disease which can be controlled by regular application of 1% Bordeaux mixture. Rose plants flower during March-April in the plains and May-June on hills. The peak flowering period is about 45 days. Sporadic flowering may continue throughout the year. Flowers are harvested from 5 a. m. to 9 a. m. in the early morning, when they begin to open. The average yield of flowers is 2000-3000 kg/ha/year.

Importance and uses: Rose oil is a colourless liquid, but on aging develops an amber colour. It has specific gravity (30 C) 0.8845, refractive index (30 C) 1.4657, optical rotation 2.5, Acid value 2-6, ester value 18.7 and total alcohol content 81.6%. The essential oil contains various alcohols, aldehydes, ketones, esters, phenols, terpenes and acids. The major components are citronellol 38%, paraffins 16%, geraniol 14%, nerol 7%, ß-phenyl ethanol 3%, eugenol methyl ester 3%, linalool 2%, ethanol 2% and farnesol 1%. Rose is a perennial erect shrub produces beautiful sweet scented flowers valued for worship, making garlands and preparation of rose oil, rose water, rose attar and rose otto. Rose oil is one of the oldest and most valuable perfumery raw materials. It imparts characteristic fragrant top notes to perfumes. The extracted absolute adds lasting notes. A mixture of distilled oil and extracted absolute combines the advantages of both the products. The distilled oil is employed in cases where solubility in dilute alcohol is important. The absolute is soluble only in high-proof alcohol and therefore used only in handkerchief perfumes or in cosmetics where solubility plays no role. In creams and powders, cheap concrete is used. Bulgarian rose oil is used for flavouring certain types of tobacco, particularly snuff and chewing tobacco and in a number of fruit flavours. Limited quantities of otto are employed in flavouring soft drinks and alcoholic liquors. Rose water has been valued from ancient times for use in making syrups and medicinal preparations. At marriages and other social functions rose water is sprinkled on the guests. Rose jam of Unani medicine is used as a mild laxative and tonic.

Scientific name	:	*Jasminum grandiflorum*
Family	:	Oleaceae
Common name	:	Jasmine

Description: Jasmines belong to the family Oleaceae and the genus *Jasminum*. Though more than 2000 species are known, three species, viz, *J. grandiflorum* L., *J. sambac* Ait. and *J. auriculatum* Vahl. are commercially cultivated and only *J. grandiflorum* L. is grown for use in perfume. *J. grandiflorum* is a twining or nearly erect growing shrub; branches are ribbed, drooping, annular; leaves are opposite, imparipinnately compound, rachis flattened or winged; leaflets, 5-7, elliptic, round or oval; flowers are borne on lax, axillary or terminal cymes longer than leaves, white, often tinged purple on the outside, fragrant; bracts ovate to spathulate, oblong, foliaceous; calyx glabrous, 5-lobbed, subulate; corolla 5-lobbed, star shaped, elliptic or obovate; corolla tube encloses 2 stamens borne on short slender filaments; ovary bicarpellary; fruit is berry. *J. sambac* Ait. is a dwarf spreading bushy shrub, 0.5-1.0 m high with attractive glabrous leaves producing attractive, white, sweet scented flowers in great profusion in the hot season. *J. auriculatum* Vahl. is a twining scandent shrub growing to 5-7 m with small opposite leaves; flowers are white and sweet smelling with calyx fine notched having round firm glandular process on the outside, light weight; around 26,000 flowers weighing a kilogram.

Cultivation: Jasmines are sun loving plants and prefer warm humid climate for successful growth. They perform well at elevations ranging from 600 to 1200 m. Areas with a warm summer, bright sunshine and mild winter almost throughout the year are considered the best. Jasmines can be grown on a wide range of soils, but well drained rich sandy loam to clay loam soils with a soil pH of 6.5-7.5 are preferred best. Layers and cuttings can be used for propagation. Improved varieties of *J. grandiflorum* ('Jaji mallige' and 'Ajjige') and *J. auriculatum* ('Vasantha mallige' 'Parimullai' and 'CO -1 mullai') are used for commercial cultivation. Cuttings for planting should be 20-25 cm long with 3-4 eyes and are dipped in seradix-B (or a solution of 4000 ppm of IBA) and planted under intermittent mist for rooting during January to March. Almost 90% rooting is achieved and the cuttings will be ready for transplanting in 4-5 months. Cuttings, are planted in pits dug upto 1 m in depth, and filled with top soil, cowdung and compost. Plants are spaced at 1 m in rows, 1.5 m apart. Irrigation is given if soil moisture is inadequate. Jasmine plant needs 15-30 kg FYM, 60-120 g N, 120-240 g P_2O_5 and 120-240 g K_2O/plant/year which are given in 3-4 split doses. Pruning is needed to get high flower yield and to keep the bushes to manageable size. Plants pruned between 17th December and 7th June produce maximum number of branches and the highest yield of flowers. The shoots are cut at 30 cm height. Diseased leaves and dry shoots are also removed. After pruning, the soil around the plant should be stirred upto a depth of 15 cm and repeated every 2-3 months. The field should be kept clean by removing weeds as and when necessary. Red scale, mealy bug, jasmine bug, leaf eating caterpillar, white flies and mites are the common insect pests of jasmine. Red mite can be controlled by spraying trithion while others by any common insecticide. Black leaf spot, mildew and rootknot (nematode) are the diseases of this crop. The plant flowers from the second year of planting. The flowering period ranges between April and May and from August to November. In Egypt, the plant flowers almost 10 months in a year. Harvesting is done during early morning because the flowers contain maximum perfume at this time. Flowers gathered at noon and in the afternoon yield lesser flower oil than those collected very early in the day. Warm weather and ample sunshine yield a crop of heavily scented flowers than in cool or rainy weather. The annual yields of flowers have been reported to be 750-1000 kg/ha in India, 2000-4000 kg in France, 4500-5500 kg in Sicily and Italy and upto 6000 kg/ha in Morocco and Egypt. A jasmine plantation gives economic yield for 10-15 years after which the crop is removed and crop rotation is followed for some years before establishing a new jasmine plantation.

Importance and uses: Jasmine concrete is a yellowish brown waxy mass with the characteristic odour of jasmine flowers. It has melting point 50-51°C, congealing point 54-55°C, acid value 0.23-0.27 and it is partly soluble in 95% alcohol. The approximate composition of jasmine flower oil obtained by enfleurage is benzyl acetate 65.0%, d-linalool 15.5%, linalyl acetate 7.5%, benzyl alcohol 6.0%, jasmone 3.0%, indole 2.5% and methyl anthranilate 0.5%. Benzyl benzoate, geraniol, nerol, terpineol, farnesol, nerolidol and p-cresol are also present in traces. Jasmine absolute is a viscous clear yellowish brown liquid possessing a delicate odour of fresh jasmine flowers. The absolute contains many of the above compounds. Jasmines are commercially grown for their fragrant flowers and essential oil production. The bulk of the flowers is used as such in garlands and decorative bunches for religious offerings and a small quantity for the production of oils and attars. Jasmine concrete and absolute are used in high grade perfumes, ranking next to the rose in the order of importance. Jasmine oil blends with every floral scent and extensively used as an important perfumery item throughout the world. Almost all high quality perfumes contain at least a small amount of jasmine oil.

The absolute, though expensive, also blends with any floral scent imparting smoothness and elegance to the perfume composition.

Scientific name : *Polyanthes tuberosa*

Family : *Amaryllidaceae*

Common name : Tuberose

Description: Tuberose *Polyanthes tuberosa* L. belongs to the family Amaryllidaceae. There are single as well as double flowered varieties. Single flowered type is mostly cultivated for the extraction of its perfume while the double flowered variety usually goes to the cut flower trade. The flowers on top of the long stalk are grouped in spike-shaped clusters 15 to 20 cm long. The flowering period begins in July reaching its maximum toward the middle of August and lasting to the end of September when a secondary blooming takes place.

There are four groups of cultivars of tuberose.

(i) Single: Most widely cultivated variety. Flower is pure white and has got a single row of corolla segments. Eg. 'Calcutta single', 'Mexican Single', 'Rajat Rekha' and 'Suvarna Rekha'.

(ii) Double : Flowers are white, tinged with pinkish red. Petals are in several whorls. Eg. 'Pearl' and 'Calcutta double'.

(iii) Semi-double: Similar to double but with only 2-3 rows of corolla segments.

(iv) Variegated: This has got variegated leaves with yellow margins.

Cultivation: Although tuberose can be grown under a wide range of climatic conditions, a mild climate with an average temperature ranging from 20°C to 30°C is considered ideal. Loam and sandy loam soils having a pH range of 6.5 to 7.5 with good aeration and drainage are best suited for its cultivation. The plant is propagated by bulbs. Spindle shaped disease free bulbs having a diameter of 1.5-3.0 cm are used for planting. Mother bulbs are the best for planting as they flower early. Finger or side bulbs take 2-3 years to come to flowering. Of the 4 types of tuberose, viz, single, double, semi-double and variegated, the single type has the maximum fragrance and is popular among the growers for the production of essential oil. Best time of planting is from April to June. Land is ploughed 2-3 times and soil is brought to fine tilth. Well-rotten FYM at 20-30 t/ha is applied and mixed well. Furrows are opened 25-30 cm apart and bulbs are planted at 25 cm spacing in furrows. About 1.25 lakhs (800-900 kg) of bulbs are required for planting a hectare of land. A fertilizer dose of 100:200:200 kg N, P_2O_5, K_2O/ha is generally recommended. Half the dose is applied basally and the other half as topdressing when the flower spikes start appearing. Weekly irrigation and regular weeding are required for best yield. Thrips are reported to attack the crop. The flowering season is between June and October. Flowers will be ready for harvest in 3-3.5 months time. They are harvested by cutting the fully opened spikes from the base during the cool hours of the day either in the morning or evening. From single planting, 2-3 ratoons can be taken for which the flower stalks of the main crop are headed back and the plot is manured and irrigated. The average yield comes to 5-10 tonnes/ha for planted crop, 9-12 tonnes/ha for first ratoon and 4-6 tonnes/ha each for subsequent ratoons.

Importance and uses: The absolute of enfleurage is a brown, semi-solid, alcohol soluble liquid possessing a characteristic odour of tuberose flowers with a fatty by-note. Concrete of tuberose is a light to dark brown waxy mass which is only partially soluble in high-proof alcohol. The chemical constituents of tuberose flower oil include geraniol, nerol, farnesol, benzyl alcohol, methyl benzoate, benzyl benzoate, methyl salicylate, methyl anthranilate, eugenol and butyric acid. Among the flowering plants which are valued much by the aesthetic world for beauty and fragrance of their flowers, tuberose occupies a very special position because of its prettiness, elegance and sweet pleasant fragrance. This bulbous plant is the source of tuberose oil of commerce which is very expensive and used in high grade perfumery. It is also cultivated for cut flowers and for preparing bouquets and garlands. The long flower spikes are excellent as cut flowers for table decoration. The individual florets are used for making garlands, floral ornaments, bouquets and button-holes.

Scientific name	:	*Tagetes spp.*
Family	:	Asteraceae
Common name	:	Marigold

Description: Marigolds belong to the family Asteraceae (Compositae) and the genus *Tagetes*. Five species have been recognized in *Tagetes*.

(i) *Tagetes erecta* L., African or Aztec Marigold: They are erect and tall growing (upto 90 cm) plants having large globular flowers of diameter 15 cm and above, in shades of lemon yellow, bright yellow, golden yellow, orange and near white. The common cultivated varieties are 'Guinea Gold', 'Apricot', 'Primrose', 'Sun Giant', 'Fiesta', 'Golden yellow', 'Glitters', 'Happiness', 'Hawai', 'Crown of Gold', 'Honeycomb' and 'Cupid'.

(ii) *Tagetes patula* L., French Marigold: These are bushy plants, 30-45 cm tall which flower profusely in singles or doubles. The flowers are in the colour range of deep scarlet, red, primrose, yellow, golden yellow, orange and their combinations. Important cultivars are 'Rusty Red', 'Flame', 'Spry', 'Naughty', 'Marietta', 'Spanish Brocade', 'Star of India' and 'Harmony'.

(iii) *Tagetes lucida* Cav., Sweet Scented Marigold: This is a perennial plant 30-40 cm tall with oblong-lanceolate, serrulate leaves. The flower heads are orange yellow and are borne in dense terminal clusters. They are often planted in gardens as a border plant.

(iv) *Tagetes minuta* L. (*T. glandulifera* Schrank.), Stinking Roger: It is a highly aromatic annual 1-2 m tall, leaves 7-15 cm long, pinnatisect, leaflets 11-19. 4 cm long linear or lanceolate, flower heads pale yellow in corymbose clusters and black achenes.

(v) *Tagetes tenuifolia* Cav. (*T. signata* Bartl.), Stripped Marigold: It is a branching annual, 30-60 cm high with pinnately compound serrate leaves and sweet smelling, bright yellow flowers.

Interspecific hybrids between African and French marigolds are also developed in USA. They are 50-60 cm tall with double flowers of 5-7 cm. Eg: 'Nugget', 'Show Boat', 'Red Seven Star', 'Red' and 'Gold Hybrid'.

Cultivation: African and French marigolds are more widely cultivated as compared to other species. Marigolds, in general, grow well a mild climate between elevations of 700-1500 m. They come up well on well drained rich loam or sandy loam soils. They are propagated by seeds and cuttings; the former is preferred for establishing tall, vigorous and heavy yielding plantations. The land is cultivated well incorporating 20-30 tonnes/ha of well decomposed FYM or compost. Seeds are sown during May-June on nursery beds. One month old seedlings are transplanted at 30-45 cm spacing. A fertilizer dose of 400 kg N, 200 kg P_2O_5 and 200 kg K_2O/ha is suggested for higher yield. Regular irrigation, weeding and hoeing are required to obtain more of large flowers. Initial flower buds are disbudded to obtain bushy and compact growth. Flower heads are harvested when they have attained full size. Regular plucking of flowers increases the flower production. The flower yield is 8-12 tonnes/ha for French marigold and 11-18 t/ha for African marigold.

Importance and uses: African marigold flower oil is reddish yellow in colour, possessing characteristic marigold odour and polymerizing readily in air. The oil contains approximately ocimene 8.5%, limonene 14.03%, linalool 21.14%, linalyl acetate 13.75% and tagetone 40.38%. Leaf and stem oils are greenish yellow in colour. French marigold oil is amber coloured, having a characteristic powerful fruity top note, reminiscent of green apples, with specific gravity (31°C) 0.8917, refractive index 1.492, optical rotation -3.4 , acid value 6.1 and ester value 25. Tagetone 40.4%, linalool 22.1%, limonene 14.0%, linalyl acetate 13.8% and ocimene 8.5% are the major constituents. Oil of *T. lucida* is greenish yellow with specific gravity (15°C) 1.5218, acid value 6.0 and ester value 22. The major chemical constituent is estragol. Marigolds gained popularity among florists on account of its easy culture, and wide adaptability. Its habit of free flowering, short duration to produce marketable flowers, wide spectrum of attractive colour, shape, size and good keeping quality attracted special attention. There are African, French, Sweet scented, Stinking Roger and Stripped marigolds. African marigold flower oil is a fly repellent and has been suggested as a modifier in hair lotions of the bay-sum type. French type finds use in perfumery particularly in certain types of herbaceous fragrances like fougere and lavender and also in florals, such as jasmine, gardenia and violet. This oil is also employed in aldehyde oriented tabac bases. Oil of stinking roger finds applications in germicidal and micro-biocidal preparations due to the presence of tagetone which is toxic. Planting of marigold is highly effective in controlling nematode population in the soil.

Scientific name : *Michelia champaca*

Family : *Magnoliaceae*

Common name : Champak

Description: The genus *Michelia* belonging to the family Magnoliaceae comprises about 50 species of evergreen trees or shrubs. The most popular among them are *Michelia champaca* L. and *M. figo* L. The evergreen tree, *Michelia champaca* grows upto a height of 8 m with ovate to lanceolate leaves of size 20 x 6.5 cm. Blooms appear during April-May and again during September-October. Buds are silky. Flowers are white, sovereign red or creamy light yellow in colour. 'Simhachalam' golden orange kind is the most sweet scented and is the most favoured of champas. The white flowered champa, though very sweet scented, lacks in substance and hence the fragrance does not last quite long. *Michelia figo* is a bushy shrub growing upto 5 m with narrow, oval, dark green, smooth shining leaves. The flowers are brownish yellow with edges of carmine colour, about 4 cm in diameter, highly fragrant and borne in axillary and solitary manner.

Cultivation: Champa requires a mild climate and an elevation of 100-1000 m with partial shade for good growth. It can be grown on a wide variety of soils and well drained rich sandy loam soils are the best. The trees are propagated both by seeds and vegetatively by grafting. The creamy yellow variety is propagated by grafting on stocks of ordinary golden orange flowered variety which produces seeds in bunches and takes 7-8 years to flower. Seedlings of 'Simhachalam' golden orange type bear flowers in about 3 years and produce taller trees than grafts which bear fruits in a short period. Though large scale commercial cultivation of champa is not very common, group planting is generally undertaken, particularly in informal gardens and homesteads. Care is to be taken till the grafts or seedlings are initially established in the field and thereafter not much attention is needed. The trees flower during April-May and again during September-October once they start blooming. A well grown tree yields 50-100 flowers daily during the peak season and 375 to 425 flowers weigh one kilogram.

Importance and uses: The champa flower concrete is a waxy solid having a low melting point of 29°C. The champa absolute has a very sweet smooth floral and velvety odour, closely resembling that of the live flowers. Champa oil has low saponification value and low solubility in 90% alcohol. The oil does not resinify during fractionation. It contains important perfumery constituents such as cineole, iso-eugenol, phenyl ethyl alcohol, benzaldehyde, methyl anthranilate, benzyl alcohol, p-cresol and its methyl ether. The tree is a great favourite in Hindu gardens, the exquisitely scented flowers being used for *Pooja* particularly of the Lord Krishna. Besides, the ladies are very fond of champa flowers because of its pleasant fragrance. Champa attars are produced in India which are used in hair oils as a head coolant. The flowers also yield a dye which is used as a base for other colours and for dyeing silk and cotton fabrics. By virtue of the refreshing appearance of its foliage, champa looks elegant even when it is out of flowers.

Scientific name	:	*Lavandula angustifolia*
Family	:	Lamiaceae
Common name	:	Lavender

Description: It is a strongly aromatic shrub, 1 to 2 m tall. The leaves are evergreen, 2–6 centimetres long, and 4–6 millimetres broad. The flowers are pinkish-purple (lavender-coloured), produced on spikes 2–8 cm long at the top of slender, leafless stems 10–30 cm long.

Cultivation: The plants perform well in aerated dry and calcareous soil rich in nutrients, pH 7.0-8.4; cold winter and cool summer; rainfall 550-700 mm in the form of rain or snow. It can also be grown on gentile and steep slopes, poor eroded soils. The plants are propagated by seeds during November-December. Besides, they are also propagated through rooted cutting. Seeds germinate in April at 14-15°C. Cuttings of 10 cm length are taken from 1yr old plants and firmly planted. Beds are covered with black polythene on protect from cold weather. A basal dose of 20 kg N and 40 kg P_2O_5 are needed. The field requires regular weeding and hoeing. Old lavender plantations are regenerated during winters. Harvesting of flowers is done in dry sunny days during Aug/Sept. Oil is extracted by steam distillation. The yield is 15 kg oil/ha.

Importance and uses: The flowering tops of lavender plant yield 1.2% -1.4% valuable essential oil, mainly composed of linalool and linaly acetate. The oil is used in high-grade perfumery. Lower quality oil goes for preparation of lavender water, toilet water and soaps. Lavender is used medicinally for centuries as a remedy for digestive issues, headaches, grief and stress.

Scientific name	:	*Majorana hortensis*
Family	:	Labiatae
Common name	:	'Sweet marjoram'

Description: It is an annual herb. This species is cultivated in Germany, Southern France, Tunisia, America and Hungary. It is commercially cultivated in Tamil Nadu. These plants grow in bunches or tufts and attain an ultimate height of 60 cm. Leaves are oval, small, whitish and soft, and highly aromatic. Dotted oil glands are present on the leaf. Flowers are greenish white in colour, small in size. This species is cultivated in Indian gardens particularly at hill-stations, as a savoury plant. Seeds are very small, dark brown in colour.

Cultivation: This plant is propagated from seeds. Although it is a perennial herb, but is cultivated as an annual. Ordinary rich soil and sunny situations are ideal. In plains, seeds are sown in October and in hills seed sowing is done during March to June. The crop is ready for harvesting within 100 days time. Essential oil content will be more if harvested before seed formation. The shoots when coming into flower are gathered and dried in shade. The dried leaves with or without flowering tops constitute the marjoram of commerce. The colour of the dried herb is light green with greyish tint. From one hectare of land, 16 to 18 tonnes of herb can be obtained which may yield 50 kg of essential oil.

Importance and uses: Oil is extracted from leaves by steam distillation. The dried herbs contain 1 to 1.5 per cent of volatile oil. The chemical compounds identified from the oil are terpinene, sabinene, myrcene, limonene, terpineol, trans-sabinene hydrate, cis sabinene hydrate, linalool, terpinen 4-ol, terpineol, linayl acetate, caryphyllene, bicaclogermacrene, etc. Seeds also yield essential oil. It is extensively used for flavouring tobacco, alcoholic beverages and food material. The oil is also used in soaps and ointments. The plant possesses medicinal properties and is used for the treatment of asthma, paralysis and hysteria. For seasoning of food, fresh and dried leaves are used as a condiment.

Scientific name	:	*Pandanus fascicularia*
Family	:	Pandanaceae
Common name	:	'Kewda', 'Fragrant Screwpine'

Description: It is a perennial shrub or a slender tree attaining a height of 6m, distributed in Tropical Asia. In India, it is distributed in the coastal districts of Orissa, Gujarat, Andhra Pradesh and Tamil Nadu. There are thick bushy forest tracts on the river banks and sea side in Ganjam, Balasore, Cuttack and Puri districts of Orissa. The oil of kewda is commercially extracted in Ganjam district of Orissa; Kotabameli of Tamil Nadu; Junapur, Kannauj, Ghazipur of Uttar Pradesh and Alwar and Pushkar of Rajasthan. It is a gregarious plant, stem much branched, resting on strong aerial roots. Leaves are bright green, in drooping tufts, terminating in short branches, 90 to 150 cm long, armed with short white spines. Male inflorescence exhale odour, and it is economically important part of the plant which is highly scented. Male spadix is solitary up to 5 cm in diameter enclosed in yellow spathe. Spadix takes nearly 15 days to develop to its full size. Fruit is yellow or red when ripe, it bears numerous drupes. Drupes are 5 cm long, oblong or elliptic, fruit is 15 to 25 cm in diameter. From 5 years onwards plants starts bearing flowers and continues up to 40 to 50 years. Maximum flower production is in between May and October and maximum flowers can be obtained during rainy season. In Orissa, two morphologically distinct types of Pandanus are found in nature, for example, 'Kewda' and 'Ketaki'. Flowers of 'Ketaki' are golden yellow in colour and strongly scented while 'Kewda' flowers are cream coloured and sweetly scented. Ketaki flower spikes are longer and heavier than 'Kewda' flowers.

Cultivation: Kewdas are propagated through ground suckers. These suckers are planted at 5 m distance each way during rainy season. As the age of the plant increases, flower production improves. Five-year-old plants on an average produce 4 flowers/plant, while 20 to 25-year-old plants may give more than 20 flowers/plants. Nearness to the sea and leaf colour affects the flower yield. Flower production is more if the plantation is within 2 km from the sea. The clones with dark green foliage produce more number of flowers. Warm and moist weather is ideal for flower production. Flowers are plucked by breaking the stalks during morning hours for distillation.

Importance and uses: The male flowers yield 0.01 to 0.03 per cent essential oil. The chief constituents of the oil is methyl ether of phenyl ethyl alcohol. Other chemical constituents identified are d-linalool, dependence, phenyl-ethyl acetate, citral, caproic acid, stearoptene, phenylethyl acetate and pthalic acid. Two other perfumery products obtained from this flower by distillation are kewda attar and kewda water. Essential oil of kewda as well as its attar highly valued in perfumery trade. These are most powerful of all perfumes. The attar is used for flavouring hair oils, soaps, bouquets, cosmetics, snuff and incense. Kewda scented cold drinks are very popular. Kewda water is used for flavouring soft drinks, syrups; the fragrance is soothing and cooling. This aroma is also used in pan masala (betel) and zarda (tobacco). The oil possesses a stimulant and antispasmodic properties. To impart lasting fragrance, male flowers are kept in packets within wardrobes and boxes along with garments. The oil relieves headache and are used for treatment of rheumatism. In India, kewda attar and water are used for toilet purposes. Strong fibre is obtained from long aerial roots of these plants, which are used for making ropes and baskets. Kewda leaves are used to prepare decorative goods like bags, hats and mats. The strong fragrance of these flowers invite snakes and it is reported that small cobras visit the flowers and lie hidden in the inflorescence, hence, the concerned persons must be careful in handling the flowers in the field.

Scientific name : *Eugenia caryophyllus*

Family : Myrtaceae

Common name : Clove

Description: Clove plant *Eugenia caryophllus* (Sprengel) Bullock & Harrison belongs to the family Myrtaceae. *Caryophllus aromaticus* L., *Eugenia aromatica* Kuntze, *Eugenia caryophyllata* Thumb., and *Syzygium aromaticum* (L.) Merr. & Perr. are often used synonymously. It is a small evergreen tree growing to a height of 15 m, conical in shape when young, later becoming cylindrical. Leaves are simple, opposite, exstipulate, glabrous and aromatic, petiole swollen, lamina lanceolate; inflorescence terminal corymbose, trichotomous panicle; bracts and bracteoles are narrow and acute; flowers are bisexual with a fleshy hypanthium surmounted by sepals; calyx, 4-lobbed, fleshy, triangular; petals 4, imbricate, red tinged, rounded; stamens numerous, grouped in four; ovary 2-celled, multiovulate, inferior, embedded in the top of the hypanthium; fruits are single seeded, oblong, fleshy, drupes, reddish purple in colour and are called *mother of cloves*. Flowers are visited by bees and are probably cross pollinated. Some botanists opine that the bisexual flowers and their structure ensure complete self-pollination.

Cultivation: Cloves grow best with insular, maritime climates in the tropics upto an elevation of 1000 m. Annual rainfall of 1500-3000 mm and a temperature of 25-32°C are ideal. Drier weather is desirable for harvesting and drying the produce. Well drained, deep, sandy, red or acid loams with high humus content are best suited. Waterlogging is fatal. Cloves are propagated by seeds, though vegetative propagation through layering, approach grafting and budding has been met with occasional success. Fully ripened, freshly fallen fruits are collected, soaked in water and heaped under wet sacks for fermentation for 3 days. The seeds are then hulled with fingers, washed and sown in nursery at 15-20 cm spacing and watered regularly. Fresh seeds give a germination over 90%, but the seed viability diminishes sharply in storage. 1-1.5 years old seedlings are transplanted to the mainfield at 6-7 m spacing in pits of 60-75 cm cube. Cattle manure or compost is applied at 15 kg/tree/year. A well grown tree, of 15 years or more, is applied with N, P_2O_5 and K_2O at 300:250:750 g/plant/year. Application of coconut meal, bone meal or fish meal at 2-5 kg/plant is beneficial. Irrigation is to be provided during summer months. Young plants are usually ring weeded. Mulching, partial shading and protection from heavy rains and winds are required during early stages of growth. Cloves can be intercropped in coconut, arecanut, nutmeg, coffee and banana plantations. Termites, the coccid-*Saissetia eugeniae* and shoot borer (*Sinoxylon* sp.) are the common insect-pests of clove. Leaf spot, twig blight and flower bud shedding are caused by different fungal pathogens. Die back caused by *Cryptosporella eugeniae* and sudden death by *Valsa eugeniae* are other diseases of the plant. These diseases except, sudden death can be controlled by suitable fungicides. The trees begin to flower in 6 years. Full bearing is achieved by about 20 years and the production continues for 80 years or more. Bearing between years shows much variation. Clove clusters are hand-picked when the buds reach full size and turn pink but before they open. They are spread thinly on mats and stirred frequently for uniform drying. Well dried cloves will snap cleanly with a sharp click across the thumb nail and weigh about one-third of the green weight. On an average, a clove tree yields 3.5-7.0 kg/year which depends upon the age, size and condition of the tree. Yields upto 80 kg/tree/year have also been recorded.

Importance and uses: Clove bud oil contains eugenol 80-90%, caryophellene 4-8%, -humulene 0.5%, -terpinyl acetate 0.1%, methyl eugenol 0.2%, humulene epoxide 0.2%, chavicol 0.3% and traces of several other compounds. In trade, cloves are the dried unopened flower buds of the plant which have been used in the Orient for over 2000 years as a spice, to check tooth decay and to counter halitosis. They are used as a table spice, in the preparation of curry powders, to flavour the betel quid in *panmasalas* and to season sausages and puddings. Clove buds, stems and leaves on steam distillation yield essential oils which are used in the manufacture of perfumes, soaps, in flavouring, medicine, dentistry and as a clearing agent in microscopy. The major constituent, eugenol is converted into iso-eugenol and a very high grade vanillin. In medicine, cloves are stimulative, antispasmodic and carminative. In dentistry, eugenol in combination with zinc oxide is used for temporary filling of cavities.

Scientific name	:	*Cinnamomum camphora*
Family	:	Lauraceae
Common name	:	Camphor

Description: Camphor tree *Cinnamomum camphora* (L.) Presl. belongs to the family Lauraceae. It is a large handsome evergreen tree, 8-18 m tall. Leaves are opposite or alternate, usually triple nerved, leathery, 5-7 cm long and aromatic. Leaves are shed every year during February-March, but simultaneously new leaves appear. Flowers are small, bisexual, borne on axillary, lateral or sublateral cymes or panicles; perianth tube short, funnel shaped, enlarged in fruit, lobes 6, subequal, persistent and partly truncate; perfect stamens 9 or fewer with introse 4-celled anthers; ovary is sessile, at the bottom of the perianth tube narrowed into thick style; fruit seated on the enlarged perianth tube. Camphor is formed in the oil cells, distributed in all parts of the tree. These cells begin to form early in the growth of the plant organs and are filled with a yellow oil from which camphor is slowly deposited. The oil progressively becomes colourless and volatile and irregular masses of camphor appear. The formation of camphor is brought about by an enzyme present in the growing parts of the tree, particularly in the tissue within the cambium region. Each layer of wood, as it is formed, is enriched by camphor.

Cultivation: Camphor tree grows best at an altitude of 1350-1500 m with temperature not going below 33°C. In the Nilgiris it does well upto 2100 m above MSL. It withstands an annual rainfall of 1000 mm where it can be successfully cultivated. Fertile well drained sandy loam soils are best suited for the cultivation of camphor tree. Deeply tilled clayey soils are also suitable, provided rendered porous by mixing leaf mould and sand supplemented with artificial fertilizers. It is chiefly propagated through seeds and rarely through layers, branches, cuttings, root cuttings and root suckers. Fresh ripe fruits are collected either direct from the tree or soon after they fall. Removal of the pulpy seed coat and pre-soaking of the seeds in water for 24 hours enhances seed germination. Seeds are sown in nursery at a spacing of 6-8 cm in rows 25-30 cm apart and irrigated regularly. Seeds start germinating after 3 months of sowing. The nursery is maintained weed free. 12-16 months old seedlings are transplanted in the mainfield in 60 cm cube pits, 2-3 m apart. Application of organic manures and inorganic fertilizers has proved beneficial. Plants are trimmed to a height of 1.5-2 m and maintained as bushes to facilitate picking of leaves. Leaf blight disease in camphor is caused by *Glomerella singulata* which can be controlled by spraying difolatan and benlate. Leaves and twigs are harvested every year and distilled to produce camphor oil. Wood over 50 years of age are also used for distillation. Bushes are harvested 3-4 times a year.

Importance and uses: The essential oil distilled from branches, wood and root is obtained as a semi solid mass. The yield is generally 1-1.2% . Leaf oil contains sabinene 1.47%, -phellandrene 0.17%, -terpinene 0.24%, terpinolene 0.30%, furfural 0.16%, piperitone 2.4%, geranyl acetate 0.22%, cuminaldehyde 0.15%, safrole 13.4%, eugenol 0.12%, cinnamyl alcohol 0.18% and traces of more than twenty compounds. Camphor plant is an evergreen tree valued for the presence of camphor in its wood and leaves. Camphor is chiefly a pharmaceutical product though it is used in the preparation of artificial essential oils like lavender and lavendin. It is used as an incense and in balms of various kinds. It is employed as a masking agent in perfumery and sometimes used to achieve lift in perfume blends. Redistilled brown oil is used directly in soap perfumes for a masking effect. The oil is a source of safrole which is a starting material for the production of various perfumery chemicals.

Scientific name	:	*Apium graveolens* Linn.
Family	:	Umbelliferae
Common name	:	Celery

Description: Two varieties are recognized under the species.

(1) *Apium graveolens L.* var. *dulce* (Mill.) Pers. the leaves and flowering stems of which are used as appetizer and salad at table and as a flavouring agent in soups.

(2) *A. graveolens L.* var. *rapaceum* (Mill.) Gandich or *turnip-rooted celery*. This is smaller with dark green leaves, less developed stalks and swollen roots (5-6 cm in diameter) which are eaten after cooking.

Apium graveolens is an annual in the plains producing seeds in the very first year but in colder climates and on the hills it becomes biennial and produces seeds only in the second year. It is an erect herb of 60-90 cm height. Roots are succulent and numerous. Stems are branched, angular or fistular, conspicuously jointed. Leaves are pinnate, deeply divided into three leaflets which are ovate to suborbicular and 3-lobbed; inflorescence is a compound umbel; flowers are small, white; calyx teeth obsolete; petals 5, ovate, acute with tip inflexed; carpels semiterete, filiform; fruits schizocarp with two mericarps, aromatic and slightly bitter.

Cultivation: Celery grows best in a moist, cool climate . It grows as an annual in the plains but as a biennial at higher elevations with cooler climate. It flourishes well on fertile, well drained, sandy and silt loam soils. Clayey soils are not suited. The plant is propagated through seeds. Seeds obtained from primary umbels are heavy and produce better seedlings in comparison to seeds obtained from quarternary umbels. On hills or higher elevations, seeds are sown during March-April, transplanted in May and harvested in November. In the plains, seeds are sown during September-October, transplanted in January and harvested in May. High temperature pre-treatment of seeds increases germination rate. Healthy seedlings are obtained by incubating the seeds at 90% relative humidity and 15-20°C for 8-10 days. For transplanting one hectare, 1.5 kg seeds are sown in a nursery area of about 1000 m². The main field is thoroughly prepared incorporating organic manures at 10-20 tonnes/ha according to availability and transplanting is done in moist soil at 30-40 cm spacing. Fertilizer application of 200 kg N and 40 kg P_2O_5/ha is much remunerative. The crop needs plenty of water and the field is irrigated every 10-15 days during non-rainy period. The crop may lodge when strong winds blow after irrigation which may be prevented by providing wind breaks. The field is kept weed free by 2-3 hoeings, first 3 weeks after transplanting and subsequently at 2 weeks interval. A leaf miner, *Liriomyza trifolii* Burgess and a fungus *Septoria apicola* Speg. attack the crop. They can be controlled by suitable insecticide and fungicide, respectively. The crop is harvested when the white flowers start turning reddish. The harvested crop is thrashed with sticks the next day and the seeds are taken. Average seed yield is about 1-1.5 tonnes/ha.

Importance & uses: Celery seed oil is a light pale-yellow coloured liquid with persistent characteristic odour of the plant. The composition of celery seed oil is as limonene 80.0%, -p-dimethyl styrene 0.9%, n-pentyl benzene 1.0%, caryophyllene 0.5%, -selinene 0.5%, n-butyl phthalide 1.0% and sedanolide 0.5%. The other constituents found are sabinene, ß-elemene, trans-1, 2-epoxy limone, linalool, iso valeric acid, cis and trans dihydro carvone, terpinen-4-ol, cis and trans p-menth-2, 8-dien-1-ol, -terpineol, carvone, cis and trans, carveol, trans anethole, trans-carvyl acetate, cis and trans-p-menth-1(7), 8-dien-2-ol, perillaldehyde and thymol. Celery leaf oil is richer in mono and sesquiterpenes in comparison to celery seed oil. It is used in the flavouring of all kinds of prepared foods such as soups, meats, pickles, vegetable juices and in the preservation of meat sauces. In pharmacy, the oil is used in certain preparations having sedative effect. It is highly priced for fixative purposes and as an ingredient of novel perfumes. It has a powerful odour and imparts a pleasant warm note. The oil is used in compounding ayurvedic formulations. Fruits yield 17% of a fatty oil which is used as an antispasmodic and nerve stimulant. Seeds of celery are rich in vitamin B.

❏❏❏

13

EXTRACTION TECHNOLOGIES OF ESSENTIAL OILS

Pharmacological activity in plant-based drugs is centred on the presence of chemical compounds which are called as active components. The major chemical groups are heterosides, polyphenols, terpenoids, and alkaloids.Active constituents are substances which are found in different parts and organs of plants, and which change or modify the functions of human and animal organs and systems.

Active constituents are classified in groups according to their chemical structure.

Products resulting from primary metabolism: glucides, lipids, amino acid derivations.

Products derived from secondary metabolism: **Heterosides** (Anthraquinones, cardiac glycosides, cyanogenics, coumarines, phenols, flavonics, ranunculosides, saponosides, sulphurides).**Polyphenoles** (Phenolic acids, coumarins, flavonoids, lignans, tannins, quinones **Terpenoids** (Essential oils, iridoids, lactones, diterpones, saponins,alkaloids).

Glycosides or heterosides: These are compounds which are composed of two parts: one is a sugar moiety (e.g. glucose) and the other a non-sugar moiety or aglycone. The bond between the two parts (moieties) involve a hydroxyl group, which can form an O-glycoside, (C-glycosides), (N-glycoside) or (S-glycoside). Glycosides are broken down upon hydrolyses with enzymes or acids, releasing the aglycone, the truly active constituent.

Type	Properties	Species
Anthraquinones	Purgatives	Holy peel, Sen
Cardiac glycoside	Diuretic. Heart tonic	Digitalis
Cyanogenics	Anaesthetics. Anti-spasmodic. Hypo-tenser.	Cherry tree. Almond tree.
Coumarin	Antibacterial. Anti-	Oats.

Contd....

Contd....

Type	Properties	Species
	coagulant. Protection from the sun.	
Phenol	Anti-pyretic and febrifuges.	Pear tree, willow.
Flavonoids	Weak capillaries. Vitamin C.	Sunflower, rude.
Ranunculosides	Irritation of the skin.	Berry plants.
Saponins	Haemolytic anaemia. Emolients. Dermatitis.	Fir tree, Maize. Liquorice, Saponaria, Violet.
Sulphurates	Antibiotics	Garlic, Onion, Radish.

The most important groups are the Anthraquinonics, Cyanogenics, Cardiotonics, Phenolics and Coumarinics.

Anthraquinonic Heterosides: Anthraquinones are yellow-brown pigments, most commonly occurring as O-glycosides or C-glycosides. Their aglycone portions consist of two or more phenols linked by a quinine ring (anthracene derivative). They are used as laxatives and purgatives. They are commonly found in Aloes (aloe juice, especially liliacs), Rhubarb (rhizomes from Rheum), Senna (leaves and pods of Cassia), Buckthorns (bark of the *Rhamnus catharticus*and *R.frangula*) and *Cascara sagrada* (bark of the *Rhamnus pursiana*).

Cardiac glycosides /Heterosides: This group contains a number of active constituents which act directly on the cardiac muscle, and thus exert therapeutic influence on congestive cardiac insufficiency, or on cardiac arrhythmia. They are found in *Digitalis purpurea*, *Digitalis lanata*, *Adonis vernalis*, *Urginea maritima* and *Nerium oleander*.

Cyanogenic Heterosides: Some plants give off hydrocyanic acid. A number of amino acids are known precursors of these glycosides (amydalin and prunasin).They are found in bark of *Prunus serotina*, roots of *Manihot utilissima*, leaves of *Sambucus nigra* and seeds of *Prunus dulcis*.

Coumarinic Heterosides: Coumarin is an aromatic. These plants contain vitamins, diminish capillary permeability, and increase the resistance of capillary walls. Some have sedative properties, such as angelican. They may have hypnotic properties. Commonly found in Indian Chestnut and *Flexinus excelsior*.

Polyphenols

These are substances with a benzene nucleus supporting a hydroxyl group. They usually bond with sugars to form heterosides, but they may be found separate. They range from very simple substances to very complex ones such as lignins and tannins. The main groups in this category are phenolic acids or phenols, coumarins, flavonoids, lignans, tannins, and quinones.

Phenolic acids: These are aril-carboxilics, and contain one or more OH groups in the aril. They have various pharmacological properties and uses: anti-oxidant, analgesics, chloretic. These include eugenol, estragol, miristicine, apiol, and anetol.

Flavonoids: Flavonoids are yellow pigments derived from phenyl-benzo γ pyrone or phenyl cromome. They are frequently found in the vegetable kingdom, usually in the form of heterosides. Their molecular structure is of the type C6-C3-C6. They are a very varied family of compounds, although all the final products are typically polyphenolic and soluble in water. There are 6 main classes, chalcones, flavones, flavonols, anthocyanidins, and condensed tannins, as well as two others, xantones, and aurones. They are found in passion flower (*Passiflora incarnata)* with approximately 2% of flavonoids; roman camomile (*Chamaemelum glabra*) and aquilea (*Achillea millefolium*); liquorice (*Glycyrrhiza glabra*) and gingko (*Ginkgo biloba)*; the Marian thistle (*Sylibum marianum*) and the white thorn (*Crataegus*).

Coumarins: These are benzo - α pyrones. Practically all coumarins, with the exception of coumarin itself, have a hydroxylic substitute (OH) at number 7, be it free, as in umbelipherone or in combination (methyle, sugars, etc...). Found in *Melilotus officinalis*, *Aesculus hippocastanum*.

Lignanes: These are molecules whose structure is the union of two units of phenyl propane (C6-C3). They are found in *Podophyllum peltatum* and *Sylibum marianum*..

Tannins: They are non- nitrogenised polyphenolic hydro-soluble substances of vegetable origin, having a molecular weight between 500-3000. Tannin is found mainly in the roots, bark, and sometimes in the leaves of plants. These compounds have antibacterial, astringent, and antiseptic properties. They are found especially in the Ericaceae, Leguminosae, Rosaceae, and Salicaceae familles.

Quinones: These are aromatic dicetones which come from the oxidising of phenols. There are several types: Para-benzoquinones, Naphthoquinones (walnut, henna, alfalfa) Anthracicflaxnes, Anthraquinones and Fenanthraquinones.

Terpenoids

Terpenoids are made up of a whole number of isoprene (C5) units.

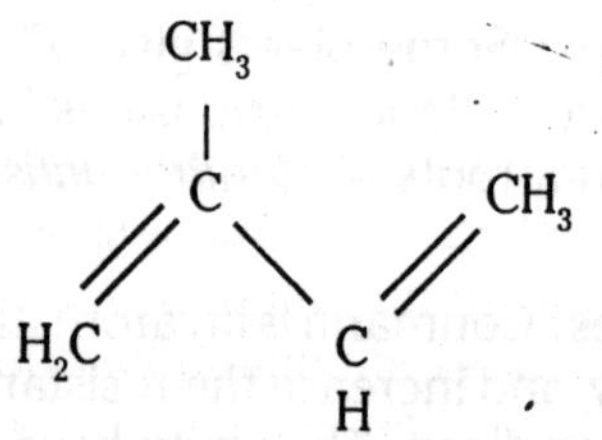

Depending on this number, they are classified as: *Isoprene unit (C5)*

- *Monoterpenes (2 units of isoprene = C10)*
- *Sesquiterpenes (3 units of isoprene=C15)*
- *Diterprenes (C20)*
- *Triterpenes (C30)*
- *Carotenes (C40)*
- *Polyterpenes (Cn)*

They include essential oils, iridoids, lactones, sesquiterpenics, saponins, and cardiotonic heterosides.

Iridoids: These are monoterpenic compounds and the name comes from a type of Australian ant *(Iridomirmex)*. They are found in *Valeriana officinalis, Harpagophytum procumbens, Olea europea* and *Gentiana lutea*.

Sesquiterpenic lactones: These are found in abundance in the Composite, Lauraceae and Magnoliaceae families, and are responsible for the bitter taste of many drugs; the holy thistle (*Cnicus benedictus)*, absinthe *(Artemisia absinthium)*, or dandelions (*Taraxacum officinale*).

Saponins: They are substances which produce foam in water solutions, and are naturally tenso-active. Many have haemolytic properties and are highly toxic when injected into the blood.

Alkaloids

This is a group of natural products of particular interest to pharmacology and are toxic even in small doses. The first alkaloid to be isolated was morphine. They are sensitive to light and to heat and can be stabilized with inorganic acids. In nature, they are found in the form of salts, although they can also be found free. In plants, they were thought to be products obtained by extraction and not particularly soluble in typical extraction solvents because of their polarity. They undergo isomerisation (lysergic and isolysergic acid) and racemization (hyoscyamine and atropine). There are several types, depending on the molecule from which they derive (tropane, quinoleine or isoquinoleine).

Essential Oils

Essential oils deals with the volatile part of the study of aromatic and medicinal plants. Most essential oils consist of hydrocarbons, esters, terpenes,lactones, phenols, aldehydes, acids, alcohols, ketones, and esters. Among these, the oxygenated compounds (alcohols, esters, aldehydes, ketones, lactones,phenols) are the principal odour source. They are more stable against oxidizing and resinifying influences than other constituents. On the other hand,unsaturated constituents like monoterpenes and sesquiterpenes have the tendency to oxidize or resinify in the presence of air and light.

Essential oils are used in a wide variety of consumer goods such as detergents, soaps, toilet products, cosmetics, pharmaceuticals, perfumes, confectionery food products, soft drinks, distilled alcoholic beverages (harddrinks) and insecticides. Essential oils are generally derived from one or more plant parts,such as flowers (e.g. rose, jasmine, carnation, clove, mimosa, rosemary,lavander), leaves (e.g. mint, *Ocimum* spp., lemongrass, jamrosa), leaves and stems (e.g. geranium, patchouli, petitgrain, verbena, cinnamon), bark (e.g.cinnamon, cassia, canella), wood (e.g. cedar, sandal, pine), roots (e.g. angelica,sassafras, vetiver, saussurea, valerian), seeds (e.g fennel, coriander, caraway, dill, nutmeg), fruits (bergamot, orange, lemon, juniper), rhizomes(e.g. ginger, calamus, curcuma, orris) and gums or oleoresin exudations(e.g. balsam of Peru, balsam of Tolu, storax, myrrh, benzoin). Depending upon the plant family, essential oils may occur in specialized secretary structures such as glandular hairs(Labiatae, Verbenaceace, Geraniaceae), modified parenchymal cells (Piperaceae),resin canals (conifers), oil tubes called vittae (Umbelliferae), lysigenous cavities (Rutaceae), schizogenous passages (Myrtaceae, Graminae,Compositae) or gum canals (Cistacae, Burseraceae).

Measuring essential oils: The basic methods for analysing essential oils are as follows.

- Physical measurements (aroma, specific weight, refractory index, optical deviation (rotational capacity, solubility in alcohol-water mixtures)
- Chemical measurements
- free acidity index
- saponisation and ester indices
- aldehyde and ketone measures (phenilhydrazone formation, oxyme formation, semi-carbazone formation and bisulphite method)
- acetyl index
- chromatographic techniques (FLC, GC, HPLC)
- spectroscopic methods (UV, IR, -MS, NMR)

Extraction is the separation of medicinally active portions of plant or animal tissues using selective solvents through standard procedures. Such extraction techniques separatethe soluble plant metabolites and leave behind the insoluble cellular marc. The products so obtained from plants are relatively complex mixtures of metabolites, in liquid or semisolid state or (after removing the solvent) indry powder form, and are intended for oral or external use. These include classes of preparations known as decoctions, infusions, fluid extracts, tinctures,semisolid extracts or powdered extracts. Such preparations have been popularly called galenicals, named after Galen, the second century Greek physician. The purpose of standardized extraction procedures for crude drugs (medicinal plant parts) is to attain the therapeutically desired portions and to eliminate unwanted material by treatment with a selective solvent known as menstruum. The extract thus obtained, after standardization,may be used as medicinal agent as such in the form of tinctures or fluid extracts or further processed to be incorporated in any dosage form such as tablets and capsules. These products all contain complex mixture of many medicinal plant metabolites, such as alkaloids, glycosides, terpenoids, flavonoids and lignans. In order to be used as a modern drug, an extract may be further processed through various techniques of fractionation to isolate individual chemical entities such as vincristine, vinblastine, hyoscyamine,hyoscine, pilocarpine, forskolin and codeine.The industrial processing of MAPs starts with the extraction of the active components using various technologies. The general techniques of medicinal plant extraction include maceration, infusion, percolation, digestion,decoction, hot continuous extraction (Soxhlet), aqueous-alcoholic extraction by fermentation, counter-current extraction, microwave-assisted extraction, ultrasound extraction (sonication), supercritical fluid extraction,and phytonic extraction (with hydrofluorocarbon solvents). For aromatic plants, hydro distillation techniques (water distillation, steam distillation,water and steam distillation), hydrolytic maceration followed by distillation,expression and enfleurage (cold fat extraction) may be employed. Some ofthe latest extraction methods for aromatic plants include headspace trapping,solid phase micro-extraction, protoplast extraction, microdistillation,thermomicrodistillation and molecular distillation.

Solvents used for active component extraction: Common solvents are water, ethanol, methanol, chloroform, ether and acetone.

Table 1. Solvents used for active component extraction

Water	Ethanol	Methanol	Chloroform	Ether	Acetone
Anthocyanins	Tannins	Anthocyanins	Terpenoids	Alkaloids	Phenol
Starches	Polyphenols	Terpenoids	Flavonoids	Terpenoids	Flavonols
Tannins	Polyacetylenes	Saponins		Coumarins	
Saponins	Flavonols	Tannins		Fatty acids	
Terpenoids	Terpenoids	Xanthoxyllines			
Polypeptides	Sterols	Totarol			
Lectins	Alkaloids	Quassinoids			
		Lactones			
		Flavones			
		Phenones			
		Polyphenols			

General Methods of Extraction of Medicinal Plants

Maceration: In this process, the whole or coarsely powdered crude drug is placed in a stoppered container with the solvent and allowed to stand at room temperature for a period of at least 3 days with frequent agitation until the soluble matter is dissolved. The mixture then is strained, the damp solid material is pressed, and the combined liquids are clarified by filtration or decantation after standing.

Infusion: Fresh infusions are prepared by macerating the crude drug fora short period of time with cold or boiling water. These are dilute solutions of the readily soluble constituents of crude drugs.

Digestion: This is a form of maceration in which gentle heat is used during the process of extraction. It is used when moderately elevated temperatureis accepted. The solvent efficiency of the menstruum is thereby increased.

Decoction: In this process, the crude drug is boiled in a specified volume of water for a defined time; it is then cooled and strained or filtered. This procedure is suitable for extracting water-soluble and heat-stable constituents.This process is mainly used in preparation of Ayurvedic extracts called"Quath" or "Kawath". The starting ratio of crude drug to water is fixed at 1:4 or 1:16; the volume is then brought down to one-fourth its original volume by boiling during the extraction procedure. Then, the concentrated extractis strained and used as such or processed further.

Percolation: This is the method used most frequently to extract active ingredients in the preparation of tinctures and fluid extracts. A percolator having a narrow, cone-shaped vessel open at both ends is generally used (Figure1). The solid ingredients are moistened with an appropriate amount of the specified menstruum and allowed to stand for approximately 4 h in a well closed container, after which the mass is packed and the top of the percolatoris closed. Additional menstruum is added to form a shallow layer above the mass, and the mixture is allowed to macerate in the closed percolator for 24 h. The outlet of the percolator then is opened and the liquid contained therein is allowed to drip slowly. Addifional menstruum is added as required,until the percolate measures about three-quarters of the required volume of the finished product. The marc is then pressed and the expressed liquid is added to the percolate. Sufficient menstruum is added to produce therequired volume, and the mixed liquid is clarified by filtration or by standing followed by decanting.

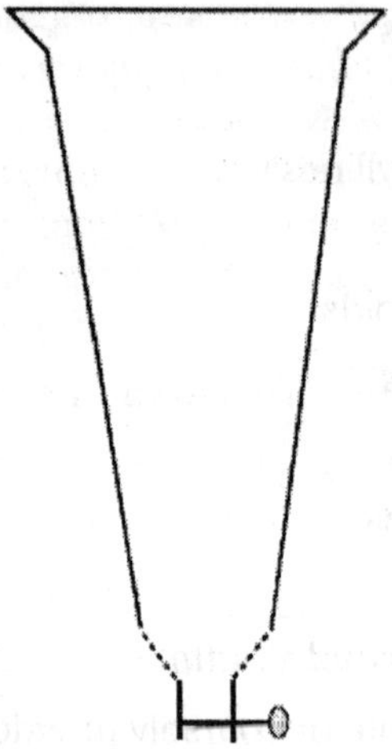

Figure 1 : Percolator

Hot Continuous Extraction (Soxhlet): In this method, the finely ground crude drug is kept in a porous bag or "thimble" made of strong filter paper, which is placed in chamber E ofthe Soxhlet apparatus (Figure 2). The extracting solvent in flask A is heated, and its vapours condense in condenser D. The condensed extractant drips into the thimble containing the crude drug, and extracts it by contact. When the level of liquid in chamber E rises to the top of siphon tube C, the liquid contents of chamber E siphon into flask A. This process is continuous and is carried out until a drop of solvent from the siphon tube does not leave residue when evaporated. The advantage of this method, compared to previously described methods, is that large amounts of drug can be extracted with a much smaller quantity of solvent. This influences tremendous economy in terms of time, energy and consequently financial inputs. At small scale, it is employed as a batch process only, but it becomes much more economical and viable when converted into a continuous extraction procedure on medium or large scale.

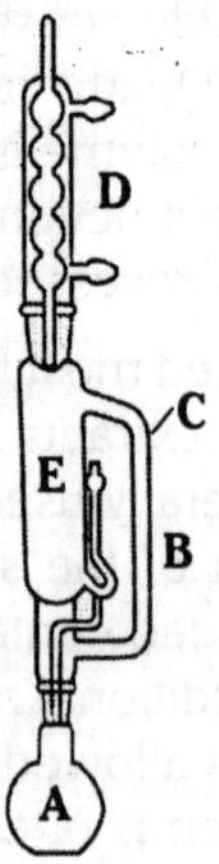

Figure 2: Soxhlet apparatus

Aqueous Alcoholic Extraction by Fermentation: Some medicinal preparations of Ayurveda (like *asava* and *arista*)adopt the technique of fermentation for extracting the active principles. The extraction procedure involves soaking the crude drug, in the form of either a powder or a decoction (*kasaya*), for a specified period of time, during which it undergoes fermentation and generates alcohol *in situ*; this facilitates the extraction of the active constituents contained in the plant material. The alcohol thus produced also serves as a preservative. If the fermentation is to be carried out in an earthen vessel, it should not be new: water should first be boiled in the vessel. In large-scale manufacture, wooden vats, porcelain jars or metal vessels are used in place of earthen vessels. Some examplesof such preparations are *karpurasava, kanakasava, dasmularista*.

Counter-current Extraction: In counter-current extraction (CCE), wet raw material is pulverized using toothed disc disintegrators to produce a fine slurry. In this process, thematerial to be extracted is moved in one direction generally in the form of a fine slurry within a cylindrical extractor where it comes in contact with extraction solvent. The further the starting material moves, the more concentrated the extract becomes. Complete extraction is made possible when the quantities of solvent and material and their flow rates are optimized. This process is highly efficient, requiring little time and posing no risk from high temperature. Finally, sufficiently concentrated extract comes out at one end of the extractor while the marc (practically free of visible solvent) falls out from the other end.

This extraction process has significant advantages

i) A unit quantity of the plant material can be extracted with much smaller volume of solvent as compared to other methods like maceration, decoction and percolation.

ii) CCE is commonly done at room temperature, which spares the thermolabile constituents from exposure to heat whichis employed in most other techniques.

iii) Since the pulverization of the drug is done under wet conditions,the heat generated during comminution is neutralized by water. This again spares the thermolabile constituents from exposure to heat.

iv) The extraction procedure is rated to be more efficient and effective than continuous hot extraction.

Ultrasound Extraction (Sonication): This method involves the use of ultrasound with frequencie sranging from 20 kHz to 2000 kHz; which increases the permeability of cellwalls and produces cavitation. Although the process is useful in some cases, like extraction of rauwolfia root, its large-scale application is limited dueto the higher cost investments. One disadvantage of the procedure is the occasional but known deleterious effect of ultrasound energy (more than 20 kHz) on the active constituents of medicinal plants through formation of free radicals and consequently undesirable changes in the drug molecules.

Supercritical Fluid Extraction: Supercritical fluid extraction (SFE) is an alternative sample preparation method with general objectives of reduced use of organic solvents and increased sample throughout. The factors to consider include temperature, pressure, sample volume, analyte collection, modifier addition flow and pressure control, and restrictors. Generally, cylindrical extraction vessels are used for SFE and their performance is good beyond any doubt.

The collection of the extracted analyte following SFE is another important step: significant analyte loss can occur during this step, leading the analyst to believe that the actual efficiency was poor. There are many advantages to the use of CO_2 as the extracting fluid. In addition to its favorable physical properties, carbon dioxide is inexpensive, safe and abundant. But while carbon dioxide is the preferred fluid for SFE, it possesses several polarity limitations. Solvent polarity is important when extracting polar solutes and when strong analyte-matrix interactions are present. Organic solvents are frequently added to the carbon dioxide extracting fluid to alleviate the polarity limitations. Of late, instead of carbon dioxide, argon is being used because it is inexpensive and more inert. The component recovery rates generally increase with increasing pressure or temperature: the highest recovery rates in case of argon are obtained at 500 atm and 150°C.

The extraction procedure possesses distinct advantages

i) The extraction of constituents at low temperature, which strictly avoids damage from heat and some organic solvents.

ii) No solvent residues.

iii) Environmentally friendly extraction procedure.

The largest area of growth in the development of SFE is the rapid expansion of its applications. SFE have extensive application in the extraction of pesticides, environmental samples, foods and fragrances,essential oils, polymers and natural products. The major deterrent in the commercial application of the extraction process is its prohibitive capital investment.

Phytonics Process

A new solvent based on hydrofluorocarbon-134a and a new technology to optimize its remarkable properties in the extraction of plant materials offer significant environmental advantages and health and safety benefits over traditional processes for the production of high quality natural fragrant oils, flavours and biological extracts. Advanced Phytonics Limited (Manchester, UK) has developed this patented technology termed "Phytonics process". The products mostly extracted by this process are fragrant components of essential oils and biological or phyto-pharmacological extracts which can be used directly withoutfurther physical or chemical treatment. The properties of the new generation of fluorocarbon solvents have been applied to the extraction of plant materials. The core of the solventis 1,1,2,2-tetrafluoroethane, better known as hydrofluorocarbon-134a(HFC-134a). This product was developed as a replacement for chlorofluorocarbons. The boiling point of this solvent is -25° C. Unlike chlorofluorocarbons, it does not deplete the ozone layer. It has a vapour pressure of 5.6 bar at ambient temperature. By most standards this is a poor solvent. For example, it does not mix with mineral oils or triglycerides and it does not dissolve plant wastes. The process is advantageous in that the solvents can be customized by using modified solvents with HFC-134a, the process can bemade highly selective in extracting a specific class of phyto-constituents. Similarly, other modified solvents can be used to extract a broader spectrum of components. The biological products made by this process have extremely low residual solvent. The residuals are invariably less than 20 parts perbillion and are frequently below levels of detection. These solvents are neither acidic nor alkaline and, therefore, have only minimal potential reactioneffects on the botanical materials. The processing plant is totally sealed so that the solvents are continually recycled and fully recovered at the endof each production cycle. The only utility needed to operate these systems is electricity and, even then, they do no consume much energy. There is no scope for the escape of the solvents and therefore pose no threat to the ozone layer. The waste biomass from these plants is dry and "Ecofriendly" to handle.

Advantages of the Process

- Unlike other processes that employ high temperatures, the phytonics process is cool and gentle and its products arenever damaged by exposure to excessive temperatures
- No vacuum stripping is required which, in other processes,leads to the loss of precious volatiles.
- The process is carried out entirely at neutral pH and, in the absence of oxygen, the products never suffer acid hydrolysis damage or oxidation.
- The technique is highly selective, offering a choice of operating conditions and hence a choice of end products.
- It is ecofriendly.
- It requires a minimum amount of electrical energy.
- It releases no harmful emissions into the atmosphere and the resultant waste products are innocuous and pose no effluent disposal problems.
- The solvents used in the technique are not flammable, toxic or ozone depleting.
- The solvents are completely recycled or utilized within the system.

Applications

The phytonics processes are used for extraction in biotechnology(e.g for the production of antibiotics), in the herbal drug industry, in the food, essential oil and flavour industries, and in the production of other pharmacologically active products. In particular, it is used in the production of top quality pharmaceutical-grade extracts, pharmacologically active intermediates,antibiotic extracts and phyto-pharmaceuticals. The technique is being used in the extraction of high-quality essential oils, oleoresins, natural food colours, flavours and aromatic oils from all manner of plant materials. The technique is also used in refining crude products obtained from other extraction processes. It provides extraction without waxes or other contaminants. It helps remove many biocides from contaminated biomass.

Parameters for Selecting an Appropriate Extraction Method

i) Authentication of plant material is must before performing extraction. Any foreign matter should not be allowed.

ii) Use the right plant part and, for quality control purposes, record the proper age of plant and the time, season and place of collection.

iii) Conditions used for drying the plant material largely depend on the nature of its chemical constituents. Hot or cold blowing air flow for drying is generally recommended. If a crude drug with high moisture content is to be used for extraction, suitable weight corrections should be incorporated.

iv) Grinding methods must be specified and techniques that generate heat should be avoided as much as possible.

v) Powdered plant material should be passed through suitable sieves to get the required particles of uniform size.

vi) Nature of constituents:

a) If the therapeutic value lies in non-polar constituents, a non-polar solvent may be used. For example, lupeol is theactive constituent of *Crataeva nurvala* and, for its extraction,hexane is generally used. Likewise, for plants like *Bacopa monnieri* and *Centella asiatica*, the active constituents are glycosides and hence a polar solvent like aqueous methanol may be used.

b) If the constituents are thermolabile, extraction methods like cold maceration, percolation and CCE are preferred.For thermostable constituents, Soxhlet extraction and decoction are useful.

c) Suitable precautions should be taken when dealing with constituents that degrade while being kept in organic solvents, e.g. flavonoids and phenyl propanoids.

d) In case of hot extraction, higher than required temperature should be avoided. Some glycosides are likely to break upon continuous exposure to higher temperature.

e) Standardization of time of extraction is important, as:

- Insufficient time means incomplete extraction.
- If the extraction time is longer, unwanted constituents may also be extracted. For example, if tea is boiled for long time, tannins are extracted which impart astringency to the final preparation.

f) The number of extractions required for complete extractionis as important as the duration of each extraction.

vii) The quality of water or menstruum used should be specified and controlled.

viii) Concentration and drying procedures should ensure the safety and stability of the active constituents. Drying under reduced pressure (e.g. using a Rotavapor) is widely used. Lyophilization,although expensive, is increasingly employed.

ix) The design and material of fabrication of the extractor are also to be taken into consideration.

x) Analytical parameters of the final extract, such as TLC andHPLC fingerprints, should be documented to monitor the quality of different batches of the extracts.

Steps Involved in the Extraction of Medicinal Plants

In order to extract medicinal ingredients from plant material, the following sequential steps are involved:

1. Size reduction
2. Extraction

3. Filtration
4. Concentration
5. Drying

Size Reduction: The dried plant material is disintegrated by feeding it into a hammer mill or a disc pulverizer which has built-in sieves. The particle sizeis controlled by varying the speed of the rotor clearance between the hammers and the lining of the grinder and also by varying the opening of the discharge of the mill. Usually, the plant material is reduced to a size between 30 and 40 mesh, but this can be changed as per requirement. The objective for powdering the plant material is to rupture its organ, tissue and cell structures so that its medicinal ingredients are exposed to the extraction solvent. Furthermore, size reduction maximizes the surface area, whichin turn enhances the mass transfer of active principle from plant material to the solvent. The 30-40 mesh size is optimal, while smaller particles maybecome slimy during extraction and create difficulty during filtration.

Extraction

Extraction of the plant material is carried out in three ways:

i) Cold aqueous percolation
ii) Hot aqueous extraction (decoction)
iii) Solvent extraction (cold or hot)

Cold Aqueous Percolation: The powdered material is macerated with water and then poured into a tall column. Cold water is added until the powdered material is completely immersed. It is allowed to stand for 24 hours so that water-soluble ingredients attain equilibrium in the water. The enriched aqueous extract is concentrated in multiple-effect evaporators to a particular concentration. Some diluents and excipients are added to this concentrated extract, whichis then ready for medicinal use.

Hot Aqueous Extraction (Decoction): This is practiced in an open-type extractor. The extractor is a cylindrical vessel made from type 316 stainless steel and has a diameter (D)greater than the height (H), i.e. the H/D ratio is approximately 0.5. The bottom of the vessel is welded to the dished end and is provided with an inside false bottom with a filter cloth. The outside vessel has a steam jacket and a discharge valve at the bottom.

One part powdered plant material and sixteen parts demineralized water are fed into the extractor. Heating is done by injecting steam into the jacket. The material is allowed to boil until the volume of water is reduced to one-fourth its original volume. By this time the medicinal ingredients present in the plant material have been extracted out.

Filtration: The extract so obtained is separated out from the exhausted plant material by allowing it to trickle into a holding tank throughthe built-in false bottom of the extractor, which is covered with a filter cloth.The exhausted plant material is retained at the false bottom, and the extract is received in the holding tank. From the holding tank, the extract is pumped into a sparkler filter to remove fine or colloidal particles from the extract.

Spray Drying: The filtered extract is subjected to spray drying with a high pressure pump at a controlled feed rate and temperature, to get dry powder.The desired particle size of the product is obtained by controlling the inside temperature of the chamber and by varying the pressure of the pump. The dry powder is mixed with suitable diluents or excipients and blended in a double cone mixer to obtain a homogeneous powder that can be straight away used, for example, for filling in capsules or making tablets.

Solvent Extraction: The principle of solid-liquid extraction is that when a solid materialcomes in contact with a solvent, the soluble components in the solid material move to the solvent. Thus, solvent extraction of plant material results in themass transfer of soluble active principle (medicinal ingredient) to the solvent, and this takes place in a concentration gradient. The rate of mass transfer decreases as the concentration of active principle in the solvent increases, until equilibrium is attained, i.e. the concentrations of active principle in the solid material and the solvent are the same. Thereafter, there will no longer be a mass transfer of the active principle from plant material to the solvent. Since mass transfer of the active principle also depends on itssolubility in the solvent, heating the solvent can enhances the mass transfer. Moreover, if the solvent in equilibrium with the plant material is replaced with fresh solvent, the concentration gradient is changed. This gives rise to different types of extractions: cold percolation, hot percolation and concentration.

Cold Percolation: The extraction of plant material is carried out in a percolator which is a tall cylindrical vessel with a conical bottom and a built-in false bottom with a filter cloth. The percolator is connected to a condenser and a receiver for stripping solvent from the marc.The powdered material is fed into the percolator along with a suitable solvent (ethyl alcohol or another non-polar solvent). The materialis left in contact with the solvent until equilibrium of the active principle is achieved. The solvent extract, known as *miscella*, is taken out from the bottom discharge valve of the percolator. Fresh solvent is added into the percolator and the miscella is drained out after acquiring equilibrium. Overall,the plant material is washed four to five times until it gets exhausted. All washes from the percolator are pooled and concentrated.

The solvent in the marc is stripped out by passing steam from the bottom of the percolator. The solvent and steam vapors rise and are condensed in a tubular condenser. The condensate, a mixture of alcohol and water, is collected in a receiver and then subjected to fractional distillation to get 95% pure ethyl alcohol which is again used as a fresh solvent. This type of percolation is not efficient as it takes a long time to reach equilibrium due to the slow mass transfer rate. The mass transferrate can be increased if some sort of movement is created between the particles and the solvent. This can be achieved either by providing inside agitation with a mechanical stirrer or by repeated circulation of the extract back to the percolator. The first method is cumbersome and power intensive whereas the latter has been successful. A circulation pump that continuously circulates the miscella back to the top of the percolator gives a better mass transferrate and reduces the equilibrium time considerably. Still, this type of percolation is energy-consuming as large amounts of miscella from multiple washes must be concentrated to remove the solvent. To overcome this problem, a battery of percolators can be connected in series. If three washes are required for completion of the extraction,four percolators are connected in series with their respective miscella storage tanks.

At a particular time, one percolator is out of circuit, for charging and discharging the material and also for stripping solvent from the marc, whereas the other three percolators are in operation. Material is fed into all the percolators and the solvent is fed into the first percolator. When the equilibrium in the first percolator is reached, the extract from the first percolator is sent to the second percolator. The first percolator is again filled with fresh solvent. The extract of second percolator is transferred to the third, the extract of first is transferred to second, and fresh solvent is added to the first. The extract of the third percolator is transferred to the fourth percolator.

After reaching equilibrium, the extract from the fourth percolator is drained off. The extract of the third percolator goes to fourth, the extract of second goes to third, and the extract of first goes to second percolator. The material of the first percolator, which has received three washes, is completely exhausted. This percolator is taken out of the system for stripping the solvent and discharging the extracted marc. This is again filled with fresh plant material and the sequence is repeated with other percolators. In this way, solvent of each percolator comes in contact three times with solid material and gets fully enriched with active principle. The enriched extract is sent for solvent recovery and concentration. Thus, instead of concentrating three volumes of solvent, only one volume has to be concentrated; and therefore, saves energy and the process is efficient.

Hot Percolation: Increasing the temperature of the solvent increases the solubility of the active principle, which increases the concentration gradient and therefore enhances the mass transfer of active principle from solid material to the solvent, provided the active principle is not heat sensitive. This is achieved by incorporating a heat exchanger between the circulation pump and the feed inlet of the percolator. The extract is continuously pumped into a tubular heat exchanger which is heated by steam. The temperature of the extract in the percolator is controlled by a steam solenoid valve through a temperature indicator controller. This sort of arrangement can be incorporated in single percolators or in a battery of percolators as needed.

The percolators that are tall cylindrical towers must be housed in sheds of relatively great height. Tall towers are difficult to operate, especially when charging material and discharging the marc from the top and bottom man holes, which are time-consuming and labour-intensive procedures.Tall towers have been replaced by extractors of smaller height for which theH/D ratio is not more than 1.5. These extractors have perforated baskets in which the material to be extracted is charged. These perforated baskets, when loaded outside, can be inserted into the extractor with a chain pulley block and, after the extraction, they can be lifted out from the extractor for discharging the marc. Some extractors have an electrical hoist for the charging the material and discharging the marc, which makes the operation less labour-intensive, quick and efficient. The other type of instrument for extraction of medicinal ingredients from plant material is the Soxhlet apparatus, which consists of an extractor, a distillation still, a tubular condenser for the distillation still, a tubular condenser for the recovery of solvent from the marc, a receiver for collecting the condensate from the condenser, and a solvent storage tank. The plant material is fed into the extractor, and solvent is added until it reaches the siphon point of the extractor. Then, the extract is siphoned out into the distillation still, which is heated with steam. The solvent vapours go to the distillation condenser, get condensed and return to the extractor. The level of the solvent in the extractor again rises to the siphon point and the extractis siphoned out into the distillation still. In this way, fresh solvent comes incontact with the plant material a number of times, until the plant material is completely extracted. The final extract in the distillation still, which is rich inactive principle, is concentrated and the solvent is recovered.

Concentration: The enriched extract from percolators or extractors, known as miscella, is fed into a wiped film evaporator where it is concentrated under vacuum to produce a thick concentrated extract. The concentrated extract is further fed into a vacuum chamber dryer to produce a solid mass free from solven'. The solvent recovered from the wiped film evaporator and vacuum chamber dryer is recycled back to the percolator or extractor for the next batch of plant material. The solid mass thus obtained is pulverized and used directly for the desired pharmaceutical formulations or further processed for isolation of its phyto-constituents.

Aromatic Plant Extracts

The types of volatile isolates that are obtained commercially from aromatic plants are essential oils, concretes, absolutes, pomades and resinoids. Essential oils are isolated from plant material by distillation whereas other volatile isolates are obtained by solvent extraction.

Concrete: This is an extract of fresh flowers, herbs, leaves and the flowering tops of plants obtained by the use of a hydrocarbon solvent such as butane,pentane, hexane and petroleum ether. Concrete is rich in hydrocarbon soluble material and devoid of water-soluble components. It is generally a waxy, semi-solid, dark-coloured material free from the original solvent.In practice, concretes are produced in static extractors. These extractors are fitted with numerous perforated trays so that the flowers do not get compressed by their own weight. Each perforated tray has a spacer so the number and distance between them are predetermined.The set of perforated trays can be within a removable cylindrical basket. In the centre of the lower tray, there is a rod on which the spacers and the perforated trays are fitted while at the top there is a ring or a hook so that the entire contents of the extractor can be readily removed by a chain pulley block. While stacking the flowers on these trays, care should be taken to minimize bruising and damage of the flowers, because such damage can result in the release of enzymes in the flower juice which deteriorates the quality of concrete. The basket stacked with flowers is inserted into the extractor and the solvent of choice is introduced from the bottom into the extractor until the material on the perforated disc assembly is completely immersed.

Four to five such washes are done until the material is exhausted.The enriched solvent from the extractor is pumped into an evaporator for solvent recovery and the solvent content is reduced to about one-tenth the original volume. The recovered solvent is pumped to the solvent tanks to be used again. The concentrated material from the evaporator is pumped into a vacuum evaporator where the solvent is removed more carefully under high vacuum and the recovered solvent is returned to the solvent tanks for repeated use. The resultant concrete has an odour similar to but stronger than the material from which it was extracted. In concrete manufacturing, it is a normal practice to circulate fresh solvent through a battery of extractors. At each cycle, the solvent becomes more enriched with the flower volatiles until extraction is complete. The number of extractors has to be synchronized with the number of solvent washes.

Absolutes: Concretes are generally converted into an alcohol-soluble volatile concentrate known as an absolute, i.e. they have to be extracted with alcohol.To make an absolute, the concrete is mixed with absolute alcohol and agitated thoroughly in a vessel with an agitator. During agitation, the temperature is kept at 40°-60° C and the concrete is immersed in the solution. The solution is cooled down to -5° to -10° C to precipitate out the wax, since waxes are normally insoluble in alcohol below -1° C. The precipitated wax is removed by passing the solution through a rotary filter. The filtrate from the rotary filter is pumped into a primary evaporator, where it is concentrated to about 10% alcohol content. Finally, the concentrated extract is pumped into an agitating-type evaporator, where the alcohol is carefully removed under high vacuum.

Resinoids: Resinoid is an extract of naturally resinous material, made with a hydrocarbon solvent. Resinoids are usually extracted from dry materials.The extraction process is same as that of concrete production, except that perforated discs are not used for stacking the material; instead powder from dry plant material is fed into the extractor.

Pomades: Pomades are obtained by a process known as enfleurage,which is a cold fat extraction method. The fat is spread out on glass plates contained in wooden frames, leaving a clear margin near the edges. The absorptive surface of the fat is increased by surface grooves made with a wooden spatula. Fresh flowers are spread out on the surface of the fat and the frames are stacked in piles. After the perfume oils have been absorbed from the flowers, the spent flowers are removed by hand. Fresh flowers are again spread on the fat surface. This is repeated until the fat surface is completely enriched with perfume oils. The pomade so obtained is ready for cold alcoholic extraction.

Essential Oils: Essential oils are used in a wide variety of consumer goods such as detergents, soaps, toilet products, cosmetics, pharmaceuticals, perfumes, confectionery food products, soft drinks, distilled alcoholic beverages (hard drinks) and insecticides. The world production and consumption of essential oils and perfumes are increasing very fast. Production technology is an essential element to improve the overall yield and quality of essential oil. The traditional technologies pertaining to essential oil processing are of great significance and are still being used in many parts of the world. Water distillation,water and steam distillation, steam distillation, cohobation, maceration and enfleurage are the most traditional and commonly used methods. Maceration is adaptable when oil yield from distillation is poor. Distillation methods are good for powdered almonds, rose petals and rose blossoms, whereas solvent extraction is suitable for expensive, delicate and thermally unstable materials like jasmine, tuberose, and hyacinth. Water distillation is the most favoured method of production of citronella oil from plant material. Most essential oils consist of hydrocarbons, esters, terpenes,lactones, phenols, aldehydes, acids, alcohols, ketones, and esters. Among these, the oxygenated compounds (alcohols, esters, aldehydes, ketones, lactones,phenols) are the principal odour source. They are more stable against oxidizing and resinifying influences than other constituents. On the other hand,unsaturated constituents like monoterpenes and sesquiterpenes have the tendency to oxidize or resinify in the presence of air and light. The knowledge ofindividual constituents and their physical characteristics, such as boiling point, thermal stability and vapour-pressure-temperature relationship, is of paramount importance in technology development of oxygenated compounds.

Methods of Producing Essential Oils

Regarding hydrodistillation, the essential oils industry has developed terminology to distinguish three types: water distillation;water and steam distillation; and direct steam distillation.Originally introduced by Von Rechenberg, these terms have become established in the essential oil industry. All three methods are subject to the same theoretical considerations which deal with distillation of two-phase systems. The differences lie mainly in the methods of handling the material.

Some volatile oils cannot be distilled without decomposition and therefore are usually obtained by expression (lemon oil, orange oil) or by other mechanical means. In certain countries, the general method for obtaining citrus oil involves puncturing the oil glands by rolling the fruit over a troughlined with sharp projections that are long enough to penetrate the epidermisand pierce the oil glands located within outer portion of the peel (*ecuelle* method). A pressing action on the fruit removes the oil from the glands, anda fine spray of water washes the oil from the mashed peel while the juice is extracted through a central tube that cores the fruit. The resulting oil-water emulsion is separated by centrifugation.

A variation of this process is to remove the peel from the fruit before the oil is extracted. Often, the volatile oil content of fresh plant parts (flower petals)is so small that oil removal is not commercially feasible by the afore mentioned methods. In such instances, an odourless, bland, fixed oil or fat is spread ina thin layer on glass plates. The flower petals are placed on the fat for a few hours; then repeatedly, the oil petals are removed, and a new layer of petals isintroduced. After the fat has absorbed as much fragrance as possible, the oilmay be removed by extraction with alcohol. This process, known as enfleurage,was formerly used extensively in the production of perfumes and pomades. In the perfume industry, most modern essential oil production is accomplished by extraction, using volatile solvents such as petroleum etherand hexane. The chief advantages of extraction over distillation is that uniform temperature (usually 50° C) can be maintained during the process, As a result,extracted oils have a more natural odour that is unmatched by distilled oils, which may have undergone chemical alteration by the high temperature. This featureis of considerable importance to the perfume industry; however, the established distillation method is of lower cost than the extraction process. Destructive distillation means distilling volatile oil in the absence of air. When wood or resin of members of the Pinaceae or Cupressaceaeis heated without air, decomposition takes place and a number of volatile compounds are driven off. The residual mass is charcoal. The condensed volatile matter usually separates into 2 layers: an aqueous layer containing wood naptha (methyl alcohol) and pyroligneous acid (crude acetic), and atarry liquid in the form of pine tar, juniper tar, or other tars, depending on the wood used. This dry distillation is usually conducted in retorts and, if the wood is chipped or coarsely ground and the heat is applied rapidly, the yield often represents about 10% of the wood weight used.

Hydrodistillation

In order to isolate essential oils by hydrodistillation, the aromatic plant material is packed in a still and a sufficient quantity of wateris added and brought to a boil; alternatively, live steam is injected into the plant charge. Due to the influence of hot water and steam, the essential oil is freed from the oil glands in the plant tissue. The vapour mixture of water and oil is condensed by indirect cooling with water. From the condenser, distillate flows into a separator, where oil separates automatically from the distillate water.

Mechanism of Distillation

Hydrodistillation of plant material involves the following main physicochemical processes:

i) Hydro-diffusion

ii) Hydrolysis

iii) Decomposition by heat

Hydro-diffusion: Diffusion of essential oils and hot water through plant membranes is known as hydro-diffusion. In steam distillation, the steam does not actually penetrate the dry cell membranes. Therefore, dry plant material can be exhausted with dry steam only when all the volatile oil has been freed from the oil-bearing cells by first thorough comminution of the plant material. But, when the plant material is soaked with water, exchange of vapours within the tissue is based on their permeability while in swollen condition. Membranes of plant cells are almost impermeable to volatile oils. Thus, in the actual process, at the temperature of boiling water, a part of volatile oil dissolves in the water present within the glands, and this oil-water solution permeates, by osmosis, the swollen membranes and finally reaches theouter surface, where the oil is vapourized by passing steam. Another aspect of hydro-diffusion is that the speed of oil vaporizationis not influenced by the volatility of the oil components, but by their degree of solubility in water. Therefore, the high-boiling but more water-soluble constituents of oil in plant tissue distil before the low-boiling but less water-soluble constituents. Since hydro-diffusion rates are slow, distillation of uncomminuted material takes longer time than comminuted material.

Hydrolysis: Hydrolysis in the present context is defined as a chemical reaction between water and certain constituents of essential oils. Esters are constituents of essential oils and, in the presence of water, especially at high temperatures, they tend to react with water to form acids and alcohols.However, the reactions are not complete in either direction and the relationship between the molal concentrations of various constituents at equilibrium is written as:

$$K = \frac{(\text{alcohol}) \times (\text{acid})}{(\text{ester}) \times (\text{water})}$$

where K is the equilibrium constant.

Therefore, if the amount of water is more, the amounts of alcohol and acid will also be more, resulting in a decreased yield of essential oil. Furthermore,since this is a time-dependent reaction, the extent to which hydrolysis proceeds depends on the time of contact between oil and water. This is one of the disadvantages of water distillation.

Three Types of Hydro-distillation

Three are three types of hydro-distillation for isolating essential oils from plant materials:

1. Water distillation
2. Water and steam distillation
3. Direct steam distillation

Water Distillation: In this method, the material is completely immersed in water,which is boiled by applying heat by direct fire, steam jacket, closed steamjacket, closed steam coil or open steam coil. The main characteristic of this process is that there is direct contact between boiling water and plant material.

When the still is heated by direct fire, adequate precautions are necessary to prevent the charge from overheating. When a steam jacketor closed steam coil is used, there is less danger of overheating; with opensteam coils this danger is avoided. But with open steam, care must be taken to prevent accumulation of condensed water within the still. Therefore, the still should be well insulated. The plant material in the still must be agitated as the water boils, otherwise agglomerations of dense material will settle on the bottom and become thermally degraded. Certain plant materials like cinnamon bark, which are rich in mucilage, must be powdered so that the charge can readily disperse in the water; as the temperature of the water increases, the mucilage will be leached from the ground cinnamon. This greatly increases the viscosity of the water-charge mixture, thereby allowing it to char. Consequently, before any field distillation is done, a small-scale water distillation in glassware should be performed to observe whether any changes take place during the distillation process. From this laboratory trial,the yield of oil from a known weight of the plant material can be determined.The laboratory apparatus recommended for trial distillations is the Clevenger system (Fig 3.).

During water distillation, all parts of the plant charge must be kept in motion by boiling water; this is possible when the distillation materialis charged loosely and remains loose in the boiling water. For this reason only, water distillation possesses one distinct advantage, i.e. that it permits processing of finely powdered material or plant parts that, by contact with live steam, would otherwise form lumps through which the steam cannot penetrate. Other practical advantages of water distillation are that the stills are inexpensive, easy to construct and suitable for field operation. These are still widely used with portable equipment in many countries.The main disadvantage of water distillation is that complete extraction is not possible. Besides, certain esters are partly hydrolyzed and sensitive substances like aldehydes tend to polymerize. Water distillation requires a greater number of stills, more space and more fuel. It demands considerable experience and familiarity with the method. The high-boiling and somewhat water-soluble oil constituents cannot be completely vaporized or they require large quantities of steam. Thus, the process becomes uneconomical. For these reasons, water distillation is used only in cases in which the plant material by its very nature cannot be processed by water and steam distillation or by direct steam distillation.

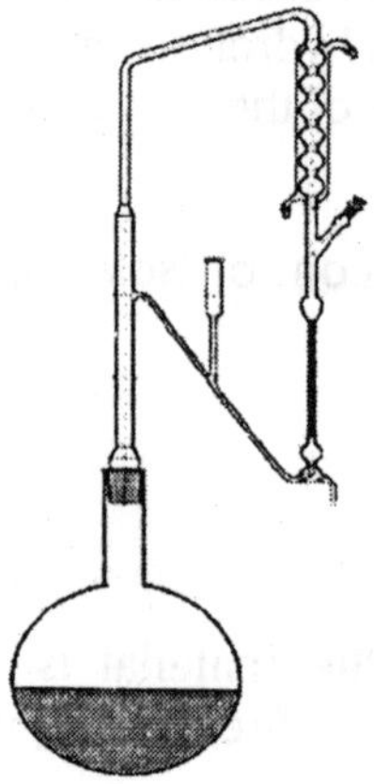

Fig 3: *Clevenger-type laboratory-scale hydrodistillation apparatus*

Traditional Method of Producing Attar using Hydro-distillation

Floral attars are defined as the distillates obtained by hydro-distillation of flowers (such as saffron, marigold, rose, jasmine, pandanus) in sandal wood oil or other base materials like paraffin.

Attar manufacturing takes place in remote places because the flowers must be processed quickly after collection. The apparatus and equipment used to manufacture attar are light, flexible, easy to repair, and have a fair degree of efficiency. Keeping in view these facts, the traditional "Deg and Bhapka" process has been used for centuries and is used evennow with the following traditional equipment (Fig 4).

- Deg (still)
- Bhapka (receiver)
- Chonga (bamboo condenser)
  - Traditional bhatti (furnace)
- Gachchi (cooling water tank)
- Kuppi (leather bottle)

Fig 4: *Traditional process of water distillation for making attar*

Disadvantages of Water Distillation

- Oil components like esters are sensitive to hydrolysis while others like acyclic monoterpene hydrocarbons and aldehydes are susceptible to polymerization.
- Oxygenated components such as phenols have a tendency to dissolve in the still water, so their complete removal by distillation is impossible.
- As water distillation tends to be a small operation, it takes a long time to accumulate much oil, so good quality oil is often mixed with bad quality oil.
- The distillation process is treated as an art by local distillers,who rarely try to optimize both oil yield or quality.
- Water distillation is a slower process than either water and steam distillation or direct steam distillation.

Water and Steam Distillation: In water and steam distillation, the steam can be generated either in a satellite boiler or within the still, although separated from the plant material. Like water distillation, water and steam distillation is widely used in rural areas. Moreover, it does not require higher capital expenditure than water distillation. Also, the equipment used is generally similar to that used in water distillation, but the plant material is supported above the boiling water on a perforated grid. In fact, it is common that persons performing water distillation eventually progress to water and steam distillation. It follows that once rural distillers have produced a few batches of oil by water distillation, they realize that the quality of oil is not very goodbecause of its subdued aroma. As a result, some modifications are made. Using the same still, a perforated grid or plate is fashioned sothat the plant material is raised above the water. This reduces the capacity of the still but affords a better quality of oil. If the amount of water is not sufficient to allow the completion of distillation, a cohobation tube is attached and condensate water is added back to the still manually, thereby ensuring that the water, which is being used as the steam source, will never run out. It is also believed that this will, to some extent, control the loss of dissolved oxygenated constituents in the condensate water because the re-used condensate water will allow it to become saturated with dissolved constituents,after which more oil will dissolve in it.

Cohobation: Cohobation is a procedure that can only be used during water distillation or water and steam distillation. It uses the practice of returning the distillate water to the still after the oil has been separated from it so that it can be re-boiled. The principal behind it is to minimize the losses of oxygenated components, particularly phenols which dissolve to some extent in the distillate water. For most oils, this level of oil loss through solution in water is less than 0.2%, whereas for phenol-rich oils the amount of oil dissolved in the distillate water is 0.2%-0.7%. As this material is being constantly re-vapourized, condensed and re-vapourized again, any dissolved oxygenated constituents will promote hydrolysis and degradation of themselves or other oil constituents. Similarly, if an oxygenated component is constantly brought in contact with a direct heat source or side of a still, which is considerably hotter than 100° C, then the chances of degradation are enhanced. As a result, the practice of cohobation is not recommended unless the temperature to which oxygenated constituents in the distillate are exposed is no higher than 100° C. In steam and water distillation, the plant material cannot be in direct contact with the fire source beneath the still; however, the walls of the still are good conductors of heat so that still notes can also be obtained from the thermal degradation reactions of plant material that is touching thesides of the still. As the steam in the steam and water distillation process is wet, a major drawback of this type of distillation is that it will make the plant material quite wet. This slows down distillation as the steam has to vapourize the water to allow it to condense further up the still. One way to prevent the lower plant material resting on the grid from becoming waterlogged is to use a baffle to prevent the water from boiling too vigorously and coming in direct contact with the plant material.

Advantages of Water and Steam Distillation over Water Distillation

- Higher oil yield.
- Components of the volatile oil are less susceptible to hydrolysis and polymerization

- If refluxing is controlled, then the loss of polar compounds is minimized.
- Oil quality produced by steam and water distillation is more reproducible.
- Steam and water distillation is faster than water distillation,so it is more energy efficient.

Many oils are currently produced by steam and water distillation,for example lemongrass is produced in Bhutan with a rural steam and water distillation system.

Disadvantages of Water and Steam Distillation

Due to the low pressure of rising steam, oils of high-boiling range require a greater quantity of steam for vaporization -hence longer hours requirement of distillation.

- The plant material becomes wet, which slows down distillation as the steam has to vaporize the water to allow it to condense further up the still.
- To avoid that the lower plant material resting on the grid becomes waterlogged, a baffle is used to prevent the water from boiling too vigorously and coming in direct contact withthe plant material.

Direct Steam Distillation: As the name suggests, direct steam distillation is the process of distilling plant material with steam generated outside the still in as atellite steam generator generally referred to as a boiler. As in water and steam distillation, the plant material is supported on a perforated grid abovethe steam inlet. A real advantage of satellite steam generation is that the amount of steam can be readily controlled. Because steam is generated in a satellite boiler, the plant material is heated no higher than 100°C and,consequently, it should not undergo thermal degradation. Steam distillation is the most widely accepted process for the production of essential oils on large scale. Throughout the flavour and fragrance supply business, it is a standard practice.

An obvious drawback to steam distillation is the much higher capital expenditure needed to build such a facility. In some situations, such as the large-scale production of low-cost oils (e.g. rosemary, Chinese cedar wood,lemongrass, *litsea cubeba*, spike lavender, eucalyptus, citronella,corn mint), the world market prices of the oils are barely high enough to justify their production by steam distillation without amortizing the capital expenditure required to build the facility over a period of 10 years or more.

Advantages of Direct Steam Distillation

- Amount of steam can be readily managed.
- No thermal decomposition of oil constituents.
- Most widely accepted process for large-scale oil production,superior to the other two processes.

Disadvantage of Direct Steam Distillation

- Much higher capital expenditure required to establish this activity than for the other two processes.

Essential Oil Extraction by Hydrolytic Maceration Distillation

Certain plant materials require maceration in warm water before they release their essential oils, as their volatile components are glycosidically bound. For example, leaves of wintergreen (*Gaultheria procumbens*)contain the precursor gaultherin and the enzyme primeverosidase;when the leaves are macerated in warm water, the enzyme acts on the gaultherin and liberates free methyl salicylate and primeverose. Other similar examples include extractions in brown mustard (sinigrin), bitter almonds (amygdalin)and garlic (alliin).

Essential Oil Extraction by Expression

Expression or cold pressing, is only used in the production of citrus oils. The term expression refers to any physical process in which the essential oil glands in the peel are crushed or broken to release the oil. One method that was practiced many years ago, particularly in Sicily (*spugna* method), commenced with halving the citrus fruit followed by pulp removal with the aid of sharpened spoon-knife (known asa *rastrello*). The oil was removed from the peel either by pressing the peel against a hard object of baked clay (*concolina*) which was placed undera large natural sponge or by bending the peel into the sponge. The oil emulsion absorbed by the sponge was removed by squeezing it into the *concolina* or some other container. It is reported that oil produced this way contains more of the fruit odour character than oil produced by any other method. A second method known as equaling (or the *scodella* method),uses a shallow bowl of copper (or sometimes brass) with a hollow central tube; the equaling tool is similar in shape to a shallow funnel. The bowl is equipped with brass points with blunt ends across which the whole citrus fruit is rolled by hand with some pressure until all of the oil glands have burst. The oil and aqueous cell contents are allowed to dribble down the hollow tube into a container from which the oil is separated by decantation.Obviously, hand pressing is impractical because it is an extremely slow process, e.g. on average only 2-4 lbs oil per day can be produced by a single person using one of these hand methods. As a result, over the years a number of machines have been designed to either crush the peel of a citrus fruit or crush the whole fruit and then separate the oil from the juice.

Pelatrice Process

In the *pelatrice* process, citrus fruits are fed from a hopper into the abrasive shell of the machine. The fruits are rotated against the abrasive shell by a slow-moving Archimedian screw whose surface rasps the fruit surfaces causing some of the essential oil cavities on the peel to burst andrelease their oil-water emulsion. This screw further transports the fruit into a hopper in which rollers covered with abrasive spikes burst the remaining oil cavities. The oil and water emulsion is washed away from the fruit by a fine spray of water. The emulsion next passes through a separator where any solids are removed, after which it passes through two centrifugal separators working in series to yield the pure oil. Most bergamot oil and some lemonoil are produced this way in Italy.

Sfumatrice Process

The *sfumatrice* equipment consists of a metallic chain that is drawn by two horizontal ribbed rollers. The peels are conveyed through these rollers during which time they are pressed and bent to release their oil. Asin *pelatrice*, the oil is washed away from the *sfumatrice* rollers by fine sprays of water. Again, the oil is initially passed through a separator prior to being sent to two centrifuges in series, so that purified oil can be produced. At onetime, *sfumatrice* was the most popular process for citrus oil isolation in Italy;however, today the *pelatrice* method appears more popular.

Essential Oil Extraction with Cold Fat (Enfleurage)

Despite the introduction of the modern process of extraction with volatile solvents, the old fashioned method of enfleurage, as passed on from father to son and perfected in the course of generations, still plays an important role. Enfleurage on a large scale is today carried out only in theGrasse region of France, with the possible exception of isolated instances in India where the process has remained primitive. The principles of enfleurage are simple. Certain flowers (e.g.tuberose and jasmine) continue the physiological activities of developing and giving off perfume even after picking. Every jasmine and tuberose flower resembles, so to speak, a tiny factory continually emitting minute quantities of perfume. Fat possesses a high power of absorption and, when brought in contact with fragrant flowers, readily absorbs the perfume emitted. This principle, methodically applied on a large scale, constitutes enfleurage. During the entire period of harvest, which lasts for eight to ten weeks, batches of freshly picked flowers are strewn over the surface of a specially prepared fat base (corps), let there (for 24 hours in the case of jasmine and longer in thecase of tuberose), and then replaced by fresh flowers. At the end of theharvest, the fat, which is not renewed during the process, is saturated with flower oil. Thereafter, the oil is extracted from the fat with alcohol and then isolated.

Enfleurage and Defleurage

Every enfleurage building is equipped with thousands of so called chassis, which serve as vehicles for holding the fat corps during the process. A chassis consists of a rectangular wooden frame. The frame holds a glass plate upon both sides of which the fat corps is applied with aspatula at the beginning of the enfleurage process. When piled one above the other, the chassis form airtight compartments, with a layer of fat on the upper and lower side of each glass plate.Every morning during the harvest the freshly picked flowers arrive,and after being cleaned of impurities, such as leaves and stalks, are strewn by hand on top of the fat layer of each glass plate. Blossoms wet from dew or rain must never be employed, as any trace of moisture will turnthe corps rancid. The chassis are then piled up and left in the cellars for 24 hours or longer, depending upon the type of flowers. The latter rest in direct contact with one fat layer (the lower one), which acts as a direct solvent whereas the other fat layer (beneath the glass plate of the chassis above)absorbs only the volatile perfume given off by the flowers. After 24 hours, the flowers have emitted most of their oil and start to wither, developing an objectionable odour. They must then be removed from the corps, which process, despite all efforts to introduce labor-saving devices, is still done by hand. Careful removal of the flower (defleurage) isalmost more important than charging the corps on the chassis with fresh flowers (enfleurage) and, therefore, the persons doing this work must be experienced and skilled. Most of the exhausted flowers will fall from the fat layer on the chassis glass plate when the chassis is struck lightly against the working table, but since it is necessary to remove every single flower and every particle of the flower, tweezers are used for this delicate operation. Immediately following defleurage, i.e., every 24 hours, the chassis are recharged with fresh flowers. For this purpose the chassis are turned over and the fat layer, which in the previous operation formed the top (ceiling) of the small chamber, is now directly charged with flowers. In the case of jasmine, the entire enfleurage process lasts about 70 days: daily the exhausted flowers are removed and the chassis are recharged with fresh ones. At the beginning of, and several times during, the harvest, the fat on the chassis is scratched over with metal combs and tiny furrows are drawn in order change and increase the surface of absorption.

At the end of the harvest, the fat is relatively saturated with flower oil and possesses the typical fragrance. The perfumed fat must then be removed from the glass plates between the chassis. For this purpose,it is scraped off with a spatula and then carefully melted and bulked in closed containers. The final product is called pomade (pomade de jasmine,pomade de tuberous, pomade de violet, etc.). The most highly saturated pomade is pomade no. 36, because the corps on the chassis have been treated with fresh flowers 36 times during the whole process of enfleurage. At the beginning of the harvest, every chassis is charged with about 360g fat corps on each side of the glass plate, in other words, with 720 g per chassis. Every kilogram of fat corps should be in contact with about 2.5 kg(preferably with 3.0 kg) of jasmine flowers for the entire period of enfleurage,which lasts from 8 to 10 weeks. The quantities differ somewhat for different flowers. At the end of enfleurage, the fat corps has lost about 10% of its weight because of the various manipulations.

Hot Maceration Process

In this process, the long enfleurage time is reduced by the immersion of petals in molten fat heated at 45°-60° C for 1 to 2 h, depending upon the plant species. After each immersion, the fat is filtered and separated from the petals. After 10 to 20 immersions, the fat is separated from waste flowers and water. Absolute of maceration is then produced from fat containing oil through the process of extraction and concentration under reduced pressure. It is mainly used for highly delicate flowers whose physiological activities are lost rapidly after their harvest, such as lily of valley.

Modern (Non-traditional) Methods of Extraction of Essential Oils

Traditional methods of extraction of essential oils are the methods most widely used on commercial scale. However, with technological advancement, new techniques have been developed which may not necessarily be widely used for commercial production of essential oils but are considered valuable in certain situations,such as the production of costly essential oils in a natural state without any alteration of their thermosensitive components or the extraction of essential oils for micro-analysis. These techniques are as follows:

Headspace Trapping Techniques

Headspace trapping techniques are used to capture the odour emitted by aromatic materials. Odour can be sampled either directly or trapped on an adsorbent material. The trapped odorous components can be freed by solvent extraction or thermal desorption prior to analysis by modern instrumental techniques.

In the static headspace sampling technique, analyte is kept in a closed vial and the air (headspace air) above the solid or a liquid sample is sampled by a gas syringe or directed on to the gas chromatography column or more usually first concentrated on an adsorbent trap.

Vacuum headspace sampling technique involves suction of the headspace air via a vacuum pump through condensers cooled with liquid nitrogen to condense odorous principles. This technique is also used by some perfumery companies for commercial scale production of fragrances.

Dynamic headspace sampling involves sweeping the analyte with a stream of air or gas and adsorption of the volatiles from the gas stream on an adsorbent trap. Dynamic headspace sampling techniques can be applied in one of the following ways: Closed-loop Stripping Method and Direct Sampling Method.

The three main historical techniques are static headspace, dynamic headspace (purge and trap) and solid-phase microextraction (SPME). Repeatability, detection limit, and elimination of water before the chromatographic run are the three principal variables. Static headspace has an excellent repeatability [coefficients of variation (CV) < 5%], with no interference by water, but sometimes the achieved LOD is not enough to comply with some environmental regulation. For example, it cannot be used to measure concentrations of benzene below 0.5–1 μg/L. SPME has a good LOD (below 0.5 μg/L), but a worse repeatability (CV > 10 %) when using an autosampler (10). The purge and trap technique has a better LOD than static headspace, but the elimination of water may be difficult, and the system is prone to carry-over effects. This makes it is necessary to clean the system using reagent water between every sample. The device must be washed, purged with detergent solution, rinsed with distilled water, and dried for analyses of samples containing large amounts of water-soluble materials, suspended solids, high boiling point compounds, or high levels of volatile compounds.

Solid Phase Micro-extraction (SPME)

Sample preparation is an essential step in analysis, greatly influencing the reliability and accuracy of resulted the time and cost of analysis. Solid-Phase Microextraction (SPME) is a very simple and efficient, solventless sample preparation method, invented by Pawliszyn in 1989. SPME has been widely used in different fields of analytical chemistry since its first applications to environmental and food analysis and is ideally suited for coupling with mass spectrometry (MS). All steps of the conventional liquid–liquid extraction (LLE) such as extraction, concentration and transfer to the chromatograph are integrated into one step and one device, considerably simplifying the sample preparation procedure. It uses a fused-silica fibre that is coated on the outside with an appropriate stationary phase. The analytes in the sample are directly extracted to the fibre coating. The SPME technique can be routinely used in combination with gas chromatography, high-performance liquid chromatography and capillary electrophoresis and places no restriction on MS. SPME reduces the time necessary for sample preparation, decreases purchase and disposal costs of solvents and can improve detection limits. The SPME technique is ideally suited for MS applications, combining a simple and efficient sample preparation with versatile and sensitive detection.

Supercritical Fluid Extraction (SFE)

Supercritical fluid extraction is the most effective way to extract valuable constituent botanicals. Supercritical Fluid Extraction (SFE) is the process of separating one component (the extractant) from another (the matrix) using supercritical fluids that is CO_2 as the extracting solvent. CO_2 is the king of extraction solvents for botanicals. Extraction conditions for supercritical CO_2 are above the critical temperature of 31°C and critical pressure of 74 bar. Supercritical fluids are highly compressed gases, which have combined properties of gases and liquids in an intriguing manner. Supercritical fluids can lead to reactions, which are difficult or even impossible to achieve in conventional solvents. It is a fast process completed in 10 to 60 minutes. A supercritical fluid can be separated from analyte by simply releasing pressure, leaving almost no trace and yields a pure residue.

Phytosol (phytol) Extraction

The method called "advanced Phytonics" use 1,1,1,2- tetrafluoroethane with or without modifiers. The solvents are named "Phytosols®". Phytosol A consists only of the gas. Phytosol B is a mixture of the gas with butane/isobutene. Phytosol C consists of the gas with dimethylether as modifier. Boiling point of 1,1,1,2-tetrafluoroethane or HFC 134 as known in trade is –26.2°C. It is a non-toxic, odourless gas whose pressure in liquid state at room temperature is 5 bar which can be compared to the pressure of a bottle of Champagne (6 bar).

The process is aptly applicable to those compounds that are sensitive to thermal or UV energy, oxygen and acidic conditions, pharmaceutical, cosmetic, medical, food, flavors and fragrances industries.

Poroplast Technique

Liquid-liquid extraction can be defined as a partition technique where the constituents of a solution are separated due to their different partition coefficients between the two immiscible liquids. Porous support which holds the stationary phase is teflon powder made into a sponge through a special heat treatment. Aqueous dispersion is pumped into the column. Aromatic constituents are transferred from the aqueous phase to the organic phase.

Simultaneous Distillation Extraction (SDE)

SDE was designed by Likens and Nickerson to diminish limitations of liquid liquid extraction (LLE)36, regardless of its limitation in time and solvent consumption. SDE allows simultaneous extraction, concentration and isolation of flavor constituents within a short time due to continuously recycling the two immiscible solvents. Only a small amount of solvent is used for extraction so the volatile compounds can be easily concentrated more than tenfold. During the initial extraction step, both the sample and solvent are simultaneously heated in two different flasks, which are separately connected to arms of SDE apparatus, from which both vapors mix at the upper part of the arms and condense on the cold finger. Then, the liquid mixture flows into the separator to form two immiscible phases that flows back individually to their original containers because of the different liquid density. The sample and solvent bottles should be placed at the correct place on the extractor in order to avoid the separated phases flowing back into their opposite bottles. Overall, SDE not only possesses the advantages of LLE, by which many solvents, such as: hexane, dichloromethane (DCM) and n-pentane, can be used for volatiles extraction, but also shortens the extraction time and improves the extraction efficiency.

Microwave Distillation

Application of microwaves in separation and extraction processes has shown to reduce both extraction time and volume of solvent required, minimizing environmental impact by emitting less CO_2 in atmosphere and consuming only a fraction of the energy used in conventional extraction methods such as steam distillation, SD. Advances in microwave-assisted extraction (MAE) have led in the development of various techniques such as compressed air microwave distillation (CAMD), vacuum microwave hydro distillation (VMHD), microwave hydrodistillation (MWHD), solvent-free microwave extraction (SFME), microwave accelerated steam distillation (MASD), microwave by hydro -diffusion and gravity.

Controlled Instantaneous Decomposition (CID)

The process involves subjecting the partially humidified plant material for a short period of time to a steam pressure varying from 0.5 to 3 bar followed by a rapid decompression to a vacuum (about 15 mbar) for 200 msec each time. The vapour in the plant material created by autovaporization produces a mechanical strength which ruptures the oil cells. Due to pressure differences in the extractor and vacuum chamber, in each opening of the pressure valve oil-rich vapour is instantaneously sucked into the vacuum chamber where it instantly condenses.

Microdistillation

Microdistillation is a micro scale distillation technique which allows rapid programmable distillation. The system has 6 heating and 6 cooling units for parallel operations. Small amount of material to be distilled (100-500 mg) is placed in a 15 mL capacity vial with water (10 mL) and sealed with a cap having a small slit on top requires less than 1 gr or 1 ml of material, which makes it excellent for working with herbarium samples. Fast preparation and set-up, fast isolation and readiness for analysis, cheaper and more efficient in long term use and easy operation.

Molecular Spinning Band Distillation

Spinning band distillation uses a rotating helical band to create a high number of theoretical plates. The spinning bands can be made of Teflon or metal. Teflon spinning bands are used for distillations below 225°C. Metal bands are used for higher temperature distillations where Teflon would become soft. Spinning band distillation gives a very efficient separation in a short distillation column with High Efficiency, Low Column Hold Up and Low pressure drop.

Membrane Extraction

In all living systems the transfer of solutes is achieved through membrane processes. In membrane technology, the membrane used is not always a solid. Liquid Membrane Extraction consists of three liquid systems, Feed, Membrane and Sripping. There are several membrane extraction and distillation techniques such as Pervaporation and Liquid membrane extraction. The basic principle of pervaporation for the removal of volatile organic compounds (VOC) from aqueous media is as follows: Hot aqueous feed flows alongside a nonporous hydrophobic membrane which consists of a thin film of silicone rubber on teflon.

□□□

14

Processing, Storage and Utilization of Drugs

Medicinal principles are usually present in different parts of the plant like root, stem, bark, heartwood, leaf, flower, fruit or plant exudates. These medicinal principles are separated by different processes where the most common being extraction. Technology for the manufacture of standardised extracts and phytochemicals is available and there are many extracts already in the international market as drugs. A drug such as an extract of *Centella asiatica* can be manufactured as an extract containing a standard quantity of asiaticoside. Similarly for senna a standardised extract of which, containing a standard quantity of sennosides a and b could easily be produced with equipment that can be designed and constructed in most developing countries.

The promotion and development of processing of medicinal and aromatic plants have gained importance recently in many developing countries. Green consumerism and resurgence of interest for plant based products, liberalised and free market economy, increasing awareness about biodiversity conservation and sustainable use of natural resources coupled with poor socio-economic conditions of native populations are ground realities for planning and harnessing the low-cost and purpose oriented process technologies.

UNIDO has already developed a Polyvalent Pilot Plant with a view to enabling developing countries to upgrade their technology for the processing of medicinal and aromatic plants.

This plant incorporates all salient features of a low cost, efficient, small capacity factory which can carry out solvent extraction, solvent percolation, concentration of miscella, solvent recovery, steam distillation and oil separation. The design and fabrication of the process equipment need not be over emphasised, as even if a good design is available for adaptation, it must be done to fit the given situation.

The polyvalent plant is characterised by simplicity of design, installation, operation,maintenance and repair. Some of its features are:

1. Modular construction so as to permit increase in capacity and function by duplicating or adding modules.
2. Simultaneous processing for more than one product, such as extraction at one end, production of solid extract or oleoresins at the other.
3. Standardised or optimised process control and measuring units, pumps and other ancillaries can be easily replaced.
4. All plumbing and electrical wiring are simple and easily accessible.
5. Multipurpose uses. E.g. Solvent/aqueous extraction, continuous extraction, preparation of solid extract and oleoresins, essential oil distillation, fractionation of essential oils and production of absolutes and concretes or even processing of other phytoproducts.

Processing Technologies Developed by Institutions

1. **National Institute of Science Communication and Information Resources (NISCAIR), New Delhi.**

   Traditional Knowledge Digital Library (TKDL): A joint venture of CSIR and Department of AYUSH (Ayurveda, Yoga and Naturopa-thy, Unani, Siddha and Homeopathy) inventorising traditional knowledge, formulations developed in India over several centuries in the above mentioned traditional systems of medicine. TKDL has over 200,000 traditional medicine formulations codified in 30 mil-lion pages in different languages.

2. **Central Food Technological Research Institute (CFTRI), Mysore.**

   Preparation of annatto dye; Natural colours; Deterpenation of cit-rus oils; Monsooning of cherry coffee; Cocoa: bean curing, cocoa mass, cocoa butter, cocoa powder; Compounded asafoetida; Cardamom-fixation of green colour; Coriander dhal, supari; Encapsulated food flavors; Garlic powder; Dehydration/ drying of ginger; Kokum: concentrate and powder; Mustard powder; Pepper: white, green and dehydrated green pepper; Drying (including dipsol formulation, fractionation) of red chillies; Spice oils: pepper, ginger, turmeric, cardamom; Spice oleoresins: pepper, ginger, turmeric, chillies; Tamarind: concentrate and powder; Turmeric: curing and polishing; Walnut processing; Product of phycocyanin from *Spirulina;* Rural-based biotechnological production of *Spirulina*; Technol-ogy for manufacturing plant growth hormones from agricultural wastes; Technology for manufacturing plastic pouches; Protein foods; Fruit/vegetable products; Beverage products; Convenience foods; Microbial and fermentation products; Cereal and pulse products; Microbial and fermentation products; Bakery products; Meat, fish and poultry products.

3. **Central Drug Research Institute (CDRI), Lucknow.**

   Centchroman (contraceptive); Centbucridine (local anesthetic); arteether (anti-malarial); Bulaquin (anti-malarial); gugulipid (hypolipidaemic); standardized Herbal Remedy (memory enhanc-er); Isaptent (abortion); Centpropazine (anti-depressant); Consap (spermicidal cream); Picroliv (hepato-protective); CDRI-99/373 (anti-osteoporosis, anti-resorptive agent).

4. **Institute of Himalayan Bioresource Technology (IHBT), Palampur.**

   Tea wines and concentrates; Theaflavin (anti-oxidant); Vanillin from vanilla beans; Vinyl guaicol (flavoring agent); Herbal tooth-picks; Himalyan delight (perfume); Viral diagnostic kits; Geprot-edTM for biote :hnology; HerbostillTM (mini distillation unit); Steri-flowTM (mini laminar flow cabinet); Tea withering machine; Natural colours; Genes and corresponding enzymes; Gulkand; Rose oil and rose water (3 grades); Beta aescin from *Aesculus indica*; Stevioside from *Stevia*.

5. **Indian Institute of Chemical Biology (IICB), Kolkata.**

   Prostalyn (herbal formulation remedy for prostate problem); Herbal composition for treating asthma; Herbal products for controlling gastric ulcer; *Murraya koenigii* extracts for treating asthma; Herbal composition for treating chronic myeloid leukemia; Anti-leishmanial activity of betel leaf extract; Herbal molecule as poten-tial anti-leukemic drug; Herbal extract and lupinoside as potential anti-diabetic type- II drug from *Pueraria tuberosa*; An anti-leukemic composition comprising withaferina-A; Process for the isolation of a carbohydrate fraction from *Feronia limonia* possessing anti-tumor activity; Process for the isolation of active principle from *Azadirachta indica* useful for controlling gastric hyperacidity and gastric ulceration; Process for isolation of fraction consisting (-) frullanolide from *Sphaeranthus indicus* possessing antifungal, antibacterial and anti-protozoal activities; Anti-peptic ulcer activity of *Woodfordia fruticosa* flower extract; Anti-leishmanial activity of *Centella asiatica* leaf extract; Anti-leukaemic activity of withanolide-D; Development of vaginal contraceptive with clove oil; *Pueraria tuberosa* extract and lupinoside and its analogues as anti-diabetic Type- II drugs; *Piper betel* and *Murraya koenigii* formulation for blocking 5-lipoxygenase activity; Leishmanicidal activity of night jasmine leaf extract; Methanolic extract of *Piper betel* leaves for the treatment of human malignancies.

6. **Indian Institute of Integrative Medicine (IIIM) (formerly Regional Research Laboratory), Jammu.**

   Piperine (bio-enhancer) and more potent molecules; Oral care rinse products; Hepato-protective herbal product; *Hypericum per-foratum* and *Tinospora cordifolia* standardized extracts; *Rhizobium* biofertilizers and other plant growth promoting bio-inoculants; Bio-control agents for plant disease management; Industrial enzymes viz. glucose oxidase, dehydrogenases, amylases, lipases and esterases; Bioreactor cultivation of *Swertia chirata*; Chemo-profiling of 40 medicinal and aromatic plants through GLC/ HPLC/ HPTLC; An anti-osteoporosis compound from *Boerhaavia diffusa*; Herbal drug for the treatment of liver disorders and as prophylactic treatment against viral diseases such as common cold and fever; Standardization and testing of herbal drugs; Indian herbal pharma-copoeia;

   Cultivars/chemotypes of *Ocimum*, Ashwagandha, Lemongrass, *Cymbopogon nardus* var. *confertiflorus*, *Mentha longifolia*, *Picrorhiza kurroa*, *Podophyllum hexandrum*, *Swertia chirata* etc.; Agro-technology for *Echinacea angustifolia*,

Clarysage, *Hypericum perforatum*, *Valeriana*, hops, Clocimum and many other plants; Micro propagation of medicinal and aromatic plants.

7. **National Institute of Interdisciplinary Science and Tech-nology (NIIST) (formerly Regional Research Laboratory), Thiruvananthapuram.**

   Oil and oleoresin from ginger; Oil, oleoresin and piperine from pepper; Colour and capsaicin fraction from red chillies and oleo-resin from fresh and dry chillies; Oleoresin and curcumin from turmeric; Micro encapsulated flavours and oleoresins.Extraction and separation of phyto-chemicals from plants for healthcare applications: carotenes, tocopherol/ tocotrienols, sterols, oryzanol, ferulic acid, phospholipids, high purity wax fractions; Supercritical fluid extraction of phytochemicals, formulations for healthcare applications; Separation of active principles from spices and herbals for nutraceutical and functional food supplements; Natural and nature identical bio-active molecules from bio-resources for healthcare applications.

8. **North East Institute of Science and Technology (NEIST) (formerly Regional Research Laboratory), Jorhat.**

   Agrotechnologies for important medicinal and aromatic plants (Citronella, Lemongrass, Palmarosa, Medicinal yam, *Plumbago zeylanica* etc.); High yielding varieties of medicinal and aromatic plants (BLI-Arun of Lemongrass); Tissue culture protocols for medicinal and aromatic plants; Isolation and characterization of insecticidal compounds of plant origin; Arteether; Liquid deodorant and cleaner; Herbal room freshener; Caffeine from tea-waste; Diosgenin from *Dioscorea* tuber; 16-DPA from diosgenin; Pesticides (Phosphamidon, Quinalphos, Chlorfenvinphos).

9. **Institute of Minerals and Materials Technology (IMMT) (formerly Regional Research Laboratory), Bhubaneswar.**

   Agrotechnologies : Cultivation of aromatic, medicinal and other economic plants (Palmarosa, Lemongrass, Citronella, Patchouli, *Ocimum*, Cinnamon, Screw pine, Ambrette, *Cola nitida*, Vetiver, *Catharanthus roseus*, Rubber, Simarouba and fast growing tree species.

   Natural Products: Strychnine and brucine from seeds of *Strychnos nux-vomica*; Solasodine from Berries of *Solanum khasianum*; Development of dyes of various colors for fabric colouring.

   Herbal drugs: Standards for 100 single and compound herbal drugs; Optimization of manufacturing process for 18 Ayurvedic formulations; HPT-LC/ HPLC quality evaluation methods for well known herbal drugs; Monographs and Pharmacopoeia.

10. **National Botanical Research Institute (NBRI), Lucknow**

    Luvstick-herbal lipstick; Lip balm; NBIRA-fermented drink; NBIRA-SOFT; Herbal gulal; Instant glow face pack for teenagers; Anti- ageing cream for wrinkle-free, younger looking healthy skin; Anti-inflammatory and anti- psoriatic cream; Anti-ageing face pack for normal skin; Facial scrubs; Anti-ageing face pack for oily skin; Instant glow face pack for acne-prone skin; Anti-bacterial, anti-inflammatory toothpicks for healthier gums; Hair tonics and shampoos for controlling hair fall and graying; Anti-oxidant, anti-septic swabs; After-shave lotion; Face and body wash; Mouth wash; Dental care and cure product from neem extract; 150 single herbal drugs and 18 herbal formulations have been standardized following WHO guidelines. High yielding varieties of Opium and grain amaranth; Cultivation package of *Rosa damascena* and rose water extraction; Extraction and standardization of neem active principles.

11. **Central Institute of Medicinal and Aromatic Plants (CIMAP), Lucknow**

    Agro and processing technologies: High yielding varieties of medicinal and aromatic plants; Agrotech-nologies for medicinal and aromatic plants; Distillation and rectification technologies for aromatic plants; Aroma chemicals extraction technologies; Extraction technologies for medicinal plants; Books, Periodicals publication; Services to farmers, entrepreneurs and industry on all aspects of medicinal and aromatic plants.

    Technologies, processes, products: Bio-enhancers: lysergol (*Ipomoea muricata*), glycyrrizin (*Glycyrrhiza glabra*), niaziridin; Antifungals; Anti-bacterials: oenostacin (*Oenothera biennis*), thymol (*Trachysermum ammi*); Anti-cancer: docetaxel (taxotere) (*Taxus wallichiana*), camptothecin (*Mappia foetida*); Hepato-protective: silymarin (*Silybum maria-num*), Cliv-92 (*Cleome viscosa*); Andrographolides (*Andrographis paniculata*).

    Herbal products: Cracknil (anti-crack cream); Mosrep (mosquito repellant agarbat-ties); Mospray (mosquito repellant spray); Mosaway (mosquito repellant cream); Mosex (mosquito repellant lotion); Mosnobite (mosquito repellant vaporizer); Pain chhoo (pain balm for head-ache & sprain); Swabee (surface disinfectant); Hankool (hand disinfectant); Skinpro (antifungal cream); Myconil (antifungal cream); Herbal tooth powder (for plaque & gingivitis); Kleenzie (Hand & face wash with *Aloe vera*); Rose water; Essential oil kit (Organic); CIM-Phalse (nutraceutical); Herbisoft and Geranium Active (Shampoos); CIM-Poshak; Haloe skin.

Storage of Medicinal Plants and Crude Drugs

Dry extracts are usually very hygroscopic and should therefore be ground, mixed under conditions, which exclude moisture as much as possible. Intermediate and end products must also be stored under dry conditions. Annealing or sealing of the products in suitable moisture tight synthetic foils has proved a good method for this.

Requirement of Packaging Materials

The general requirements of packaging materials are as follows:

1. Economical or low cost
2. Impermeable as glass or metal or of acceptable permeability to moisture, gases, volatile solvents etc.
3. Non- reactive-relatively inert with no extraction, exchange or interaction
4. Easy to manufacture in a wide range of shapes, preferably by a number of manufacturing processes.

5. Easy to decorate and /or print by a range of processes
6. Good production line efficiency-performance, with the minimum of rejects or wastage.
7. Effective as a pack(container and closure), i.e., easy to open and reclose and use if multidose; or open if single dose, whilst meeting any special requirements such as child resistance, tamper evidence or resistance, etc.
8. Easy to produce and maintain clean
9. Preferably readily available both in terms of source of supply for raw materials and as a converted item component from several suppliers.
10. Environmentally friendly
11. Able to optimise use of space when stacked or during transportation.

Types of Packing Materials

The common types of packaging materials currently available are given below

1. Glass: It can be found as several variants such as treated soda glass, soda glass and non parenteral.
2. Metals: A variety of metals including tin plate(tin coated mild steel) tin free steel, aluminium, aluminium alloys are widely used in packaging, being found as rigid containers, collapsible containers, aluminium foils, metalised coatings etc.
3. Plastics: There are five economical materials for rigid type of containers i.e., those based on polyethylene (PE), polypropylene(PP), Polystyrene(PS), PVC and polyester.
4. Elastomeric materials: Elastomers can be found as a wide range of basic materials (i.e., natural rubber, synthetic polyisoprene, neoprene, nitryl, butyl, including bromo-and chloro butyl, ethylene propylene-diene modified (EPDM), acid silicone elastomers.

Good Store House for Storage of Herbs

There should be adequate space in the store house for 'Under Test'. Approved and Rejected herbs with arrangements and equipments to allow washing, cleaning, drying and orderly placement of stored herbs with controlled temperature and humidity. The store house for storage of herbs, handling of herbs and drying space etc. shall be as per the provision of GMP.

Sores under GMP1.1 (F): Proper ventilation, free from dampness, adequate space for different type of materials.

Sores under GMP1.1 (F)(A): Different stores may be used for herbs as mentioned below:

- Fresh herbs
- Dry herbs
- Plant concentrates/ extracts

Steps Involved for Proper Storage of Herbs

- Identification by naked eyes
- Cleansing of herbs: flowers, fruits etc may be cleaned
- Washing of crude drugs like stem barks, roots, leaves, stems with potable water
- Drying of raw crude drugs
- Removal of microbes from raw herbs

Sterilization of Raw Herbs

Sterilization of raw herbs needs to be stored to be done by any of the following methods:

- Dry heat sterilization
- Steam sterilization
- Sterilization by radiation like infra-red radiation, UV rays, X-rays, gamma rays, alpha radiation and beta radiations
- Chemical method – gaseous sterilization by ethylene oxide

Temperature for Storage of Raw Herbal Drugs

All the raw herbal drugs may be stored at cool place between 8-25°C. Raw herbs must be protected from moisture, freezing , light and excessive heat from preventing decomposition.

Testing of Quality Raw Herbs

Testing of raw herbs may be conducted as per guidelines of WHO.

Authentication: Taxonomical study and botanical identity may be established.

Foreign matter: Herbs should be entirely free from soils, stones, dusts, insects and other animal contaminations.

Organo-leptic evaluation: It covers the microscopic appearance of drugs, its odour, test, sound, feel of touch and colour.

Microscopic examination: It gives information on type of stomata, vessel thickenings, which is helpful for identification of species.

Volatile matter: Volatile oil in *Mentha*, *Ocimum* etc. is determined by steam distillation of plant. It plays an important role for quality control parameter.

Ash value: The presence of ash in medicinal plants is determined as total ash, acid insoluble ash etc.

Extractive values: It indicates on quality of raw herbs.

Chromatographic profile and marker component: Thin layer chromatography and high performance liquid chromatography provides information on active alkaloids of drug.

Pharmacological evaluation: Bitterness value, haemolytic activity, foaming index etc.

Toxicological pesticide residues: Medicinal plants are used for longer periods and therefore pesticide residues limit may be established for safety.

Microbial contamination: The presence of moulds, microbes, aflatoxins etc.in plant material may be tested.

Radioactive contamination: The presence of radioactive particles may also be tested.

Determination of heavy metals: In plants material show permissible limit by WHO.

Testing of raw herbs as per API: Testing of raw herbs may conducted for macroscopic examination, microscopic examination identity , purity and strength, TLC, constituents, etc. Determination of heavy metals: In plants material permissible limit by API.

Table 1. Permissible limit heavy metal cosmetic herbal products

Heavy metals	Permissible limits
Mercury	1 ppm
Cadmium	0.30 ppm
Lead	10 ppm
Arsenic	3 ppm

Table 2. Materials for specific packaging and storage of raw herbs

Sl.No	Category	Packing materials
1.	Woody in nature like stem, heart wood, bark etc.	Gunny bags and woven sacks
2.	Soft in nature like creepers, leaves etc.	High gauge high density polyethylene bags, woven sacks with low density liner, high gauge polyethylene bags
3.	Fleshy in nature like fruits, rhizomes etc.	High gauge high density polyethylene bags, woven sacks with low density liner ,wooden boxes
4.	Flowers, anthers, stigma, petals, seeds, etc.	Corrugated boxes with polypropylene woven sacks, HDPE containers, Fibre Boards Drums.
5.	Volatile contents	Air tight HDPE containers, Air tight HDPE carboys, Card board box with polyethylene liners
6.	Herbal extracts and compounds	Air tight HDPE containers, Corrugated boxes with polypropylene woven sacks, Fibre Boards Drums.\ with polyethylene bags

. There are no strict guidelines for proper storage of raw herbal material. It is only the proper handing, packaging and storage of herbs which can preserve the safety, efficacy and quality of herbs. Exposures to light, moisture, dust, etc are all harmful which should be restricted to maximum extent.

Formulations and Industrial Utilizations

Medicinal plants are utilized as raw materials for extraction of active constituents in pure form (e.g. alkaloids like quinine and quinidine from cinchona bark, emetine from ipecacuanha root, glycosides from digitalis leaves, sennosides from senna leaves), as precursors for synthetic vitamins or steroids, and as preparations for herbal and indigenous medicines. Products such as ginseng, valerian and liquorice roots are part of the herbal and health food market, as well as the food flavours, fragrance and cosmetic industries. Certain plant products are industrially exploited like liquorice in confectionery and tobacco, papaine as meat tenderiser, quinine as soft drink tonic and cinchona as wine flavour. A large quantity of medicinal plant material is used in the preparation of herbal and medicinal teas, e.g. chamomile. These herbal and food uses are of great importance, also to the exporters from developing countries. Hundreds of medicinal plants are items of commerce, however relatively small countries are used in formulated herbal remedies. Several formulations like herbal teas, extracts, decoctions, infusions, tinctures, etc are prepared from medicinal plants.

Herbal teas, Herbal remedies: Herbal tea or infusion mixtures are mixture of unground or suitably ground medicinal plants to which drug plant extracts, ethereal oils or medicinal substances can be added. Infusion mixtures should be as homogenous as possible.

Drug extracts: They are preparations obtained by extracting drugs of a certain particle size with suitable extraction agents (menstrua). The extract obtained after separation of the liquid from the drug residue is called miscella. It may already represent the final liquid dose form e.g. as a so called fluid extract, or be used as an intermediary product which is to be further processed as quickly as possible.

Aqueous drug extracts : The degrees of comminution are used for the extract depending on the type of plant parts. For leaves, flowers and herbs shredded (4000mm); woods, barks and roots shredded (2800mm); fruits and seeds (2000mm). Alkaloid containing drugs are powdered (700mm).

Decoctions: The drug in the prescribed comminution is put in to water at a temperature above 90°C. The container is suspended in a water bath and maintained at this temperature for 30 minutes, with repeated stirring. The mixture is then strained while still hot.

Infusions: One part of the comminuted drug is kneaded several times in a mortar with 3-5 parts of water and left to stand for 15 minutes. The rest of the boiling water is then poured on to the mixture, which is suspended in a container in a water bath and kept for 5 minutes, with repeated stirring at a temperature above 90°C. The mixture is covered and left to stand until cool.

Macerates: The comminuted drug is left to stand, with occasional stirring, for 30 minutes after the required quantity of water has been poured on to it at room temperature. The extract is then strained and made up to the prescribed weight with rinsings.

Tinctures: Tinctures are extracts from drug plants prepared with ethanol of varying concentration, ether or mixtures of these, perhaps with certain additives, in such a way that one part of drug is extracted with more than two parts, but at most ten parts, of extraction liquid.

Fluid extracts: Like tinctures, they are liquid preparations, the difference being that they are more concentrated.

Dry extracts: They are usually very hygroscopic and should therefore be ground and mixed under conditions which exclude moisture as much as possible. Intermediate and end product must also be stored under dry conditions.

There are also liquid, semisolid, solid and controlled release formulations or preparations. The other dose forms are injections, implants, ocular preparations, inhalations and transdermal systems. Liquid formulations may be solutions, emulsions, colloids or suspensions in the increasing order of particle size. They may be intended for administration parenterally, orally or topically including administration into body cavities. Homogeneity for the formulations is very important, particularly where the active ingredient is present in lower concentration.

A generalised production scheme include the following:

Pre-processing: Washing, particle size optimisation, moisture reduction, refinement or concentration.

Solublisation: Insolubles removal, product stabilisation.

Primary extraction: Primary contaminations removal.

Purification: Secondary contaminations removal, decolourisation, concentration, recemization

Derivatization (optional): Chemical modification.

Drying(optional): Lyophilization or spray drying

Compounding of drugs : According to the guidelines of formulations, a prescription is composed of four different component parts of ingredients as given below.

- The principal ingredient which provides the principal curative action
- The adjuvent which strengthen the principal action
- The auxillary ingredient which relieves secondary symptoms or tempers the action of principle ingredient
- The conductant which directs action to the affected conduit or site. It may also be a less significant auxillary ingredient.

□□□

15

IPR Issues in Medicinal and Aromatic Plants

Based on the British Law of 1852, then the Government of India introduced the Act of Protection of Invention in 1883 and subsequently modified it to the Inventions and Designs Act in 1888. And on August 15, 1947, Indian patents and designs came to be managed by the Controller of Patents and Designs. In 1970, the Indian Patents Act was passed. Over the years, sublime and novel ideas, a design or an invention or a manuscript were defined as 'intellectual property'. But in any case, these properties led to useful products/applications.

The tragedy is that most intellectual properties can be copied or imitated, thus lowering returns to the original inventor. The above account recognises the right of an inventor to accrue economic gains from his inventions. This right is called 'Intellectual Property Right' (IPR).

The IPR is protected in several ways: trade secrets, patents, plant breeders' right and copyright.The Paris Convention (1983) established equal protection ofindustrial IPR. This stipulated that inventions claim priority in all member States by filing a patent application in one member country. This convention had 100 members initially and India joined it in December 1998. In1978, the European Patent Convention came into being and had 17 member States. Incidentally, provisions for biotechnology inventions were introduced. In January 1995, the Trade Related Intellectual Property Rights (TRIPs) agreement,'pertinent to IPR, came into force. This is enforced and administered by the World Trade Organisation (WTO), Geneva. Each member country has the option to frame its patent laws as defined in the GATT agreement. On a complaint by the U. S. A. in 1997, India is required to modify its patent laws by 2004.

The World HealthOrganization (WHO) estimates that 80 percent of the world population presently uses herbal medicine for some aspect of primary health care. Herbal medicine is a major component in all indigenous peoples' traditional medicine and a common element in Ayurvedic, homeopathic,naturopathic and traditional oriental.

Substances derived from the plants remain the basis for a large proportion of the commercial medications used today for the treatment of heart disease, high blood pressure, pain, asthma, and other problems (Table 1.)

Table 1. Medicinal plants/ products

Botanical name	Common name	Medicinal use
Abroma augulata Lam.	Devil's cotton	Uterine tonic regulates menstrual flow; urinary trouble, bronchitis, broncho-pneumonia, carbuncles and poisonous boils; Leaf: used in diabetes, rheumatic pain and sinusitis.
Acorus calamus Linn.	Sweet Flag	Laxative, diuretic, carminative; improves voice and appetite; good for oral diseases, abdominal pain, epilepsy, bronchitis, hysteria, loss of memory, rat bite and worms in ear.
Aegle marmelos Corr	Wood apple	Destroys phlegm and a good remedy in fevers associated with catarrhal symptoms, also good for dropsy, bleeding piles, dysentery, diarrhoea, bowel complaints and is a good laxative.
Balanites aegyptica Linn.	Desert date, Soapberry tree, Thorntree	Bark, Leaf, Fruit and Seed: as anthelmintic and purgative; Fruit: in boils, leucoderma and other skin diseases; Fruit pulp: in whooping cough.
Butea monosperma	Flame of the forest, Parrot tree.	Gum: in diarrhoea; Bark: as haemostatic, in wounds,Flower: in eye complaints; antifertility.
Catharanthus roseus Linn.	Madagascar periwinkle,Old maid, Red Periwinkle	Root-paste: in septic wounds Leaf-juice: in blood dysentery Root: in cancerous wounds
Embelia ribes Burm	Embelia, Embelia fruit	Stomach pain and increases exothermic metabolism;useful in skin diseases including leprosy.
Ficus racemosa Linn.	Cluster fig, Country fig, Gular fig.	Useful in antifertility, pimples and wounds;
Gymnema sylvestre	Vine, Periploca of the woods	Useful in amennorrhoea, asthma, bronchitis, cardiopathy, conjunctivitis, constipation, cough, dyspepsia, haemorroids, hepatosplenomegaly, inflammations, intermittant fever, jaundice and leucoderma
Vetiveria zizanioides	Cuscusn root	Amenorrhea, antifertility
Terminalia chebula, Terminalia reticulata	Indian Gall Nut	Remunerative, tonic, astringent, laxative, nervine,expectorant, anthelmintic, alterative

Source: Somasundaram.S, Ethno botany, 2004.

Medicinal Plants are the Local Inheritance with Global Importance

The Indian Systems of Medicine, particularlyAyurveda, Siddha, Unani, & Homoeopathy medicine largely use plant base materials, minerals, metals, marine and products of animal origin. Our ancient texts had documented medicinal uses of a large number of plants.These plants are being used for preparation of medicines for centuries. The traditional Medicinal plants have successfully set an example of natural resource use incuring many complex diseases for more than 3,000 years.The plants used for various therapies are readily available, are easy to transport, and have a relatively long shelf life. The most important advantage of herbal medicine is the minimal side effects, and relatively low cost compared to the synthetic medicines. The success of medicinal plants sector mainly depends on the awareness and interest of the farmers as well as its other stakeholders, supportive government protection policies,availability of assured markets, profitable price levels, andsuitable agro-techniques. The World Health Organization(WHO) has estimated the present demand for medicinal plants is approximately US $14 billion per year. The demand for medicinal plant-based raw materials is growing at the rate of 15 to 25% annually, and according to an estimate of WHO, the demand for medicinal plants is likely to enhance more than US $5 trillion in 2050. In India, the medicinal plant-related trade is estimated to be approximately US $1 billion per year.

Bio-piracy: Bane for Developing Countries

Green forests have currently become component of a vibrant, profit-seeking economy and challenging pluralistic political principles worldwide. Medicinal plants signify not only a valuable part of India's biodiversity but also a foundation of immense conventional knowledge. Knowledge-rich companies and researchers from the industrial world have attracted to the wealth of the poorer countries have in their biodiversity and the traditional knowledge systems. Some argue that the access to such biodiversity and community knowledge by the industrially developed nations isessential for the larger welfare of mankind as this advances knowledge and leads to innovative products which contribute to the well being of global consumers. The point is that this access to the resources of the poor does not benefit in any way, while their natural resources and intellectual property continues to be appropriated and monopolized. The patenting (A patent may be granted for a new, useful, and non-obvious invention, and gives thepatent holder a right to prevent others from practicing the invention without a license from the inventor for a certain period of time, typically 20 years from the filing date of apatent application) of indigenous knowledge by foreigncorporations is a cultural threat to countries like India as well as economic loss. Developing countries increasingly face bio-piracy (Bio-piracy' describes a process in which living resources or traditional knowledge and practices are patented, thus applying intellectual property restrictions to their use) as a significant issue that immediately needs to be addressed. Since many of these countries are shelters of biodiversity and rely economically on their ability to export indigenous products and processes, they see the rising importance of protecting their traditional knowledge from unjustifiable foreign patenting.

The World TradeOrganization (WTO) isencouraging developing countries to expand their legal protection of intellectual property rights in order to be on a similar "playing field" with the developed countries. Hence India, as a biodiversity hotspot, is taking a leading role it committing to the protection of its knowledge base. The tribulations experienced in case of patents on the turmeric, neem, basmati etc also highlight the problem of bio-piracy and demand for patenting of indigenous bio-diversity- related knowledge. Further, anew trend is that foreign countries have exposed interest in medicinal plants accessible in India and well documented in our books signifying the formulation inwhich they are used. A number of medicinal plants and their uses have been patented by foreign countries. Medicinal plants are a significant target of patent claims since they have become of immense significance to the international drug and cosmetic industries. This isdue to the increasing occurrence of allergic reaction in the west to chemical drugs, and the complications arising from side effects after prolonged drug use in chronic illnesses. The global drug industry, spurred by the accomplishment of plantbased cancer cures like Vinblastine, Vincristine andnow Taxol, has intensified itssearch of the plant world for potential drugs. Because of this international awareness, illegal prospecting for medicinal plants has been carried out in developing countries on a large and unsustainable scale. But, India is behind the rest of the world in patents both quantitatively and qualitatively, even when comparison is made with our neighbour China. The persistent illiteracy and confusion about patents is a serious matter. Our pool of knowledge that is protected by patents, even in areas where we have a competitive advantage is rather poor. Take the area of herbal products, where so much emotion has been raised. The number of herbal patents (1995-1998) was 1889, out of which China had a share of 889, and the Indian share was next to nothing. Different ways and systems for awarding patents on the medicinal plants in India, United States, Europe, Canada and other countries have widened the confusion. In many countries, the plants and inventions directed to the plants and the plant products (i.e., seeds, flowers, gums, and resins) are not eligible for filing a patent. In United States, however, any living organism derived by human invention, such as by breeding or by laboratory-based manipulation, can befiled for awarding patent. The Indian Protection of Plant Varieties and Farmers Rights Act of 2001 recognize the contribution of farmers who actively participate in the breeding programs. Furthermore, this act contains provisions for benefit sharing whereby local communities are acknowledged as a contributor of plants. Now that areasonable level of consciousness has been generated concerning the sort of IPR regime that the GATT/TRIPs have heaped upon developing countries. It necessitates defining our positions on all our biological resources like microorganisms and medicinal plants which along with seeds will be in the forefront of IPR claims. Unfortunately,there is a wide gap between developed and developing nations such as India on patenting the products. For example, out of the 3,125,603 patents filed in 91countries, only 301,177 or 9.6% are registered indeveloping countries while the rest is in developed countries. Of these, only 0.2% of the total and 2.3% ofthose registered in developing countries belong to residents. In addition, 97.7% of the total patents filed thus far are in the name of non-residents, who apply solely to control export markets in developing countries.

Developing nations and many scientists who want to exploit medicinal plants demand more specific rules about the recording of nationality of samples and sharing of their benefits between the nations of origin, the inventor, and the commercial sponsors. Some developed nations are not inclined to accept such provisions. These conflicts have frustrated many scientists who believe that natural products remain the most promising source for new drugs. Besides government agencies, there are numbers of stakeholders ranging from herb gatherers, local middlemen, urban traders, wholesalers, manufacturers,exporters and herbal healers in the medicinal plants trade sector. The marketing system in medicinal plants sector islargely unregulated and inequitable. The medicinal plant collectors are generally the marginal farmers and labourers. They get cash income to meet their basic requirements for food, health and children education by selling medicinal plants. They are often unaware about the real market prices of many medicinal plant species.Generally, in medicinal plants sector, there is a top down approach and even the many stakeholders at the bottom are not aware of the rising demand of their product and the availability of its market. For example, in some villages of Chamoli district of Uttaranchal, the farmers had cultivated Kut (*Saussurea costus*) and Dolu (*Rheum emodi*) but they were unable to sell them due to lack of knowledge on the marketing system. Conversely, many medicinal plant species are traded through illegal ways.

The other Constraints in The Medicinal Plants Sector

1) Slow rate of production of many medicinal plants, 2)long gestation period, 3) shortage of suitable cultivation techniques, 4) production of small quantity, 5) unscientific harvesting, 6) paucity of research on the high yielding varieties, 7) inefficient processing techniques, 8) fluctuation in demand and supply, 9) poor quality control procedures, 10) scarcity of good manufacturers, (11) poor marketing infrastructure, and 12) poor coordination among different stakeholders and (13)the medicinal plant sector is largely less documented and inadequately regulated.

Avenues in Support of Medicinal Plants Sector

A huge diversity of medicinal plants would be helpful for further scientific research on exploring their medical efficacy, value addition, and use in curing various old and new diseases. India has already established a reputation as a low-cost manufacturer of high quality generic drugs in the global market. This fact can be used as an important tool for the marketing of herbal products produced in India.

Existing Policies

In the National Five Year Plans of India,the medicinal plants sector has been identified as an integral part of the Indian System of Medicine and Homeopathy In 1983, the National Health Policy recognized that the large stock of health manpower in Ayurveda, Siddha, Unani, Homeopathy and Naturopathy had not been adequately utilized; therefore, steps need to be taken to move towards a meaningful integration of the indigenous and modern systems of medicine. The Planning Commission and the National Medicinal Plants Board (NMPB) of the Government of India have prepareda policy document on the promotional and commercial aspects of the medicinal plants. The various policies at national and state level and their subsequent implementation will provide an opportunity in the advancement of medicinal plants sector.

Institutional Support

In India, many government and nongovernment organizations have had the focused attention on improving the medicinal plants sector (Table 2). The successful establishments of medicinal plants sector may help in raising rural employment, boost commerce around the world, and contribute to the health of million.

Table 2. Major institutions involved in funding projects to the medicinal plants research in India

Institutions	Funding for major areas in medicinal plants research
National Medicinal Plants Board, NMPB	Survey, documentation, cultivation, marketing,Conservation
Department of Science & Technology, DST	Taxonomy, ecology, pathology, survey, propagation,documentation, cultivation, conservation
Council for Scientific &Industrial Research, CSIR	Ecology, taxonomy, biochemistry, survey,documentation, cultivation, genetics, agro-technology,conservation
Indian Council of Medical Research, ICMR	Breeding, value addition
All India Council for Technical Education, AICTE	Management technology
Department of Biotechnology, DBT	Agro-technology, molecular biology, biochemistry,rural bio-technology
Defence Research &Development Organization,DRDO	Agro-technology, survey, documentation, conservation
Indian Council of Agricultural Research, ICAR	Breeding, pathology, molecular biology
Ministry of Environment & Forest, MoEF	Survey, documentation, conservation, management,ecological impact assessment, cultivation
National Bank for Agriculture and Rural Development,NABARD	Cultivation, marketing
University Grant Commission, UGC	Ecology, biochemistry, survey, documentation
Herbal Research and Development institute, HRDI	Survey, documentation, nursery development
G.B. Pant Institute of Himalayan Environment & Development, GBPIHED	Survey, documentation, cultivation, conservation

Source: Journal of Ethno biology and Ethno medicine 2006.

Traditional Knowledge Digital Bank (TKDL)

The Government of India has set up a Traditional Knowledge &Digital Library (TKDL), namely, an electronic database of traditional knowledge in the field of medicinal plants. Such a database would enable the Patent Officers all over theworld to search and examine any prevalent use/prior art,and thereby prevent incorrect grant of patent based on knowledge in public domain, including knowledge associated with medicinal plants. The issue had also been taken up at the international level in the Inter Government

Committee of the World Intellectual Property Organization to ensure that TKDL is prescribed as a non patent literature and minimum PCT documentation to ensure that patent examiners are duty bound to search the said database for any prior art.

Conservation of Conventional Knowledge

India ought to set into action the drafting of suitable legislations to guarantee that the laws we devise permitus to keep hold of power over our biological resources and be acquainted with the privileges of local and indigenous communities as owners of these bioresources. In countries like Australia and the Philippines, strong national laws have been passed for the protection of local genetic resources. These laws establish the ownership of communities over the biological resources of their region and recognize their right to share in the profits of products derived from indigenous knowledge.There is a need to document the indigenous knowledge related to Indian herbs and plants and their medicinal and other uses and convert it into easily navigable computerize data base for easy access and to secure patenting rights; to discourage other countries for patenting Indian heritage. Along with the documentation of the bioresources, we must document the local/community knowledge that exists about the various uses of these resources. This documentation which should be compiled as a National Bio-resource Register will provide several functions. The first is that of a data bank for people seeking access to information. This access should be made available for a fee accompanied by the conditions governing the use of this information.

MEDICINAL TREES

Aegle marmelos

Alstonia scholaris

Averrhoea carambola

Azadirachta indica

Butea monosperma

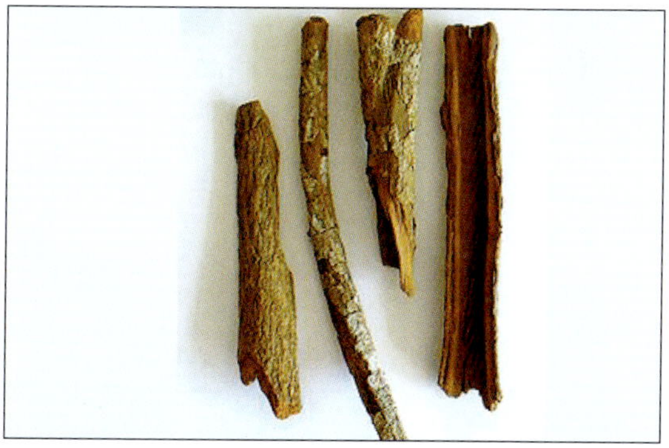

Cinchona officinalis

Emblica officinalis

Eugenia jambolana

MEDICINAL TREES

Ficus microcarpa

Mallotus philippensis

Mesua nagassarium

Mimusops elengi

Moringa oleifera

Saraca asoca

Strychnosnux vomica

Terminalia arjuna

MEDICINAL SHRUBS

Adhatoda vasica

Artemisia maritima

Calotropis gigantea

Cassia senna

Commiphora wightii

Datura metel

Desmodium gangeticum

Ephedra gerardiana

MEDICINAL SHRUBS

Hibiscus rosa sinensis

Holarrhenna antidysenterica

Leptadenia reticulata

Plumbago zeylanica

Premna integrifolia

Rauvolfia serpentina

Vitex negundo

Withania somnifera

MEDICINAL CLIMBERS

Antigonon leptopus

Asparagus racemosus

Cissus quadrangularis

Clitorea ternatea

Coccnia indica

Dioscorea bulbifera

Gloriosa superba

Gymnema sylvestre

MEDICINAL CLIMBERS

Hemidesmus indicus

Ipomoea hederacea

Merremia hederacea

Mucuna pruriens

Paederia foetida

Tinospora cordifolia

Tylophora indica

Vigna trilobata

MEDICINAL HERBS

Aconitum heterophyllus

Acorus calamus

Aloe vera

Alpinia galanga

Ammi majus

Andrographis paniculata

Anthemis nobilis

Anthriscus cerefolium

MEDICINAL HERBS

Atropa belladonna

Bacopa monnieri

Boerhaavia diffusa

Catharanthus roseus

Centratherum anthelminticum

Cephaelis ipecacuanha

Chlorophytum arundinaceum

Colchicum luteum

MEDICINAL HERBS

Coleus barbatus

Coptis teeta

Costus speciosus

Curculigo orchioides

Curcuma caesia

Dendrobium nobile

Eclipta prostrata

Enhydra fluctuans

MEDICINAL HERBS

Eulophia campestris

Euphorbia hirta

Gentiana kurroo

Glycyrrhiza glabra

Habenaria intermedia

Hyoscyamus niger

Inula racemosa

Kaempferia rotunda

MEDICINAL HERBS

Lycopodium clavarum

Malaxis muscifera

Malva parviflora

Nardostachys grandiflora

Panax gingseng

Picrorhiza kurroa

Plantago ovata

Podophyllum emodi

MEDICINAL HERBS

Polygonum recumbens

Rheum emodi

Rhyncostylis retusa

Ruta chalepensis

Saussurea lappa

Solanum indicum

Solanum khasianum

Swertia chirayita

MEDICINAL HERBS

Trianthema monoegyna

Trillium erectum

Typha latifolia

Urtica dioica

Valeriana officinalis

Vanda tessellata

Verbascum thapsus

Viola serpens

AROMATIC PLANTS

Apium graveolens

Artemisia pallens

Cymbopogon flexuosus

Cymbopogon martinii

Cymbopogon nardus

Eucalyptus citriodora

Jasminum grandiflorum

Kaempferia galangal

AROMATIC PLANTS

Lavandula angustifolia

Majorana hortensis

Matricaria chamomilla

Mentha arvensis

Michelia champaca

Ocimum sanctum

Pandanus fascicularia

Pelargonium graveolens

AROMATIC PLANTS

Pogostemon patchouli

Polyanthes tuberosa

Rosa damascena

Rosmarinus officinalis

Salvia sclarea

Thymus vulgaris

Vanilla planifolia

Vetiveria zizanioides

References

Agarwal, Poonam. 2013. Study of useful climbers of Fatehpur, Uttar Pradesh, India. *International Journal of Pharmacy and Life Sciences*, 4(9): 2957-2962.

Akshay KR, Sudharani N, Anjali K B and Deepak TM. 2014. Biodiversity and strategies for conservation of rare, endangered and threatened medicinal plants. Research and reviews : *Journal of Pharmacognosy and Phytochemistry.* 2(3): 2332-2347.

Alexander, M. P., Ganeshan, S.1993. 'Pollen storage'. In: *Advances in Horticulture.* Vol. I. *Fruit Crops-part I.*(Eds.). Chadha, K. L. and Pareek, O. P. New Delhi: Malhotra Publishing House, pp. 481-496.

Anjalika, M.R. and Banerjee, N. and Mandal, S. 2005. Structural and Functional parameters of mating and seedgermination in mutants of *Catharanthus roseus*(Apocynaceae). *J. Appl. Biosciences,31*(2):145-149.

Asolkar, L.V., Kakkar, K. K. and Chakre, O. J. 1992. *Second Supplement to Glossary of Indian MedicinalPlants with Active Principles Part I (A-K).* (1965-81). Publications and Informations Directorate(CSIR), New Delhi. pp.18-20.

Atal C.K. et al., 1977: *Cultivation and Utilization of Aromatic Plants.* Regional Research Laboratory, Council of Scientific & Industrial Research, Jammu-Tawi (India), 568.

Atal, C. K. and Kapur, B. N. 1982. *Gultivation and Utilisation of Medicinal Plants.* CSIR, RRL, Jammu-Tawi, India. 727p.

Baishya, R.A., Sarma, J. and Begum, Ajima. 2013. Forest-based medicinal plants rendering their services to the rural community of Assam, India. *International Journal of Applied Biology and Pharmaceutical Technology*, 4(4): 10-20.

Bajaj, Y. P. S. (Ed.). 1989. *Biotechnology in Agriculture and Forestry: 7. Medicinal and Aromatic plantsII.* Springer-Verlag, Berlin Heidelberg, New York. 545 p.

Bajaj, Y. P. S. (Ed.). 1991. *Biotechnology in Agriculture and Forestry: 15. Medicinal and Aromatic plants III.* Springer-Verlag, Berlin Heidelberg, New York. 502 p.

Baskaran, K., Srinivas, KVNS and Kulkarni, R.N. 2013. Two induced macro-mutants of periwinkle with enhanced contents of leaf and root alkaloids and their inheritance. *Industrial Crops & Products*,43:701-703.

Bhattacharjee, S. K. and De, L. C. 2005. *'Medicinal Herbs and Flowers'*. Pp-434. Aavishkar Publishers & Distributors, Jaipur , Rajasthan.

Bhattacharya, Rohit, Reddy KRC and Mishra AK. 2014. Export strategy of Ayurvedic Products from India. *International Journal of Ayurvedic Medicine,* 2014, 5(1),125-128.

Bhau, B.S. 2012. Molecular Markers in the Improvement of the Medicinal Plants. *Med Aromat Plants* 1:e108. doi:10.4172/2167-0412.1000e108.

Biswas, K. and Chopra, R. N. 1982. *Common Medicinal Plants of Darjeeling and the Sikkim Himalayas*.Periodical Experts Book Agency, D-42, Vivek Vihar, Delhi-110032. 157p.

Brown, E. and Matthews, W. S. A. 1951. Notes on the aromatic grasses of commercial importance. *Col. Pl.Ani. Prod.*, **2**:174-187.

Chadha K.L. et al., 1995: *Advances in Horticulture: Medicinal and Aromatic Plants*. Vol.11. Malhotra Pub. House, New Delhi, India.

Chatterjee S.K. 1978. The cultivation of *Catharanthusroseus* in India: In: Alfermann AW and Reinhard E(eds), Production of natural compounds by cellculture methods. Proc. Int. Symp. Plant Cell Culture,Munchen FRG. 74-85.

Chaudhuri, S. P. R. 1991, 1992. *Recent Advances in Medicinal, Aromatic and Spice Crops*. Vol. I-II. Todayand Tomorrow's Printers and Publishers New Delhi.

Chomchalow, N. and Henle, H. V. (Ed.). 1995. *Medicinal and Aromatic Plants in Asia*. Oxford & IBH, NewDelhi. 196 p.

Chopra, R. N., Nayar, S. L. and Chopra, I. C. 1980. *Glossary of Indian Medicinal Plants*. CSIR, New Delhi.

CSIR. 1953. *The Wealth of India*. Part I-III. Council of Scientific and Industrial Research, New Delhi.

Curtis, S. and Fraser, R. 1992. *Natural Healing for Women: Caring for yourself with herbs, homeopathy and essential oils*. Pandora Press, UK. 418 p.

Das, Sudipta and Choudhury Dutta Manabendra. 2012. Ethnomedicinal uses of some traditional medicinal plants found in Tripura, India, *Journal of Medicinal Plants Research*, Vol. 6(35), pp. 4908-4914.

De, L.C, Rao, A.N., Rajeevan, P.K., Promila Pathak and Singh, D.R. 2015. Medicinal and Aromatic Orchids -An Overview. *International Journal of Current Research,* 7(9): 19931-19935.

De, L. C. De 2011. *Value Addition in Flowers and Orchids*, Pp.294, New India Publishing Agency, Pitampura, New Delhi.

De, L.C. De 2014. *Production Technology of Commercial Flowers*: In 2 volumes Pp. 599. Pointer Publisher, Jaipur, Rajasthan.

De, L.C. 2014. *Production of Seed and Planting Materials of Horticultural Crops* pp. 372 Aavishkar Publishers, Distributors, Jaipur, Rajasthan.

De, L. C. Pathak,P., Rao, A.N. and Rajeevan, P.K. 2014. *Commercial Orchids*. De Gruyter Open, Polland, Pp.300.

Dey, A. C. 1980. *Indian Medicinal Plants Used in Ayurvedic Preparations*. Bishen Singh, Mahendra Pal Singh, Dehra Dun-202p.

Dey, R. B. K. L.1984. *The Indigenous Drugs of India*. International Book Distributors, Dehradun. India.387p

Dhar, M. L., Dhar, M. M., Dhawan, B. N. and Ray, C., 1979. Screening of Indian plants for biological activityPart- 1. *Ind. J. Exp. Biol.*, **6**: 232.

Dhar, M. L., Dhar, M. M., Dhawan, B. N., Mehrotra, B. N. and Ray, C., 1968. Screening of Indian Plants for biological activity, Part I. *Indian J. Exp. Biol.*, **6**:234-247.

Duke, J. A. and Wain, K. K. 1981. *Medicinal Plants of the World. Computer index with more than 85,000entries*. 3 Vols. 1654 p.

Edison, S., Johny, A. K., Babu, N. and Ramadasan, A. 1991. *Spices Varieties: A compendium of morphological and agronomic characters of improved varieties of spices in India*. National Research Centre for Spices, P.O. Marikunnu, Calicut, Kerala, India. 68 p.

Efterpi Christaki,, Eleftherios Bonos , Ilias Giannenas and Panagiota Florou-Paneri. 2012. Aromatic plants as source of bio-active compounds. *Agriculture*, 2: 228-243.

Farooqi, A. A. and Khan, M. M. 1991. *Cultivation Practices for Medicinal and Aromatic Crops*. Division of Horticulture, University of Agric. Sci., Bangalore. 77 p.

Fischer-Rissi, S. 1992. *Complete Aromatherapy Handbook: Essential oils for radiant health*. Sterling Publishers, UK. 240 p.

Foster, S. 1992. *Herbs of Commerce*. American Herbal Products Association. 3 p.

Graves, G. 1996. *Medicinal Plants-An illustrated guide to more than 180 herbal plants*. Bracken Books,London. p.91.

Gupta, R. 1980. Medicinal and aromatic plants. In *Handbook of Agriculture*. ICAR, New Delhi. pp.1206-1225.

Haq, N. 1993. Breeding and improvement of medicinal and aromatic plants. In FAO, 1993.

Harrington, J. F. 1970. 'Seed and pollen storage for conservation of plant genetic resources'. In: *Genetic Resources in Plants. Their Exploration and Conservation.* (Eds.). Frankel, O. H. and Bennet, E. Edinburgh, U.K: Blackwell, pp. 501-521.

Hussain, A., Virmani, O. P., Sharma, A., Kumar, A. and Misra, L. N. 1988. *Major Essential Oil Bearing Plants of India*. Central Institute of Medicinal and Aromatic Plants, Lucknow, India. 237p.

Ilyas, M. 1978. Spices of India. Part II. *Econ. Bot.*, **32**(3):239-263.

Ilyas, M. 1980. Spices of India. *Econ. Bot.*, **34**(3):236-259.

Jackson, W. P. 1999. 'Plenary address, BGCI. 4th International Congress on Education in Botanic Gardens held in Thiruvanthapuram, Kerala.

Jain, S., (Ed.). 1981. *A Perspective of the perfumes and flavours industry in India*. Perfumes and Flavours Association of India, Bombay. 190 p.

Jain, H. C., Mehta, S. P., Mehtani, S., Singh, G., Mukherjee, T. K., Doreswamy, R. and Sharma, D. 1994.Report of the workshop on Information management for the medicinal and aromatic plants industry. *J. Sci.Ind. Res.*, **53**(1):43-46.

Joshi, Pravin R , Harisha C.R. and Patel, Bupesh R. 2011. Regionally accepted popular source of Ayurvedic medicinal plants in Southern India. *International Journal of Pharmacy & Life Sciences*, 2 (10); 1123-1132.

Kaikini, N. S. 1969. Cultivation of essential oil yielding plants in Mysore state. *Indian Perfumer*,**13**(1):1-6.

Kapoor, L. D. and Krishnan, R. (Eds.). 1977. *Advances in Essential Oil Industry*. Today and Tomorrow's Printers and Publishers, New Delhi. 250 p.

Khory, R. N and Katrak, N. N. 1985. *Materla Medica of India and their Tnerapeutics*. Neeraj Publ. House, Delhi 809 p.

Khumbongmayum, Ashalata Devi, Khan, M.L. and Tripathi, R.S. 2005. Ethnomedicinal plants in the sacred grooves of Manipur. *Indian Journal of Traditional Knowledge*, 4(1): 21-32.

Kirtikar, K. R. and Basu, B. D. 1987. *Indian Medicinal Plants*. Part I-IV. International Book Distributors,Dehradun-248 001.

Kokate, C. K. and Varma, K. C. 1971. Cultivation of lemongrass (*Cymbopogon citratus* (D.C.) Stapf.) and citronella grass (*C. nardus*) at Sagar, M. P. *Indian J. Agric. Sci.*, **41**(4):382-385.

Kraisintu, K. 1997. Industrial exploitation of indigenous medicinal and aromatic plants: Formulation and industrial utilisation. In UNDP, 1997.

Kulkarni, R.N. and Baskaran, K. 2008. Inheritance of Pollen-less Anthers and "Thrum" and"Pin" Flowers in Periwinkle. *J. of Hered.* 99(4): 426-431.

Kulkarni, R.N., Bhaskaran, K, Chandrasekhara, R.S. and Kumar, S. 1999. Inheritance of morphological traits of periwinkle mutants with modified contents andyields of leaf and root alkaloids. *Plant Breed.*,118(1): 71-74.

Kulkarni, R.N. and Suresh, N. 1999b. Inheritance of characters in periwinkle: leaf pubescence and corolla colour. *J.Herbs Spices Med.*, 6: 85-88.

Kumar, Neelum.2014. Some plants used in ayurvedic and unani systems of medicine, tehsil joginder nagar, district mandi, H.P., India. *International Journal of Food, Agriculture and Veterinary Sciences, 2014 Vol. 4 (1)* January-April, pp.73-80.

Kumar, S. Rai, S.P., Kumar, S.R., Singh, D., Srivastava, S. and Mishra, R.K. 2007. Plant variety of *Catharanthus roseus* named 'lli', United States Patent PP18315.

Kumar, N., M. Abdul Khader, J. B. M., Rangaswami, P. and Irulappan, I. 1997. *Introduction to Spices,Plantation Crops, Medicinal and Aromatic Plants*. Oxford and IBH Publishing Co. Pvt. Ltd., NewDelhi.

Kumar, S., Shukla, Y. N., Lavania, U. C., Sharma, A. and Singh, A. K. 1997. Medicinal and Aromatic Plants:Prospects for India. *J. Med. Arom. Pl. Sc.* **19** (2):361-365.

Kurup, P. N. V., Ramdas, V. N. K. and Joshi, P.1979. *Handbook of Medicinal Plants*, New Delhi.

Lawless, J. 1992. *Encyclopaedia of Essential Oils*. Element Books Ltd., UK. 232 p.

Lawrence, B. M. 1979, 1980, 1982, 1984. Progress in essential oils. *Perfumers Flavorist,* **4**:54-56,**5**(1):55, **7**:45, **9**:31.

Lee, W. H. and Lee, L. 1993. *The Book of Practical Aromatherapy*. Keats. 200 p.

Leung, A. Y. 1980. *Encyclopaedia of Common Natural Ingredients Used in Food, Drugs and Cosmetics*.pp. 261-262

Lewis, Y. S., Nambudiri, E. S. and Natrajan, C. P. 1967. Studies on some essential oils. *Indian Perfumer*,**11**:5-9 Lewis, Y. S., Mathew, A. G., Nambudiri, E. S. and Krishnamurthy, N. 1972. Oleoresin ginger.*Flavour Industry*, **3**(2):78-81.

Lindley, J. 1984. *Medical and Economical Botany.* Bishen Singh, Dehra Dun, 912 p.

Maheswari, J.2011. Patenting Indian medicinal plants and products. *Indian Journal of Science and Technology*, 4(3): 298-301.

Maithy Roy, A. Banerjee, N. and Mandal, S. 2004. Environmental effects of mutation on *Catharanthus roseus* (L.) G.Don. In: Proc. Impact of Civilization on Environment. 156-159.

Masand, Sameet, Swati Madan and Balian, S.K. 2014. Modern concept of storage and packaging of raw herbs used in Ayurveda. *Int.J.Res.Ayurveda Pharm.*, 5(2): 242-245.

Misra, A. 1992. Nutritional requirements of some important medicinal and aromatic plants for higher productivity. *Curr. Res. Med. Arom. Plants,* 14(2):121-132.

Mishra, P. and Kumar, S. 2003. Manifestation of heterostyled character by induction of recessive hsf mutation responsible for thrum type herkogamous flowers in *Catharanthus roseus*. JMAPS, 28: 2.

Mukherjee, D. 2013. Improved agricultural practices for high value medicinal plant under mid to high altitude situation. *Journal of Crop and Weed,* 9(1):201-206.

Pala, Nazir A., Negi, A. K. and Todaria, , N. P. 2010. Traditional uses of medicinal plants of Pauri Garhwal, Uttrakhand. *Nature and Science*, 8(6): 57-61.

Pandey, Amita and Tripathi, S. 2014. Concept of standardization, extraction andpre phytochemical screening strategies forherbal drug. *Journal of Pharmacognosy and Phytochemistry*, 2 (5): 115-119.

Pank, Friedrich. 2007. Use of breeding to customise characteristics of medicinal and aromatic plants to postharvest processing requirements. *Stewart Postharvest Review* 4:1.

Paroda, Raj, Dasgupta, S. Bhag Mal, Ghosh, S.P. and Pareek, S.K. 2014. *Expert Consultation on Promotion of Medicinal and Aromatic Plants in the Asia-PacificRegion: Proceedings*, Bangkok, Thailand; 2-3 December, 2013. 259 p.

Patel, D.K. 2015. Introduction of Some Tree Species of Medicinal and Aromatic Pants MAPs) in Herbal Garden for Their Ex-situ Conservation. *Universal Journal of Plant Science* 3(1): 9-12.

Patel, D.K. 2014. Some Medicinal and Aromatic Shrubs introducedin Herbal Garden for their Propagation and Ex-situ Conservation*Journal of Medicinal Plants Studies*, 2014; 2(6): 9-12.

Patra, N. K. and Sharma, J. R. 1989. Genetic improvement of aromatic grasses: A review. *Curr. Res. Med.Arom. Plants*, 11(1):32-39.

Purohit S.S and Vyas, S.P. 2005. *Medicinal Plant Cultivation: A Scientific Approach*, Agrobios, Jodhpur, India. pp:1-8, 236-237.

Rai, Nitin , Tiwari, Lalit and Sharma, Rajeev Kr. 2012. Agronomic Practices for High Trade Value Indian Aromatic Plants. International Journal of Advanced Agricultural Science and Technology 2012, Volume 1, Issue 1, pp. 1-11, Article ID Sci-16.

Rai, S.P. and Kumar, S. 2000. Induced mutation tomonocotyledony in periwinkle, *Catharanthus roseus*, and suppression of mutant phenotype bykinetin. *Indian Aca. of Sci.*,79(3): 97-104.

Rajeswara Rao B.R. 2012. Medicinal and aromatic plants processing technologies of council of scientific and industrial research (CSIR) institutions. *Journal of Pharmacognosy*, 3(2):138-141.

Rama Shankar and Ramesh Babu Devalla. 2012. Conservation of folk healing practices and commercial medicinal plants with special reference to Nagaland. *International Journal of Biodiversity and Conservation,* 4(3): 155-163.

Ramesh Kumar, M. and Janagam, D. 2011. Export and import pattern of medicinal plants in India. *Indian Journal of Science and Technology*, 4(3): 245-248.

Robiya Mol and Binu Thomas. 2015. Some medico-potential climbers of Kottayam district, Kerala, india. *International journal of phytopharmacy research*, 5(4): 397-401.

Roberts, E. H. 1975. Problems oflong-term storage and seed and pollen of genetic resources conservation'. In: *Crop Genetic Resources for Today and Tomorrow.* (Eds.). Frankel, O. H. and Hawks, J. G. Cambridge, U.K.: Cambridge University Press, pp. 226-296.

Roberts, E. H. 1972. *Viability of Seeds.* London: Chapman and Hall, 1972.

Samy, Ramar Perumal , Peter Natesan Pushparajand Ponnampalam Gopalakrishnakone. 2008. A compilation of bioactive compounds from Ayurveda. *Bioinformation* 3(3): 100-110.

Sharma M., Sharma C.L. and Marak, P.N. 2014. Indigenous uses of medicinal plants in North Garo Hills, Meghalaya, NE India. *Research Journal of Recent Sciences,* Vol. 3(ISC-2013), 137-146.

Sharma, S. and Thokchom, R. 2014. A review on endangered medicinal plants of India and their conservation. *Journal of Crop and Weed,* 10(2):205-218.

Sharma, S. C., Goel, A. K. 1994. 'Role of botanic gardens in modern times'. *Indian* J. *Forestry* 17: 230-238.

Silva, T. D. 1997. Industrial utilisation of aromatic plants in developing countries. In UNDP, 1997.

Singh, S. 2005. Medicinal plants – A Natural gift to mankind. *Agriculture Today.* 3(3): 58-60 pp.

Singh, S. R., Phurailatpam, Wangchu, L., Ngangbam, P. and Chanu, T.M. 2014.Traditional medicinal knowledge of underutilized minor fruits as medicine in Manipur. *International Journal of Agricultural Sciences* Vol. 4 (8), pp. 241-247.

Sivarajan, V. V. and Balachandran, I. 1994. *Ayurvedic Drugs and their Plant Sources.* Oxford and IBH Publishing Co. Pvt . Ltd., New Delhi. p.570.

Subramoniam, A. 2014. Present scenario, challenges and future perspectives in plant based medicine development *Annals of Phytomedicine*, 3(1): 31-36.

Thakur, R. S., Puri, H. S. and Husain, A. 1989. *Major Medicinal Plants of India*, CIMAP, Lucknow, India.

Thomas, J. 1997. Medicinal and aromatic plants research in India. In UNDP, 1997.

Tilling, Rupa, Bharali, Pankaj, Dutta, Priyanka,Gogoi, Gaottham, Paul, Ashish and Das, Arup Kumar. 2015. Ethnomedicinal plants used by apatani tribe of ziro Valley of Arunachal Pradesh. *International journal of conservation science*, Volume 6, Issue 3, July-September 2015: 411-418.

Towill, L. E. 1985. 'Low temperature and freeze/vacuum-drying preservation of pollen'. In: *Cryopreservation of Plant Cells and Organs.* (Ed.). Kartha, K. K. Boca Raton, Florida, U. S. A.: CRC Press, Pp'. 171-198.

Trivedi, P.C. 2009. *Medicinal Plant-Utilization and Conservation.* Aavishkar Publishers, Distributors, Jaipur, Rajasthan, Pp. 542.

Varshney, S. C. 1992. *Essential Oils by Steam Distillation.* S. C. Varshney, Anand Vihar, Delhi-110092. 41 p.

Verghese, J. (Ed.). 1986. *On Essential Oils.* Synthite. Industrial Chemicals Pvt. Ltd., Kolenchery-682 311.428 p.

Vijayalatha, S.J. 2004. An Ornamental garden with medicinal plants : an indirect approach for conservation of medicinal plants. *Indian J. Arecanut, Spices and Medicinal Plants.* 6(3): 98-107.

Warrier, P. K., Nambiar, V. P. K. and Ramankutty, C. (Eds.). 1993-'95. *Indian Medicinal Plants: A compendium of 500 species.* Vol 1-5. Vaidyaratnam P. S. Varier's Arya Vaidya Sala, Kottakkal. OrientLongman, Madras, India.

Wijesekera, R. O. B. (Ed.).1991. *The Medicinal Plant Industry.* CRC Press, Boca Raton. Pp 1-17.

Withers, L. A. 1991. 'Biotechnology and plant genetic resources conservation'. In: *Plant Genetic Resources Conservation and Management.* (Eds.). Paroda, R. S. and Arora, R. K. New Delhi: IBPGR Regional Office, pp. 273-297.

Zu, L., Li, Y., Li, B., Lu, B., and Xia, N. 1993. *Aromatic Plants and Essential Oil Constituents.* SouthChina Inst. Bot., China Academy of Sciences, 344 p.

□□□

INDEX

D